D0615569

FUNCTION OF QUINONES IN ENERGY CONSERVING SYSTEMS

FUNCTION OF QUINONES IN ENERGY CONSERVING SYSTEMS

Edited by
BERNARD L. TRUMPOWER
Department of Biochemistry
Dartmouth Medical School
Hanover, New Hampshire

1982

ACADEMIC PRESS
A Subsidiary of Harcourt Brace Jovanovich, Publishers
New York London
Paris San Diego San Francisco São Paulo Sydney Tokyo Toronto

ACADEMIC PRESS, INC.
111 Fifth Avenue, New York, New York 10003

United Kingdom Edition published by
ACADEMIC PRESS, INC. (LONDON) LTD.
24/28 Oval Road, London NW1 7DX

Library of Congress Cataloging in Publication Data

Main entry under title:

Function of quinones in energy conserving systems.

Includes bibliographies and index.
1. Ubiquinone--Physiological effect. 2. Energy metabolism. I. Trumpower, Bernard L.
QP801.Q5F86 1982 574.19'127 82-8846
ISBN 0-12-701280-X AACR2

PRINTED IN THE UNITED STATES OF AMERICA

82 83 84 85 9 8 7 6 5 4 3 2 1

Assunta Baccarini-Melandri
1940–1981

The scientific community has been struck by the death of Assunta Baccarini-Melandri, who died in April 1981 after a tragic traffic accident—a great loss for her family and for the people in bioenergetics to whom she increasingly dedicated her activities in the past years.

Born in 1940 and raised in Faenza, Italy, she studied biology in Bologna and received her doctoral degree in 1962. While spending two postdoctoral years in Bloomington, Indiana, she and her husband were introduced to photosynthetic bacteria by Howard Gest. The two young scientists were immediately fascinated with the versatility of these respiring and photosynthesizing organisms, as well as the possibilities for research on bacterial bioenergetics (work on *E. coli* was almost purely molecular biology at that time). They recognized the possible evolutionary implications for respiratory and photosynthetic processes in higher organisms. In the intervening years, the photosynthetic bacteria remained important in research by the Melandris, who maintained intensive contact and experimental collaboration with other laboratories, especially with the groups of Baltscheffsky, Crofts, Marrs, and Jackson. I personally feel very fortunate to have participated in this experimental collaboration and stimulating exchange of ideas.

In Bloomington, Assunta started her work on photosynthetic bacteria with a study on the coupling factor ATPase, demonstrating that the same enzyme complex functions in oxidative as well as in photosynthetic ATP formation. This result suggested that both processes take place on the same membrane. Back in Bologna, the respiratory chain of *Rhodopseudo-*

monas capsulata was investigated in detail. As in mitochondria, three coupling sites were identified between NADH and O_2. However, the chain was found to be branched not only on the dehydrogenating side but also on the oxidizing end: two cytochrome *b* type oxidases were characterized, one of them being reduced by cytochrome c_2, much as cytochrome *c* reduces cytochrome oxidase in mitochondria. Since cytochrome c_2 was known also as the electron donor for the reaction center in these bacteria, it was concluded that respiratory and photosynthetic electron transport share common carriers. Assunta established this for cytochrome c_2 in further experiments. In an additional study, cytochrome c_2 was found to be located in the periplasmic space, where it proved to be topographically and functionally analogous to cytochrome *c* in mitochondria, and to algal cytochrome *c* or plastocyanin in chloroplasts. From these observations, a basically universal organization of energy conserving electron transport in biomembranes was derived.

Another extensive study by the research group in Bologna was of the electrochemical proton gradient maintained in illuminated chromatophores of photosynthetic bacteria and of the associated photophosphorylation. Although this system behaved in many respects according to the chemiosmotic hypothesis, there were some inconsistencies, one of them being that ATP formation stopped very rapidly after illumination, although the H^+-potential decayed only slowly, remaining above the threshold for phosphorylation for a rather long time. This pointed to a direct effect of electron transport on the coupling factor ATPase.

The last major investigation by Assunta was to discover the role of ubiquinone in photosynthetic bacteria. She contributed to the belief that there are four different sites of ubiquinone involvement in cyclic electron transport: two lying in series on the donor side of the reaction center, one lying between cytochromes *b* and *c*, and one constituting the large ubiquinone pool—the function of which still remains unclear. In extraction–reconstitution studies, Assunta demonstrated that the length of the isoprenoid side chain determines the differential interaction with these reaction sites. Moreover, she established that the whole quinone pool was required for ATP formation in continuous light; however, as had been demonstrated by others, over 90% of the pool could be removed without effect on the turnover of cyclic electron transport in flashing light. It was found later that the ATP yield per single turnover flash did not depend on the presence of the quinone pool if flashes were optimally spaced with time, but it did so if the frequency of the flashes was increased. This interesting result could not be explained in simple chemiosmotic terms and again pointed to a regulation of the coupling factor ATPase by electron transport, possibly by the quinone pool.

It is obvious that Assunta contributed substantially to our current view of bioenergetic processes. But it was not only research which made her important to our field: it was also her charming personality, basically shy but quite determined, which ideally qualified her as an ambassador for bioenergetics far beyond the borders of her country, on numerous visits to the United States, the countries within Europe, and the Soviet Union. Another facet of her character was her ability as an organizer, revealed upon her election, in 1977, to the presidency of the Italian Bioenergetics Group, a position she held for four years. Many of us also remember vividly the delightful First European Bioenergetics Conference which she organized in Urbino, Italy. To her memory we dedicate this book.

Günter Hauska
Regensburg, West Germany

CONTENTS

CONTRIBUTORS

Numbers in parentheses indicate the pages on which the authors' contributions begin.

Brian A. C. Ackrell (319), Department of Biochemistry and Biophysics, University of California, San Francisco, and Molecular Biology Division, Veterans Administration Medical Center, San Francisco, California 94121

Adolfo Alexandre (541), Department of Physiological Chemistry, The Johns Hopkins University School of Medicine, Baltimore, Maryland 21205

Hiroyuki Arata[1] (199), Department of Biochemistry, University of Washington, Seattle, Washington 98195

C. J. Arntzen (443), MSU-DOE Plant Research Laboratory, Michigan State University, East Lansing, Michigan 48824

A. Baccarini-Melandri[2] (285), Istituto e Orto Botanico, Università di Bologna, 40126 Bologna, Italy

W. F. Becker (351), Institut für Physikalische Biochemie, Universität München, Munich, Federal Republic of Germany

Helmut Beinert (227), Institute for Enzyme Research, University of Wisconsin, Madison, Wisconsin 53706

Jan A. Berden (153, 235), Laboratory of Biochemistry, B.C.P. Jansen Institute, University of Amsterdam, 1018TV Amsterdam, The Netherlands

Charles L. Bering[3] (35), Department of Biochemistry and Molecular Bi-

[1] Present address: Department of Biochemistry and Biophysics, University of Pennsylvania, Philadelphia, Pennsylvania 19104.

[2] Deceased.

[3] Present address: Department of Natural Sciences, State University of New York, College of Technology, Utica, New York 13502.

ology and Department of Chemistry, Northwestern University, Evanston, Illinois 60201

Enrico Bertoli (111), Istituto di Chimica Biologica, Università di Bologna, 40126 Bologna, Italy

Haywood Blum (247), Department of Biochemistry and Biophysics, University of Pennsylvania, Philadelphia, Pennsylvania 19104

Bernadette Bouges-Bocquet[4] (409), Institut de Biologie Physico-Chimique, 75005 Paris, France

John R. Bowyer[5] (365, 377, 477), Department of Biochemistry, Dartmouth Medical School, Hanover, New Hampshire 03755

F. Capuano (527), Institute of Biological Chemistry, Faculty of Medicine and Center for the Study of Mitochondria and Energy Metabolism, University of Bari, 70124 Bari, Italy

Claudio Casali (111), Istituto e Orto Botanico, Università di Bologna, 40126 Bologna, Italy

Therese M. Cotton[6] (35), Department of Biochemistry and Molecular Biology and Department of Chemistry, Northwestern University, Evanston, Illinois 60201

Antony R. Crofts (477), Department of Physiology and Biophysics, University of Illinois, Urbana, Illinois 61801

Robert Crowley (453), Department of Plant and Soil Biology, University of California, Berkeley, California 94720

David Crowther[7] (499), Biology Department, Brookhaven National Laboratory, Upton, New York 11973

S. C. Darr (443), MSU-DOE Plant Research Laboratory, Michigan State University, East Lansing, Michigan 48824

R. J. Debus (299), Department of Physics, University of California, San Diego, La Jolla, California 92093

Mauro Degli Esposti (111), Istituto e Orto Botanico, Università di Bologna, 40126 Bologna, Italy

Simon de Vries (153, 235), Laboratory of Biochemistry, B.C.P. Jansen Institute, University of Amsterdam, 10181 TV Amsterdam, The Netherlands

[4] Present address: UPMTG-Institut Pasteur, 28 Rue du Dr. Roux, 75724 Paris, cedex 15 France.

[5] Present address: Sittingbourne Research Centre, Sittingbourne, Kent, England, ME9 8AG.

[6] Present address: Illinois Institute of Technology, Department of Chemistry, Chicago, Illinois 60616.

[7] Present address: Department of Botany and Microbiology, University College London, London WC1E 6BT, England.

P. Leslie Dutton (29, 265, 271, 277), Department of Biochemistry and Biophysics, University of Pennsylvania, Philadelphia, Pennsylvania 19104

Carol A. Edwards[8] (377), Department of Biochemistry, Dartmouth Medical School, Hanover, New Hampshire 03755

W. D. Engel (351), Institut für Physikalische Biochemie, Universität München, Munich, Federal Republic of Germany

Romana Fato (111), Istituto di Chimica Biologica, Università di Bologna, 40126 Bologna, Italy

G. Feher (299), Department of Physics, University of California, San Diego, La Jolla, California 92093

N. Gabellini (285), Istituto e Orto Botanico, Università di Bologna, 40126 Bologna, Italy

Peter B. Garland (465), Department of Biochemistry, Medical Sciences Institute, University of Dundee, Dundee DD1 4HN, Scotland, England

Asher Gopher (511), Department of Biochemistry, Tel-Aviv University, Tel-Aviv, Israel

A. C. F. Gorren (213), Department of Biophysics, Huygens Laboratory of the State University, 2300RA Leiden, The Netherlands

F. Guerrieri (527), Institute of Biological Chemistry, Faculty of Medicine and Center for the Study of Mitochondria and Energy Metabolism, University of Bari, 70124 Bari, Italy

M. R. Gunner (29, 265), Department of Biochemistry and Biophysics, University of Pennsylvania, Philadelphia, Pennsylvania 19104

Menachem Gutman (511), Department of Biochemistry, Tel-Aviv University, Tel-Aviv, Israel

Charles R. Hackenbrock (125), Laboratories for Cell Biology, Department of Anatomy, School of Medicine, University of North Carolina at Chapel Hill, Chapel Hill, North Carolina 27514

Günter Hauska (87), Institute of Botany, University of Regensburg, Regensburg, Federal Republic of Germany

Geoffrey Hind (499), Biology Department, Brookhaven National Laboratory, Upton, New York 11973

Eduard Hurt (87), Institute of Botany, University of Regensburg, Regensburg, Federal Republic of Germany

G. Izzo (527), Institute of Biological Chemistry, Faculty of Medicine and Center for the Study of Mitochondria and Energy Metabolism, University of Bari, 70124 Bari, Italy

[8] Present address: Laboratory for Plant Molecular Biology, The Rockefeller University, New York, New York 10021.

U. Johanningmeier (425), Department of Biology, Ruhr-University, 4630 Bochum, Federal Republic of Germany

Robert W. Jones (465), Noyes Laboratory, School of Chemical Sciences, University of Illinois, Urbana, Illinois 61801

Edna B. Kearney (319), Department of Biochemistry and Biophysics, University of California, San Francisco, California 94143, and Molecular Biology Division, Veterans Administration Medical Center, San Francisco, California 94121

Tsoo E. King (3), Department of Chemistry and Laboratory of Bioenergetics, State University of New York at Albany, Albany, New York 12222

D. Kleinfeld (299), Department of Physics, University of California, San Diego, La Jolla, California 92093

Albert L. Lehninger (541), Department of Physiological Chemistry, The Johns Hopkins University School of Medicine, Baltimore, Maryland 21205

John J. Lemasters (125), Laboratories for Cell Biology, Department of Anatomy, University of North Carolina at Chapel Hill, Chapel Hill, North Carolina 27514

Giorgio Lenaz (111), Istituto e Orto Botanico, Università di Bologna, 40126 Bologna, Italy

Paul A. Loach (35), Department of Biochemistry, Molecular and Cell Biology, Northwestern University, Evanston, Illinois 60201

M. Lorusso (527), Institute of Biological Chemistry, Faculty of Medicine and Center for the Study of Mitochondria and Energy Metabolism, University of Bari, 70124 Bari, Italy

Richard Malkin (453), Department of Plant and Soil Biology, University of California, Berkeley, California 94720

Sergio Mascarello (111), Istituto di Chimica Biologica, Università di Bologna, 40126 Bologna, Italy

K. Matsuura (277), Department of Biochemistry and Biophysics, University of Pennsylvania, Philadelphia, Pennsylvania 19104

Steve W. Meinhardt (477), Department of Physiology and Biophysics, University of Illinois, Urbana, Illinois 61801

B. A. Melandri (285), Istituto e Orto Botanico, Università di Bologna, 40126 Bologna, Italy

Peter Mitchell (553), Glynn Research Institute, Bodmin, Cornwall PL30 4AU, England

Larry E. Morrison[9] (35), Department of Biochemistry and Molecular Biology, and Department of Chemistry, Northwestern University, Evanston, Illinois 60201

[9] Present address: Standard Oil Company, Warrenville Road and Mill, Naperville, Illinois 60540.

Jennifer Moyle (553), Glynn Research Institute, Bodmin, Cornwall PL30 4AU, England

P. Mueller (277), Department of Molecular Biology, Eastern Pennsylvania Psychiatric Institute, Philadelphia, Pennsylvania 19129

J. E. Mullet (443), MSU-DOE Plant Research Laboratory, Michigan State University, East Lansing, Michigan 48824

W. Oettmeier (425), Department of Biology, Ruhr-University, 4630 Bochum, Federal Republic of Germany

Tomoko Ohnishi (247), Department of Biochemistry and Biophysics, University of Pennsylvania, Philadelphia, Pennsylvania 19104

M. Y. Okamura (299), Department of Physics, University of California, San Diego, La Jolla, California 92093

Daniel P. O'Keefe (271), Department of Biochemistry and Biophysics, University of Pennsylvania, Philadelphia, Pennsylvania 19104

N. K. Packham (277), Department of Molecular Biology, Eastern Pennsylvania Psychiatric Institute, Philadelphia, Pennsylvania 19129

S. Papa (527), Institute of Biological Chemistry, Faculty of Medicine and Center for the Study of Mitochondria and Energy Metabolism, University of Bari, 70124 Bari, Italy

Giovanna Parenti-Castelli (111), Istituto di Chimica Biologica, Università di Bologna, 40126 Bologna, Italy

William W. Parson (199), Department of Biochemistry, University of Washington, Seattle, Washington 98195

K. Pfister[10] (443), Botanical Institute I, University of Würzburg, D-87 Würzburg, Germany

Veronica M. Poore (141), Department of Biochemistry, University of Southampton, Southampton SO9 3TU, England

Roger C. Prince (29, 265, 271), Department of Biochemistry and Biophysics, University of Pennsylvania, Philadelphia, Pennsylvania 19104

C. Ian Ragan (141), Department of Biochemistry, University of Southampton, Southampton SO9 3TU, England

Rona R. Ramsay (319), Department of Biochemistry and Biophysics, University of California, San Francisco, California 94143 and Molecular Biology Division, Veterans Administration Medical Center, San Francisco, California 94121

Peter R. Rich (73), Department of Biochemistry, University of Cambridge, Cambridge CB2 1QW, England

Frank J. Ruzicka (227), Institute for Enzyme Research, University of Wisconsin, Madison, Wisconsin 53706

John C. Salerno (247), Department of Biology, Rensselaer Polytechnic Institute, Troy, New York 12181

[10] Present address: CIBA-Geigy Co., Basel, Switzerland.

H. Schägger (351), Institut für Physikalische Biochemie, Universität München, Munich, Federal Republic of Germany

Jean E. Schelhorn[11] (35), Department of Biochemistry and Molecular Biology, and Department of Chemistry, Northwestern University, Evanston, Illinois 60201

Heinz Schneider[12] (125), Laboratories for Cell Biology, Department of Anatomy, University of North Carolina at Chapel Hill, Chapel Hill, North Carolina 27514

Thomas P. Singer (319), Department of Biochemistry and Biophysics, University of California, San Francisco California 94143, and Molecular Biology Division, Veterans Administration Medical Center, San Francisco, California 94121

E. C. Slater (153, 235), Laboratory of Biochemistry, B.C.P. Jansen Institute, University of Amsterdam, 1018TV Amsterdam, The Netherlands

K. E. Steinback (443), MSU-DOE Plant Research Laboratory, Michigan State University, East Lansing, Michigan 48824

A. J. Swallow (59), Paterson Laboratories, Christie Hospital and Holt Radium Institute, Manchester M20 9BX, England

A. P. G. M. Thielen (213), Department of Biophysics, Huygens Laboratory of the State University, 2300RA Leiden, The Netherlands

G. Denis Thorn (319), Research Institute, Agriculture Canada, London, Ontario N6A 5B7, Canada

D. M. Tiede (265, 277), Department of Biochemistry and Biophysics, University of Pennsylvania, Philadelphia, Pennsylvania 19104, and Department of Molecular Biology, Eastern Pennsylvania Psychiatric Institute, Philadelphia, Pennsylvania 19129

A. Trebst (425), Department of Biology, Ruhr-University, 4630 Bochum, Federal Republic of Germany

H. J. van Gorkom (213), Department of Biophysics, Huygens Laboratory of the State University, 2300 RA Leiden, The Netherlands

B. R. Velthuys (401), Martin Marietta Laboratories, Baltimore, Maryland 21227

André Verméglio[13] (169), Service de Biophysique, Département de Biologie, Centre d'Etudes Nucléaires de Saclay, BP 2, 91190 Gif-sur-Yvette, France

G. von Jagow (351), Institut für Physikalische Biochemie, Universität München, Munich, Federal Republic of Germany

[11] Present address: Owens-Corning Fiberglas, Technical Center, Granville, Ohio 43023.

[12] Present address: Department of Pharmaceutical Research, F. Hoffmann-La Roche & Co., Ltd., CH-4002 Basel, Switzerland.

[13] Present address: Département de Biologie, Service de Radioagronomie Centre d'Etudes Nucléaires de Cadarache, BP 1, 13115 St Paul-lez-Durance, France.

Gordon A. White (319), Research Institute, Agriculture Canada, London, Ontario N6A 5B7, Canada

C. A. Wraight (181), Department of Physiology and Biophysics and Department of Botany, University of Illinois, Urbana, Illinois 61801

C.-A. Yu[14] (333), Department of Chemistry, State University of New York at Albany, Albany, New York 12222

L. Yu[15] (333), Department of Chemistry, State University of New York at Albany, Albany, New York 12222

[14] Present address: Department of Biochemistry, Oklahoma State University, Stillwater, Oklahoma 74078.

[15] Present address: Department of Biochemistry, Oklahoma State University, Stillwater, Oklahoma 74078.

PREFACE

Our appreciation for the importance of quinones in biological systems has grown considerably in the years since the isolation of "Köfler's quinone," the discovery of ubiquinone, and the subsequent identification of plastoquinone as the quinone first isolated by Köfler. It is now widely recognized that quinones fulfill a universal and possibly unique function in electron transfer and energy conserving systems. At the same time, there is an equally wide awareness that the molecular basis of quinone function is probably much more complex than is implied by the relative structural simplicity of these compounds. We are now beginning to address, by experimentation, the questions thus raised. In this sense, the chapters herein acknowledge that quinones have come of age.

It is virtually certain that the specific reactivity and thermodynamic stability exhibited by quinones in biological electron-transfer systems are governed by "quinone proteins." In the first chapter, Tsoo King discusses the progression and current status of his research on ubiquinone proteins in mitochondria. Such quinone–protein interactions are a central theme, at least implicitly, in almost every chapter in this book and form the basis for the longest chapter.

A second recurring theme deals with how the biological reactivity of quinones relates to their reactivity as simple organic molecules. Apparently, biological systems, and quinone proteins in particular, function by selectively amplifying some inherent reactivities of quinones and by suppressing others. Thus, papers in the second chapter, on the physical chemistry of biologically active quinones, have special importance. These include data and discussion of the electrochemical and spectral properties of quinones and semiquinones, and a model for quinone–cytochrome electron-transfer reactions in which Rich points out the kinetic advantages that derive from altering the ionizability (pK') of ubiquinol.

In all species examined to date, there is a substantial molar excess of quinone relative to other electron-transfer components, even allowing for multiple sites of quinone action. This stoichiometry and the freely soluble nature of isoprenoid quinones in hydrophobic media, such as the lipid bilayer of energy transducing membranes, led to the early hypothesis that quinones, such as ubiquinone, function as a homogeneous pool for the collection and shuttling of reducing equivalents between otherwise noninteracting electron-transfer chain complexes. It is now recognized that there are multiple quinone binding sites and species of thermodynamically stable semiquinones in both the quinone reductase and quinol oxidase complexes of the photosynthetic and respiratory electron-transfer chains. This, in turn, has led to consideration of the possibility that there may be direct interaction between laterally mobile electron-transfer complexes. The papers in the third chapter thus deal with the properties and possible function of the quinone pool, and the relationship and possible exchange between bulk phase and bound quinone.

Although detectable levels of ubisemiquinone have been reported in mitochondria, the functional importance of thermodynamically stable semiquinones in mitochondria and photosynthetic systems has received attention only recently. In reading the papers in the fourth chapter, which are devoted to the possible function of these species, it is intriguing to consider whether the stable semiquinone pair in the mitochondrial succinate–ubiquinone oxidoreductase may have a "gating" function not unlike the more thoroughly characterized primary and secondary acceptor quinones in the plant and bacterial photosystems.

The fifth chapter is devoted to "quinone binding proteins." Although numerous papers elsewhere in the book could also have been included in this grouping, these were chosen to illustrate the types of experimental questions currently being addressed and the approaches being used to address these questions. The latter include the use of isolated electron-transfer complexes in combination with resolution and reconstitution of the endogenous quinones and individual polypeptides. The chapter also illustrates the application of inhibitory analogs and photoactivatable derivatives to identify the proteins that may interact directly with quinones. In some instances the proteins thus identified have recognizable redox functional groups, suggesting that these may be acceptors/donors for the redox-active quinone species. In other instances, the reputed quinone-binding proteins have no detectable prosthetic group, suggesting that these may be "quinone apoproteins," the function of which remains to be elucidated.

The papers in Chapter VI include results on the function of plastoquinone in the photosynthetic reaction centers and the *b-f* complex. In the plant system, primary emphasis to date has been on the function of plastoquinone in the photosystems and the interaction of herbicides at these quinone sites. However, plastoquinone is also involved at a DBMIB-sensitive site in the *b-f* complex, most likely the Rieske type iron–sulfur protein therein, and the recent isolation of a *b-f* complex opens the way to further study of this aspect of plastoquinone function.

Quinones are the only components of energy transducing electron-transfer chains that are (1) universally occurring, (2) soluble in hydrocarbon solvents, and (3) oxidized and reduced through reactions that unequivocally involve release and uptake of protons. These considerations support Mitchell's original suggestion that transmembranous reduction and oxidation of quinones constitute a direct ligand conduction mechanism for energy conserving proton translocation. The papers in the last chapter are thus devoted to pathways of electron transfer in energy transducing membranes, with particular emphasis on protonmotive mechanisms of quinone function.

This volume is not a record of the proceedings of a conference, although many of the papers herein were presented for comment and criticism at an international conference on the "Function of Quinones in Energy Conserving Systems," held at Dartmouth College, Hanover, New Hampshire, in October 1980. The conference was sponsored by the Department of Biochemistry and Office of the Dean of Dartmouth Medical School, the National Institute of General Medical Sciences, the National Science Foundation, and the New Hampshire Heart Association. In soliciting and editing papers for this volume, I have attempted to provide a representative, but not necessarily exhaustive, account of current research on quinone function in energy conserving systems.

Any collection of research articles will include some material which eventually will be outdated. This cannot be avoided if one is to obtain an accurate view of a rapidly developing research area. In selecting these papers, however, I have tried to emphasize material of enduring usefulness, in the hope that *Function of Quinones in Energy Conserving Systems* will remain a valuable resource for students and researchers for years to come.

I

ERIC REDFEARN MEMORIAL LECTURE

The Eric Redfearn Memorial Lecture traditionally has been held at Leicester University, where Eric Redfearn was Professor of Biochemistry. We are grateful to Professor William Brammer and his colleagues in the Department of Biochemistry at the University of Leicester for the opportunity to hold the Redfearn Lecture as the plenary lecture for "Function of Quinones in Energy Conserving Systems."

Ubiquinone Proteins in Cardiac Mitochondria 1

TSOO E. KING

INTRODUCTION

I am privileged to give the Eric Redfearn Memorial Lecture. It is especially pleasurable because Professor Redfearn and I once worked together so enthusiastically. In this paper I will briefly review our studies of the quinone-binding proteins in cardiac mitochondria and report some recent findings.

Ubiquinone (coenzyme Q) was discovered by two schools (*1*), one led by Professor David Green of Madison and the other by Professor Richard Morton of Liverpool, and was named compound 275, and compound SA, respectively. There is no dispute concerning the structure, which has been confirmed in synthesis by Green in collaboration with Merck in America, and by Morton in collaboration with Hoffman-La Roche in Switzerland. However, electron transport[1] in the ubiquinone regions of the respiratory chain is still somewhat unclear.

Studies of electron transport may be pursued *in situ* or in purified systems [see, e.g., Green *et al.* (*3*) and Hatefi *et al.* (*4*, *5*); see also this Chapter]. Preparations that can transfer electrons from NADH or succinate to

[1] Electron transport (electron transfer, electron pathway, or related terms) as used here is actually an abbreviation including the transport of electron, hydrogen, or any other species if those species indeed occur in the respiratory chain. The term respiratory chain is employed for convenience and must not be considered literally as a fixed chain, i.e., "a chain of beads." It should not be considered that those "segments" termed succinate-*c* reductase, NADH-Q reductase, etc., exist statically as such in mitochondria. Actually, the essence of the respiratory chain concept is only that electron or hydrogen transfer in mitochondria takes place sequentially from one carrier to the next and eventually to oxygen (*2*).

Function of Quinones in Energy Conserving Systems

ISBN 0-12-701280-X

cytochrome *c* are usually called NADH- or succinate-cytochrome *c* reductase, respectively. These reductases are devoid of cytochrome *c* and cytochrome oxidase, but contain ubiquinone (Q),[2] phospholipids, and other respiratory carriers. Both phospholipids and ubiquinone are essential for function.

Green and co-workers (*3*) and Hatefi and co-workers (*4*) originally proposed that cytochrome *c* and ubiquinone were mobile components. More recently Kroger and Klingenberg (*6*, *7*) elaborated the concept of mobile ubiquinone and introduced the hypothesis of the Q-pool. It is true that mitochondria contain a molar excess of ubiquinone relative to cytochrome a_3 or c_1. But examination of spacefilling models of phospholipid and ubiquinone, especially when ubiquinone is bound to protein, does not suggest that ubiquinone-protein can swim freely in the mitochondrial bilayer.

In our laboratory, using methods different from those of Green and Hatefi (*3*–*5*), we have been successful in resolving submitochondrial particles to cytochrome oxidase, succinate-cytochrome *c* reductase (*8*), and the NADH dehydrogenase complex. The succinate-cytochrome *c* reductase can be further cleaved into SDH and the cytochrome b-c_1 complex. The b-c_1 complex catalizes electron transfer from QH_2 to cytochrome *c*, i.e., QH_2-cytochrome *c* reductase. The SDH shows properties identical to that directly isolated from submitochondrial particles. Further purification of the dehydrogenase by a method identical to the Keilin–Hartree preparation, yielded an almost pure and completely reconstitutively (>95%) active SDH [(*9*) and unpublished observation of this laboratory].

The succinate dehydrogenase thus prepared contains 1 FAD, 4 Fe, and 4-labile sulfide in the larger subunit (70,000), and 4 Fe and 4-labile sulfide in the smaller subunit (29,000), similar to the reconstitutively active (*10*) preparation obtained by Davis and Hatefi (*11*). The larger subunit contains flavin, 4 Fe, and 4-labile sulfide in two 2 Fe–2 S clusters whereas the smaller subunit contains one 4 Fe–4 S cluster, the so-called center S-3 [see (*12*, *13*) and references cited therein]. However, Albracht (*14*) has challenged our interpretation and the ratio of flavin to iron to sulfide that we first reported in 1964 (*16*). [Although the Dutch laboratory previously claimed a different ratio(s) (*15*), other workers have found the same ratio as ours (*17*–*19*).]

[2] Abbreviations: P-HMB, *p*-hydroxymercuribenzoate; Q, ubiquinone, a subscript indicates the number of isoprenoid units, while Q without a subscript refers to ubiquinone in a general sense; QP, ubiquinone bound to specific proteins; QP-C, a ubiquinone protein that occurs in cytochrome b-c_1 region and has not been isolated; QP-N, a ubiquinone protein that exists in the NADH dehydrogenase segment and has not been isolated; QP-S, a ubiquinone protein that accepts electrons directly from SDH; QH_2, fully reduced Q; SCR, succinate-cytochrome *c* reductase; SDH, soluble, reconstitutively active succinate dehydrogenase; TTFA, thenoyltrifluoroacetone.

MORTON'S REPORT AND OUR EARLY FAILURES

In 1964, when Dr. Eric Redfearn was working in our laboratory, we debated (1) whether ubiquinone was on the main chain of electron transport, and (2) the question of ubiquinone-binding protein(s). Dr. Redfearn was active in the Liverpool laboratory's discovery of ubiquinone. In his earlier papers, Dr. Redfearn clearly expressed the view that ubiquinone might not be on the main chain. Even after he left our laboratory, as late as 1966 and not long before his untimely death, he claimed:

> its (ubiquinone's) rate of reduction is insufficient to account for the total electron flux through the (mitochondrial) system (*20*).

He further asserted

> kinetic data of ubiquinone reactions could not be reconciled with a position on a single pathway and a more complicated system containing branched pathways may operate (*21*).

He seemed to indirectly yet carefully imply that ubiquinone might react at more than one site in the respiratory chain by the way of the branches. All of the carriers in the respiratory chain are proteins except ubiquinone, if it is indeed a member of the respiratory chain. Phospholipids are required for electron transport, but do not participate in the oxidation–reduction. Is ubiquinone really unique? Did this uniqueness unconsciously influence Professor Redfearn's attitude toward the role of ubiquinone as a respiratory carrier?

In London in the late 1950s while I was working in Keilin's laboratory at Cambridge, and sometime later at Sheffield (*23*, *24*), I heard Professor Morton's discourse on SA in which he and co-workers proposed that ubiquinone might function in the respiratory chain (*22*). At that time Green and co-workers advocated, even more strongly than the Liverpool group, that ubiquinone was a carrier in the chain [see, e.g. (*2*) and references cited therein]. These views were natural because Professor Morton was essentially an organic chemist, whereas Professor Green was a pioneer in mitochondrial biochemistry. At the earliest possible chance after the London meeting I asked Professor Keilin: "Do you think ubiquinone can really serve as a respiratory carrier, whereas all others are proteins?" He was silent for a long while. Indeed, he did not answer my question but continued other discussions until the end of our usual conversation in the Molteno Library. Eventually, he simply said, "Why don't you try to find out [that ubiquinone protein problem]?" His tone of voice was more an order than a statement or a question. Because of other work, I did not take up the problem until 1964 with Dr. Redfearn.

Obviously, the technology and facilities available then were not as sophisticated as now. Dr. Redfearn and I failed to find any apoenzyme for coenzyme Q. We carefully reread a paper by Green (*25*) presented at the Faraday Society Discussion at Nottingham. In his published paper, "Structure–Function Interrelationships in Mitochondrial Electron Transport and Oxidative Phosphorylation," Green stated:

> the lipoprotein in which coenzyme Q is localized (the Q lipoprotein) has the chemical composition summarized in Table 2 [entitled 'composition of the Q-lipoprotein' which contains only protein and lipid; the lipid consists of about 90% phospholipids, 8% neutral lipid, cholesterol, etc. with 0.42% Q] . . . The lipoproteins which have been isolated thus far are associated, respectively, with coenzyme Q, cytochrome c_1 and f_D [flavin protein required in the oxidation of NADH]. These will be referred to, respectively, as Q, c_1, and f_D lipoproteins.

To my knowledge, only in this reference paper did he (or anyone else) mention that ubiquinone was in a lipoprotein complex. However, since then Green seemingly has been interested in the mobile characteristics of ubiquinone functioning in electron transport rather than ubiquinone in a lipoprotein. I wonder why he christened the compound, which Crane and Widmer initially isolated, as coenzyme Q without giving too much thought to apoenzyme(s)? Perhaps ubiquinone is so freely soluble in phospholipid that not much attention has been paid to its interaction with protein. Nevertheless, we continued intermittently to search for an apoenzyme for ubiquinone. Not until the development of our sequential resolution technique [see (*8*) and subsequent publications] and the observation of a definitive pattern of SDS gel electrophoresis of the cytochrome b-c_1 complex (*26*) did we more seriously and systematically pursue the problem.

UBIQUINONE PROTEINS

After Dr. S. Takemori and I (*27*, *28*) applied sequential fragmentation (or sequential resolution) to the respiratory chain and succeeded in reconstituting succinate-cytochrome *c* reductase using SDH and the b-c_1 particle (the b-c_1–I complex), we found that ubiquinone and phospholipid not only stimulated electron transport activity in the reconstituted system, but also in the parent preparation (succinate-cytochrome *c* reductase). Although our laboratory still used the original method for preparing succinate-cytochrome *c* reductase, it proved inconvenient for the b-c_1 particle because the product was rather insoluble. Drs. C.-A. Yu and L. Yu developed a method (*29*) for preparing a soluble cytochrome b-c_1 complex (the b-c_1–II complex) based on reversible dissociation of SDH from submitochondrial particles at alkaline pH. Almost no denaturation of the cytochromes occurs when this process is performed anaerobically, in the cold (0–4 °C) and in the presence of succinate (*30*, *31*).

From experiments with SDH and the b-c_1–II complex, we gained valuable information. As shown in Table I, SDH can reconstitute with the b-c_1–II complex but cannot reconstitute with Green's complex III (*3*), although Green's complex III contains cytochromes b and c_1 and also possesses QH_2–cytochrome c reductase activity. However, SDH does not reconstitute with our cytochrome b-c_1–II complex when the complex is subjected to a limited chymotrypsin digestion or reacted with P-HMB equivalent to approximately half of the SH content of the complex (Table I).

These observations suggest that a compound(s) in our b-c_1–II complex is essential for reconstitution and is P-HMB reactive. This compound is most probably a protein because of its lability toward chymotrypsin, and apparently contains an accessible SH group. Also, this compound is not required in the oxidation of reduced Q.

TABLE I

Comparison of the Oxidation of Succinate by Q and Oxidation of QH_2 by *c* and the Reconstitutive Activity (with SDH) of Green's Complex III and the b-c_1–II Complex before and after Certain Treatments

Compound	Succinate → Q	$QH_2 \rightarrow c$	Reconstitutive activity (succinate → c)
(1) SDH	–	–	–
(2) b-c_1–II complex	–	+	–
(3) (1) + (2)	+	+	+
(4) Complex II of Green	+	–	–
(5) Complex III of Green	–	+	–
(6) (4) + (5)	+	+	+
(7) (1) + (5)	–	+	–
(8) (2) treated by $\frac{1}{2}$ P-HMB[A]	–	+	–
(9) (1) + (8)	–	+	–
(10) (2) controlled digestion[B] by chymotrypsin	–	+	–
(11) (1) + (10)	–	+	–

[A] A sample of the b-c_1–II complex was treated with an amount of P-HMB equivalent to slightly less than 50% of the total SH group content of the complex.

[B] The b-c_1–II complex was digested with 5 μg chymotrypsin/mg complex and the reaction was stopped with phenylmethylsulfonyl fluoride. The $QH_2 \rightarrow c$ activity decreased less than 20% after 60 min incubation and did not change after 10 min; whereas 50% of the reconstitutive activity was lost within 15 min.

TABLE II

Composition of the b-c_1–III Complex of Beef Heart, Complex III from Beef Heart, and Complex III from Yeast[A]

Compound	b-c_1–III	Complex III [King (*2*)]	Complex III [Palmer (*35*)]
Cytochrome b	10–10.5	8.0–8.5	9.3
Cytochrome c_1	5.7–6.0	4.0–4.2	4.6
Nonheme iron	6.6–7.2	8–10	4.6[D]
Flavin	0.0	[B]	0.0
Q	2–3	1–2	5.8
Phospholipids	250	[C]	[B]

[A] Values given in nmole or natom/mg protein.

[B] Not available.

[C] 0.4 mg/mg.

[D] The total iron was 26.5 natom/mg protein. Palmer (*35*) added an entry to his table called "excess iron" [see original explanation (*27*)].

With this information, we began direct isolation of this compound.[3] From the cytochrome b-c_1–II complex we sequentially resolved QH_2–cytochrome c reductase, or the b-c_1–III complex, and a Q binding protein which we call QP-S (here "S" indicates succinate system) (*32–34*). As such, this b-c_1–III complex is now no longer reconstitutable with SDH, although it still catalyzes oxidation of QH_2 by cytochrome c. Table II

TABLE III

Composition of QP-S

Compound	Old method	New method
Q (nmole/mg)	~10	~15–25
b (nmole/mg)	1	6
Phospholipids (total %)	30	52

[3] A note of warning may be appropriate. The preparation of QP-S using ammonium acetate fractionation does not show as sharp a separation as described (*32*, *34*); it depends upon the sample of the b-c_1 complex employed. Indeed, in all such fractionations of membranous protein components, one cannot take the percentage of saturation of salt described in the literature too strictly. The concentration of salt required to fractionate a desired component depends on the content of lipid in the system, among others.

TABLE IV

Electron Transfer from Succinate to Q through SDH and QP-S

Addition	Specific activity[A]
(1) QP-S, as isolated	0.0
(2) SDH	0.04
(3) (1) + (2)	30.5

[A] Sixty μg of QP-S as isolated by the old method were added in excess to 2 mg SDH and diluted to 0.5 ml. The final medium was 50 m*M* phosphate buffer, pH 7.4. Aliquots were then taken and assayed with Q_2 as electron acceptor. The specific activity is expressed in μmole of succinate oxidized per min per mg of QP-S, at 22°C.

compares the composition of the b-c_1–III complex thus obtained with Green's complex III from beef heart (*3*) and complex III from yeast recently reported by Palmer (*35*). Further purification of crude QP-S cleaved from the b-c_1 complex eventually yields a protein which, by SDS polyacrylamide gel electrophoresis, shows one band with some contamination of cytochrome *b*, and has an apparent monomeric molecular weight of 15,000. This method probably slightly damages the protein, because samples obtained with another isolation technique we developed (*33*) show a higher specific activity, but have significantly more cytochrome *b* "impurity." The composition of both QP-S preparations is shown in Table III.

When SDH is mixed with QP-S, Q can be reduced as shown in Table IV. This type of assay is not unique and is similar to methods used with pyridine nucleotide-linked enzymes. The reconstitution of succinate-Q reductase is not only a functional coupling but also a genuine physical reconstitution in which SDH and QP-S are structurally combined. After binding with QP-S, succinate dehydrogenase is no longer as unstable ($t_{1/2} \cong 30$ min) as the free form and loses its low ferricyanide activity (*36*). The reconstituted succinate-Q reductase consists only of 3 sizes of protein subunits; namely, 70 K and 29 K from SDH, and 15 K from the new protein or QP-S.

It is well known that TTFA does not inhibit SDH-catalyzed succinate oxidation by phenazine dyes, ferricyanide (*11*) or Wurster's blue (*37*). However, TTFA inhibits succinate oxidation by DCIP or by cytochrome *c* in systems catalyzed by the reconstituted as well as the intact succinate-Q reductase or succinate-cytochrome *c* reductase. The reappearance of TTFA inhibition was observed for succinate oxidation by DCIP when SDH binds with QP-S.

UBIQUINONE-PROTEIN IN THE NADH DEHYDROGENASE SEGMENT (QP-N)

While working with QP-S, another idea came to mind, since the NADH dehydrogenase segment contains a significant and relatively constant amount of Q. The Q-pool concept (*6*, *7*) introduced in 1973 seemed insufficient to explain many observations, therefore we asked if the NADH segment could contain another Q-protein, analogous to QP-S in the succinate system?

Widger obtained an NADH dehydrogenase segment of the respiratory chain which can catalyze oxidation of NADH by Q and ferricyanide. It is prepared by a method similar to that of Green and associates for preparing complex I (*3*, *4*). This segment, called Widger's particulate NADH dehydrogenase, contains all of the respiratory carriers, Q and phospholipid, required to transfer electrons from NADH to Q. Also, it is rotenone sensitive, rotenone insensitive NADH oxidation is almost completely absent. By most sensitive assays, this dehydrogenase segment appears devoid of any cytochrome, QP-S, or SDH, both structurally and in some tests functionally.

I must admit two points: the NADH dehydrogenase segment is more complicated than the succinate dehydrogenase segment and the background information about the possible existence of a Q-protein is even more meager than for the succinate dehydrogenase segment, consisting mostly of a vague notion that these two chains should be somewhat symmetric. At that time however, the stable ubisemiquinone in the b-c_1 region had already been unambiguously demonstrated by Nagaoka.

The demonstration of Q-radical(s) in the b-c_1 region prompted us to try to detect any similar ubisemiquinone formation after addition of NADH, without an electron acceptor, by the EPR method at room temperature. Luckily, in the first attempt a sharp signal was obtained (*34*). This signal shows a resonance absorbance at $g = 2.00$ with a line width of 8 gauss. Upon addition of rotenone, the signal is abolished, but antimycin has no effect. This system is not as clear as that demonstrating QP-C (see below) because the particulate dehydrogenase also contains FMN. Radicals of flavoproteins can also be demonstrated at room temperature.

Circumstantial evidence favors the view that the radical(s) is a ubisemiquinone attached to a specific protein. Removal of Q from Widger's dehydrogenase by a gentle method such as fractionation by ammonium sulfate in the presence of cholate, abolishes the signal, even after repletion of phospholipids to the system. In contrast, rotenone-insensitive NADH to ferricyanide activity is not affected. Nevertheless, phospholipids are necessary in the radical formation.

It has been reported that the flavin radicals generally show a line width of 12–16 gauss for the anionic form and of 19–24 gauss for the neutral form (*38*). Indeed, after photoreduction of the FMN in the dehydrogenase a flavin radical was observed in the system used to demonstrate the QP-N signal. This flavin radical has a line width nearly three times that of the radical of ubisemiquinone (*34*).

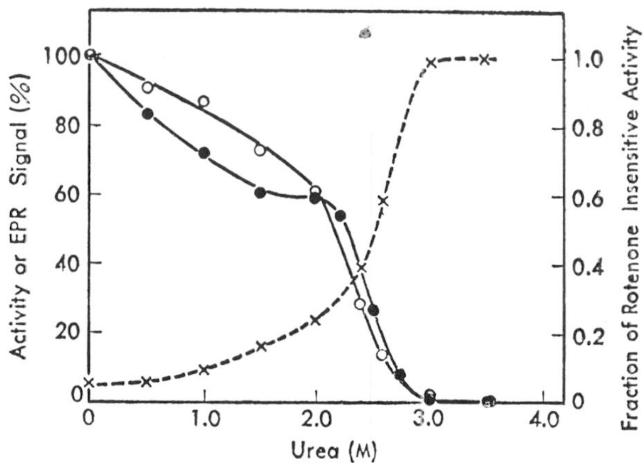

Fig. 1. Correlation of EPR amplitude of rotenone-sensitive NADH → Q and NADH → FeCy (ferricyanide) activities (○) of the particulate NADH dehydrogenase by urea treatment. NADH → FeCy activity (X) is represented as a fraction of rotenone-insensitive NADH-Q reductase. The NADH → Q activity was determined in a system containing 75 μM Q_2, 25 m*M* Tricine buffer, pH 8.0, in a total volume of 1.0 ml and a suitable amount (usually 40–60 μg protein) of the particulate NADH dehydrogenase. When the spectrophotometric readings remained constant, the reaction was started by addition of 0.1 m*M* NADH and monitored at 340 nm. After 30 to 45 sec 5 mg rotenone was added. The rotenone-sensitive NADH oxidation represents the initial rate minus the rate in the presence of rotenone. An extinction coefficient of 6.81 $mM^{-1}cm^{-1}$ was used in calculation because oxidized Q shows absorption at 340 nm with $\epsilon = 0.59\ mM^{-1}cm^{-1}$. The NADH → FeCy activity was determined in the system as reported for "30° enzyme" (*39*) using 50 m*M* Tricine, pH 8.0 instead of triethanolamine, and ferricyanide at a concentration of one m*M*. The EPR measurement was in Varian E-4 at a microwave frequency of 9.5 GHz, field modulation frequency 100 kHz, modulation amplitude 6.3 gauss, microwave power 20 mW, time constant one sec, scanning rate 50 gauss/min, receiver gain 4×10^3 (●). All activities and EPR signal were determined at room temperature. The urea treatment was conducted at 0 °C.

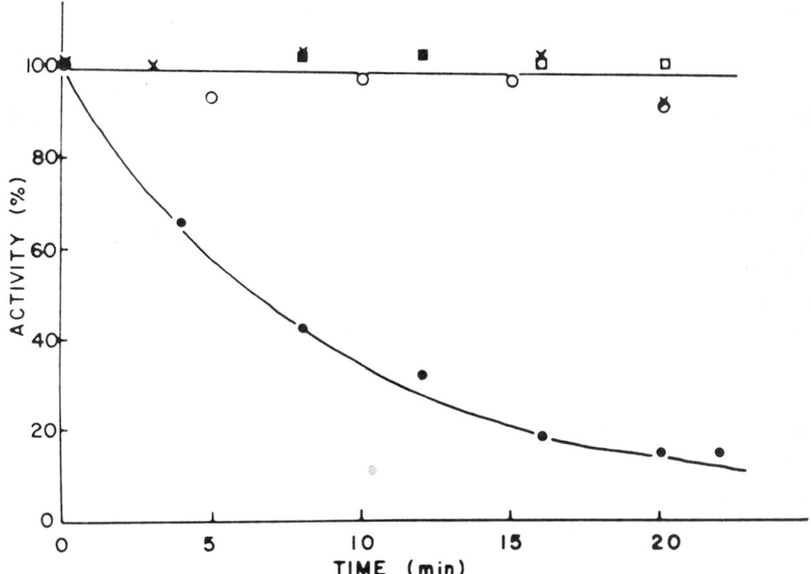

Fig. 2. Effect of chymotrypsin on rotenone-sensitive NADH → Q and NADH → FeCy (ferricyanide) activities and EPR amplitude of the particulate NADH dehydrogenase. The digestion was conducted at one mg of chymotrypsin per 50 mg of the dehydrogenase at room temperature. At indicated time periods, aliquots were removed for the determination of enzymic properties and EPR signal amplitude. These determinations were the same as Fig. 1. The curves designated as "+ chymotrypsin" indicate these samples after the proteolytic digestion, whereas those without the sign are for controls or blanks. The electron transfer activities and EPR amplitudes are represented by -○-, NADH → Q; -●-, NADH → Q + chymotrypsin; -□-, NADH → FeCy; -×-, NADH → FeCy + chymotrypsin.

The signal amplitude as shown in Fig. 1 is decreased by urea treatment. At the same time, the decrease of rotenone-sensitive NADH oxidation is practically parallel to the signal disappearance, while the rotenone-insensitive NADH oxidation by ferricyanide appears almost as a mirror image of the other two parameters. Controlled digestion of the particulate dehydrogenase with chymotrypsin also decreases, and eventually abolishes the signal (cf., Fig. 2). Interestingly, the rotenone-sensitive oxidation of NADH by Q is also gradually decreased and finally disappears. Conversely, rotenone-insensitive electron transfer from NADH to ferricyanide appears. These experiments show an invariable correlation between EPR signal and rotenone-sensitive oxidation of NADH by Q on the one hand and the rotenone-insensitive oxidation of NADH by ferricyanide on the other, thus supporting the idea that the radical is indeed a ubisemiquinone.

UBIQUINONE–PROTEIN IN THE CYTOCHROMES *b* AND c_1 REGION (QP-C)

From evidence accumulated in various experiments, we have deduced that at least one ubiquinone protein must exist in the cytochromes *b* and c_1 region. This hypothetical protein was temporarily called QP-2 and is now known as QP-C, where C stands for cytochrome. The cytochrome *b*-c_1–II complex shows seven bands by conventional SDS polyacrylamide gel electrophoresis experiments (*26*). Likewise, several laboratories also obtained seven or eight bands for Green's complex III or related preparations (*40*). We do not think these seven bands represent all the components in this region (*26*) as each band may contain more than one polypeptide or protein as respiratory components in the *b*-c_1 complex. Another significant finding was that activity of the phospholipid-and Q-deficient succinate-cytochrome *c* reductase could be restored only by reincorporation of Q and then phospholipid, but not phospholipid and then Q (*41*). Concurrently, we attempted, but so far have failed, to demonstrate the existence of a ubisemiquinone under various conditions in the reconstituted succinate-Q reductase with SDH and QP-S. Although the literature has reported or proposed the existence of free radical(s) of Q in mitochondria or submitochondrial particles the concentration is low when compared with the Q content of these samples (*42–46*). Of course, the protonmotive Q cycle requires that a Q-radical(s) exists, regardless of its half-life (*47, 48*). Equally important is the claim by certain investigators that dismutation of ubisemiquinone favors the non-free radical forms (*6, 7, 49*).

A few months after our isolation of QP-S by the "old" method (*37*), we found that succinate-cytochrome *c* reductase also can be reconstituted with the *b*-c_1–III complex plus catalytic amounts (in the order of 10^{-9} *M*) of SDH and QP-S. Naturally, these reconstituted systems show a much slower rate of electron transfer from succinate to cytochrome *c* than the parent reductase. Similarly, in the absence of an external acceptor, the rates of reduction of cytochromes *b* and c_1 are also slower but increase with additional SDH and QP-S. However, unlike the rate of reduction the extent of reduction of *b* and c_1 is not affected by the amount of SDH and QP-S added (*50*).

One night in early spring of 1978, Nagaoka of the Medical College of Gifu University, while working with the *b*-c_1–III complex plus catalytic amounts of SDH and QP-S, found a huge radical at $g = \sim 2$. First he used low temperature, about 77 °K. After my insistance that experiments be done at room temperature under aerobic conditions, the same sharp signal was obtained with a *g*-value of 2.004 and a line width of 8 gauss (*50*). The signal is not readily saturated even at microwave power as high as

200 mW at room temperature. Although I was certain that the signal was not due to the triplet state because of its line behavior, I sent the tracings to Peisach for his opinion. I received a call from him; with an ambivalent feeling he intoned that it was just a "garden variety(!)" organic radical. The signal must come from Q since the system is so simple and contains no components which can form an organic radical; QP-S and SDH added are only in the order of 10^{-9} *M*. Moreover, addition of exogenous Q_2 increases the signal amplitude (*51*), because in purifications of the b-c_1 complex a part of Q_{10} attached to the QP-C inevitably has been removed.

The radical(s) is abolished by addition of TTFA. This effect is expected because the inhibitor obstructs electron flow from succinate to QP-S. The signal is also abolished by antimycin A; this action is more complicated and is discussed elsewhere [see (*51*) and references cited therein]. Controlled treatment of the b-c_1 complex with α-chymotrypsin diminishes or abolishes the radical. Consideration of these results collectively indicates the existence of a ubiquinone protein (QP-C) in the b-c_1–III complex (*50*, *51*).

We have not yet isolated QP-C. Unpublished results show that the protein is perhaps one of the smaller proteins in the b-c_1 complex, at least when it is in the monomeric form. Recent SDS gel electrophoresis results (*52*) indicate QP-C may be either a 17,000 or 37,000 polypeptide(s) or both; the two polypeptides are at the positions we have suggested to be *b* cytochromes (*53*).

EPR STUDIES ON THE b-c_1–II COMPLEX

We have endeavored to demonstrate formation of ubisemiquinone from QP-S in the system containing SDH and QP-S with electron transfer function from succinate to DCIP or Q under various conditions, but attempts so far have failed. There may be many reasons for failing to demonstrate the radical.

Recently, Ohnishi and Trumpower (*54*) reported demonstrating at least two populations of ubisemiquinone radicals in succinate-cytochrome *c* reductase. Perhaps their work was inspired by the results of Konstantinov and co-workers (*55–57*). The succinate-cytochrome *c* reductase used by Ohnishi and Trumpower is somewhat different from our reductase. Their reductase contains a ratio of SDH to the b-c_1 complex varying from 0.3 to 1.0 with approximately 4–6 mole Q/mole c_1, in contrast to our reductase preparation, which contains about 1 flavin and a little more than 2 Q per b-c_1 complex. Ohnishi and Trumpower found a heterogeneous distribution of ubisemiquinone radicals. The one exhibits 0.1 spin (Q radical) per c_1 at 50 °K and 10 μW. Antimycin destabilizes and decreases the signal which originally shows a band width of 10 gauss. They called this ubisemiquinone SQ_c.

On the other hand, they measured at 50 °K and 100 mW and found 0.3 spin of Q-radical per c_1 in the absence of antimycin. Antimycin stabilizes the intermediate redox state (0.45 spin per c_1) of the spin-coupled ubisemiquinone. The bandwidth was 12 and 10 gauss in the absence and presence of antimycin, respectively. The investigators have claimed that the ubisemiquinone which is more difficult to saturate with increasing microwave power is SQ_s, which has been proposed to be very near center S-3 of SDH.

These results seem to support the existence of our QP-S and QP-C, and more importantly, demonstrate that the ubisemiquinone radical of our QP-S, (or SQ_s of Ohnishi and Trumpower), is indeed formed. However, there are several less desirable points in employing SCR to demonstrate two kinds of ubisemiquinone radicals at lower temperatures. SCR contains a great deal of flavin, whose radicals also show a g value of approximately 2.00. Moreover, the signal intensity from some 4 iron–sulfur centers will adversely affect the accuracy of the determination. Although clamping or poising the voltage at particular values can diminish these objections to a certain extent, as can the determination of the difference in relaxation behavior and EPR line shapes, a detailed analysis is still hampered at great intensities by interference from the nearby signal. Moreover, the existence of so many paramagnetic centers makes the use of antimycin to differentiate the semiquinone and iron–sulfur center signals quantitiatively tedious and less accurate. For example, Fig. 2 in Ohnishi and Trumpower (*54*) shows that the difference between these signals is minute, less than one mm as reproduced, in the presence and absence of antimycin.

Before publication of Ohnishi and Trumpower (*54*) we already thought that experiments might be performed at room temperature in systems free of flavin and with as few iron–sulfur clusters as possible. Ratios of fumarate and succinate should easily control the potentials of the system. With these considerations in mind, the b-c_1–II complex immediately came to our attention; it contains a full amount of QP-S (although a part of Q may have been lost during the isolation) and all respiratory carriers including the Rieske iron–sulfur protein (*58, 59*) required for electron transfer from succinate to cytochrome c, if SDH is added. Addition of catalytic amounts of SDH should demonstrate the reduction of cytochromes b and c_1 and the appearance of the ubisemiquinone radical(s). If QP-S would show a radical(s), then the observed overall radical would give indications

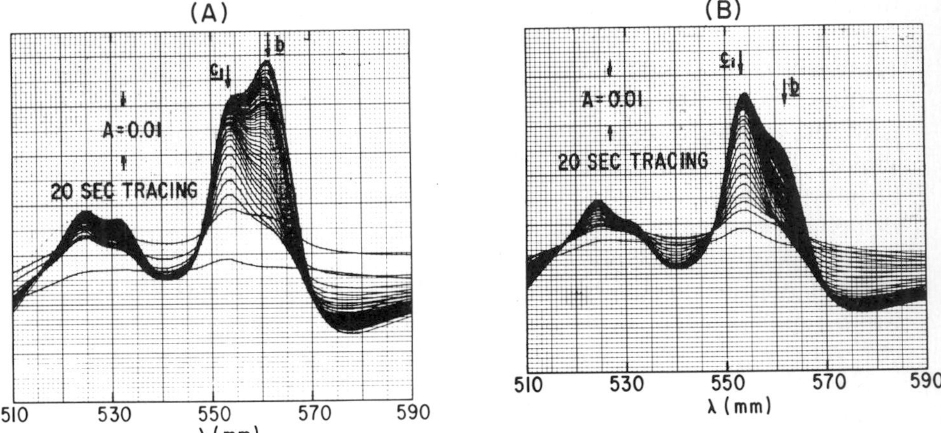

Fig. 3. Kinetics of cytochrome *b* and c_1 reduction by succinate. (A) The system contained 1.5 mg of *b*-c_1–II complex which contained 5.3 nmole cytochrome c_1, *b* to c_1 ratio of about 1.7, 7×10^{-9} M SDH and 21 nmole Q_2 in total volume of 0.8 ml. The first tracing was taken about 20 sec after addition of one μl 60 m*M* succinate without fumarate, and subsequent tracings were made at 20 sec intervals. The spectra were recorded at Aminco DW-2 spectrophotomer using an identical sample without substrate in the reference cell. (B) Same as (A) except the fumarate to succinate ratio used was 4.

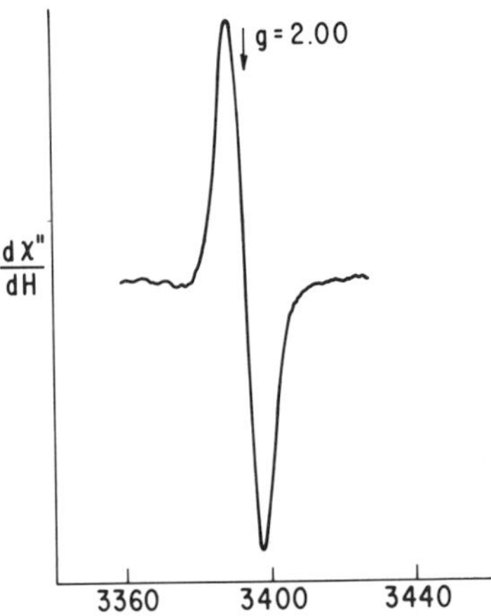

Fig. 4. A typical Q radical signal at room temperature (23 °C). The system contained 11.4 mg of *b*-c_1–II complex (40 nmole cytochrome c_1 and *b*-c_1 ratio of 1.7), 0.15 nmole SDH, and 168 nmole Q_2 in total volume of 0.20 ml. The spectrum was recorded 90 min after addition of fumarate/succinate in total concentration of 9.3 m*M*. The EPR settings were the same as those of Fig. 1, except microwave power 10 mW; time constant 3.0 sec; and scan rate 25 gauss per min. The spectrum was recorded at a Varian E-4 spectrometer using a 0.2 ml quartz EPR tube (flat cell) at room temperature.

of being heterogeneous. A more advantageous point is that the amount of SDH added can adjust the rate of reduction as well as the rate of the appearance of the Q-radical(s).

The reduction of cytochromes b and c_1 is depicted in Figs. 3A and 3B, and clearly shows that most of the c_1 is reduced prior to b reduction. Obviously, this is because the E_m of c_1 is higher than that of b. This figure also shows that the fumarate to succinate ratio affects the rates of reduction of the two cytochromes. An important observation, however, is that the Q-radical did not appear until more than half of the c_1 was reduced. This effect is very similar to what has been observed in the demonstration of the QP-C-radical(s) in the b-c–III complex (*34*, *50*) although radical appearance occurs before the reduction of b cytochromes.

A typical radical at room temperature is presented in Fig. 4. The signal height (cf., Fig. 5) is dependent on the fumarate to succinate ratio, similar to that of QP-C ubisemiquinone demonstrated in the b-c_1–III complex (*34*, *50*, *51*). Likewise, the kinetic behavior of the Q radical is a function of the molar ratio of fumarate to succinate used as substrate. An increase in the fumarate to succinate ratio causes an increase in the maximal radical concentration but decreases the rate of its formation and final concentra-

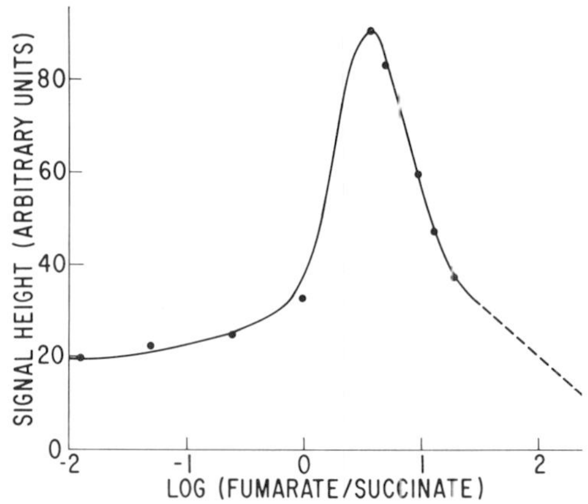

Fig. 5. Dependence of Q radical signal height of the b-c_1–II complex on the fumarate to succinate ratio. Different ratios of succinate to fumarate were added to the b-c_1–II complex (9.9 nmole c_1 with b to c_1 ratio of about 1.7) plus 1.0 nmole SDH, and 41 nmole Q_2 in 0.9 ml total volume. The EPR signals of the Q radical were monitored at machine settings of: 9.492 GHz microwave frequency, 100 mW microwave power, 200 kHz field modulation of 8.2 gauss amplitude, 3.0 sec time constant, and scan rate 25 gauss/min. The maximal signal height of each sample was plotted against the logarithm of fumarate to succinate ratio; the total concentration of the substrates was about 15 m*M*.

tion. However, there are differences between the b-c_1–II complex described here and the b-c_1–III complex reported elsewhere. In the b-c_1–III complex at a range of fumarate to succinate ratio of 80 to 100, radical(s) is formed slowly and remains constant for several hours, even at room temperature after the maximal concentration is reached (*50*, *51*). In contrast, a similar phenomenon is demonstrated in the b-c_1–II complex only at a fumarate to succinate ratio of 5 or slightly higher (*34*, *50*, *51*).

We also made a careful measurement of the Q-radical(s) of the b-c_1–II complex at approximately 0 °C (274 °K). The band width (peak–to–trough) is 8.25 ± 0.05 gauss as shown in Fig. 6. Upon freezing at 254 °K, the spectra becomes much more complicated, coupling with some unknown species occurs, and the ubisemiquinone signal decreases drastically. Upon warming to 274 °K again, the original features as well as signal intensity completely reappeared. The Q radical(s) was further examined in a Q-band EPR spectrometer. By using a computerized signal averager a spectrum has been obtained with g_1, g_2, and g_3 values of 2.0064, 2.0054, and 2.0021, respectively. The prominent g anisotropy observed prompts us to conclude that the Q is definitely bound, most probably by a protein(s), and cannot tumble freely; otherwise, only one g value would be observed.

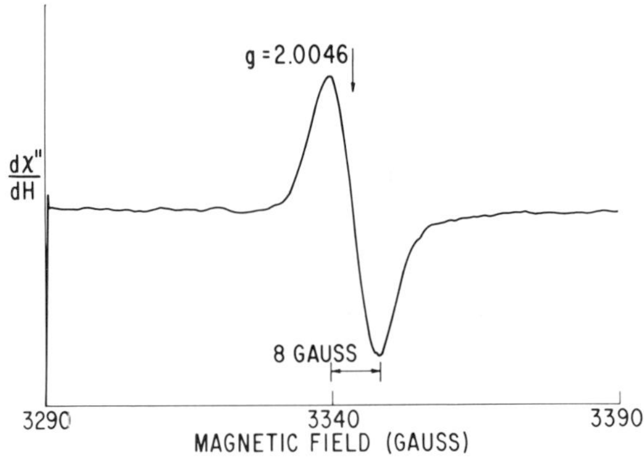

Fig. 6. The behavior of Q radical(s) signal. The system contained 5.8 mg b-c_1–II complex (containing 20.3 nmole c_1 with b to c_1 ratio of about 1.7), 0.05 nmole SDH, and 82 nmole Q_2 in 0.1 ml total volume (0.1 ml EPR flat cell). The Q radical was generated by addition of 2 μl 60 mM succinate and 8 μl 60 mM fumarate, and the experiments were carried out 30 min after addition of the substrate, with T = 274 °K. The EPR spectra were taken in a Bruker ER-420 EPR spectrometer with a computerized signal averager with 8 sweeps. The machine setting was: microwave frequency 9.367 GHz, microwave power 5 dB (~30 mW), 100 kHz field modulation of 5.0 gauss amplitude, time constant of 0.1 sec, scan rate at 230 gauss/min, gain 4.0×10^{-4}.

Figure 7 shows that TTFA inhibits formation of the ubisemiquinone radical. Spectrum A is the Q-radical(s) generated from the b-c_1–II complex with a catalytic amount of SDH. Subsequent spectra were obtained by adding increased amounts of TTFA. The final spectrum was obtained in the presence of antimycin and TTFA. Notice that in most of the experiments containing b-c_1–II complex, an excess of Q_2 was added. Both TTFA and antimycin were found to be competitive with Q in the system (*27*, *28*).

Figures 8A and 8B summarize power saturation curves of the Q-radical(s) of the b-c_1–II complex at room temperature and 77 °K in the absence of any inhibitor. Addition of antimycin in the presence of excess Q exhibits a signal magnitude higher than with addition of both antimycin and TTFA (Fig. 8C). This phenomenon is also observed when SCR is used, as Ohnishi and Trumpower reported (*54*). Nevertheless, by increasing the gain of the EPR spectrometer, the power saturation can still be measured in the presence of antimycin and TTFA (cf., Fig. 8C). In other words, under these conditions TTFA does not completely inhibit the radi-

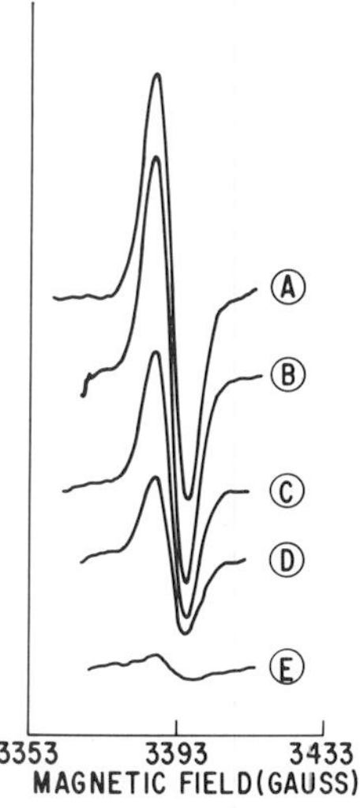

Fig. 7. Inhibition of Q radical signal by TTFA and antimycin. The system contained the b-c_1–II complex with 35.5 nmole cytochrome c_1 and b to c_1 ratio of 1.7, 0.04 nmole SDH and 382 nmole Q_2 in a total volume of 0.22 ml. The signal was generated by addition of 40 μl 60 mM fumarate and 1.5 μl 0.6 M succinate to the system. Curves A to D were recorded after 0, 0.8, 1.8, 2.6 nmole TTFA were added to the system and curve E was taken after further addition of 40 nmole antimycin A. The Varian E-4 settings are identical for all the tracings as those used in Fig. 5.

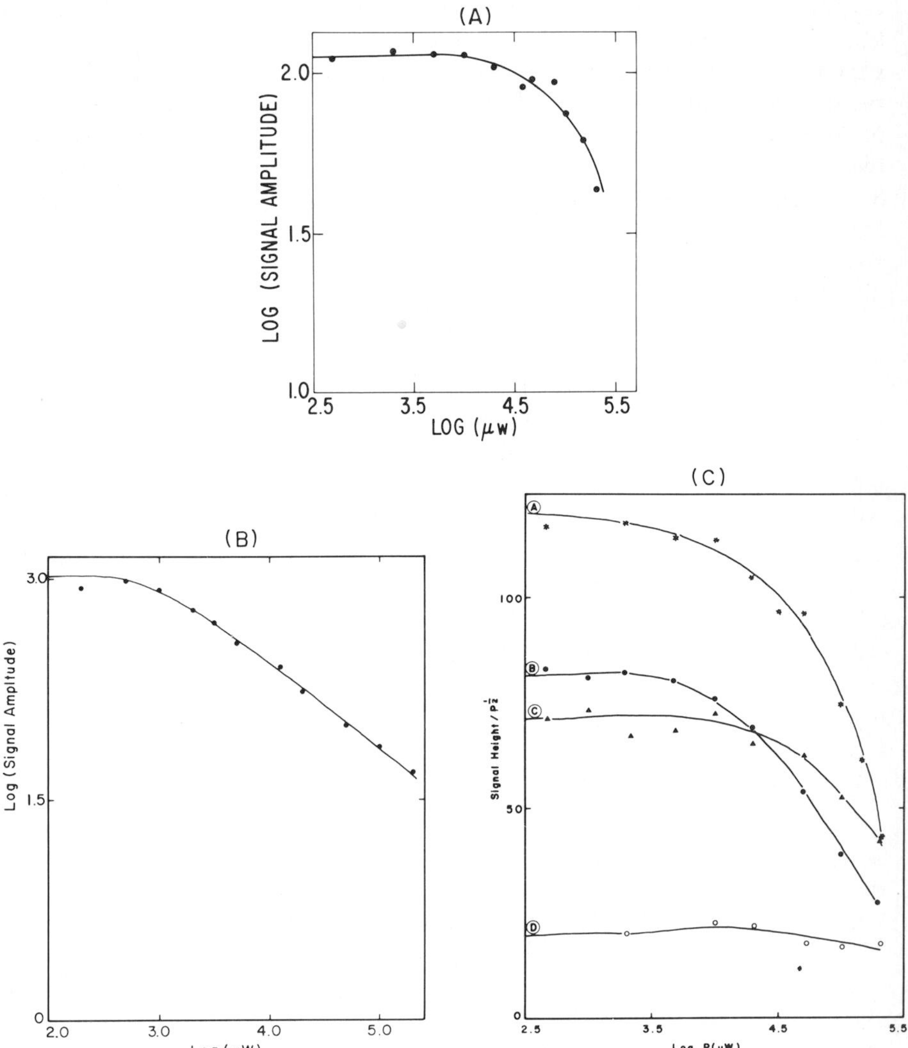

Fig. 8. The power saturation curves of Q radical of the b-c_1–II complex at room temperature and at 77 °K in the presence and absence of inhibitors. (A) At room temperature in the absence of inhibitors, the system was the same as that of Fig. 4 except that the microwave power varied from 0.5 mW to 200 mW. The experiment was carried out 45 min after addition of 30 μl 60 mM fumarate and one μl 600 mM succinate. The data was taken when the Q-radical had reached equilibrium and the signal stabilized. (B) At 77 °K in the absence of inhibitor, the system contained the b-c_1–II complex (35 nmole c_1 with b to c_1 ratio of 1.7), 0.04 nmole SDH and 448 nmole Q_2 in a total volume of 0.22 ml. The Q radical was generated by addition of 40 μl of 60 mM fumarate and 1.5 μl 600 mM succinate to the system. The power saturation was done after the signal reached equilibrium. The EPR machine settings

cal formation. This observation agrees with the functional activity [see (*8*), Fig. 4, p. 177]. Further addition of antimycin greatly decreased the signal amplitude and may have been caused by the depression of the QP-C radical produced in b-c_1–II complex in the presence of a catalytic amount of SDH at room temperature. One of the reasons for the difference between the QP-S and SQ_s signals (*54*) may be due to the coupling of center S-3 of SDH and the ubisemiquinone in this region, since only a catalytic amount of SDH is present. Because of factors which will be discussed elsewhere, the QP-S radical is difficult to demonstrate by admixing catalytic amounts of SDH and QP-S alone.

CONCLUDING REMARKS

Finally, one may ask why all that I have attributed to the radicals from QP-C and QP-N are not due to free Q, but to ubiquinone proteins. The answer is simple, because in the free system the dismutation is much in favor of the non-radical form with a predicated stability constant of 10^{-10} (*48*). **Specific Q proteins stabilize the radicals and dictate the specificity of QP.** As to the existence of specific quinone–binding proteins, very recently Hatefi and Galante (*60*) claimed cytochrome b-560 (in complex II), with equimolar amounts of two polypeptides of the sizes 13,500 and 15,500 can reconstitute with SDH to form succinate-Q reductase. If this is true, then the dual role that a cytochrome plays in electron transport is

---

are the same as those of Fig. 4 except microwave frequency, 9.021 GHz, and power varied from 0.2 to 200 mW. (C) Different power saturation behaviors of the g = 2.00 EPR signals at room temperature. Curve A: the b-c_1–II complex. The system contained 11.4 mg b-c_1–II complex (40 nmole c_1 with the b to c_1 ratio of about 1.7), 0.015 nmole SDH, and 168 nmole Q_2, 30 μl 60 mM fumarate and 1 μl 60 mM succinate in a total volume of .2 ml in 50 mM Tris–chloride–0.25 M sucrose–1 mM EDTA, pH 7.8. EPR settings were the same as those in Fig. 8B except for the gain, 6.2 × 10^3. Curve B: SCR. The system was in 0.24 ml total volume with 14.8 mg SCR containing 26.1 nmole c_1 in 50 mM Tris-chloride–0.25 M sucrose, pH 7.8, 40 μl 60 mM succinate, and 5 μl 60 mM fumarate. EPR settings were the same as curve A except the gain at 5 × 10^3. Curve C: the b-c_1–II complex in the presence of antimycin A. The system contained 11.5 mg b-c_1–II complex (c_1 40.6 nmole with the b to c_1 ratio of about 1.7), 0.014 nmole SDH, 380 nmole Q_2, 80 μl 60 mM fumarate and 1.0 μl 1.2 M succinate in 0.2 μl of total volume. After the radical signal was equilibrated, 50 nmole antimycin was added to the system. EPR settings were the same as curves A and B. Curve D: The b-c_1 complex in the presence of TTFA and antimycin. The system was the same as curve C except that power saturation tracing was done after addition of 2.2 nmole TTFA, 50 nmole antimycin to the system. EPR settings were the same as curve B except field modulation at 8.2 gauss and the gain was 8 × 10^3.

significant. I wonder then about QP-N and QP-C: do they also require *b*-560, or a similar cytochrome, or other carriers for stabilization of ubisemiquinone?

It must be emphasized, regardless of how attractive and plausible the evidence, the existence of QP-C and QP-N must be considered as an explanation for the stabilization of Q–radicals. At any rate, proof of the existence of these two Q-proteins (QP-C and QP-N) must wait their isolation, rigorous characterization of the isolated proteins and, of course, their eventual reconstitution. If Q-proteins exist, investigation of their relation with the protonmotive Q cycle (*47*, *48*) would certainly yield valuable information (*61*).

Note added on December 1, 1981: Since the original manuscript was submitted in the fall of 1980, a number of new results have been found. The more significant ones are: (1) the QP-S radical(s) has been demonstrated, and is not affected by the presence of antimycin A to the b-c_1–II complex; (2) more characteristics on QP-C radicals have been found, for example, in Q-band EPR spectrometer, a prominent anisotropic spectrum is observed at room temperature, or 0 °C and 77 °K. It shows a field separation between derivative extrema of 26 ± 1 gauss with g values of 2.0064, 2.0054, and 2.0021 at about 0 °C. These observations indicate that QP-C radical is strongly immobilized, since freely tumbling ubisemiquinone radicals would not show g anistrophy, resolved hyperfine structure, and substantial line shape change upon change of temperature (Wei, Y. H., Scholes, C. P., and King, T. E. (1981) *Biochem. Biophys. Res. Comm.* **99,** 1411, and also **101,** 326); (3) QP-C has been isolated from the b-c_1–III complex. It reconstitutes with the QP-C deficient b-c_1–III complex or the b-c_1–IV complex to recover the ubiquinone–cytochrome c reductase activity and restore the QP-C radical(s) (Wang, Y. T. and King, T. E., *Biochem. Biophys. Res. Comm.,* in press); (4) evidence suggests that there may be two QP-C's corresponding to Q_{out} and Q_{in} for proton motive force Q cycle (Kim, C. H. and King, T. E. (1981) *Biochem. Biophys. Res. Comm.* **101,** 609).

ACKNOWLEDGMENT

I am grateful for the collaboration of my research associates, present and past in this experimental work—especially, Drs. C. H. Kim, S. Nagaoka, Y. H. Wei, W. R. Widger, C.-A. Yu, and L. Yu. I am greatly appreciative of the collaborative work done with Drs. Y. H. Wei and C. P. Scholes on some of the EPR experiments. Experimental work was generously supported by grants from NIH (HL-12576 and GM-16767) and the American Cancer Society.

REFERENCES

1. Wolstenholme, G. E. W., and O'Connor, C. M., eds. (1961). "Quinones in Electron Transport," pp. xii and 453. Churchill, London.
2. King, T. E. (1978). *In* "Membrane Proteins" (P. Nicholls, J. V. Møller, P. L. Jørgensen, and A. J. Moody, eds.), pp. 17–31. Pergamon, Oxford.
3. Green, D. C., Wharton, D. C., Tzagoloff, A., Rieske, J. S., and Brierly, G P. (1965). *In* "Oxidases and Related Redox Systems" (T. E. King, H. S. Mason, and M. Morrison, eds.), pp. 1032–1076. Wiley, New York.
4. Hatefi, Y.; Hatefi, Y., and Stiggall, D. L.; Galante, Y. M., and Hatefi, Y.; Errede, B., Kamen, M. D., and Hatefi, Y. (1978). *In* "Biomembranes: Part D: Biological Oxidations, Mitochondrial and Microbial Systems" (S. Fleischer and L. Packer, eds.), Methods in Enzymology, Vol. 53, pp. 3–54. Academic Press, New York.
5. Hatefi, Y. (1966). *Compr. Biochem.* **14,** 199–231.
6. Kröger, A., and Klingenberg, M. (1973). *Eur. J. Biochem.* **34,** 358–368.
7. Kröger, A., and Klingenberg, M. (1973). *Eur. J. Biochem.* **39,** 313–323.
8. King, T. E. (1966). *Adv. Enzymol. Related Subj. Biochem.* **28,** 155–236.
9. Yu, C.-A., and Yu, L. (1980). *Biochim. Biophys. Acta* **59,** 409–420.
10. Cited in Singer, T. P., *et al.* (1978). *In* "Biomembranes: Part D: Biological Oxidations, Mitochondrial and Microbial Systems" (S. Fleischer and L. Packer, eds.), Methods in Enzymology, Vol. 53, p. 478. Academic Press, New York; Hanstein, W. G., Davis, K. A., Ghalambov, M. A., and Hatefi, Y. (1971). *Biochemistry* **10,** 2517–2521.
11. Davis, K. A., and Hatefi, Y. (1971). *Biochemistry* **10,** 2509–2516.
12. King, T. E., Ohnishi, T., Winter, D. B., and Wu, J. T. (1976). *In* "Iron and Copper Proteins" (K. T. Yasunobu, H. F. Mower, and O. Hayaishi, eds.), pp. 182–227. Plenum, New York.
13. Salerno, J. C., Lim, J., King, T. E., Blum, H., and Ohnishi, T. (1979). *J. Biol. Chem.* **254,** 4828–4935.
14. Albracht, S. P. J. (1980). *Biochim. Biophys. Acta* **612,** 11–28.
15. Zeylemaker, W. P., Dervartanian, D. V., and Veeger, C. (1965). *Biochim. Biophys. Acta* **99,** 183.
16. King, T. E. (1964). *Biochem. Biophys. Res. Commun.* **16,** 511.
17. Ackrell, B. A., Kearney, E. B., and Singer, T. P. (1978). *In* "Biomembranes: Part D: Biological Oxidations, Mitochondrial and Microbial Systems". (S. Fleischer and L. Packer, eds.), Methods in Enzymology, Vol. 53, pp. 466–483. Academic Press, New York.
18. Baginsky, M. L., and Hatefi, Y. (1969). *J. Biol. Chem.* **244,** 5313–5319.
19. Ackrell, B. A. C., Kearney, E. G., and Cole, C. J. (1977). *J. Biol. Chem.* **252,** 6963–6965.
20. Redfearn, E. R., and Burgos, J. (1966). *Nature (London)* **209,** 711–713.
21. Redfearn, E. R. (1966). *Vitam. Horm. (N.Y.)* **26,** 465–488.
22. Morton, R. A., Wilson, G. W., Lowe, J. S., and Leat, W. M. F. (1958). *Biochem. J.* **68,** 16P.
23. Falmy, N. I., Hemming, F. W., Morton, R. A., Paterson, J. Y. F., and Pennock, J. F. (1958). *Biochem. J.* **70,** 1P.
24. Pumphrey, A. M., Redfearn, E. R., and Morton, R. A. (1958). *Biochem. J.* **70,** 1P.
25. Green, D. E. (1959). *Discuss. Faraday Soc.* **27,** 206–216.
26. King, T. E., Yu, C.-A., Yu, L., and Chiang, Y. L. (1975). *In* "Electron Transfer Chains

and Oxidative Phosphorylation'' (E. Quagliariello, S. Papa, F. Palmieri, E. C. Slater, and N. Siliprandi, eds.), pp. 105–118. North-Holland Publ., Amsterdam.
27. Takemori, S., and King, T. E. (1964). *J. Biol. Chem.* **239,** 3546–3558.
28. King, T. E., and Takemori, S. (1964). *J. Biol. Chem.* **239,** 3559–3569.
29. Yu, C.-A., Yu, L., and King, T. E. (1974). *J. Biol. Chem.* **249,** 4905–4910.
30. King, T. E. (1962). *Biochim. Biophys. Acta* **58,** 375–377.
31. King, T. E. (1963). *J. Biol. Chem.* **238,** 4037–4051.
32. Nagaoka, S., Yu, C.-A., Yu, L., and King, T. E. (1979). *Fed. Proc., Fed. Am. Soc. Exp. Biol.* **38,** 638. (Abstr.)
33. Yu, C.-A., and Yu, L. (1980). *Biochemistry* **19,** 3579–3585.
34. King, T. E., Yu, L., Nagaoka, S., Widger, W. R., and Yu, C.-A. (1978). *In* ''Frontiers of Biological Energetics'' (L. Dutton, J. Leigh, and T. Scarpa, eds.), Vol. 1, pp. 174–182. Academic Press, New York.
35. Palmer, G. (1978). *In* ''Biomembranes: Part D: Biological Oxidations, Mitochondrial and Microbial Systems'' (S. Fleischer and L. Packer, eds.), Methods in Enzymology, Vol. 53, pp. 113–121. Academic Press, New York.
36. Yu, C.-A., Yu, L., and King, T. E. (1977). *Biochem. Biophys. Res. Commun.* **79,** 939–946.
37. Yu, C.-A., Yu, L., and King, T. E. (1977). *Biochem. Biophys. Res. Commun.* **78,** 259–265.
38. Palmer, G., Muller, F., and Massey, V. (1971). *In* ''Flavins and Flavoproteins'' (H. Kamini, ed.), pp. 123–136. Univ. Park Press, Baltimore, Maryland.
39. King, T. E. (1967). ''Oxidation and Phosphorylation'' (R. W. Estabrook and M. E. Pullman, eds.), Methods in Enzymology, Vol. 10, pp. 216–225. Academic Press, New York.
40. Slater, E. C., Marres, C. A. M., de Vries, S., and Albracht, S. P. J. *In* ''Oxidases and Related Redox Systems (ISOX III)'' (T. E. King, H. S. Mason, and M. Morrison, eds.), in press.
41. Yu, C.-A., Yu, L., and King, T. E. (1974). *J. Biol. Chem.* **249,** 4905–4910.
42. Bäckström, D., Norling, B., Ehrenberg, A., and Ernster, L. (1970). *Biochim. Biophys. Acta* **197,** 108–111.
43. Wikström, M. K. F., and Berden, J. A. (1972). *Biochim. Biophys. Acta* **283,** 403–420.
44. Baum, H., Rieske, J. S., Silman, H. I., and Lipton, S. H. (1967). *Proc. Natl. Acad. Sci. U.S.A.* **57,** 798–805.
45. Ruzicka, F. J., Beinert, H., Schepler, K. L., Dunham, W. R., and Sans, R. H. (1975). *Proc. Natl. Acad. Sci. U.S.A.* **72,** 2886–2890.
46. Ingledew, W. J., Salerno, J. C., and Ohnishi, T. (1976). *Arch. Biochem. Biophys.* **177,** 176–184.
47. Mitchell, P. (1975). *FEBS Lett.* **56,** 1–12.
48. Mitchell, P. (1976). *J. Theor. Biol.* **62,** 327–367.
49. Kröger, A. (1976). *FEBS Lett.* **65,** 278–280.
50. Yu, C.-A., Nagaoka, S., Yu, L., and King, T. E. (1978). *Biochem. Biophys. Res. Commun.* **82,** 1070–1078.
51. Nagaoka, S., Yu, L., and King, T. E. (1981). *Arch. Biochem. Biophys.* **208,** 334–343.
52. Yu, C.-A., and Yu, L. (1980). *Biochem. Biophys. Res. Commun.* **96,** 286–292.
53. Yu, C.-A., Yu, L., and King, T. E. (1975). *Biochem. Biophys. Res. Commun.* **66,** 1194–1200.
54. Ohnishi, T., and Trumpower, B. L. (1980). *J. Biol. Chem.* **255,** 3278–3284.
55. Ruuge, E. K., and Konstantinov, A. A. (1976). *Biofizika* **21,** 586–587.

56. Tikhonov, A. N., Burbayer, D. S., Grigolva, I. V., Konstantinov, A. A., Ksenzenko, M. Y., and Ruuge, E. K. (1977). *Biofizika* **22,** 734–735.
57. Konstantinov, A. A., and Ruuge, E. K. (1977). *FEBS Lett.* **81,** 137–141.
58. Rieske, J. S., Hansen, R. E., and Zaugg, W. S. (1964). *J. Biol. Chem.* **239,** 3017–3022.
59. Trumpower, B. L., and Edwards, C. A. (1979). *J. Biol. Chem.* **254,** 8697–8706.
60. Hatefi, Y., and Galante, Y. M. (1980). *J. Biol. Chem.* **255,** 5530–5537.
61. King, T. E. (1980). *In* "New Horizons in Biological Chemistry, A Festschrift to Professor Y. Yagi (1979)" (M. Koike *et al.*, eds.), pp. 121–134. Japan Scientific Societies Press, Tokyo.

II PHYSICAL CHEMISTRY OF BIOLOGICALLY ACTIVE QUINONES

1 Quinones of Value to Electron-Transfer Studies: Oxidation–Reduction Potentials of the First Reduction Step in an Aprotic Solvent

ROGER C. PRINCE
M. R. GUNNER
P. LESLIE DUTTON

INTRODUCTION

The realization that a single chemical species of ubiquinone can be the prosthetic group of electron and hydrogen transfer proteins with widely differing electrochemistries has posed the question of how this is achieved. One possibility is that the accessibility of protons to the quinone may be moderated by the protein to impose different redox characteristics on the prosthetic group. Another influence may be the shape of the "pocket" in the protein in which the quinone sits. A powerful experimental approach for studying these effects is by extraction of the endogenous quinones, and replacement with other compounds. For example, this allows testing not only of the varied electrochemical properties, but also of the role of shape, size, charge and chemical constituents in the function of quinones in biological macromolecules. This approach is already yielding important results in the study of the bacterial photochemical reaction center, both in describing the nature of the quinone binding site, and in

Function of Quinones in Energy Conserving Systems

ISBN 0-12-701280-X

TABLE I

$E_{1/2}$ Values of Benzoquinones in Dimethylformamide[A]

1,4-Benzoquinones	$E_{1/2}$	1,4-Benzoquinones	$E_{1/2}$
Unsubstituted (10)[B]	−400	2,5-Dimethyl- (14)	−557
2,3-Dicyano-	+383	2,6-Dimethyl- (15)	−557
2,3-Dicyano-5,6-dichloro-	+608	Trimethyl- (20)	−636
2,5-Dichloro- (5)	−65	Tetramethyl- (22)	−738
2,6-Dibromo- (6)	−58	2,3-Dimethoxy- (12)	−460
2,5-Dibromo-3-methyl-		2,5-Dimethoxy- (21)	−654
6-Isopropyl-	−249	2,6-Dimethoxy- (19)	−630
2,6-Dichloro- (4)	−30	2,3-Dimethoxy-5-methyl-	
2-Bromo-6-chloro- (7)	−62	6-Decaisoprenyl- (18)	−600
2-Bromo-6-methyl-	−281	2,3-Dimethoxy-5-methyl-	
2,6-Diiodo- (8)	−62	6-Decyl-	−622
Trichloro- (3)	+35	*t*-Butyl- (13)	−473
Tetrafluoro- (1)	+100	2,6-Di-*t*-butyl-	−618
Tetrachloro- (1)	+140	Phenyl- (9)	−390
Tetrabromo- (2)	+106	2,5-Diphenyl-	−376
Methyl- (11)	−478	2,5-Diisooctyl-	−632
Trifluoromethyl-	−119	2,5-Diamino-3,6-dibromo-	−732
2-Methyl-5-isopropyl- (17)	−561	2,3-Thiophene-	−438
2,3-Dimethyl- (16)	−538		
1,2 Benzoquinones			
Tetrachloro-	+213	Tetrabromo-	+213

[A] mV, referred to the saturated calomel electrode.

[B] Numbers in parentheses refer to compounds used by Gunner *et al.* (see paper No. 1, Chapter V, this volume).

determining the thermodynamic constraints which affect the rates of electron transfer within the protein [see (*1–3*); see also Gunner *et al.*, Chapter V, this volume]. In this paper we present the oxidation–reduction potentials of a variety of quinones which may have a use in extraction–reconstitution experiments in this and other systems. While many of these compounds have been studied before (*4*) the conditions under which such measurements were made were not always comparable. We have measured all the quinones using cyclic voltammetry under conditions such that the results are directly comparable, and have restricted ourselves to the $Q/Q^{\overline{\cdot}}$ couple.

EXPERIMENTAL RESULTS

All measurements were made in dimethylformamide (Baker Reagent Grade) dried over molecular sieves, with 50 m*M* tetrabutylammonium fluoroborate as electrolyte (*5*). Between 0.5 and 5.0 mg of quinone were dissolved in 7 ml dimethylformamide, and the solution was bubbled with nitrogen before the measurements were made. The reference electrode was a saturated calomel electrode, and the working electrode was made of platinum (surface area 64 mm²).

Tables I, II, and III present the $E_{1/2}$ values of some benzoquinones, naphthoquinones, and anthraquinones; the numbers in parentheses refer to the compounds used by Gunner *et al.* (see Paper No. 1, Chapter V, this volume). Most compounds were obtained commercially, and measured without further purification [see, however (*3*)]. The measurements were made at 22 ± 1 °C, over the range + 1040 to − 2440 mV with respect to the calomel electrode. No efforts were made to reduce any junction potentials, but ferrocene had a value of + 526 mV under the conditions used here. Values are reported as measured, but the experimental error is

TABLE II

$E_{1/2}$ Values of Naphthoquinones in Dimethylformamide[A]

1,4-Naphthoquinones	$E_{1/2}$	1,4-Naphthoquinones	$E_{1/2}$
Unsubstituted (24)[B]	−583	2,7-Dimethyl- (29)	−688
2,3-Dichloro-	−300	2-*t*-Butyl-	−654
2,3-Dibromo- (23)	−300	2-Cyclohexyl-	−667
2-Methyl- (28)	−658	2-Methylthio- (25)	−596
2,3-Dimethyl- (31)	−741	2-Methyl-3-methylthio- (27)	−628
2,6-Dimethyl- (30)	−732	2-Methyl-3-ethylthio- (26)	−616
	1,2-Naphthoquinones		
Unsubstituted	−473	4 Sulfonate	−428
	Acenaphthenequinones		
Unsubstituted	−841		

[A] mV, referred to the saturated calomel electrode.

[B] Numbers in parentheses refer to compounds used by Gunner *et al.* (see Paper No. 1, Chapter V, this volume).

TABLE III

$E_{1/2}$ Values of Anthraquinones in Dimethylformamide[A]

9,10-Anthraquinones			
Unsubstituted (35)[B]	−816	1-Nitro-	−593
2-Chloro- (34)	−695	1-Nitro-2-methyl-	−609
2-Iodo-	−700	1-Amino-	−895
2-Bromo-3-methyl- (37)	−815	2-Amino-	−1021
1,5-Dichloro- (32)	−702	1-Amino-8-chloro- (38)	−867
1,8-Dichloro- (33)	−703	1-Amino-2,4-dibromo- (36)	−750
2-Methyl- (40)	−836	1-Amino-6,7-dichloro-	−777
1,4-Dimethyl-	−931	1-Amino-4-chloro-	
2,3-Dimethyl- (42)	−868	2-Methyl- (45)	−905
2-Ethyl- (39)	−829	1-Amino-4-bromo-	
2-*t*-Butyl- (41)	−829	2-Sulfonate (Na^+)-	−860
Benz[*a*]-	−673	1-Amino-4-methylamino-	−1070
(2-Methylbenz[*a*])5-methyl-	−719	1,2-Diamino-	−1087
Dibenz[*b*,*e*]-	−1036	1,4-Diamino- (48)	−1124
2-Sulfonate (Na^+)-	−809	1,5-Diamino- (46)	−1034
1,5-Disulfonate (Na^+)-	−828	2,6-Diamino- (49)	−1207
2,6-Disulfonate (Na^+)-	−796	1-Methylamino- (44)	−868
		1,4-Diisopropylamino- (47)	−1060
Phenanthrenequinones			
Unsubstituted	−579	1-Methyl-7-isopropyl-	−686

[A] mV, referred to the saturated calomel electrode.

[B] Numbers in parentheses refer to compounds used by Gunner *et al.* (see Paper No. 1, Chapter V, this volume).

probably of the order of ±20 mV in these somewhat preliminary measurements.

DISCUSSION

Quinone–hydroquinone couples are perhaps the most studied organic redox systems, and a vast literature has accumulated on them. Their electrochemistry in aprotic solvents has been discussed by Chambers (*4*) and in various aqueous solutions by Clark (*6*), Moriconi *et al.* (*7*), and Dewar and Trinajstic (*8*), among others. Electrochemical reduction of quinones is usually reversible, and so polarographic half-wave potentials ($E_{1/2}$), as measured here, are good approximations to the equilibrium oxidation–reduction midpoint potentials (E_m) (*4*). The half-wave potentials of several series of compounds have been shown to correlate with substituent and positional constants in modified Hammett (*4*) or Taft (*9*) equations, and our results generally agree, at least qualitatively, with such predictions.

ACKNOWLEDGMENTS

This work was supported by grants from the National Science Foundation (PCM 79-09042) and the Department of Energy (DE-AC02-80-ER 10590).

REFERENCES

1. Cogdell, R. J., Brune, D. C., and Clayton, R. K. (1974). *FEBS Lett.* **43,** 344–348.
2. Okamura, M. Y., Isaacson, R. A., and Feher, G. (1975). *Proc. Natl. Acad. Sci. U.S.A.* **72,** 3491–3495.
3. Dutton, P. L., Gunner, M. R., and Prince, R. C. (1982). "Trends in Photobiology" (C. Helene *et al.,* eds.), pp. 561–570. Plenum, New York.
4. Chambers, J. Q. (1974). *In* "The Chemistry of the Quinonoid Compounds" (S. Patai, ed.), Part II, pp. 737–791. Wiley, New York.
5. House, H. O., Feng, E., and Peet, N. P. (1971). *J. Org. Chem.* **36,** 2371–2375.
6. Clark, W. M. (1960). "Oxidation-Reduction Potentials of Organic Systems." Williams & Wilkins, Baltimore, Maryland.
7. Moriconi, E., Rakoczy, B., and O'Connor, W. F. (1962). *J. Am. Chem. Soc.* **27,** 2772–2776.
8. Dewar, M. J. S., and Trinajstic, N. (1969). *Tetrahedron* **25,** 4529–4534.
9. Zuman, P. (1962). *In* "Progress in Polarography" (P. Zuman, and I. M. Kolthoff, eds.), Vol. 1, pp. 319–332. Wiley (Interscience), New York.

2

Electrochemical and Spectral Properties of Ubiquinone and Synthetic Analogs: Relevance to Bacterial Photosynthesis

LARRY E. MORRISON
JEAN E. SCHELHORN
THERESE M. COTTON
CHARLES L. BERING
PAUL A. LOACH

INTRODUCTION

Evidence is accumulating that ubiquinone (or plastoquinone in chloroplasts) plays a multiple role in electron transport and in the primary photochemical events of photosynthesis. In addition to its traditionally assumed role as a two-electron, two-proton acceptor, it is also now known to function in one-electron transfer reactions. For example, the first stable electron acceptor in bacterial photosynthesis in *Rhodospirillum rubrum* and *Rhodopseudomonas sphaeroides* is known to be a protein bound ubiquinone molecule (*1–7*). Good evidence also exists that plastoquinone plays a similar role in the Photosystem II of oxygen evolving organisms (*8*). In each case, initial transfer of one electron normally occurs, possibly followed by protonation and transfer of a second electron (*4, 5*). Currently we are on the threshold of discovering a wide variety of protein-bound quinones with as much variation and importance as the flavoproteins, NAD dehydrogenases, iron–sulfur proteins, heme proteins, etc. Thus,

Function of Quinones in Energy Conserving Systems

ISBN 0-12-701280-X

the need to understand more fully the electrochemical properties of this important redox group is evident.

The midpoint potential for the two-electron reduction of the ubiquinone pool in preparations of mitochondria and photosynthetic bacteria has been determined in a number of laboratories with reasonably good agreement. However, it has not yet been clearly demonstrated that this two-electron redox reaction is directly involved in any reaction of either photosynthesis or mitochondrial electron transport. There is good evidence for the one-electron reduction of ubiquinone in preparations from photosynthetic bacteria (*1–7*, *9–11*) and mitochondria (*12–14*). Unfortunately, there is considerable variation in the reported values of the midpoint potential for such equilibria and their respective pH dependencies (*4*, *15–17*). Since three oxidation states and several states of protonation for both reduced states may play a role in biological systems, it is not surprising to find differences in interpretation.

Surprisingly little information concerning the solution electrochemistry and spectroscopy of ubiquinone has been reported, although the electrochemistry of quinones in general has been studied extensively (*18*, *19*). The polarographic half-wave potentials of the two-electron reductions of ubiquinone-6, ubiquinone-7, and ubiquinone-4 in aqueous ethanol at various pH values have been reported. (*20*). The absorbance spectra of ubiquinones and ubihydroquinones are well known in ethanol (*21*), and the transient spectrum of neutral ubisemiquinone-6 and its anionic form produced by pulse radiolysis in methanol have been reported (*22*). In general, as stated by Chambers (*18*), quinone electrochemistry in protic environments presents a "non-textbook" complexity. The 3×3 array of possible reactants, intermediates, and products which are interrelated by electron and proton transfer steps must be taken into consideration. In addition, the effect of binding to protein-active sites and the distribution of local charges in these sites add a third dimension to the complexities that need to be evaluated.

In an effort to increase our knowledge in this area, we have undertaken *in vitro* studies of ubiquinone-10 to determine electrochemical properties and absorbance spectra of the various oxidation and protonation states. These studies were performed in both protic and aprotic solvents so that comparison of spectra and midpoint potentials with those determined *in vivo* might be helpful. Because it is often true that a systematic examination of simpler, but related structures will provide a better understanding of the properties of a particular molecule of interest, we have also recorded the spectral and electrochemical properties of several related molecules such as 2,3-dimethyl-5,6-dimethoxybenzoquinone. In addition, we have studied the electrochemical properties of a variety of benzoquinone analogs which have been used as inhibitors of electron transport, especially in photosynthesis. For example, 2,5-dibromo-3-methyl-6-isopropylbenzoquinone (DBMIB)[1] was found to block electron flow between photosystems I and II, while in bacteria strong inhibition of photophosphorylation and electron transport was observed (*23*, *24*). In addition, higher concentrations of DBMIB were reported to inhibit the primary photochemistry in photosynthetic bacteria (*25*). The electrochemical and spectral information reported here for these analogs should provide additional insights into the mechanism by which their inhibitory action occurs.

MATERIALS AND METHODS

The benzoquinone analogs used in this study were obtained from a variety of sources as follows: *p*-Benzoquinone* from City Chemical Corporation (New York, New York); 2,5-dimethyl-*p*-benzoquinone*, 2,5-di-*t*-butyl-*p*-benzoquinone, and 2,6-dichloro-*p*-benzoquinone* from Eastman Organic Chemicals (Rochester, New York); thymoquinone* from ICN (Plaineville, New York); methoxyquinone from Pfaltz and Bauer, Inc. (Flushing, New York); ubiquinone-10 from Sigma Chemical Co. (St. Louis, Missouri); 2,3-dimethyoxy-5,6-dimethyl-*p*-benzoquinone from Biochemical Laboratories, Inc. (Redondo Beach, California); duroquinone from Aldrich Chemical Co. (Milwaukee, Wisconsin). All other quinones were the generous gift of Dr. A. Trebst and Dr. W. Oettmeier of Ruhr-Universität, Bochum, West Germany. Tetrabutylammonium perchlorate was obtained from Southwest Analytical Chemicals Inc. (Austin, Texas) and dried under vacuum at 90 °C for at least 24 hours before use. Distilled water was demineralized further and freed of organics with a Bion Exchanger System (Pierce Chemical Co., Rockford, Illinois). Triply distilled mercury, platinum wire, gauze, and foil were purchased from D. F. Goldsmith Chemical Co. (Evanston, Illinois). Distilled in glass grade acetonitrile and dimethylformamide were obtained from Burdick and Jackson Laboratories (Muskegon, Michigan) which specified water con-

* Sublimed before use.

[1] Abbreviations: DBMIB, 2,5-dibromo-3-methyl-6-isopropylbenzoquinone; DDQ, 2,3-dimethyl-5,6-dimethoxybenzoquinone; UQ or UQ_{10}, ubiquinone-10 where the 10 indicates the number of isoprenoid units in the tail; SHE, standard hydrogen electrode; SCE, saturated calomel electrode; $E_{1/2}$, polarographic half-wave potential; RR, resonance raman; λ_{max}, wavelength of maximal absorbance; ϵ, molar extinction coefficient; TBAP, tetrabutylammonium perchlorate.

tents of no more than 0.01% and 0.03%, respectively. For experiments involving ubiquinone and DBMIB, acetonitrile was dried further by repeated shaking with phosphorous pentoxide. The acetonitrile was then decanted, shaken with additional phosphorous pentoxide and distilled under high vacuum at room temperature (*26*). For routine cyclic voltammetry, acetonitrile was used without further purification. Dimethylformamide was dried further by stirring with barium oxide followed by vacuum distillation at temperatures below 40 °C (*27*). Solutions of the dried aprotic solvents were prepared in plastic glove bags (Instruments for Research and Industry, Cheltenham, Pennsylvania) purged with dry nitrogen. Protic solutions were 80% ethanol–water by weight. Buffers consisted of lysine, arginine, tris-(hydroxymethyl)aminomethane, bis-(2-hydroxymethyl)-imino-tris-(hydroxymethyl) methane, 3-(cyclohexylamino)-propanesulfonic acid, all obtained from Sigma Chemical Co., and acetic acid from Mallinckrodt Chemical Co. (Paris, Kentucky) as specified in the text. Adjustment of pH was made with potassium hydroxide or hydrochloric acid.

Electrolysis experiments and cyclic voltammetry were performed with an "adder" type operational amplifier potentiostat (*28*) constructed at the electronics facility of the Chemistry Department, Northwestern University. The potentiostat was used together with a voltage generator which provided a reference voltage of either a constant voltage or a triangular wave form with 1 mV incremental steps. In cyclic voltammetry experiments the sample solution and electrodes were contained in a Metrohm titration vessel assembly (Brinkmann Instruments, Westburg, New York). The glass bottom portion of this assembly held samples between 5 and 10 ml and was sealed with a Penton top containing five ground joint openings for introduction of electrodes and degassing apparatus. Matheson ultra high purity argon gas (Joliet, Illinois), passed over copper filings at 475 °C to remove oxygen further, was then bubbled through the samples for thirty minutes prior to recording cyclic voltammograms. The working electrode was a Metrohm hanging mercury drop electrode (Brinkmann Instruments). A platinum foil was used as the counter electrode and an aqueous saturated calomel electrode was employed as the reference electrode. The reference made contact with the sample solution through a Princeton Applied Research model 9361 reference electrode bridge (Princeton, New Jersey). The bridge contained the same solvent and supporting electrolyte as that in the test solution to which it made a liquid junction through a porous glass plug. Sodium chloride was used in place of potassium chloride since perchlorate in the supporting electrolyte tends to precipitate with potassium ions.

Quinones in protic and aprotic solutions were bulk electrolyzed at constant preset potentials with the potentiostat described above. The anaerobic titration vessel, patterned after a cell designed by Bull and Hoffman for oxidation–reduction titrations, was modified for three electrode potentiostatic electrolysis (*29*, *30*). The electrolysis vessel was comprised of a glass chamber opening into a horizontally oriented one centimeter pathlength quartz cuvette cemented to the wall of the chamber with Torr Seal epoxy (Varian Associates, Palo Alto, California). Four openings at the top of the chamber allowed entry of the three electrodes and an inlet/outlet for Matheson ultra-high purity argon. A Gilmont microburette (Cole Parmer Instrument Co., Chicago, Illinois) and Radiometer model GK 2392C combination pH electrode could be added when desired. The reference saturated calomel electrode made contact with the sample solution through the salt bridge described above. A cylindrical platinum-gauze working electrode which fit snugly in the bottom portion of the cell was constructed from a $1 \times 3\frac{1}{2}$ inch piece of platinum gauze with ends folded over and spot welded. A square opening was cut in this gauze to match the entrance to the cuvette. A platinum wire auxiliary electrode was isolated from the main solution by an ultra-fine porosity glass frit sealed in the end of a glass tube. Solutions were degassed one hour prior to recording spectra, or initiating electrolysis, and were continuously stirred with a Kel-F coated magnetic stir bar which allowed rapid equilibration between the solution in the cuvette portion of the cell with that in the main body.

Differential pulse polarography was performed with a Princeton Applied Research Corporation Model 374 Polarographic Analyzer. This microprocessor-controlled unit contained a dropping mercury working electrode, platinum wire counter electrode, and silver–silver chloride reference electrode (saturated KCl). Samples were purged 10 minutes with Matheson prepurified nitrogen prior to recording polarograms.

For Resonance Raman spectra, an Ar^+ laser was used as the excitation source. The monochromator, detector, and other aspects of the experiment have all been described (*31*). Anion radicals were generated by exhaustive, controlled potential coulometry in a high-vacuum cell (*32*). The procedure used in these experiments was identical to that previously used to prepare other radical ions for RR spectroscopy (*31*, *33*).

RESULTS

Ubiquinone in 80% Ethanol–water

The peak reduction potentials of the ubiquinone-10/ubihydroquinone-10 couple in 80% ethanol–water (w/w) were determined by differential pulse polarography. The peak potential on a differential pulse polarogram corresponds to the point of inflection on the wave of a dc polarogram and should closely correspond to the half-wave potential The peak potentials of ubiquinone-10 recorded at various PH values between 6 and 10 are plotted in Fig. 1 (solid circles). Least squares analysis determined the standard potential to be 0.51 V versus SHE ($E_{1/2}{}^0 = E^0$) with a pH dependence equal to 58 mV per pH unit, identical to the dependence expected for a two-electron reduction involving one protonation per electron transferred at 22 °C. Thus, the reaction measured was

$$UQ + 2e + 2H^+ \rightleftharpoons UQH_2. \quad (1)$$

The potentials recorded here compare well with those reported by Moret *et al.* for ubiquinone-6 and ubiquinone-7 in 74% ethanol–water and ubiquinone-4 in 50% ethanol–water (*20*). For example, at pH 7.0 we find E_{peak} = 0.110 V versus SHE while a value of 0.133 V was interpolated from data reported by Moret *et al.* for the several quinones tested. In their data and ours, there is only a very small effect of the isoprenoid chain length on the electrochemical potential of the quinone, and a relative insensitivity to the proportions of ethanol and water in the solvent.

The absorbance spectra of ubiquinone-10 and neutral ubihydroqui-

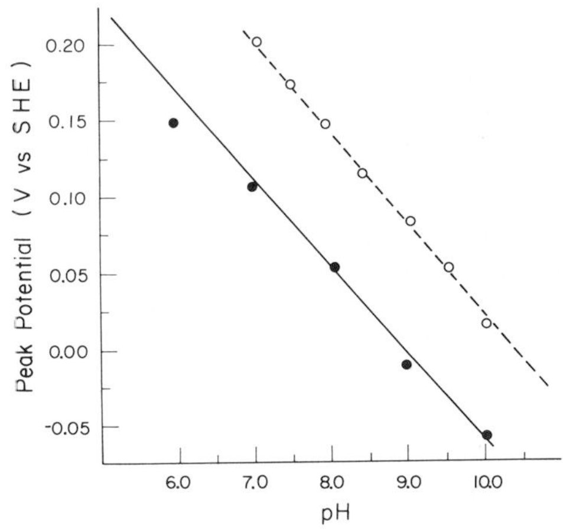

Fig. 1. The pH dependence of peak potentials from differential pulse polarography of ubiquinone-10 (solid line, solid circles) and DBMIB (dashed line, open circles) in 80% ethanol–water (w/w). The buffer system was composed of 0.02 *M* tris-(hydroxymethyl) aminomethane, 0.02 *M* 3-(cyclohexylamino)-propanesulfonic acid, 0.02 *M* bis-(2-hydroxymethyl)-imino-tris-(hydroxymethyl) methane, and 0.02 *M* acetic acid. The temperature was approximately 22 °C (room temperature).

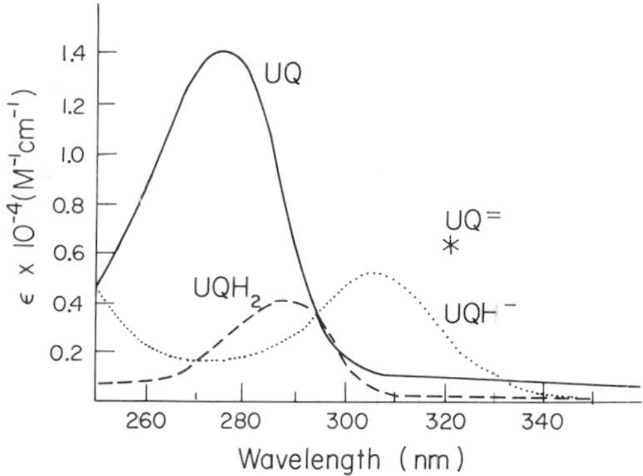

Fig. 2. Optical absorbance spectra of ubiquinone-10 (solid line), neutral ubihydroquinone (dashed line), ubihydroquinone-10 anion (dotted line), and the predicted λ_{max} and $\epsilon_{\lambda max}$ of ubihydroquinone dianion (asterisk) in 80% ethanol–water (w/w). Neutral ubihydroquinone was electrolytically generated at a platinum gauze electrode and titrated with potassium hydroxide to form ubihydroquinone-10 anion. The buffer-supporting electrolyte system was 0.01 *M* lysine (monohydrochloride) and 0.05 *M* potassium chloride. The titration vessel was kept at 25.0 °C.

none-10 were recorded in 80% ethanol–water (w/w) as shown in Fig. 2 (solid and dashed lines, respectively). The peak absorbance of ubiquinone-10 was identical to that in 100% ethanol located at 275 nm with an extinction coefficient of 1.40×10^4 M^{-1}cm^{-1} (*5*, *21*, *30*). Neutral hydroquinone was generated electrolytically on platinum at pH =8.0. The peak absorbance occurs at 288 nm with an extinction coefficient of 4.14×10^3 M^{-1}cm^{-1}. Dimethoxydimethylbenzoquinone displayed a similar absorbance at pH = 8 with λ_{max} = 288 nm. The spectrum of this species remained unaltered upon lowering the pH to a value of 2.8, supporting the assignment of this absorbance to the neutral hydroquinone. Table I contains data obtained by Baxendale and Hardy (*34*) for various hydroquinones, indicating that only hydroquinones with strong electron withdrawing substituents have pK_a' values near eight or lower for loss of one proton (*34*).

As potassium hydroxide was added to the hydroquinone, spectral changes began to occur at pH values greater than 12, resulting from formation of ubihydroquinone-10 anion. The spectrum of this species is also shown in Fig. 2 (stippled line) after substraction of a small spectral contribution from ubisemiquinone anion. At high pH values autooxidation of hydroquinones occurs readily (*34*). Although the cell was continually flushed with high purity argon gas, some semiquinone formation could not be prevented. Subtraction of ubisemiquinone anion

TABLE I

Hydroquinone pK_a Values and Spectroscopic Data[A]

Hydroquinone	pK_a (temperature °C)	λ_{max} (nm)	$\epsilon_{\lambda max} \times 10^{-3}$ (M^{-1} cm^{-1})
Benzohydroquinone			
neutral	9.85 (25.0)	288	2.30
anion	11.4 (25.9)	307	2.75
dianion		319	3.15
Methylbenzohydroquinone			
neutral	10.0 (25.1)	289	2.70
anion	11.6 (25.9)	307	3.20
dianion		317	3.70
2,6-Dichlorobenzohydroquinone			
neutral	7.30 (26.1)	294	3.50
anion	9.99 (26.1)	314	5.15
dianion		333	5.70
1,4-Naphthohydroquinone			
neutral	9.37 (26.4)	322	5.25
anion	10.9 (26.4)	346	5.70
dianion		363	6.05
Tetramethylbenzohydroquinone			
neutral	11.2 (25.0)	283	2.15
anion	12.7 (25.0)	302	3.70
dianion		317	4.95
Ubiquinone-10			
neutral	13.3 (23)	288	4.14
anion		306	5.22
dianion		321[B]	6.26[B]

[A] All data except that for ubihydroquinone were taken from Baxendale and Hardy (*34*), used with permission. The solvent for these studies was water. Ubihydroquinone data is from this study in 80% ethanol.

[B] Predicted (see text).

absorbance from spectra of ubihydroquinone anion was based on the absorbance at 445 nm where the hydroquinone anion possesses negligible absorbance. The protic spectrum and extinctions of ubisemiquinone anion recorded by Land *et al.* (*22*) in methanol were used in the procedure ($\epsilon_{445} = 6.53 \times 10^3 M^{-1}\text{cm}^{-1}$). Ubihydroquinone-10 anion was determined to possess a $\lambda_{max} = 306$ nm with an extinction coefficient of $5.22 \times 10^3 M^{-1}\text{cm}^{-1}$. Spectra of ubihydroquinone-10 were recorded at several pH values between pH 12 and 13.5. The concentrations of neutral ubihydroquinone-10 and ubihydroquinone-10 anion in each spectrum were calculated from the extinctions determined at 288 and 306 nm, respectively, and the log of their ratios plotted against pH

(Fig. 3). The linearity of the data indicates that the approximation of the semiquinone absorbance spectrum introduced little error into the calculation of the hydroquinone anion spectrum. A pK'_a of 13.3 was determined from the data in Fig. 3. This value of the pK'_a appears high when compared with those determined for other hydroquinones in pure water (*34*). However, it must be considered that both the hydroquinone and the water present in the solvent solution would have smaller dissociation constants in the presence of 80% ethanol.

The observed shift in the absorbance maximum of 18 nm from neutral ubihydroquinone-10 to ubihydroquinone anion is within the range reported for other hydroquinones as is the ratio of the peak extinction coefficients (*34*) (Table I). Considering the average values for the red shifts and extinction-coefficient increases reported in forming the dianion hydroquinone from the anion for a variety of hydroquinones, predictions may be made concerning the spectrum of ubihydroquinone-10 dianion (*34*). The predicted λ_{max} and ϵ_{max} for ubihydroquinone-10 dianion thus determined are represented by an asterisk in Fig. 2, the values being 321 nm and $6.26 \times 10^3 \ M^{-1} \ cm^{-1}$, respectively. The spectrum could not be determined directly due to difficulty in preventing formation of ubisemiquinone anion at the high pH values required to form the dianion.

At pH values near 14 or greater, reduction of ubiquinone-10 electrolytically at platinum, or chemically by borohydride addition led predominantly to the ubisemiquinone-10 anion. In Fig. 4A (solid line), the absorbance spectrum of ubisemiquinone anion generated by such borohydride reduction is displayed. Successively recorded spectra showed a

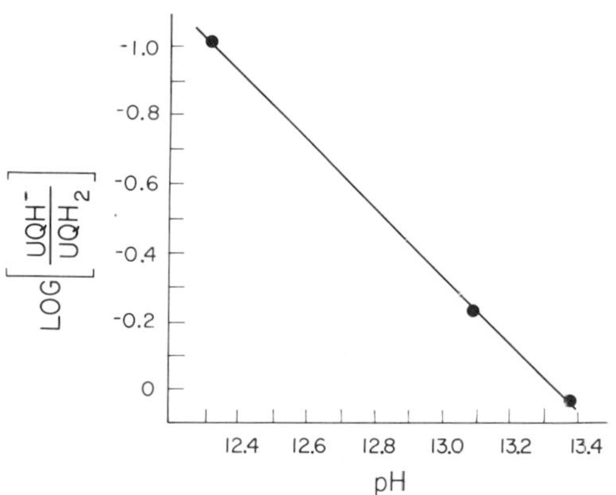

Fig. 3. The pH dependence of log ([ubihydroquinone-10 anion]/[neutral ubihydroquinone-10]). The solid line represents the theoretical relationship for $pK_a = 13.34$.

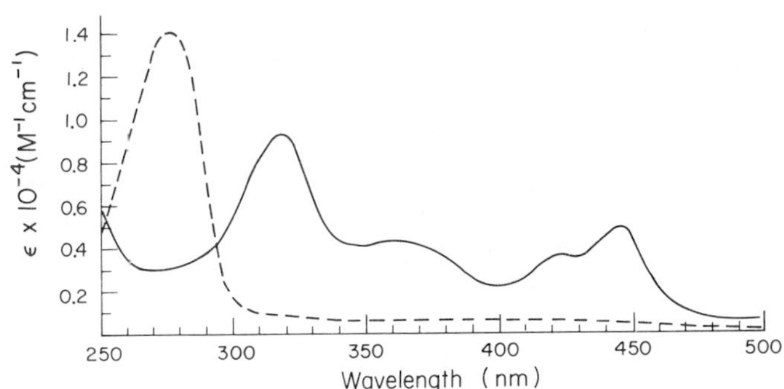

Fig. 4A. Optical absorbance spectra of ubisemiquinone-10 anion radical (solid line) and ubiquinone-10 (dashed line) in 80% ethanol–water (w/w) at 25.0 °C. The semiquinone was generated by reduction of ubiquinone with potassium borohydride and addition of potassium hydroxide (final pH >14). 0.02 *M* KCl, 0.02 *M* lysine and 0.02 *M* arginine were present as buffers for the titration of ubihydroquinone prior to generation of semiquinone. The extinction coefficients apply strictly to the ubiquinone spectrum but only approximately to that of the semiquinone due to its instability at high pH (see text). The same spectral characteristics were observed for electrolytically generated ubisemiquinone.

slowly decreasing absorbance (A_{445} decreased by several percent per minute). In addition, the exact amount of hydroquinone anion and dianion in equilibrium can not be determined since their absorbance peaks lie in the region of the strong semiquinone absorbance near 320 nm. The spectrum is presented here for a comparison to the point-by-point

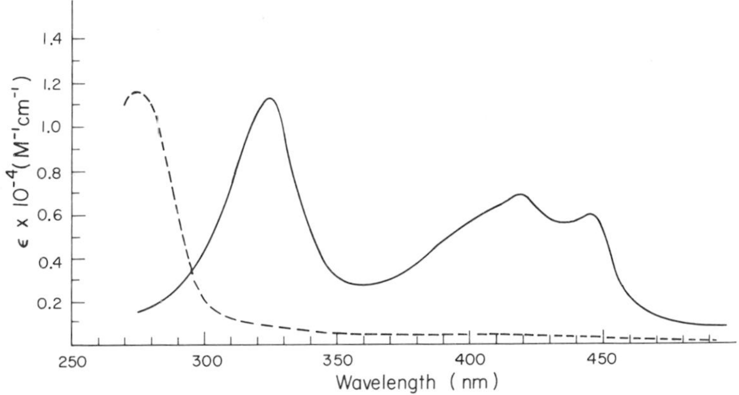

Fig. 4B. Optical absorbance spectra of ubiquinone-10 (dashed line) and ubisemiquinone-10 anion radical (solid line) in dimethylformamide at room temperature. The ubisemiquinone-10 anion radical was electrolytically generated at a platinum gauze electrode. The supporting electrolyte was 0.1 *M* TBAP.

spectrum of Land and Simic for ubisemiquinone anion produced in methanol by pulse radiolysis. The similarities between these data are very high. Maxima are observed at 420, 444, 318, 365 nm. The extinction coefficients labeled on the ordinate pertain strictly to the spectrum of ubiquinone-10 also plotted in Fig. 4 (dashed line). Although the normalized spectrum should be reasonably accurate, the extinction coefficient values given for the semiquinone may be as much as 30% below their actual extinction coefficients.

Ubiquinone-10 in Aprotic Solvents

Cyclic voltammetry was performed with solutions of ubiquinone-10 in acetonitrile and dimethylformamide. The supporting electrolyte was 0.1 *M* tetrabutylammonium perchlorate (TBAP) in all aprotic experiments. Negative peak potentials in acetonitrile were determined to be −0.73 V and −1.45 V versus the saturated calomel electrode (SCE) for the quinone–semiquinone anion and semiquinone anion–hydroquinone dianion transitions, respectively (Fig. 5). Reversibility was observed at each of the two one-electron transitions as determined by a 60 mV separation between peak potentials for the positive and negative scans (positive peak potentials = −0.67 and −1.39 V versus SCE). Additional peaks, however, appeared in the positive portion of the voltage sweep following reduction to the hydroquinone dianion. These peaks became larger and more complex as the potential at which the negative

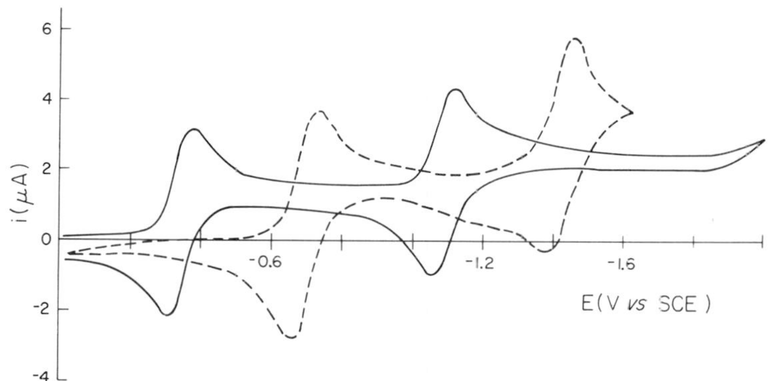

Fig. 5. Cyclic voltammogram of 5×10^{-4} *M* ubiquinone-10 (dashed curve) and 6.2×10^{-4} *M* DBMIB (solid curve) at a hanging mercury drop electrode in acetonitrile. The supporting electrolyte was 0.1 *M* TBAP, scan rate = 27 mV/sec, surface area of mercury drop $\sim 2.2\ mm^2$.

scan ends and the positive scan begins was extended to more negative values. A portion of the hydroquinone dianion evidently undergoes additional change.

The cyclic voltammetry data reported above show peak potentials significantly more positive than those reported by Marcus and Hawley (*35*) for ubiquinone-1 in acetonitrile (in the Hawley study negative peak potentials = −0.87 V and −1.74 V versus SCE). The electron affinity of both quinones should be very similar since the only difference between substituents on the ring is the length of the isoprenoid chains. Perhaps the difference might best be explained by differences in the supporting electrolytes employed and liquid junction potentials between the reference electrodes and quinone solutions.

Ubiquinone-10 was quantitatively reduced to the ubisemiquinone anion electrolytically at platinum in both acetonitrile and dimethylformamide at a potential of −1.0 V versus SCE. Very similar spectra were observed in the two aprotic solvents. The semiquinone was not exceptionally stable in acetonitrile, showing an absorbance decrease at 444 nm of several percent per minute. In dimethylformamide, though, a more stable ubisemiquinone anion spectrum was observed. This spectrum is shown in Fig. 4B in addition to the absorbance spectrum of the oxidized form. For the ubisemiquinone anion, maxima were observed at 325, 419 and 445 nm with respective extinction coefficients of 1.12×10^4, 6.8×10^3, and 6.0×10^3 M^{-1} cm^{-1}. A shoulder was observed near 390 nm.

The hydroquinone dianion could not be generated in stable form to a degree that would allow spectroscopic characterization by electrolytic reduction in either acetonitrile or dimethylformamide at platinum.

Study of 2,3-Dimethyl-5,6-dimethoxybenzoquinone (DDQ) as a Model for Ubiquinone

The absorbance spectra of 2,3-dimethyl-5,6-dimethoxybenzoquinone (DDQ) and ubiquinone-10 are very similar as are the spectra of their respective semiquinone anions. It appears then that DDQ, differing from ubiquinone by replacement of the isoprenoid chain by a methyl group, should be a good model compound for ubiquinone. The semiquinone anion of DDQ was found to be considerably more stable in acetonitrile than ubisemiquinone. The stability of this molecule allowed both the oxidized form and the semiquinone anion to be studied by Resonance Raman spectroscopy.

Resonance Raman spectroscopy of DDQ was performed using

457.9 nm excitation. This is not in resonance with the strongest electronic transition near 273 nm, but is close to a broad weak band near 405 nm (approximate extinction coefficient = 500 cm^{-1} M^{-1}). As may be seen in Fig. 6A, the strongest vibrations are 1658 and 1613 cm^{-1}. From a comparison of these bands with those of chloranil, the 1658 cm^{-1} band is tentatively assigned to the C=O stretch [1693 cm^{-1} in

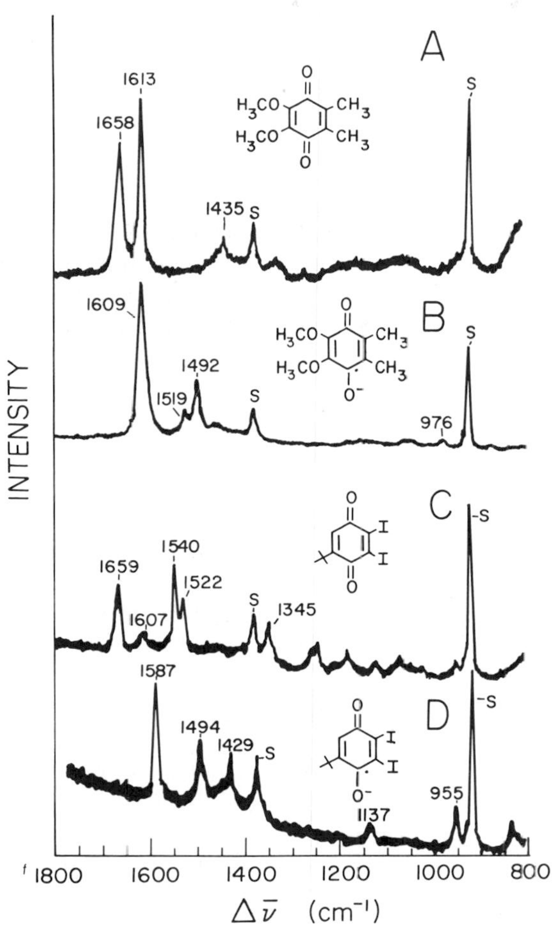

Fig. 6. Resonance Raman spectra of DDQ at 2×10^{-2} M (A) and its semiquinone anion at 1.2×10^{-3} M (B) and 2,3-diiodo-5-*t*-butyl-1,4-benzoquinone at 2×10^{-2} M (C) and its semiquinone anion at 1.1×10^{-3} M (D) in acetonitrile. Each quinone was reduced electrolytically to form the semiquinone anion, DDQ at −0.70 V versus SCE and 2,3-diiodo-5-*t*-butyl-1,4-benzoquinone at −0.35 V versus SCE at 25 °C. Laser excitation wavelength was 457.9 nm and the power was 20 mW. Bands due to solvent are marked with an S.

chloranil (*36*)] and the 1613 cm^{-1} to the C=C stretch [1609 cm^{-1} in chloranil (*36*)]. Other vibrations in the spectra are much weaker, except for a broad band near 803 cm^{-1} which is comparable in intensity to the 1658 cm^{-1} band. No assignment of this band can be made at this time.

The one-electron reduction of DDQ to its anion radical in acetonitrile causes substantial changes in its Resonance Raman spectrum. Figure 6B illustrates the spectrum of a 10 m*M* solution after electrolyzing at −0.7 V, or approximately 50 mV below the negative peak. The spectrum was recorded using 457.9 nm excitation which is close to the broad, weak electronic transition at 444 nm. The most intense band is at 1609 cm^{-1}. This may correspond to the C=C stretch, in which case this normal mode is relatively unaffected by reduction. However, as expected, the C=O stretch, observed at 1658 cm^{-1} in the oxidized molecule, is shifted considerably and corresponds to the 1519 or 1492 cm^{-1} band in the anion radical. This assignment is based on similar shifts in chloranil and bromanil on one–electron reduction (*36*). Further studies are underway using deep blue and near UV excitation in an effort to find an optimum excitation wavelength for monitoring semiquinone formation.

Other Ubiquinone Analogues and Antagonists

2,5-Dibromo-3-methyl-6-isopropylbenzoquinone (DBMIB) has been much used as an inhibitor in photosynthetic and electron transport systems. We have examined its electrochemical properties in some detail. The peak reduction potentials of DBMIB recorded in 80% ethanol–water (w/w) at various pH values by differential pulse polarography are plotted in Fig. 1 (open circles). Least squares analysis determined E^0_{peak} = 623 mV versus SHE with a pH dependence of 60 mV/pH unit. Cyclic voltammetry in acetonitrile produced two nicely reversible single electron reductions of the quinone nucleus, with negative peak potentials of −310 and −1060 mV versus SCE (Fig. 5, solid curve).

The DBMIB semiquinone anion could be generated as a stable species in acetonitrile by electrolytic reduction at platinum. The UV-visible spectrum is displayed in Fig. 7. The hydroquinone dianion could not be generated in the aprotic solvent as was the case for ubiquinone-10. No spectroscopic studies in water or ethanol were performed.

Electrochemical data were also collected for a number of other quinones which have been tested for antagonistic effects on bacterial photosynthesis. The potentials recorded by cyclic voltammetry in acetonitrile are listed in Table II. Aqueous electrochemistry for a number of

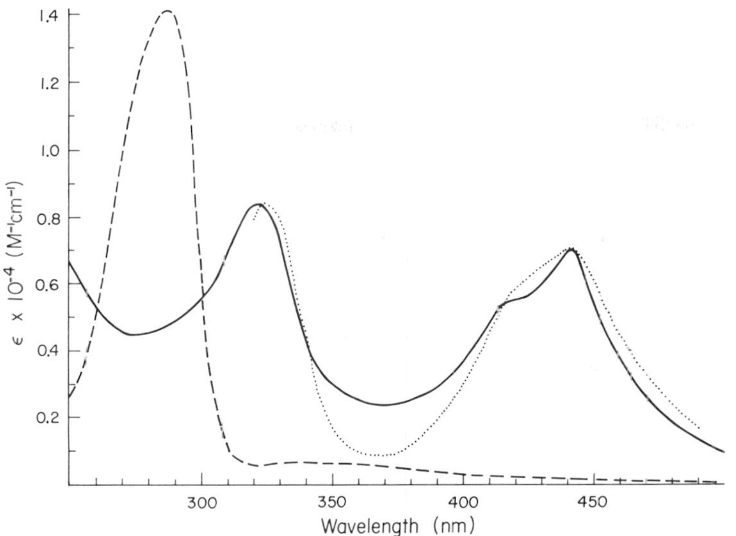

Fig. 7. Optical absorbance spectra of DBMIB (dashed line), its semiquinone anion (solid line) and the diiodo-*t*-butylbenzosemiquinone anion (dotted curve) in acetonitrile at room temperature. The semiquinones were electrolytically generated at platinum. The supporting electrolyte was 0.1 *M* TBAP. The extinction coefficients apply only to DBMIB and its semiquinone. The diiodo-*t*-butylbenzosemiquinone spectrum has been normalized to the DBMIB semiquinone spectrum.

these and for similar compounds have been reported in the literature and little differentiation in midpoint potentials was observed. The compound 2,3-diiodo-5-*t*-butylbenzoquinone has displayed some interesting inhibitor properties in photosynthetic bacteria (*37*). The spectrum of its semiquinone anion was found to be quite stable in acetonitrile and is also shown in Fig. 7. The Resonance Raman spectrum was also measured and may be compared with DDQ in Fig. 6D. Bands at 1587 cm^{-1}, 1494 cm^{-1} and 1429 cm^{-1} were observed and are similar to bands observed for the DDQ semiquinone anion.

DISCUSSION

The thermodynamic and spectroscopic data reported here for ubiquinone are summarized in Fig. 8, which also contains data from Land *et al.* (*22*). Six species which may play a role in biological systems are shown.

Although the redox equilibria in aprotic solvents show an expected

TABLE II

$E_{1/2}$ Values for Reduction of Benzoquinones[A]

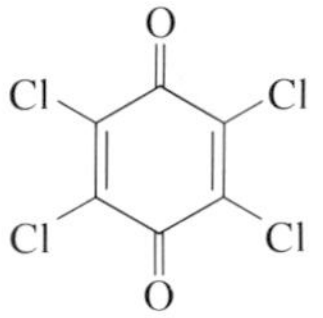

(.01, −.71)[B]	(−.03, −.76)	(−.15, −.93)	(−.15, −.95)
(−.18, −.97) (−.18, −.81)[B]	(−.31, —)	(−.31, −1.11)	(−.31, −1.12)
(−.35, −1.15)	(−.35, −1.13)	(−.41, −1.17)	(−.48, −1.09) (−.51, −1.14)[B]
(−.50, −1.25)	(−.62, —)[B]	(−.63, −1.45)	(−.65, −1.33) (−.66, —)[B]
(−.66, −1.31) (−.67, −1.27)	(−.70, −1.39)	(−.70, −1.43)	(−.84, −1.45)[B]

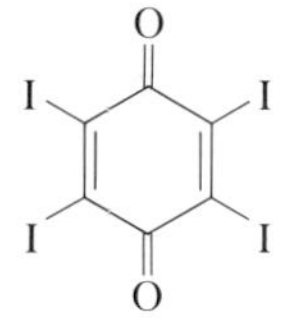

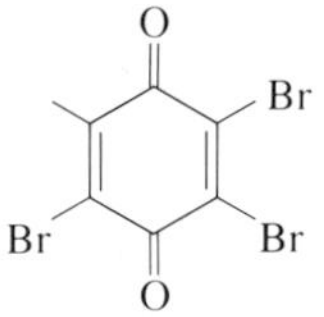

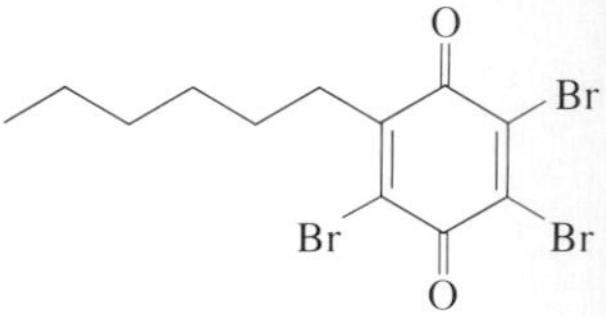

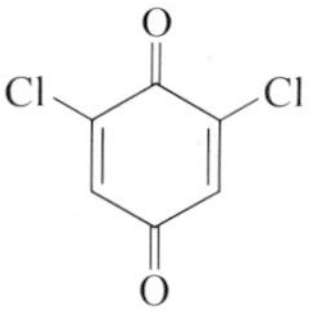

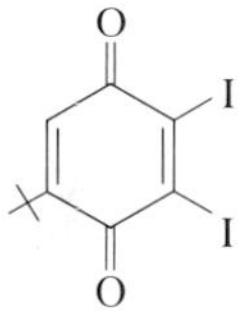

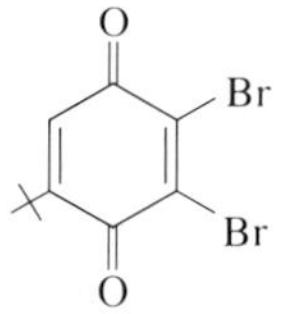

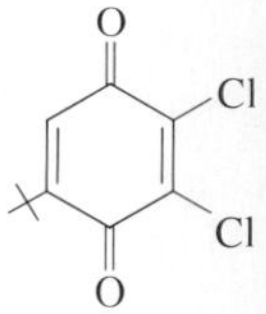

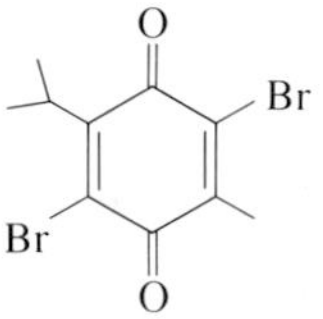

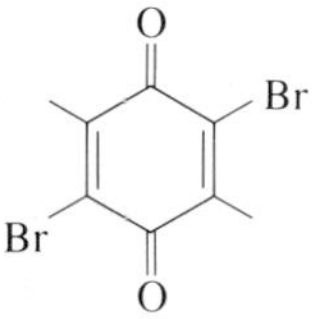

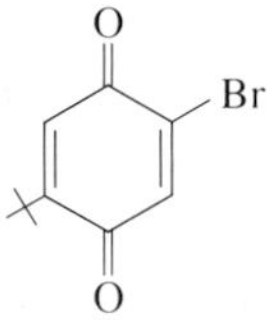

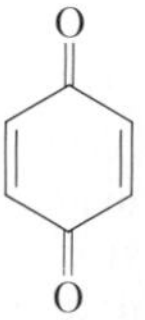

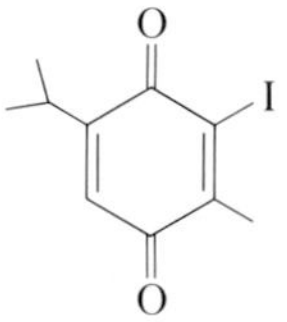

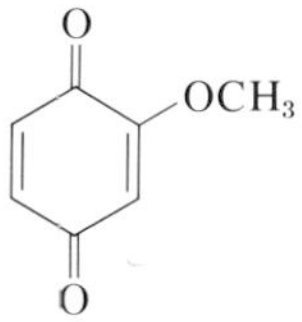

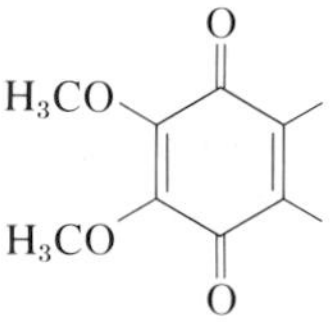

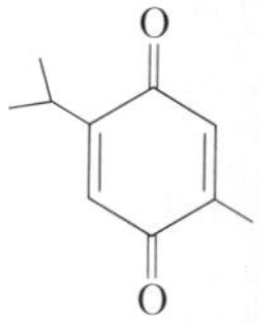

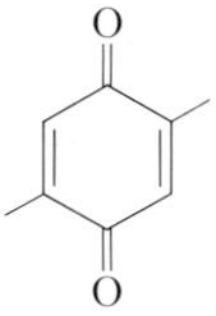

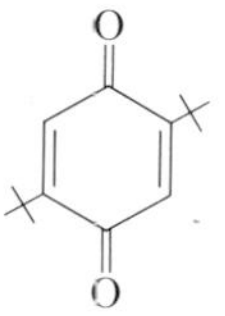

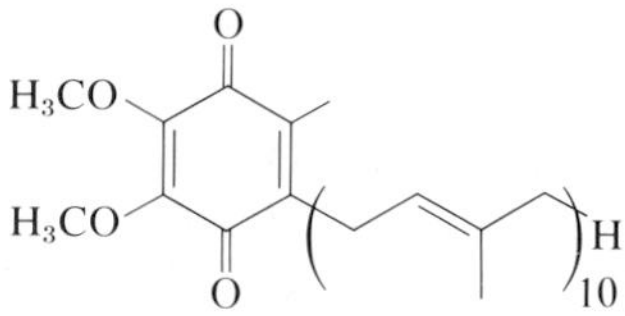

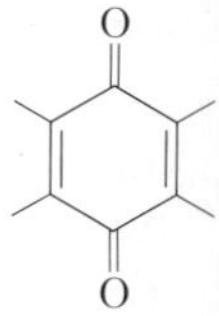

[A] Each compound was studied by cyclic voltammetry. The solvent system for these determinations was acetonitrile. All values are versus SCE at 25 °C. See text for other details.
[B] These are literature values (*18*, *19*).

APROTIC

UQ_{10} $\underset{E_{P,C} = -.73}{\overset{e^-}{\rightleftharpoons}}$ $UQ_{10}^{\dot{-}}$ $\underset{E_{P,C} = -1.46}{\overset{e^-}{\rightleftharpoons}}$ $UQ_{10}^{=}$

275 (1.16 x 10⁴)

425 (6.0 x 10³)
419 (6.8 x 10³)
325 (1.12 x 10⁴)

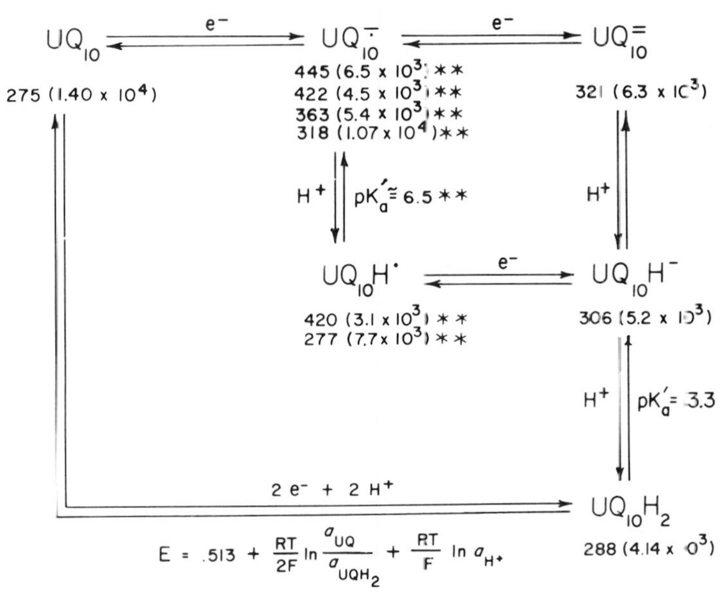

$$E = .513 + \frac{RT}{2F} \ln \frac{a_{UQ}}{a_{UQH_2}} + \frac{RT}{F} \ln a_{H^+}$$

Fig. 8. Summary of thermodynamic and spectroscopic data pertaining to ubiquinone-10 in its various states of oxidation and protonation. **Data taken from Land *et al.* (*22*) and Land and Swallow (*41*), used with permission.

major effect from the electron donating or accepting properties of substituents on the ring ($E_{1/2}$ values of compounds in Table II vary by nearly 1.0 V), much smaller effects are observed for equilibria in protic solvents. For example, if one compares ubiquinone and DBMIB data, the $E_{1/2}$ value for one-electron reduction of DBMIB in an aprotic solvent is 350 mV more positive than the $E_{1/2}$ for ubiquinone. However, in a protic solvent the difference for the two-electron and two-proton reduction is only 95 mV (see Fig. 1). As established in the literature, in aprotic solvents, the $\Delta E_{1/2}$ value for the difference between the addition of the first electron and the second is nearly constant (700 mV ± 100 mV). Also, $\Delta E_{1/2}$ is relatively independent of the electron withdrawing or donating substituents, although some compounds with more negative $E_{1/2}$ values show a decreasing trend. When considering possible roles in biological systems, notice that the ubisemiquinone anion, ubihydroquinone anion, and ubihydroquinone dianion would be very strong reductants in aprotic environments.

From the UV-visible absorbance spectra, the major band of the oxidized form shifts toward the red with addition of electron withdrawing substituents while the longest wavelength bands of the semiquinone anion in aprotic solvents are only slightly affected (Fig. 7). In contrast, the band at 420 nm in the ubisemiquinone anion is much decreased and red shifted in the complexes containing halogens. The Resonance Raman spectra of DDQ and 2,3-diiodo-5-*t*-butyl-benzoquinone were very different (Figs. 6A and 6C) while the semiquinone anions were unique but similar (see Figs. 6B and 6D).

Resonance Raman spectra have great promise for use as a diagnostic tool to determine the specific quinone species (oxidation and protonation state) that may be involved in a particular biological reaction. Our data for a close ubiquinone analogue, 2,3-dimethyl-5,6-dimethoxybenzoquinone (Figs. 6A and B) demonstrates the dependency of particular vibrations on oxidation state of the molecule. The method is becoming sufficiently sensitive and rapid to be applied to photosynthetic preparations. We are continuing our work in this area to establish more completely properties of the simple analogs and isolated ubiquinone and to examine intermediates in bacterial photosynthesis.

Many benzoquinone analogs are being used to probe the sequences of electron transfer in biological systems (*3*, *23–25*, *38*). In most cases, it is assumed that the analog competes for a site where ubiquinone is normally bound but that some property of the analog is sufficiently different to prevent proper function. The physical property of the quinone that is thought to be important in most inhibitions is its electrochemical behavior, although quenching of excited states by substituted heavy atoms may also be important in photosynthetic systems (*25*). We have measured the aprotic $E_{1/2}$ values for many benzoquinone analogs currently being studied and listed them in Table II. These data should be useful to those conducting research with such analogs, both to ascertain the reason for their inhibition and to probe structure–function relationships in some electron carrier protein complexes. We have tested most of these compounds as inhibitors of the primary photochemistry in *R. rubrum* and also for their ability to restore secondary electron transport in systems where ubiquinone was first selectively depleted by extraction. These results will be described elsewhere (*50*).

Of the biological systems in which ubiquinone plays an important role as an electron carrier, the first well defined experiments that showed direct involvement of the ubisemiquinone oxidation state were those with photosynthetic bacteria (*1*, *2*). Although much evidence is now beginning to accumulate which indicates that the semiquinone oxidation state of ubiquinone (or plastoquinone) is probably an important intermediate in all photosynthetic and aerobic electron transport systems, by far the clearest evidence was obtained using photosynthetic bacteria. Therefore, discussion of a few insights gained from the electrochemical and spectroscopic data will be limited to photosynthetic bacteria.

From measurements on membrane vesicles (also called chromatophores) and pigment protein complexes isolated from them, a tightly bound ubiquinone-10 molecule is known to be reduced by one electron during the primary photochemical events and it is thought to exist as the anion radical. Evidence exists from studies on *R. rubrum* that a second electron may be transferred to this acceptor from a second donor complex under saturating light conditions to produce two-electron-reduced ubiquinone (*4*). This special ubiquinone molecule thus serves as the first stable (life-time in reduced state longer than a microsecond) electron acceptor. This evidence was first provided from ESR data with low iron preparations (*1*) and then from ESR (*2*) and absorbance difference spectra (*9*, *10*, *39*, *40*) of reaction center preparations. Extraction experiments confirmed the importance of this molecule to the primary photochemical event (*3*–*7*). Therefore, identification of ubisemiquinone anion as the stable reduction product in reaction centers was based on ESR signal and absorbance change measurements. From the data presented here, it appears that the difference spectra observed with reaction centers are better fit by a ubisemiquinone anion in a protic (or limited protic) environment (Fig. 4A) rather than in a non-protic environment (Fig. 4B). The reaction center absorbance band at 360 nm together with the lack of a distinctive peak at about 430 nm (next to the prominent 450 nm peak) can be fit with the data of a protic environment much more closely than that of an aprotic. Such a comparison is made in Fig. 9. On the basis of p*K*a′ data on isolated ubisemiquinones one would expect that the ubisemiquinone in reaction center preparations is an anion radical. This is because a p*K*a′ of 6.5 (Fig. 8) has been measured for the equilibria shown below in methanol (*41*).

$$QH\cdot \rightleftharpoons Q^{\overline{\cdot}} + H^{+}. \qquad (2)$$

The analogous value for benzoquinone in water has been determined to be 4.0 (*47*). Thus, in the polar environment of reaction center preparations in

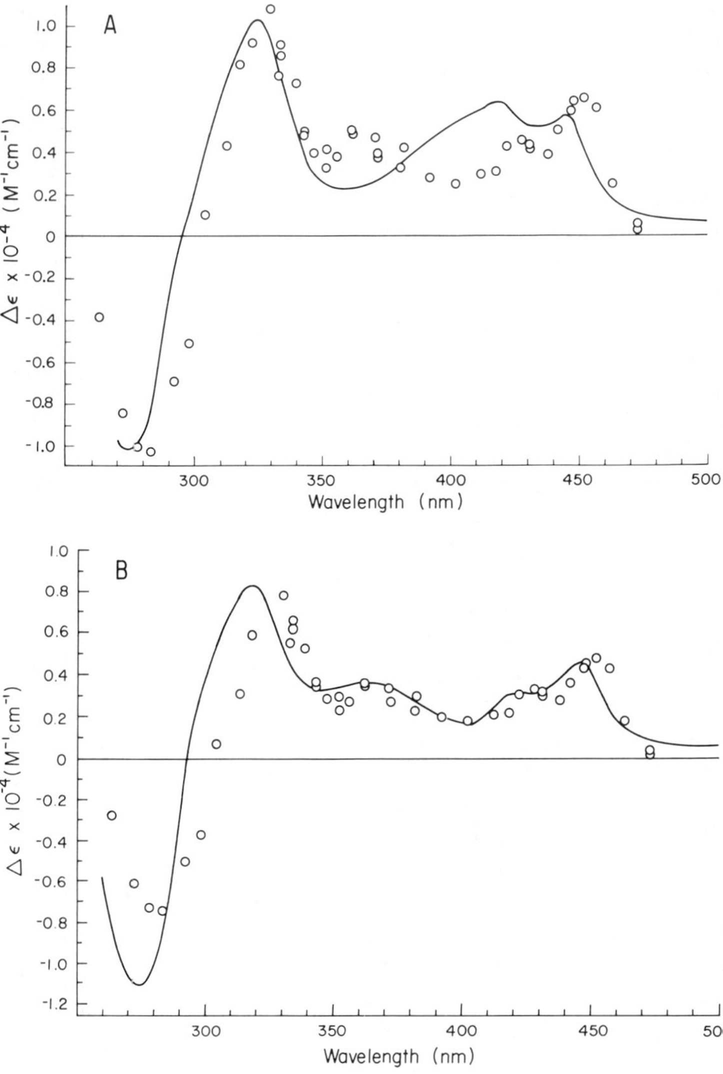

Fig. 9. Difference spectra of ubisemiquinone anion minus ubiquinone, (A), in dimethyl formamide; and (B) in 80% ethanol–water (w/w) calculated from spectra presented in Fig. 4. Data from Verméglio (*9*, used with permission) (open circles) for the light-induced absorbance changes in reaction centers of *Rps. sphaeroides* strain R-26 have been normalized to the solution spectra for comparison. Differential extinction coefficients are exact for the difference spectrum in dimethyl formamide but subject to the qualifications presented in the text for the spectrum in 80% ethanol–water.

water at pH 7, the ubisemiquinone anion would be expected to be the stable species. This is verified by comparison of the reaction center ubisemiquinone spectrum with the ubisemiquinone anion and neutral species recorded by Land and Swallow (*41*) in methanol. The neutral semiquinone displays only a single broad absorption peak (λ_{max} = 420 nm; $\epsilon = 3.1 \times 10^3\ M^{-1}cm^{-1}$) in the region between 350 and 500 nm. In simple systems the semiquinone anion would not be observed as a stable species at pH 7 because the disproportionation reaction is very fast for the neutral semiquinone and the equilibria greatly favor disproportionation. However, in the system discussed here, the ubisemiquinone anion is tightly bound to a large complex and probably not free to diffuse on a time scale that would allow the disproportionation reaction to be significant.

Results of redox control experiments on reaction center activity are consistent with the above interpretation, that is the stable form of the reduced ubiquinone is the anion radical. In these experiments, it is assumed that a vital electron carrier in the reaction center can be chemically reduced as the environmental potential is lowered and will consequently no longer be able to accept an electron from the light excited bacteriochlorophyll donor species. The rationale and methodology for conducting such experiments was developed in this laboratory many years ago (*15*, *42*, *43*). Reed *et al.* (*16*) and later Dutton *et al.* (*17*) found that the equilibrium at low potential associated with quenching normal activity in the isolated reaction center complexes of the R-26 mutant of *Rps. sphaeroides* had no pH dependency in the range of pH 6 to 9. This is now interpreted as being due to the reduction of the first stable electron acceptor ubiquinone to its anionic semiquinone form with an E_0' value at pH 7.0 of −45 mV (versus SHE). From the data of Table II, one would have expected a much more negative value (e.g., −450 mV versus SHE). If the equilibrium is correctly identified, the ubisemiquinone anion formed must be greatly stabilized by a positive charge(s) in its environment. Stabilization of semiquinone anions and anionic hydroquinones by monovalent and especially polyvalent cations has been demonstrated in solution (*44*, *49*). Also, stabilization of these species by hydrogen bonding has been studied (*48*). Because an iron atom is thought to be relatively near this ubisemiquinone anion (*3*), it may provide the needed stabilization.

Contrary to the results with reaction center preparations, measurement with intact membrane vesicles of *Rps. sphaeroides,* while yielding approximately the same E_0' value at pH 7, show a very clear pH dependency of 60 mV/pH unit between pH 6 and 9 (*17*, *45*) with a p*K*a′ of about 10 (*46*). It has been suggested that the first stable acceptor ubiquinone has been reduced to its neutral ubisemiquinone form at pH values near 7 and that the p*K*a′ of about 10 (in *Rps. sphaeroides*) should be assigned to the equilibria indicated in equation (2) above. However, notice that this value is almost 4 pK_a units above that measured for ubisemiquinone in methanol (*41*). Furthermore, if a positively charged iron atom is in the vicinity and stabilizes the negative charge of the ubisemiquinone anion as is popularly believed, such stabilization would have the effect of shifting the p*K*a′ to lower values, not higher. We would like to suggest that the data could better be interpreted as resulting from the redox equilibria

$$\mathrm{UQ} + \mathrm{e} + \mathrm{H}^+ \rightleftharpoons \mathrm{UQH}\cdot \tag{3}$$

and

$$\mathrm{UQH}\cdot + \mathrm{e} + \mathrm{H}^+ \rightleftharpoons \mathrm{UQH_2}, \tag{4}$$

where Reaction (4) occurs considerably more slowly than Reaction (3) so that the shape of the titration curve may vary between the best fit by a one-electron reduction and that of a two-electron reduction (*15*). The p*K*a′ of about 10 would then more reasonably be assigned to

$$\mathrm{UQH_2} \rightleftharpoons \mathrm{UQH^-} + \mathrm{H}^+. \tag{5}$$

See Table I for appropriate comparisons. A value for E_0' at pH 7 of −50 mV is also in the range expected for Reactions (3) and (4).

The latter interpretation is the only one of the two which is also consistent with evidence for *R. rubrum* that indicates there are conditions under which two sequential one-electron reductions of a single ubiquinone molecule may be important in the primary photochemistry (*4, 5*). The conditions referred to are actually those most consistent with the native state, either intact cells or very fresh membrane vesicles. The two one-electron reductions were suggested to have midpoints of −95 mV and < −400 mV. These may be attributed to Reactions (3) and (6), respectively,

$$\mathrm{UQH}\cdot + \mathrm{e} \rightleftharpoons \mathrm{UQH^-} \tag{6}$$

which could occur in the limited protic environment of the membrane and protein. The E_0' value associated with Reaction (3) would be −95 mV while that with Reaction (6) would be −400 mV or less. The last species, UQH^-, is presumed to (a) be very unstable at pH 7, (b) cause a significant conformational change in the protein, and (c) quickly become protonated to UQH_2. This would explain an observed hysteresis (*4*) upon reoxidation where the environmental potential must be raised above zero millivolts before activity is restored, presumably due to reaction (1). Equilibria (3) and (6) would be expected to have E_0' values in the regions measured. For suggestions of how these equilibria are related to the normal primary photochemical events, see Loach (*4*).

In this paper, we have presented additional electrochemical and spectral data which should help provide insights and experimental tools to better probe the chemical mechanism of electron transport reactions involving ubiquinone as well as the inhibition of these reactions with selected benzoquinone analogues.

ACKNOWLEDGMENTS

This investigation was supported by research grants from the National Science Foundation (PCM 74-12588 and PCM 78-16660) and the National Institutes of Health (GM 11741 and GM 27498).

REFERENCES

1. Loach, P. A., and Hall, R. L. (1972). *Proc. Natl. Acad. Sci. U.S.A.* **69,** 786–790.
2. Feher, G., Okamura, M. Y., and McElroy, J. D. (1972). *Biochim. Biophys. Acta* **257,** 222–226.
3. Okamura, M. Y., Isaacson, R. A., and Feher, G. (1975). *Proc. Natl. Acad. Sci. U.S.A.* **72,** 3491–3495.
4. Loach, P. A. (1976). *Prog. Bioinorg. Chem.* **4,** 89–192.
5. Morrison, L., Runquist, J., and Loach, P. (1977). *Photochem. Photobiol.* **25,** 73–84.
6. Cogdell, R. J., Brune, D. C., and Clayton, R. K. (1974). *FEBS Lett.* **45,** 344–347.
7. Vadeboncoeur, C., Nöel, H. Poirier, L., Cloutier, Y., and Gingras, G. (1979). *Biochemistry* **18,** 4301–4308.
8. Knaff, D. B., Malkin, R., Myron, J. C., and Stoller, M. (1977). *Biochim. Biophys. Acta* **459,** 402–411.
9. Verméglio, A. (1977). *Biochim. Biophys. Acta* **459,** 516–524.
10. Wraight, C. A. (1979). *Biochim. Biophys. Acta* **548,** 309–327.
11. Feher, G., Isaacson, R. A., McElroy, J. D., Ackerson, L. C., and Okamura, M. Y. (1974). *Biochim. Biophys. Acta* **368,** 135–139.
12. Ruzicka, F. J., Beinert, H., Schepler, K. L., Dunham, W. K., and Sands, R. H. (1975). *Proc. Natl. Acad. Sci. U.S.A.* **72,** 2886–2890.
13. Ohnishi, T., and Trumpower, B. L. (1980). *J. Biol. Chem.* **255,** 3278–3284.
14. Bowyer, J. R., and Trumpower, B. L. (1980). *FEBS Lett.* **115,** 171–174.
15. Loach, P. A. (1966). *Biochemistry* **5,** 592–600.
16. Reed, D. W., Zankel, K. L., and Clayton, R. K. (1969). *Proc. Natl. Acad. Sci. U.S.A.* **63,** 42–46.
17. Dutton, P. L., Leigh, J. S., and Wraight, C. A. (1973). *FEBS Lett.* **36,** 169–173.
18. Chambers, J. Q. (1974). *In* "The Chemistry of the Quinonoid Compounds" (S. Patai, ed.), pp. 737–792. Wiley (Interscience), New York.
19. Mann, C. K., and Barnes, K. K. (1970). "Electrochemical Reactions in Nonaqueous Systems," pp. 190–199. Dekker, New York.

20. Moret, V., Planamanti, S., and Fornasari, E. (1961). *Biochim. Biophys. Acta* **54,** 381–383.
21. Crane, F. L., and Dilley, R. A. (1963). *Methods Biochem. Anal.* **11,** 279–306.
22. Land, E. J., Simic, M., and Swallow, A. (1971). *Biochim. Biophys. Acta* **226,** 239–240.
23. Trebst, A., Harth, E., and Dräber, W. (1970). *Z. Naturforsch.* **25,** 1157–1159.
24. Baltscheffsky, M. (1975). *Proc. Int. Congr. Photosynth., 3rd, Rehovot, 1974* **1,** 799–806.
25. Bering, C. L., and Loach, P. A. (1977). *Photochem. Photobiol.* **26,** 607–615.
26. van Duyne, R., personal communication.
27. Thomas, A. B., and Rochow, E. G. (1957). *J. Am. Chem. Soc.* **79,** 1843–1848.
28. Sawyer, D. T., and Roberts, J. L., Jr. (1974). "Experimental Electrochemistry for Chemists," Chap. 5. Wiley, New York.
29. Bull, C. (1976). Ph.D. Thesis, Northwestern Univ. Evanston, Illinois.
30. Morrison, L. (1978). Ph.D. Thesis, Northwestern Univ., Evanston, Illinois.
31. Cotton, T. M., and van Duyne, R. P. (1978). *Biochem. Biophys. Res. Commun.* **82,** 424–433.
32. van Duyne, R. P., Suchanski, M. R., Lakovits, J. M., Siedle, A. R., Parks, K. D., and Cotton, T. M. (1979). *J. Am. Chem. Soc.* **101,** 2832–2837.
33. Cotton, T. M., Parks, K. D., and van Duyne, R. P. (1980). *J. Am. Chem. Soc.* **102,** 6399–6407.
34. Baxendale, J. H., and Hardy, H. R. (1953). *Trans. Faraday Soc.* **49,** 1140–1144.
35. Marcus, M. F., and Hawley, M. D. (1971). *Biochim. Biophys. Acta* **226,** 234–238.
36. Girlando, A. Z., Bozio, I., and Pecile, C. J. (1978). *J. Chem. Phys.* **68,** 22–31.
37. Schelhorn, J. E., Bering, C. L., and Loach, P. A. (1980). *Annu. Meet. Am. Soc. Photobiol., Colorado Springs, Colo.* Abstr. TAM-C7.
38. Bowyer, J. R., Tierney, G. V., and Crofts, A. R. (1979). *FEBS Lett.* **101,** 206–212.
39. Clayton, R. K., and Straley, S. C. (1970). *Biochem. Biophys. Res. Commun.* **39,** 1114–1119.
40. Sloofen. L. (1972). *Biochim. Biophys. Acta* **275,** 208–218.
41. Land, E. J., and Swallow, A. J. (1970). *J. Biol. Chem.* **245,** 1890–1894.
42. Loach, P. A., Androes, G. M., Maskin, A. F., and Calvin, M. (1963). *Photochem. Photobiol.* **2,** 443–454.
43. Kuntz, I. D., Jr., Loach, P. A., and Calvin, M. (1964). *Biophys. J.* **4,** 227–249.
44. Jaworski, J. S., and Kalinowski, M. K. (1977). *J. Electroanal. Chem.* **76,** 301–314.
45. Jackson, J. B., Cogdell, R. J., and Crofts, A. R. (1973). *Biochim. Biophys. Acta* **292,** 218–225.
46. Prince, R. C., and Dutton, P. L. (1976). *Arch. Biochem. Biophys.* **172,** 329–334.
47. Smith, I. C. P., and Carrington, A. (1967). *Mol. Phys.* **12,** 439–448.
48. Peover, M. E., and Davies, J. D. (1963). *J. Electroanal. Chem.* **6,** 46–53.
49. Fujihira, M., and Hayano, S. (1972). *Bull. Chem. Soc. Jpn.* **45,** 644–645.
50. Shelhorn, J. E., Bustamante, P. B., and Loach, P. A. (1981). *Proc. Int. Congr. Photosynth., 5th, Halkidiki, Greece, 1980,* Vol. III, 969–980.

Physical Chemistry of Semiquinones 3

A. J. SWALLOW

INTRODUCTION

Information about the physical chemistry of semiquinones is gained from experiments in which chemical, electrochemical, photochemical and radiation–chemical reactions are used to generate the semiquinone. The photochemical and radiation–chemical methods have been particularly fruitful since the 1950s and 1960s, profiting greatly by the introduction of techniques such as flash photolysis and pulse radiolysis. Approximately 280 publications based on these methods are listed in a bibliography on semiquinones published elsewhere (*1*). This paper summarizes some advances in the physical chemistry of semiquinones based principally on photochemistry and radiation chemistry.

The elements of photochemistry are relatively familiar, but those of radiation chemistry (*2*) may be less so. In experiments with ionizing radiation the samples are irradiated with either x or γ rays or fast electrons (energy about 2–10 MeV). x and γ rays are highly penetrating, but interact with matter by producing photo or Compton electrons and these produce the chemical effects which are observed. Thus the effects of x and γ rays are the same as those produced when fast electrons are introduced directly. Fast electrons are less penetrating than x or γ rays, but are still sufficiently energetic to pass right through samples of thickness 0.5 cm or more.

Fast electrons eject electrons from molecules randomly along their track, and also similarly produce excitations. In the irradiation of a dilute solution nearly all of the initial action is therefore on the solvent rather than the solute, contrasting with photochemistry where the light is nor-

Function of Quinones in Energy Conserving Systems

ISBN 0-12-701280-X

mally absorbed by the solute. It is not yet easy to explain the effects of irradiating complex systems; a significant fraction of the observed effect would be expected to result from direct action rather than from indirect action via species produced from the solvent. For the time being this prevents the application of radiation–chemical methods to organized biological or biochemical systems.

The consequences of ionization and excitation depend on the solvent. They are particularly well understood for water. In this case the positive ions react at every collision with neighboring molecules, yielding free hydroxyl radicals, while the electrons rapidly become hydrated

$$H_2O \longrightarrow H_2O^+ + e^-, \quad (1)$$

$$H_2O^+ + H_2O \longrightarrow H_3O^+ + OH, \quad (2)$$

$$e^- + aq \longrightarrow e_{aq}^-. \quad (3)$$

Minor processes give rise to small amounts of H atoms, H_2 and H_2O_2 as well. Hydroxyl radicals can be converted into reducing radicals by reaction with solutes like formate ion or isopropanol,

$$OH + (CH_3)_2CHOH \longrightarrow H_2O + (CH_3)_2\dot{C}OH. \quad (4)$$

Hydroxyl radicals can also be converted into less powerful oxidizing radicals by reactions, such as

$$OH + N_3^- \longrightarrow OH^- + N_3\cdot. \quad (5)$$

Hydrated electrons [$E(aq/e_{aq}^-) = -2900$ mV] can be converted into less powerfully reducing radicals if required, or into hydroxyl by reaction with nitrous oxide,

$$e_{aq}^- + N_2O \xrightarrow{H^+} N_2 + OH. \quad (6)$$

Irradiation of a substance in a dilute aqueous solution exposes it to the action of reducing or oxidizing radicals of chosen power depending upon the solute or solutes present. Excited water plays no significant part. Irradiation in alcohols exposes solutes to the action of reducing radicals, the overall effect of radiation on methanol (*3*), is summarized by

$$CH_3OH \longrightarrow \cdot CH_2OH + e_{MeOH}^- + H^+. \quad (7)$$

Irradiation in other solvents also exposes solutes to the action of free radicals, while in solvents of low dielectric constant excited states become quite significant.

The consequences of attack on the solute by species formed from the solvent are followed by the usual methods of detection. In pulse radiolysis the irradiation lasts about a microsecond or less, and subsequent reactions are generally observed optically as in flash photolysis and other fast reaction techniques. Conductimetric or other methods of detection can also be used.

FORMATION OF SEMIQUINONES

For physical chemistry studies, semiquinones can be formed photolytically from quinones, as in the well-known abstraction of a hydrogen atom by triplet anthraquinone,

$$Q^* + (CH_3)_2CHOH \longrightarrow QH\cdot + (CH_3)_2\dot{C}OH. \tag{8}$$

The organic radicals formed may give more semiquinone,

$$(CH_3)_2\dot{C}OH + Q \longrightarrow (CH_3)_2CO + QH\cdot. \tag{9}$$

Excited quinones can also abstract electrons from suitable donors, as in the reaction of triplet anthraquinones with hydroxide or other inorganic ions

$$Q^* + OH^- \longrightarrow Q^{\overline{\cdot}} + OH, \tag{10}$$

giving the anionic form of the semiquinone. Reduction of quinones by excited states of other molecules is of great importance as with triplet chlorophyll

$$Chl^* + Q \longrightarrow Chl^{\dot{+}} + Q^{\overline{\cdot}}. \tag{11}$$

Radiolytic reduction of quinones is achieved through the action of the solvated electron, e.g.,

$$e_{aq}^- + Q \longrightarrow Q^{\overline{\cdot}}, \tag{12}$$

or by the action of organic radicals produced as in Reactions (4) and (7) (Reaction 9).

Semiquinones can be formed from hydroquinones by photoionization,

$$QH_2 \longrightarrow QH\cdot + e_{aq}^- + H^+. \tag{13}$$

Excited dyestuffs can oxidize hydroquinone as well as other organic compounds,

$$D^* + QH_2 \longrightarrow DH\cdot + QH\cdot, \tag{14}$$

(cf., Reaction 8). Oxidizing radicals can oxidize hydroquinones. Hydroxyl radicals do this by adding to the hydroquinone, followed by elimination of water in a reaction catalyzed by acid or base

$$OH + QH_2 \longrightarrow \dot{Q}H_2OH \longrightarrow QH\cdot + H_2O. \quad (15)$$

A similar reaction produces semiquinones from methoxylated phenols by the elimination of methanol or from nitrophenols by the elimination of nitrite. Less powerful radicals (e.g., azide) act by electron transfer.

OPTICAL ABSORPTION SPECTRA

A high proportion of the work on the physical chemistry of semiquinones relies on knowledge of the optical absorption of the radicals. Consequently there are numerous determinations of absorption spectra (position of absorption bands and extinction coefficients). The spectrum of the semiquinone anion is relatively easy to determine, as the anion is fairly stable in alkaline aqueous solution or in organic solutes like DMF, but it is less easy to get quantitative spectra for the neutral radical. Both radical anion and neutral radical can be observed with flash photolysis, however flash photolysis does not lend itself to the ready determination of extinction coefficients. Pulse radiolysis can produce both the radical anion and neutral radical. The input of radiation energy to the solution can be measured within about 5% accuracy by the standard methods of radiation dosimetry. With radiation dosimetry the yield of radicals is known, especially for aqueous solutions, for which it is established that about 6–6.5 radicals (depending on conditions) are produced per 100 eV of energy absorbed. Consequently, radical concentrations can be calculated, and extinction coefficients at a series of wavelengths can be found from changes in optical absorption produced by the radiation pulse.

Semiquinone radicals absorb quite strongly at wavelengths where the parent quinone shows little absorption. Typical quantitative data obtained by means of pulse radiolysis are given in Table I.[1] The absorption maxima given in the table (λ_{max}) can be regarded as correct to within about 5 nm. Extinction coefficients (ϵ) can generally be considered accurate to within about 20%. The λ_{max} and ϵ for the radical anion are in adequate agreement with values obtained for aqueous solutions using other methods (*9*).

The quinonoid structure of the parent changes to aromatic in the semiquinone, and the absorptions in Table I are considered to represent π-π^*

[1] Based on the main on Hulme *et al.* (*4*), Patel and Willson (*5*), Rao and Hayon (*6*), Gohn and Getoff (*7*), and Richter (*8*). Mean values are given where there is more than one value in the literature.

TABLE I

Optical Absorption Data for Semiquinones in Aqueous Solution

Compound	Neutral Semiquinone λ_{max} (nm)	Neutral Semiquinone $\epsilon(M^{-1}\ cm^{-1})$	Radical Anion λ_{max} (nm)	Radical Anion $\epsilon(M^{-1}\ cm^{-1})$
Benzoquinone	415	4,700	425	6,900
Methylbenzoquinone	405	4,500	430	6,200
2,3-Dimethylbenzoquinone	415	5,100	430	6,700
2,5-Dimethylbenzoquinone	415	4,300	435	7,000
2,6-Dimethylbenzoquinone	405	4,900	430	6,100
Trimethylbenzoquinone	410	4,300	435	6,700
Duroquinone	420	4,400	445	7,400
4-Methyl-1,2-benzoquinone	285, 390	8,700, 1750	310	11,300
4-*t*-Butyl-1,2-benzoquinone	290, 390	7,700, 1850	315	12,200
1,4-Naphthoquinone	370	7,200	390	12,500
2-Methyl-1,4-naphthoquinone	370	9,500	395	12,000
2,3-Dimethyl-1,4-naphthoquinone	380	7,300	400	11,000
2-Hydroxy-1,4-naphthoquinone	370	5,900	390	6,300
1,2-Naphthoquinone	<260	16,000	265	40,000
Anthraquinone	375	11,000	395, 480	7,800, 7,300
Anthraquinone-1-sulphonate	385	12,000	400, 500	8,000, 8,000
Anthraquinone-2-sulphonate	390	12,500	400, 500	8,000, 8,000

transitions. Evidently transitions between the energy levels of the free electron in the semiquinone require less energy than those of the paired electrons in the parent. As for radicals in general, the radical anions of the benzoquinones and the naphthoquinones absorb at slightly longer wavelengths than the corresponding neutral radicals. They also have higher extinction coefficients. For the anthraquinones the wavelength of maximum absorption increases in going from the neutral radical to the radical anion, but two bands are seen instead of one. In no case, however, is the λ_{max} very different for the neutral radical and the anion, consistent with the orbitals involved being little affected by protonation. Within each series the influence of the substituents examined is quite small and no systematic effect can be discerned for the simple substituents in the compounds in Table I.

The values of λ_{max} in organic solvents as seen in flash photolysis experiments are similar to those in Table I. The neutral durosemiquinone radical has been estimated to have an extinction coefficient in ethanol at λ_{max} = 420 nm of 5,500 ± 500 $M^{-1}cm^{-1}$ compared with 4,400 $M^{-1}cm^{-1}$ in Table I (*10*). In cyclohexane λ_{max} appears at 407.5 nm, but in contrast to water or alcohol the band is very sharp, so that the apparent extinction coefficient is strongly dependent on monochromator bandwidth. A true extinction coefficient of 8850 $M^{-1}cm^{-1}$ has been estimated, but values less than half this are observed experimentally. Oscillator strengths for the transitions are taken to be independent of solvent. Values of 0.06 and 0.052 have been calculated for the transitions in Table I from experimental observations on the benzoquinone anion (*9*) and the neutral durosemiquinone radical (*10*) respectively. Both λ_{max} and oscillator strength are in adequate agreement with molecular orbital calculations (*9*).

The absorption spectra of neutral and anionic ubisemiquinone in methanol are shown in Fig. 1 (*11*). It may be noted that the neutral semiquinone shows a band at 275–280 nm (ϵ = 7400 $M^{-1}cm^{-1}$) where the parent ubiquinone also absorbs (ϵ = 15,150 $M^{-1}cm^{-1}$). With the anion, however, this band is displaced to a longer wavelength (λ_{max} = 320 nm, ϵ = 10,700 $M^{-1}cm^{-1}$). Extinction coefficients for the neutral radical at 425 nm and the anion radical at 445 nm have also been estimated by examining ubiquinone in 7 *M* isopropanol, 1 *M* acetone, and comparing the durosemiquinones in this solvent and in water (*5*). It was concluded that the neutral ubisemiquinone would have ϵ = 5300 $M^{-1}cm^{-1}$ at 425 nm if it could be prepared in simple aqueous solution, and the anionic form would have ϵ = 8600 $M^{-1}cm^{-1}$ at 445 nm. Both of these values are slightly higher than those for other benzosemiquinones in Table I and significantly higher than for the ubisemiquinones in Fig. 1. It may be that the true extinction coefficient in water is close to the other benzosemiquinones, but the extinction coefficient in methanol is higher than in Fig. 1 as also found for durosemiquinone in ethanol (*10*). This would also be consistent with the solvent affecting the shape of the absorption but not λ_{max} or oscillator strength.

Absorption spectra have also been given for the neutral and anionic forms of ubisemiquinones 0 and 10 and plastoquinone-9 in methanolic

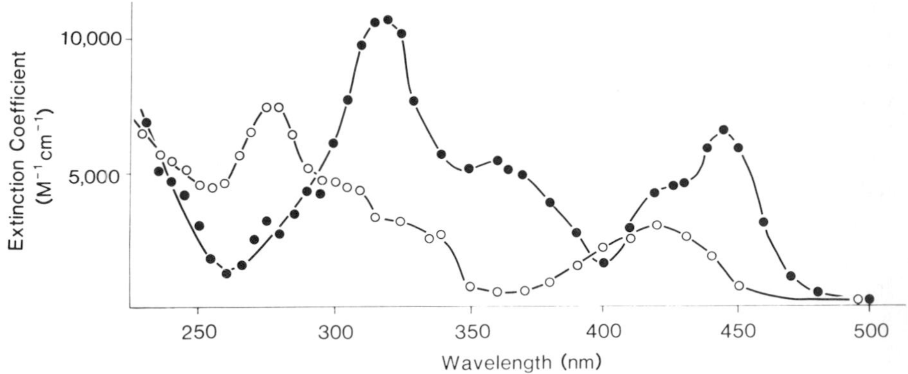

Fig. 1. Absorption spectrum of ubisemiquinone in methanol (*11*). (○), neutral form; (●), anionic form. Used with permission.

solution (*12*). The values of λ_{max} are indistinguishable from those of the ubisemiquinones and the simple quinones in Table I. Extinction coefficients appear to be uniformly on the low side, but due to the above considerations this could reflect the errors in the methanol experiments. For 2-methyl-3-phytyl-1,4-naphtoquinone the neutral radical and the anion have maxima at 380 nm (ϵ in water would be 9,900 $M^{-1}cm^{-1}$) and 400 nm (ϵ in water would be 10,200 $M^{-1}cm^{-1}$), respectively, not very different from the other 1,4-naphthoquinones in Table I.

None of the experiments so far discussed have provided any indication that the semiquinones exist other than as separated molecules. However, with alkaline methanol solutions of 2-amino-anthraquinones substituted in the 4-position with $NHCH_3$, NH_2, or OH, pulse radiolysis results in more than one quinone molecule being lost for each reducing radical introduced into the system (*13*). This suggests that quinone molecules may be aggregated in the solution, and implies that semiquinone anion radicals are associated with quinone ($Q–Q^{\overline{\cdot}}$). Further work on this phenomenon is desirable.

ACID–BASE PROPERTIES

After electron transfer to a quinone, the semiquinone anion formed may protonate rapidly by reaction with a sufficiently high concentration of a suitable proton donor such as the hydrated proton itself or a strong Lowry-Brønsted acid, attaining the equilibrium corresponding to its p*K*

$$QH\cdot \rightleftharpoons Q^{\overline{\cdot}} + H^+, \qquad (16)$$

before disappearance by mutual reaction. The neutral semiquinone radical may deprotonate similarly. Since the absorption spectrum of the uncharged form differs from that of the anionic form, measurements of optical density after a microsecond pulse of radiation are expected to show a dependence on the pH of the solution. This should correspond to the theoretical titration curve and enable the p*K* of the semiquinone to be determined. Values of p*K* have also been determined from the conductivity of solutions in which the semiquinone has been produced by action of OH on methoxylated phenols.

Table II[2] lists values of p*K* which have been determined in aqueous solution by these methods. The p*K* values can generally be considered accu-

[2] Based on Hulme *et al.* (*4*), Patel and Willson (*5*), Rao and Hayon (*6*), Gohn and Getoff (*7*), Richter (*8*), Adams and Michael (*14*), Willson (*15*), Hayon *et al.* (*16*) and Steenken and O'Neill (*17*). Selection has been made in cases where there is doubt.

TABLE II

p*K* of Semiquinones in Aqueous Solution

Compound	p*K*
Benzoquinone	4.0
Methylbenzoquinone	4.45
2,3-Dimethylbenzoquinone	4.65
2,5-Dimethylbenzoquinone	4.6
2,6-Dimethylbenzoquinone	4.75
Trimethylbenzoquinone	4.95
Duroquinone	5.1
1,2-Benzoquinone	5.0
3-Methoxy-1,2-benzoquinone	5.0
4-Methyl-1,2-benzoquinone	4.5
4-*t*-Butyl-1,2-benzoquinone	5.2
1,4-Naphthoquinone	4.1
2-Methyl-1,4-naphthoquinone	4.5
2,3-Dimethyl-1,4-naphthoquinone	4.25
2-Hydroxy-1,4-naphthoquinone	4.7
1,2-Naphthoquinone	4.8
Anthraquinone	5.3
Anthraquinone-1-sulphonate	5.4
Anthraquinone-2-sulphonate	3.4
Anthraquinone-2,6-disulphonate	3.2

rate to within ± 0.2 units. For the 1,4-benzoquinones there are sufficient values to see that introduction of methyl groups increases the p*K* by 0.25 units per group, consistent with electron donation into the ring. Methyl or hydroxy groups in the quinone ring of naphthoquinone also increase the p*K* of the semiquinone. Electron-withdrawing SO_3^- groups decrease the p*K* of anthrasemiquinones especially in the 2-position.

Ubisemiquinone had a p*K* value of 5.9 in 7 *M* isopropanol/1 *M* acetone; 2-methyl-3-phytyl-1,4-naphthosemiquinone in 5 *M* isopropanol/2 *M* acetone had a p*K* of 5.5 (*5*). In the 7 to 1 solvent system, durosemiquinone exhibited a p*K* one pH unit higher than in water containing only one *M* isopropanol and acetone. It could be expected that if ubisemiquinone could be prepared in aqueous solution its p*K* would be about 4.9 (cf., duroquinone, 5.1) and the p*K* of 2-methyl-3-phytyl-1,4-naphthosemiquinone would be about 4.5 (cf., 2,3-dimethyl-1,4-naphthosemiquinone, 4.25). These values are therefore consistent with the electron donating properties of the side groups in the two cases. The p*K* of ubisemiquinone has also been determined in methanol (*18*) by measuring the rate of deprotonation of the neutral semiquinone after formation by

$$\cdot CH_2OH + Q \longrightarrow QH\cdot + HCHO. \quad (17)$$

The rate of deprotonation by the forward reaction of Eq. (16) was $1.0 \times 10^4 sec^{-1}$. By assuming that the reverse reaction is diffusion controlled ($k = 3 \times 10^{10} M^{-1}sec^{-1}$) the p$K$ of ubisemiquinone in methanol becomes 6.45.

No pK appears to have been determined for plastosemiquinone, but it is expected that the value would be slightly less than that for ubisemiquinone because it contains one less electron-donating group. A value of about 4.7 would therefore be expected for the hypothetical aqueous solution case.

Neutral semiquinones can protonate further in highly acidic solutions: duroquinone shows a pK of −1.1 in 50% aqueous ethanol for ionization of the cation radical $QH_2^{+\cdot}$ (*19*).

REDOX PROPERTIES

The redox potentials for semiquinones consist of the potentials for formation from quinone (the one-electron reduction potential of the parent quinone, $E(Q/Q^{\overline{\cdot}})$ for reduction to hydroquinone (the reduction potential, $E(Q^{\overline{\cdot}}/Q^{2-})$, for addition of the second electron to the quinone. Because of the protonation of $Q^{\overline{\cdot}}$ and Q^{2-}, the reduction potentials are dependent on pH.

Until recently, it has not been easy to determine these reduction potentials in aqueous solution at biological pH values, although deductions could be made from experiments in strong alkali. However after the anionic form of a semiquinone has been formed at low concentration by the action of a microsecond pulse of radiation through Reactions (1–3) and subsequent reactions such as (4), (9), (12), and (16), it is able to come into reversible equilibrium with a reference redox couple $R/R^{\overline{\cdot}}$ by

$$Q^{\overline{\cdot}} + R \rightleftharpoons Q + R^{\overline{\cdot}}, \quad (18)$$

before disappearance through mutual reaction. Measurements of optical absorption can then be used to obtain the equilibrium constant K_{eq} for Reaction (18), and hence, if the reduction potential $E(R/R^{\overline{\cdot}})$ is known, the value of $E(Q/Q^{\overline{\cdot}})$ for the quinone can be found. At 25 °C we have.

$$E(Q/Q^{\overline{\cdot}}) = E\ (R/R^{\overline{\cdot}}) - 59 \log K_{eq}, \quad (19)$$

for reduction potentials in mV. The reduction potential $E(Q^{\overline{\cdot}}/Q^{2-})$ is given

TABLE III

Reduction Potentials of Quinones in Aqueous Solution[A] at 25 °C

Compound	(mV, versus NHE)		
	$E(Q/Q^{\cdot -})_7$	$E(Q/Q^{2-})_7$	$E(Q^{\cdot -}/Q^{2-})_7$
1,2-Benzoquinone	+210	+370	+530
1,4-Benzoquinone	+99	+286	+473
Methyl-1,4-benzoquinone	+23	+230	+437
1,4-Naphthoquinone-2-sulphonate	−60	+120	+300
2,5-Dimethyl-1,4-benzoquinone	−66	+180	+426
2,3-Dimethyl-1,4-benzoquinone	−74	+175	+424
2,6-Dimethyl-1,4-benzoquinone	−80	—	—
Trimethyl-1,4-benzoquinone	−165	+114	+393
2-Methyl-3-phytyl-1,4-naphthoquinone	−170	−60	+220
2-Methyl-1,4-naphthoquinone	−203	−5	+193
2,3-Dimethyl-1,4-naphthoquinone	−240	—	—
Duroquinone	−240	+57	+354
9,10-Anthraquinone-2-sulphonate	−380	−228	−76

[A] In the case of 2-methyl-3-phytyl-1,4-naphthoquinone the solution also contained isopropanol (5*M*) and acetone (2*M*). In other cases the solutions contained 1*M* (or less) added solutes.

by

$$E(Q^{\cdot -}/Q^{2-}) = 2E(Q/Q^{2-}) - E(Q/Q^{\cdot -}), \tag{20}$$

where $E(Q/Q^{2-})$ is the known two-electron reduction potential of the quinone.

Determinations of reduction potentials $E(Q/Q^{\cdot -})_7$, at pH 7 by different workers are in excellent agreement to within about 10 mV. Values obtained to date are given in Table III.[3] The table also gives values of the two-electron reduction potentials $E(Q/Q^{2-})_7$, at pH 7, based on the older literature (*26*) and of $E(Q^{\cdot -}/Q^{2-})_7$ obtained using Eq. (20). For comparison, $E(O_2/O_2^{\cdot -})_7$ for oxygen may be taken to be −330 MV (or −155 mV with reference to 1 *M* rather than 1 atmosphere), $E(O_2/O_2^{2-})_7$ to be +230 (1 atmosphere) or + 355 mV (1 *M*) and $E(O_2^{\cdot -}/O_2^{2-})_7$ to be + 865 mV (*23*).

Where only a single ionization of the semiquinone needs to be considered, $E(Q/Q^{\cdot -})$ depends on pH according to

$$E(Q/Q^{\cdot -}) = E(Q/Q^{\cdot -})_7 + 59 \log \left(\frac{K_i + [H^+]}{K_i + 10^{-7}}\right), \tag{21}$$

[3] Based on Meisel and Neta (*20*), Meisel and Czapski (*21*), Wardman and Clarke (*22*), Ilan *et al.* (*23*), Meisel and Fessenden (*24*), and Steenken and Neta (*25*).

where K_i is the ionization constant. $E(Q/Q^{\overline{\cdot}})$ is therefore independent of pH above that corresponding to the pK of the semiquinone, but rises with decreasing pH below the pK. In the pH-independent region the redox process is the simple addition of an electron, while in the acid region it is $Q + H^+ + e^- \rightarrow QH\cdot$.

Inspection of Table III shows that for the benzoquinones, introduction of each electron-donating methyl group into the ring decreases $E(Q/Q^{\overline{\cdot}})_7$ by about −85 mV regardless of position. The stability constant for a semiquinone at a given pH may be defined by

$$K_s = \frac{[Q^{\overline{\cdot}}]^2}{[Q][Q^{2-}]}, \tag{22}$$

where the terms in brackets refer to the total concentrations of each species, regardless of protonation state. It is given in terms of $E(Q/Q^{\overline{\cdot}})$ and $E(Q^{\overline{\cdot}}/Q^{2-})$ by

$$59 \log K_S = E(Q/Q^{\overline{\cdot}}) - E(Q^{\overline{\cdot}}/Q^{2-}). \tag{23}$$

Figure 2 shows the decreasing stability of the methyl-substituted benzosemiquinones at pH 7 as the number of methyl groups (electron-donating)

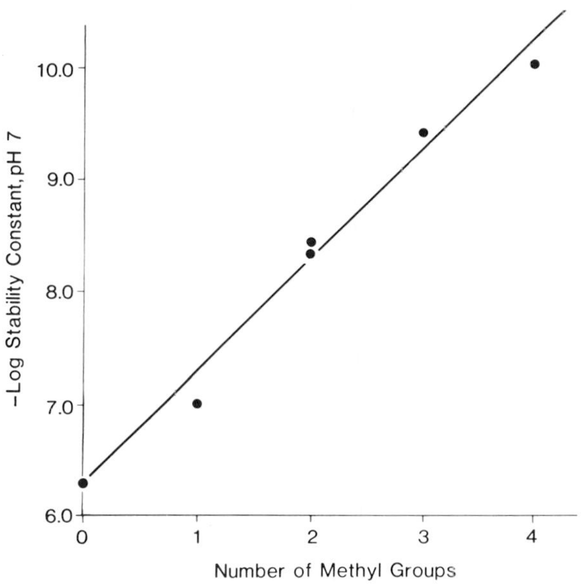

Fig. 2. Stability constant of substituted benzoquinones as a function of number of methyl groups.

increases. The main reason for the decrease is the increase in the pK values of the hydroquinone as the number of methyl groups increases. In strongly alkaline solutions where hydroquinones and semiquinones (pK ~ 10–11 and 11.5–13) are fully ionized, semiquinones are orders of magnitude more stable, K_s being in the region of 1–10.

There are numerous electrochemical measurements of half-wave potentials corresponding to the addition of both first and second electrons in non-aqueous solvents such as acetonitrile and dimethylformamide. These are for the addition of the electrons as such, uncomplicated by protonation steps. Values obtained up to the end of 1971 are tabulated elsewhere (*27*). Those quoted against the SCE can be converted to the NME scale by the addition of 241 mV. They cannot however be compared directly with the reduction potentials obtained in aqueous solution because of uncertain junction potentials and differences in solvation energies. Inspection of half-wave potentials shows that for the methyl-substituted benzoquinones, each methyl group makes the half-wave potential for addition of the first electron more negative by 65 mV in acetonitrile and 62.5 in DMF, somewhat less than noted above for water. Methoxy groups are not greatly different from methyl groups. These are examples of a more general linear correlation with Hammett substituent constants, which can be related to the energy of the lowest unoccupied orbital (*28*). It is interesting that half-wave potentials for addition of the second electron in solvents like acetonitrile and DMF are typically about 700 mV more negative than for addition of the first, implying that K_s is $\sim 10^{12}$, an exceedingly high stability of the semiquinone anion with respect to disproportionation in these solvents. No doubt this is due to strong destabilization of the double-negatively charged anion (dielectric constant ~37 for both solvents) as well as the absence of acid–base equilibria. Measurements of equilibria between oxygen and the semiquinones of trimethylbenzoquinone, 2-methyl-1,4-naphthoquinone, ubiquinones 0 and 9, tocopherylquinone and 2-methyl-3-phytyl-1,4-naphthoquinone in DMF are in reasonable agreement with half-wave potentials, but also indicate that the semiquinones form dimeric complexes with their quinones in this solvent (*29*). When this occurs, half-wave potentials for the addition of the first electron should be about 100 mV less negative.

It is not possible to measure true reduction potentials for ubiquinone and plastoquinone in aqueous solution at pH 7, but assessing the methoxy groups as rather similar to methyl, and the unsaturated side chain as less influential (*30*), and taking into account measurements of half-wave potential in acetonitrile, it may be estimated that for comparison with the figures in Table III,

Ubiquinone $E(Q/Q^{\overline{\cdot}})_7 = -230 \pm 20$ mV,

and

Plastoquinone, $E(Q/Q^{\overline{\cdot}})_7 = -130 \pm 20$ mV.

Rate constants for electron transfer reactions of semiquinones in aqueous solution can be as high as about $5 \times 10^9 M^{-1}sec^{-1}$ (*31*), aproaching the diffusion-controlled limit. Rate constants are lower when the reacting species have the same charge and when the equilibrium constant for the reaction [cf., Eqs. (18) and (19)] is small. A good correlation is found with the Marcus theory of electron transfer using reduction potentials as in Table III (*24, 32*).

In a biological environment the redox properties of semiquinones will be profoundly affected by solvation and complexing with other molecules, as well as being dominated by the availability of protons. Complexing of the semiquinone with quinone has already been mentioned. Complexes with metal ions also occur. In the reaction of benzosemiquinone with Cu^{2+}, for example, the semiquinone substitutes for a water molecule in the coordination sphere of the Cu^{2+}

$$\cdot O - C_6H_4 - O^- + Cu^{2+} \longrightarrow \left| O - C_6H_4 - O - Cu \right|^{\cdot +} . \quad (24)$$

Although the reduction potential of benzoquinone is +99 mV and that of Cu^{2+} is +153 mV, no rapid inner-sphere electron transfer is seen, and in flash photolysis experiments the complexes disappear in second order reactions (*33*). Micellar systems provide useful models for biological systems. At pH 8 the radical anions of 9,10-anthraquinone sulphonates are stable for weeks in anionic and non-ionic detergents (*34*) contrasting strongly with simple aqueous solutions. It is now beginning to be possible to construct models which increasingly imitate biological processes (*35*).

ACKNOWLEDGMENTS

The author wishes to thank his collegues, especially Professor J. H. Baxendale and Drs. J. M. Bruce and E. J. Land for helpful discussions. Some of the work on which the paper is based has been supported by grants from the Cancer Research Campaign and the Medical Research Council.

REFERENCES

1. Swallow, A. J., Ross, A. B., and Helman, W. P. (1981). *Radiat. Phys. Chem.* **17,** 127–140.
2. Swallow, A. J. (1973). "Radiation Chemistry: An Introduction." Wiley-Halsted, New York.
3. Baxendale, J. H., and Wardman, P. (1975). NSRDS-NBS 54. U.S. Dep. Commer., Washington, D.C.
4. Hulme, B. E., Land, E. J., and Phillips, G. O. (1972). *J.C.S. Faraday I* **68,** 1992–2002.
5. Patel, K. B., and Willson, R. L. (1973). *J.C.S. Faraday I* **69,** 814–825.
6. Rao, P. S., and Hayon, E. (1973). *J. Phys. Chem.* **77,** 2274–2276.
7. Gohn, M., and Getoff, N. J. (1977). *J.C.S. Faraday I* **73,** 1207–1215.
8. Richter, H. W. (1979). *J. Phys. Chem.* **83,** 1123–1129.
9. Fukuzumi, S., Ono, Y., and Keii, T. (1973). *Bull. Chem. Soc. Jpn.* **46,** 3353–3355.
10. Amouyal, E., and Bensasson, R. (1976). *J.C.S. Faraday I* **72,** 1274–1287.
11. Land, E. J., Simic, M., and Swallow, A. J. (1971). *Biochim. Biophys. Acta* **226,** 239–240.
12. Bensasson, R., and Land, E. J. (1973). *Biochim. Biophys. Acta* **325,** 175–181.
13. McAlpine, E., Sinclair, R. S., Truscott, T. G., and Land, E. J. (1978). *J.C.S. Faraday I* **74,** 597–602.
14. Adams, G. E., and Michael, B. D. (1967). *Trans. Faraday Soc.* **63,** 1171–1180.
15. Willson, R. L. (1971). *Chem. Commun.* pp. 1249–1250.
16. Hayon, E., Ibata, T., Lichtin, N. N., and Simic, M. (1972). *J. Phys. Chem.* **76,** 2072–2078.
17. Steenken, S., and O'Neill, P. (1977). *J. Phys. Chem.* **81,** 505–508.
18. Land, E. J., and Swallow, A. J. (1970). *J. Biol. Chem.* **245,** 1890–1894
19. Land, E. J., and Porter, G. (1960). *Proc. Chem. Soc., London* p. 84.
20. Meisel, D., and Neta, P. (1975). *J. Am. Chem. Soc.* **97,** 5198–5203.
21. Meisel, D., and Czapski, G. (1975). *J. Phys. Chem* **79,** 1503–1509.
22. Wardman, P., and Clarke, E. D. (1976). *J.C.S. Faraday I* **72,** 1377–1390.
23. Ilan, Y. A., Czapski, G., and Meisel, D. (1976). *Biochim. Biophys. Acta* **430,** 209–224.
24. Meisel, D., and Fessenden, R. W. (1976). *J. Am. Chem. Soc.* **98,** 7505–7510.
25. Steenken, S., and Neta, P. (1979). *J. Phys. Chem.* **83,** 1134–1137.
26. Clark, W. M. (1960). "Oxidation-Reduction Potentials of Organic Systems." Williams & Wilkins, Baltimore, Maryland.
27. Chambers, J. Q. *In* "The Chemistry of the Quinonoid Compounds" (S. Patai, ed.), pp. 737–791. Wiley, New York.
28. Zuman, P. (1967). Substituent Effects in Organic Polarography," pp. 273–308. Plenum, New York.
29. Afanas'ev, I. B., and Polozova, N. I. (1979). *Zh. Org. Khim.* **15,** 1802–1810.
30. Creed, D., Hales, B. J., and Porter, G. (1973). *Proc. R. Soc. London, Ser. A* **334,** 505–521.
31. Swallow, A. J. (1978). *Prog. React. Kinet.* **9,** 195–365.
32. Meisel, D. (1975). *Chem. Phys. Lett.* **34,** 263–266.
33. Khudyakov, I. V., Kuzmin, V. A., and Emanuel, N. M. (1978). *Int. J. Chem. Kinet.* **10,** 1005–1018.
34. Kano, K., and Matsuo, T. (1974). *Bull. Chem. Soc. Jpn.* **47,** 2836–2842.
35. Kano, K., Takuma, K., Ikeda, T., Nakajima, D., Tsutsui, Y., and Matsuo, T. (1978). *Photochem. Photobiol.* **27,** 695–701.

A Physicochemical Model of Quinone–Cytochrome *b-c* Complex Electron Transfers

4

PETER R. RICH

INTRODUCTION

A diverse range of biological electron-transfer chains are known to contain a complement of quinones that can undergo redox changes. Since the quinones are stoichiometrically in excess of electron-transfer chains, it was generally thought that they formed a mobile "pool" which electronically connected complex multiprotein components. This concept adequately explains the first-order behavior of quinone-pool oxidations (*1*, *2*), sigmoidal inhibition with tightly bound inhibitors (*2*), inhibition profiles in double-inhibitor experiments (*3*), extraction–reconstitution studies (*4*), and provides a mobile H-atom carrier as required for a protonmotive loop (*5*). In this view, the quinone species are not protein bound and are essentially freely diffusible in the lipid phase.

In contrast, extensive evidence has been gained for the existence of a specific-bound quinone species, generally termed "Z", as the immediate donor to the *b-c* complex. Evidence includes: potentiometric–kinetic data in photosynthetic bacteria (*6*, *7*); extraction–reincorporation studies (*8*); stoichiometric studies (*9*); isolation of a Z-binding protein (*10*); detection of semiquinone other than that associated with center S-3 in mitochondria (*11*); labeling of Q-binding proteins (*12*); and studies on the recovery of the slow electrochromic phase *b* in chloroplasts (*13*). Indeed, extrapolations have been made to suggest that all of the active quinone

Function of Quinones in Energy Conserving Systems

ISBN 0-12-701280-X

may be protein associated and that a pool may not be functionally involved.

The present report describes work that builds a physicochemical model of the reactions of quinols with acceptor species, both in model and biological reactions. A summary of this work to date and its biological implications will be presented.

EXPERIMENTAL RESULTS

Cytochrome *c* Reduction by Substituted *p*-Benzoquinols in Solution

Extensive studies have already appeared on this reaction (*14–18*). Essentially, two major routes of equilibration can be shown to occur, semiquinone- and anionic-quinol-mediated. Factors that contribute to the relative contributions of these paths have been discussed (*18*). The anionic-quinol-mediated route is the one of biological relevance since it is clear that inhibition of electron transport does not occur under conditions where the Q-pool is highly reduced. Reaction via this route, in the direction of quinol to cytochrome, is summarized in Fig. 1, taken from data by Rich and Bendall (*17*, *18*). Reductants are the anionic forms of quinol and semiquinone and the rate-limiting step is QH^- reduction of protonated cytochrome *c*. A stable bimolecular complex involving specific steric interactions is not considered since a plot of $\log_{10} k_1$ versus ΔG_0 of the rate-limiting reaction follows the -120 mV slope prediction of Marcus (*19*). Anaerobically, the semiquinone was able to reduce a second cytochrome *c* more rapidly than it dismutated (*18*).

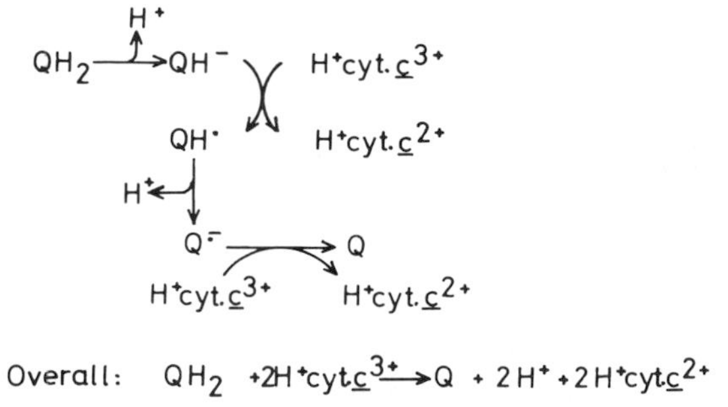

$QH_2 \xrightarrow{H^+} QH^-$; $QH^- + H^+cyt.\underline{c}^{3+} \rightarrow QH^{\cdot} + H^+cyt.\underline{c}^{2+}$

$QH^{\cdot} \xrightarrow{H^+} Q^{\cdot-}$; $Q^{\cdot-} + H^+cyt.\underline{c}^{3+} \rightarrow Q + H^+cyt.\underline{c}^{2+}$

Overall: $QH_2 + 2H^+cyt.\underline{c}^{3+} \rightarrow Q + 2H^+ + 2H^+cyt.\underline{c}^{2+}$

Fig. 1. Mechanisn of reduction of cytochrome *c* by *p*-benzoquinols in solution. Data taken from Rich and Bendall (*17, 18*).

Quinol and Quinone Equilibration in Aqueous Solution

A similar set of physicochemical criteria has been applied to the equilibration of quinols and quinones in aqueous solution. A series of quinols donating to a spectrally distinct quinone was used so that the reactions could be followed spectrophotometrically. For example, in the reduction of ubiquinone-1 by plastoquinol-1, the initial rate was linearly dependent on quinol and quinone concentrations, essentially independent of ionic strength, and showed a pH dependency inversely proportional to proton concentration. Such behavior is consistent with the equilibration mechanism illustrated in Fig. 2. In this scheme the anionic species were the major reductants and H-atom transfers did not occur to a significant extent. Furthermore, when a series of benzoquinol derivatives were used as donors to 2,5-dichloro-*p*-benzoquinone, it was found that a relationship existed between the rate constant and the ΔG_0 of the rate-limiting QH^- reduction of quinone (Fig. 3). This relationship consisted of a 10-fold increase in rate constant for each 60-mV decrease in E_0 ($QH^-/QH\cdot$) of the donor couple until a diffusion-limited rate was reached. Since this diffusion limit was reached at around ΔG_0 of 0, it was concluded that the reaction was probably diffusion-limited in the exergonic direction.

Quinol and Quinone Equilibration in Hydrophobic Media

The reaction chosen for study was the equilibration of the highly hydrophobic plastoquinol-9 and ubiquinone-10 in the solvent *n*-hexane.

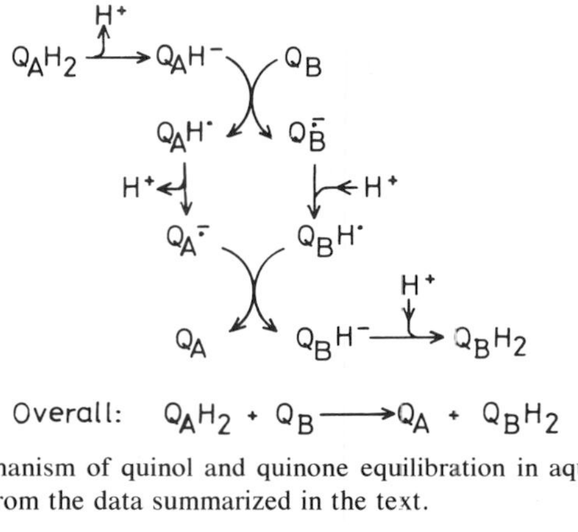

$$Q_AH_2 \xrightarrow{-H^+} Q_AH^- + Q_B \rightarrow Q_AH^\cdot + Q_B^{\cdot-}$$

$$Q_AH^\cdot \xrightarrow{-H^+} Q_A^{\cdot-} \qquad Q_B^{\cdot-} \xrightarrow{+H^+} Q_BH^\cdot$$

$$Q_A^{\cdot-} + Q_BH^\cdot \rightarrow Q_A + Q_BH^- \xrightarrow{+H^+} Q_BH_2$$

$$\text{Overall:} \quad Q_AH_2 + Q_B \longrightarrow Q_A + Q_BH_2$$

Fig. 2. The Mechanism of quinol and quinone equilibration in aqueous solution. The scheme is deduced from the data summarized in the text.

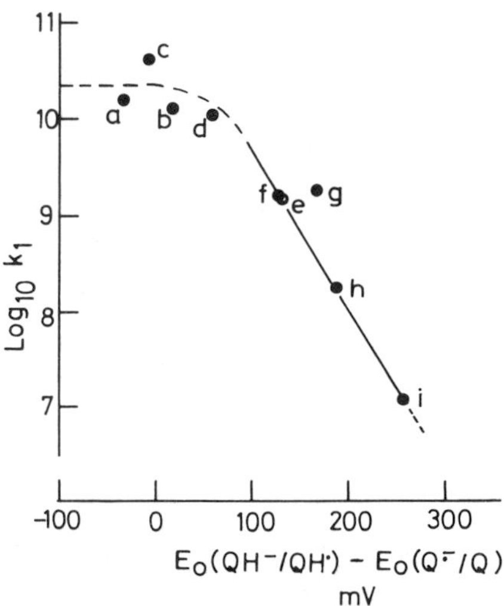

Fig. 3. Relationship between rate constant and ΔG_0 for the reduction of 2,5-dichloro-*p*-benzoquinone by anionic quinols in aqueous solution. A buffer of 100 m*M* sodium citrate plus 2 m*M* EDTA at pH 3.35 and 20 °C was used. An appropriate amount of quinol was added and the reaction monitored at 310 nm. Quinols used were as labeled in Fig. 4. k_1 refers to the reaction $QH^- + Q \rightarrow QH\cdot + Q^{\overline{\cdot}}$.

Equilibration occurred extremely slowly in such a system, with no detectable reaction even after 30 min at 20 °C. To test whether this was caused by a change in the physical properties of one of the reactants in *n*-hexane (such that the overall equilibrium constant made the reaction undetectable), the reverse reaction of ubiquinol-10 reduction of plastoquinone-9 in the same solvent was also tested. Again, the reaction rate was too slow to be detected, demonstrating that disequilibria had indeed been set up in the solvent. These results may be rationalized with reference to the aqueous solution situation: since ionic species do not exist for a significant time in aprotic media, no route of electron transfer involving a thermodynamically feasible quinone couple is available, and H-atom transfer is still too slow a route to be detected.

Reduction of Purified Cytochrome *f* by Quinols in Solution

By techniques analogous to those already outlined, the reaction mechanism of quinol reduction of purified cytochrome *f* in aqueous solution was shown to be similar to that of the reduction of cytochrome *c*. Again, no sterically fitted complex was indicated since the prediction by Marcus of a −120 mV slope in a plot of $\log_{10} k_1$ versus ΔG_0 of the rate-limiting step was followed for a wide range of quinol donors (Fig. 4).

Reduction of Cytochrome *f* by Quinols in Catalytically Active *b-f* Preparations

The initial acceptor of the *b-f* complex is presumably the Rieske-type iron–sulfur center (*20*). Equilibration of the Rieske center and cytochrome *f* is rapid; therefore, measurement of the rate of cytochrome *f* reduction gives a measure of the rate-limiting step in quinol reduction

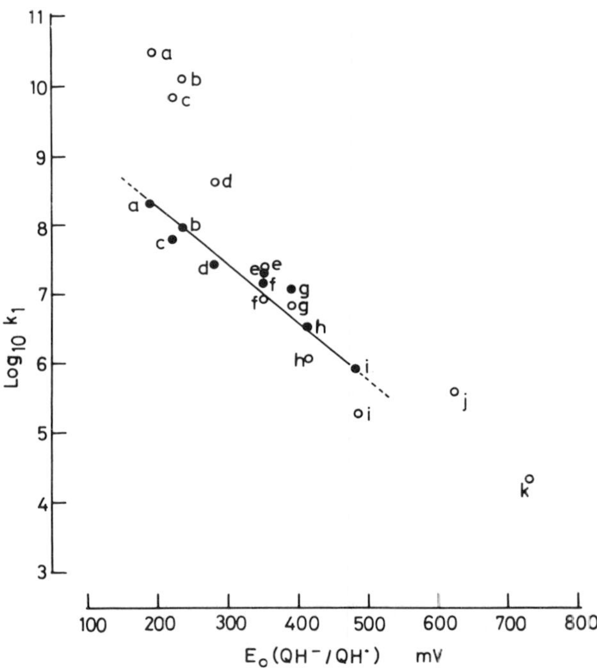

Fig. 4. Relationship between rate constant and $E_0(QH^-/QH\cdot)$ for quinol donation to purified and *b-f* complex-bound cytochrome *f*. The k_1 values were determined in 100 m*M* sodium citrate or MES, both plus 2 m*M* EDTA, in the pH range 5–6. Donors were: (a) ubiquinol-1; (b) plastoquinol-1; (c) duroquinol; (d) trimethylquinol; (e) thymoquinol; (f) 2,6-dimethylquinol; (g) 2,3-dimethylquinol; (h) methylquinol; (i) hydroquinone; (j) 2,6-dichloroquinol; (k) tetrachloroquinol. (●), purified cytochrome *f*; (○), cytochrome *f* in the *b-f* complex. The thermodynamic data was taken from Rich and Bendall (*18*).

of acceptor species. This was studied using a digitonin solubilized, catalytically active *b-f* preparation from lettuce chloroplasts (*21*), demonstrating

1. The maximum turnover of cytochrome *f* is as great as, or even exceeds, turnover in chloroplasts in the steady state.
2. A pH dependency. This may be simulated as reaction between anionic quinol and an acceptor with a redox proton on the oxidized form, and a positively charged site of interaction with a p*K* of around 6.5 (*22*). In intact chloroplasts, this p*K* may be somewhat lower.
3. A first-order dependency on quinol concentration up to the limit of solubility of the quinol used. Figure 5 illustrates the plastoquinol concentration dependency, which is linear up to 60–70 μM, the limit of solubility of plastoquinol-1 in the buffer used.
4. An ionic strength dependency which indicates that the active quinol reductant and the site of interaction on the *b-f* complex are oppositely charged (Fig. 6).
5. A relationship between rate constant and ΔG_0 of the rate-limiting step which deviates from the Marcus prediction when a series of substituted *p*-benzoquinols are used as donors to the *b-f* complex (Fig. 4). The deviation appears to be related to the hydrophobicity of the quinol.

These results are interpreted in the form of a model in the next section.

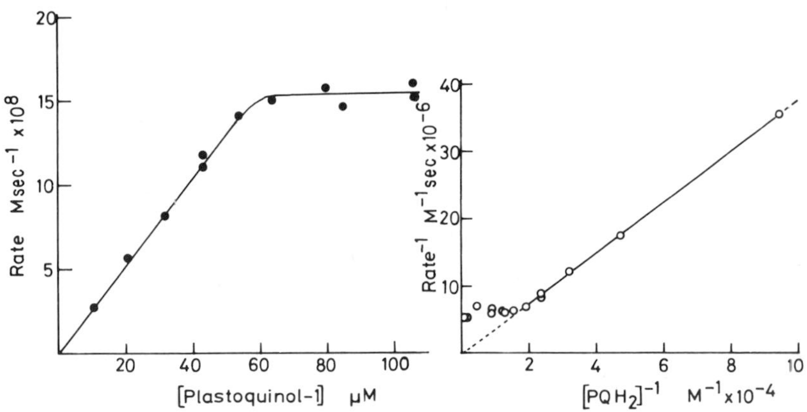

Fig. 5. Plastoquinol concentration dependency of cytochrome *f* reduction rate in *b-f* preparations. Initial reaction rate was determined in a buffer containing 100 m*M* sodium citrate and 2 m*M* EDTA at a pH of 4.3 and at 20 °C. Initial cytochrome *f* concentration was around 110 n*M*.

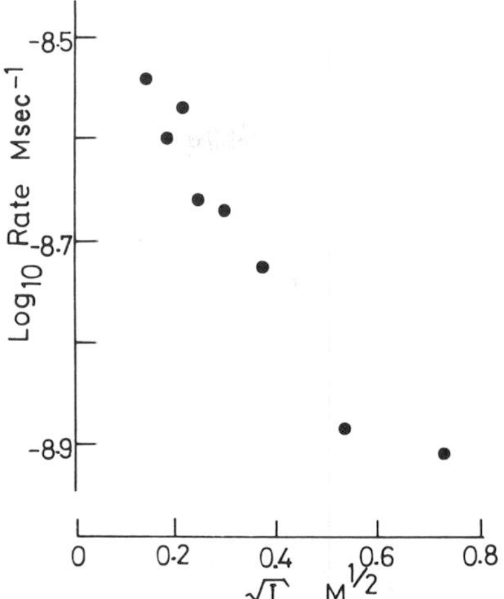

Fig. 6. Ionic strength dependency of plastoquinol-1 reduction of cytochrome *f* in *b-f* Particles. Ionic strength was varied with MES or NaCl concentration, while maintaining an EDTA concentration of 1 m*M* and a pH of 5.60. Initial cytochrome *f* concentration was around 110 n*M* and initial plastoquinol-1 around 6 μ*M*.

DISCUSSION

A Physicochemical Model of Quinol to *b-c* Complex Electron Transfer

The model that emerges from the studies described is depicted in Fig. 7. In view of the similarity of the cytochrome system present in mitochondria and chloroplasts, the model may be of general validity. The Rieske-type [2Fe–2S] iron–sulfur center is shown as possessing the useful collision site (cf., *20*), but with a positively-charged moiety at the site (with a p*K* of around 6.5 in isolated *b-f* preparations). The redox center is presumably protonated in its oxidized form at neutral pH (*23*). The anionic quinol is able to interact at this site such that a useful electron transfer can occur. Steps subsequent to formation of the initial complex are so rapid that saturation of the rate with quinol is not experimentally feasible. After electron transfer, further rapid transfer to the cytochrome *f* heme and dissociation of PQH· can occur. Alternatively, deprotonation of the semiquinone may occur before dissociation so that a second reduction of the iron–sulfur center may occur. The precise route of the electron from the semiquinone will depend upon

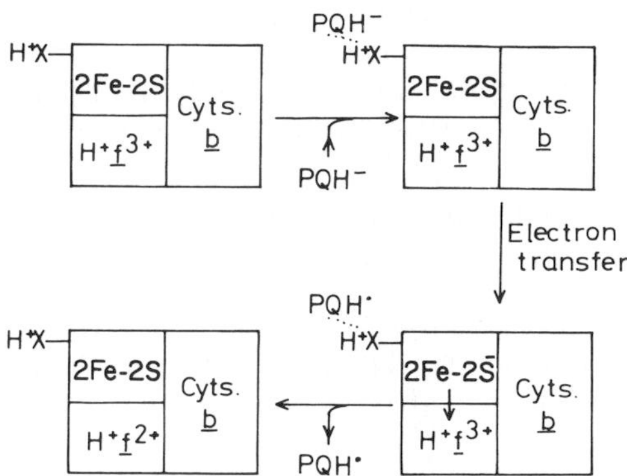

Fig. 7. Model for electron transfer from plastoquinol to acceptors in the *b-f* region of chloroplasts. The model is derived from the data presented in the text.

TABLE I

Assessment of the Thermodynamic and Kinetic Feasibility of Anionic Quinol Reduction of the *b-c* Complex *in situ*

Parameter	Submitochondrial particles	Chloroplasts
Lipid/protein (mg/mg)[A]	50/50	60/40
UQH_2 or PQH_2	6.3 nmole/mg protein, 6.3 m*M* in lipid	15 nmole/mg chlorophyll, 1 m*M* in lipid
UQH^- or PQH^- [B]	1.12×10^{-6} *M*	5×10^{-7} *M*
Acceptor (c_1, *f* or Rieske)	0.4 nmole/mg protein, 0.4 m*M* in lipid	2.3 nmole/mg chlorophyll, 0.16 m*M* in lipid
Maximum electron transport, pH 7.5	250 nmole O_2/mg/min, 0.017 molar electrons · sec^{-1}	200 μmole O_2/mg chlorophyll/hr, 0.0148 molar electrons · sec^{-1}
Required overall[C] rate constant for electron flux from QH^- to acceptor	$3.8 \times 10^7\ M^{-1} \cdot sec^{-1}$	$1.85 \times 10^8\ M^{-1} \cdot sec^{-1}$
E_m acceptor couple[D] minus E_m donor couple at rate-limiting step	~+90 mV	~+40 mV

[A] 1 gm lipid = 1 ml volume.
[B] pK_A UQH_2 = 11.25; pK_A PQH_2 = 10.8.
[C] All reactants freely soluble in lipid.
[D] E_m (UQH^-/UQH·), +190 mV; E_m (PQH^-/PQH·), +240 mV; E_m Rieske, +280 mV.

several relative rate constants and redox states. Thus, although the route *in vivo* may be well defined, when observed experimentally the route will be variable.

The deviation of the rate constants from Marcus behavior (Fig. 4) in the catalytic complex may arise from a number of sources:

1. The hydrophobic environment of the *b-f* complex may concentrate the hydrophobic quinols such that their effective concentration close to the collision site is higher, increasing the measured rate constant.
2. The intermediate complex may involve some hydrophobic binding between protein and quinol (*24*) such that

$$k_{obs} = \frac{k_{e.t.}\ K[QH^-]}{1 + K[QH^-]}$$

where $$QH^- + b\text{-}f \overset{K}{\rightleftharpoons} QH^- - b\text{-}f \text{ complex.}$$

3. The acceptor may be held in such a way that oxidized and reduced conformations are identical so activation of only the quinol need occur.

These possibilities are not distinguishable from the available data.

Relation of the Model to the Specialized Bound Quinone, Z

The model proposed is inconsistent with the notion of Z, a quinone thought to be bound to a protein in so stable a manner that it is distinguishable from the bulk quinone pool, and which forms the immediate donor to *b-c* complexes (*6–13*). One possibility of reconciliation of the two models would be if the rate-limiting step of the reaction were always mobile quinol donation to Z. Such a scheme, however, would not only add an unnecessary component, but would also be inconsistent with the observed data on Z which indicates a rate-limiting step in single turnover experiments between ZH_2 and cytochrome (*6, 7, 9, 13*).

Instead, reconciliation of the experimental data may lie in a consideration of the experimental approaches used in the studies. For example, consider the physicochemical model proposed in terms of the following simple kinetic parameters:

$$QH^- + b\text{-}c \underset{k_{-a}}{\overset{k_a}{\rightleftharpoons}} QH^-\ bc \longrightarrow QH\cdot + b\text{-}c^-$$

In a single turnover, provided that $k_a > k_{-a}$, the immediate situation after flash oxidation would consist of a significant proportion of quinol-*b-c* complex units. The observed rate of electron transfer would be given by:

$$\text{Rate} = k_b[\text{QH}^- b\text{-}c]$$

and would exhibit a first-order decay, as observed experimentally.

In the steady-state turning over condition, however, which has experimentally pointed to the involvement of a mobile quinone pool (*1–4*), the rate would be limited by the formation of the enzyme–substrate complex and the steady-state concentration of this complex would be very low. If k_b is greater than k_a or k_{-a}, the predicted kinetics would be those observed in steady state and approach to steady-state conditions, and the overall rate would approximate to

$$\text{Rate} = k_a[\text{QH}][b\text{-}c].$$

Although the above is clearly a great oversimplification and has many inherent assumptions, the general mechanism proposed is one which provides a means of rationalization of apparently dissonant data obtained from many laboratories. It therefore, for the present, forms a valid working hypothesis for the analysis of quinone-related reactions of electron-transfer chains.

ACKNOWLEDGMENTS

Many useful discussions with Drs. D. S. Bendall and R. Hill have contributed to the direction of this work. Financial support of the Science Research Council is also gratefully acknowledged.

REFERENCES

1. Haehnel, W. (1973). *Biochim. Biophys. Acta* **305,** 618–631.
2. Kröger, A., and Klingenberg, M. (1973). *Eur. J. Biochem.* **39,** 313–323.
3. Moreira, M. T. F., Rich, P. R., and Bendall, D. S. (1980). *Eur. Bioenerg. Conf., 1st, Short Rep., Urbino* pp. 61–62..
4. Ernster, L., Glaser, E., and Norling, B. (1978). *In* "Biomembranes: Part D: Biological Oxidations, Mitochondrial and Microbial Systems" (S. Fleisher and L. Packer, eds.), Methods in Enzymology, Vol. 53, pp. 573–579. Academic Press, New York.
5. Mitchell, P. (1976). *J. Theor. Biol.* **62,** 327–367.
6. Cogdell, R. J., Jackson, J. B., and Crofts, A. R. (1972). *Bioenergetics* **4,** 413–429.

7. Prince, R. C., and Dutton, P. L. (1977). *Biochim. Biophys. Acta* **462,** 731–747.
8. Takamiya, K., Prince, R. C., and Dutton, P. L. (1979). *J. Biol. Chem.* **254,** 11307–11311.
9. Prince, R. C., Bashford, C. L., Takamiya, K., van den Berg, W. H., and Dutton, P. L. (1978). *J. Biol. Chem.* **253,** 4137–4142.
10. Bodmer, S., Snozzi, M., and Bachofen, R. (1981). *Proc. Int. Congr. Photosynth., 5th, Halkidiki, Greece, 1980,* **2,** 655–663.
11. Ohnishi, T., and Trumpower, B. L. (1980). *J. Biol. Chem.* **255,** 3278–3284.
12. Yu, C.-A., and Yu, L. (1980). *Biochem. Biophys. Res. Commun.* **96,** 286–292.
13. Bouges-Bocquet, B. (1980). *FEBS Lett.* **117,** 54–58.
14. Baxendale, J. H., and Hardy, H. R. (1954). *Trans. Faraday Soc.* **50,** 808–814.
15. Williams, G. R. (1963). *Can. J. Biochem. Physiol.* **41,** 231–237.
16. Yamazaki, I., and Ohnishi, T. (1966). *Biochim. Biophys. Acta* **112,** 469–481.
17. Rich, P. R., and Bendall, D. S. (1979). *FEBS Lett.* **105,** 189–194.
18. Rich, P. R., and Bendall, D. S. (1980). *Biochim. Biophys. Acta* **592,** 506–518.
19. Marcus, R. A. (1963). *J. Phys. Chem.* **67,** 853–857.
20. Trumpower, B. L., Edwards, C. A., and Ohnishi, T. (1980). *J. Biol. Chem.* **255,** 7487–7493.
21. Rich, P. R., and Bendall, D. S. (1980). *Biochim. Biophys. Acta* **591,** 153–161.
22. Rich, P. R., and Bendall, D. S. (1980). *Eur. Bioenerg. Conf., 1st, Short Rep., Urbino* pp. 59–60..
23. Prince, R. C., and Dutton, P. L. (1976). *FEBS Lett.* **65,** 117–119.
24. Segal, M. G., and Sykes, A. G. (1978). *J. Am. Chem. Soc.* **100,** 4585–4592.

III POOL FUNCTION BEHAVIOR AND UBIQUINONE MOBILITY IN ENERGY CONSERVING MEMBRANES

Pool Function Behavior and Mobility of Isoprenoid Quinones

I

GÜNTER HAUSKA
EDUARD HURT

INTRODUCTION

Isoprenoid quinones have been recognized as constituents of electron-transport chains for many years. However, their exact role still remains unclear. Two central questions are

1. Is the pool of excess, free quinone an obligate intermediate in electron transport, or is it on a dead alley, in relatively slow exchange with protein bound, specialized quinone molecules?
2. Is the redox reaction of the quinone involved in H^+-translocation the H-carrying step of vectorial electron transport (*1*, *2*)?

The intention of this brief survey is to compile and discuss evidence related to these two questions. We will not confine ourselves to one-electron transport systems, but consider respiration and photosynthesis, both eukaryotic and bacterial, as well as membrane model systems, side by side. For further aspects of the chemistry and biochemistry of isoprenoid quinones the reader is referred to accompanying articles in this volume, as well as the following reviews. For an early but broad account, see Morton (*3*); for ubiquinone in mitochondria, see Crane (*4*, *5*) and Trumpower (*6*); for plastoquinone in chloroplasts, see Amesz (*7*), Bendall (*8*), and Witt (*9*); for ubiquinone in photosynthetic bacteria, see Parson (*10*) and Wraight (*11*).

Function of Quinones in Energy Conserving Systems

ISBN 0-12-701280-X

EXPERIMENTAL RESULTS AND DISCUSSION

The Size of Quinone Pools

Table I provides a selection of values collected mainly from former reviews, supplemented with a few more recent results. Three main types of isoprenoid quinones are employed in electron transport throughout: ubiquinone (UQ), plastoquinone (PQ), and menaquinone (MK). Ubiquinone functions in the respiration of mitochondria and gram-negative bacteria, and in photosynthesis of purple bacteria. Plas-

TABLE I

Quinone Pool Sizes in Electron Transport Systems[A]

Membrane system	Quinone species	Pool size	Reference
Chloroplasts	PQ_9	40[B]	*(12)*
Cyanobacteria			
Anacystis nidulans	PQ_9	50[B]	*(13)*
Anabaena variabilis	PQ_9	7[C]	*(13)*
Rhodospirillaceae			
photosynthetic	UQ_{10}	25–40[D]	*(10)*
aerobic	UQ_{10}	5–10[E]	*(14)*
			(15)
Chromatium vinosum	UQ_7	15[D]	*(10)*
	MK_7	6[D]	*(10)*
Mitochondria			
mammals	UQ_{10}	6–8[F]	*(16)*
plants (*Arum*)	UQ_{10}	7[G]	*(12)*
yeast	UQ_6	36[F]	*(16)*
Enterobacteria			
(*E. coli, Proteus rettgeri*)	UQ_8	6–8[E]	*(17)*
	MK_8	6–8[E]	*(18)*
			(19)
Mycobacterium phlei	MK_9 (H_2)	45[F]	*(16)*
Bacillus megatherium	MK_7	5–10[E]	*(20)*
Vibrio succinogenes	$MK_?$	6[E]	*(21)*
Desulfovibrio vulgaris	MK_6	21[E]	*(22)*

[A] UQ, PQ, and MK, stand for ubiquinone, plastoquinone and menaquinone.
[B] Related to P700, the reaction center of photosystem I (chlorophyll/P700 = 400).
[C] Related to P700 (chlorophyll/P700 = 150); Kawamura *et al.* *(23)*.
[D] Related to reaction center.
[E] Related to cytochrome *b*.
[F] Related to cytochrome aa_3.
[G] Related to the value for cytochrome aa_3 of Douce *et al.* *(24)*.

toquinone participates in oxygenic photosynthesis of chloroplasts and cyanobacteria (*7*, *12*). It occurs in three forms and is always associated with phylloquinone (*7*). A role for ubiquinone in the respiration of cyanobacteria was suggested by Peschek (*25*), although plastoquinone is more abundant. Menaquinone functions in the respiration of gram-positive bacteria (*4*, *17*), and in anaerobic fumarate respiration of gram-negative bacteria [see (*26*, *27*); see also Jones and Garland, Paper 1, Chapter VII, this volume]. *Chlorobium*-quinone (1′-oxo-MK-7) occurs in the green photosynthetic sulfur bacteria (*28*), and rhodoquinone replaces menaquinone in fumarate respiration of parasitic worms (*29*). Menaquinone also plays a role in photosynthetic electron transport next to ubiquinone in *Chromatium* (*10*).

The size of a quinone pool is often obtained by combining older data on quinone content per gram dry weight, or per milligram protein, with newer data on cytochrome or reaction-center contents; the estimations are therefore rather rough. The values determined for cytochrome *b* are underestimated compared with cytochrome aa_3 or the reaction center, because the systems usually contain more than one species of cytochrome *b*. In addition, the pool size may vary with physiological conditions [see (*4*) and Table II for plastoquinone in *Chlorella;* see (*14*, *15*) for ubiquinone in Rhodospirillaceae]. In Enterobacteria the relative

TABLE II

Requirement for a Minimum Length of the Isoprenoid Side Chain in Ubiquinone[A]

Compound	Ferricyanide reduction in liposomes	ATP formation in extracted/reconstituted chromatophores: μmole mg BChl^{-1} h^{-1}	
	k (s^{-1})	(Q_{III} present)	(Q_{III} missing)
Control	0.015	22	4
UQ_1	0.03	29	5
UQ_2	0.06	35	8
UQ_3	1.50	181	39
UQ_5	1.50	185	40
UQ_7	1.85	186	180
UQ_{10}	1.28	208	212

[A] All data are compiled from Futami *et al.* (*90*) and from Baccarini-Melandri *et al.* (*48*). For the absence and presence of Q_{III} (see Fig. 1) consult Baccarini-Melandri *et al.* (this volume).

amounts of ubiquinone and menaquinone vary extensively with growth conditions (*4*, *18*); however, the pool size for each specific electron-transport chain is probably less variable. In spite of individual variability, the average pool sizes range from about 7 up to 50 in different electron-transport systems. Chloroplasts of spinach have about 40 plastoquinones per reaction center of photosystem I, and similarly high values can be derived for other plants from the data of Redfearn (*30*). However, only about seven plastoquinones per reaction center were reported to be photoactive (*31*). At least part of the inactive quinone may not be located in the thylakoid membranes but, instead, in the so-called plastoglobuli [PQ-rich lipid particles in the chloroplast (*32*, *33*)]. Within photosynthetic bacteria, including cyanobacteria, the size of quinone pools varies. However, the value differences for Rhodospirillaceae grown in the light, and aerobically in the dark, are not real. The dark-obtained values should be multiplied by about five when relating to the respiratory chains, (e.g., cytochrome *b*-type oxidases constitute only 20% of the *b*-type cytochromes (*34*). Mitochondria of higher organisms contain about seven ubiquinones per cytochrome oxidase, 80% being redox-active (*35*). According to the same authors, in extracted mitochondrial membranes refurnished with ubiquinone, up to 300% of the original ubiquinone complement was found to be redox active. Heron *et al.* (*36*), however, reported that about six ubiquinones per cytochrome c_1 were enough to saturate the interaction of isolated NADH-dehydrogenase with cytochrome b-c_1 complex (see below), being close to the endogenous pool size. Yeast mitochondria seem to have a much larger quinone pool. In conclusion, it appears that pool size is variable from system to system, making the idea of an obligatory role of unbound quinone in electron transport questionable.

Physical Heterogeneity

The observation that different amounts of ubiquinone are required to reconstitute NADH and succinate dehydrogenase in extracted submitochondrial particles (*35*), and in membranes from a photosynthetic bacterium (*14*), might suggest that there are different forms of ubiquinone in the respiratory chain. This is corroborated by the different specificity of the two dehydrogenases for the length of the ubiquinone isoprenoid side chain (*5*, *14*, *37*). However, both observations might just reflect a difference in the binding site for ubiquinone in the two dehydrogenases. Better evidence for separate ubiquinones stems from Gutman and Silman (*38*), who observed that the two dehydrogenases operated independently when extracted submitochondrial particles were reconstituted with low amounts of ubiquinone (in contrast to particles containing the full complement). Indeed, bound ubiquinone was found in the isolated dehydrogenase as well as the isolated cytochrome b-c_1 complex of mitochondria (*36*, *39*).

An indication that ubiquinone might operate both before and after cytochrome *b* in the respiratory chain, comes from mutant studies with *Escherichia coli* (*40*). This suits the idea of a Q-cycle [see (*2*, *41*); see also Mitchell and Moyle, Paper 7, Chapter VII this volume], or of a sequential double loop of ubiquinone in electron transport (*42*). The loop concept was suggested for cyclic electron transport of *Rhodopseudomonas*, and a dual function of ubiquinone before and after cytochrome *b* was demonstrated in these bacteria by extracting membranes with organic solvents (*43*). It is now widely accepted that ubiquinone and plastoquinone exist in different forms in mitochondria, photosynthetic bacteria, and chloroplasts. These are schematically depicted in Fig. 1, which compares the various electron-transport systems. For photosynthetic electron transport in chloroplasts and bacteria, up to four forms of the isoprenoid quinone, plastoquinone or ubiquinone, are discussed. There is a remarkable similarity of cyclic electron transport in Rhodospirillaceae, photosystem II, and the cytochrome f-b_6-complex in chloroplasts (Fig. 1A). The first relatively stable electron acceptor of the photoreactions in photosynthesis is a specialized form of plastoquinone in the chloroplast photosystem II reaction center (*9*, *45*), and of ubiquinone in the reaction center of Rhodospirillaceae (*10*, *11*). In

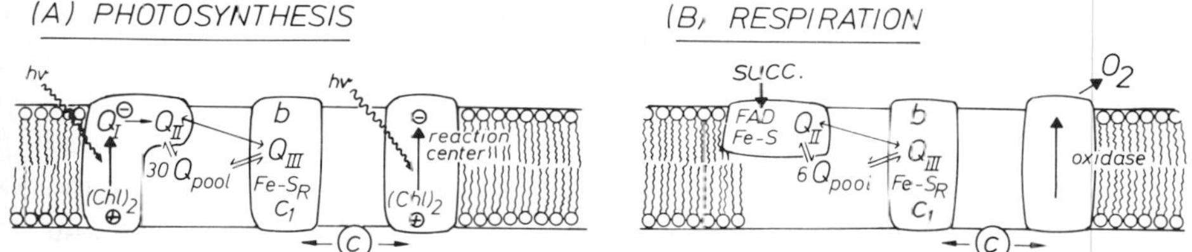

Fig. 1. The different forms of isoprenoid quinones and universal features of electron transport systems. (A) Photosynthetic electron transport in chloroplasts and Rhodospirillaceae, (B) aerobic, respiratory electron transport in mitochondria and bacteria. The schemes show cross sections through the respective membranes, the upper edges representing the surface facing the matrix of bacteria and mitochondria, or the stroma of chloroplasts. Symbols are: Q for quinone; $(Chl)_2$ for reaction center chlorophyll dimer; b for cytochrome *b*; c_1 for cytochrome c_1 in mitochondria and Rhodospirillaceae (*44*), and for cytochrome *f* in chloroplasts; c for cytochrome *c* in mitochondria, bacteria and algal chloroplasts, and for plastocyanin in chloroplasts of higher plants and algae; Fe$-S_R$ for the Rieske iron-sulfur protein; succ. for succinate.

Chromatium, Q_I the acceptor is probably a form of menaquinone (*10*), and in the green photosynthetic bacteria it may be menaquinone or *Chlorobium*-quinone (*28*). Q_I operates between the quinone and semiquinone form as a one-electron carrier and transfers electrons to a second, special form of either plastoquinone or ubiquinone (Q_{II} in Fig. 1A) that can accept two electrons. A remarkable feature of Q_{II} is the stability of its semiquinone form [for PQ_{II}, see (*8*); for UQ_{II}, see Wraight (*11*) and Paper 2, Chapter IV this volume; Evans and Rutherford (*46*)]. A third special form of either plastoquinone or ubiquinone (Q_{III} in Fig. 1A) is thought to operate in the cytochrome *b-c*-complexes, close to the Rieske Fe–S center [see Bowyer *et al.* (*47*); Baccarini-Melandri *et al.* (*48*) and Paper 4, Chapter 5, this volume; Bouges-Bocquet (*49*) and Paper 2, Chapter VI, this volume]. The fourth form is the free, excess quinone, denoted as Q_{pool} in Fig. 1A. The controversy of whether the Q pool mediates between Q_{II} and Q_{III}, or whether the special, protein bound forms interact by direct collision is not yet settled.

Another striking analogy between electron transport in chloroplasts and photosynthetic bacteria is the resemblance between the cytochrome *b-c* complexes (Fig. 1A), in chloroplasts and mitochondria (Fig. 1B). In bacteria, electrons are transferred from the b-c_1 complex to the reaction center by cytochrome c_2; in chloroplasts, electrons are transferred from the f-b_6 complex to the reaction center of photosystem I via plastocyanin (or cytochrome *c* in algae). These are peripheral, small redox proteins located on topographically equivalent membrane surfaces: the periplasmic surface in photosynthetic bacteria (*50*), and the intrathylakoid surface in chloroplasts (*51*, *52*).

Analogous to photosynthesis (Fig. 1A), the generalized scheme of aerobic respiratory chains (Fig. 1B) shows three different forms of quinone, Q_{II} bound to the dehydrogenases (Q_{II} might occur in two forms, one for each dehydrogenases), Q_{III} bound to the b-c_1 complex and the free pool. Q_I of Fig. 1A may be compared to the flavins in the dehydrogenases. Evidence for the bound quinone forms Q_{II} and Q_{III} comes from ESR studies (*53*, *54*). Q_{II} of succinate dehydrogenase seems to exist as a semiquinone dimer and could be identical to the quinone protein isolated and characterized by Yu and Yu [see (*54*) and Paper 7, Chapter V, this volume]. Q_{III}, analogous to the quinone in the cytochrome complex of photosynthetic bacteria, was also identified by redox titration of cytochrome *c* reduction kinetics after light flashes, in a hybrid system containing reaction centers from *Rps. sphaeroides* and the cytochrome b-c_1 complex from beef heart mitochondria (Matsuura *et al.,* Paper 3, Chapter V, this volume).

How real are these different, specialized forms Q_{I-III}? How fast do they exchange with the Q_{pool}? Opinions diverge: on one extreme, no exchange occurs; on the other, the "specialized quinones" just reflect the sites of interaction of Q_{pool} with its reaction partners in the electron-transport chain.

Stabilization of semiquinone alone, as visualized by ESR, does not prove a separate entity beyond a binding site, because any binding of semiquinone should result in stabilization by slowing the bimolecular collision in dismutation. A difference in the overall E_m is more conclusive. The E_{m7} for Q_{pool} in mitochondria is 65 mV (*55*), and in *Rps. sphaeroides* it is 90 mV (*56*). For chloroplasts a value of 30 mV was estimated [(*8*); see, however, (*57*)]. Q_I operates only between quinone and semiquinone at an E_{m7} of 0 to -100 mV [(*8*, *11*, *46*); see, however, (*57*) for chloroplasts] and its overall E_{m7} seems to be much more negative (*46*). The E_{m7} for $Q_{II}/Q_{II}H_2$ for *Rps. sphaeroides* is around 60 mV (*46*) [slightly more negative than for Q_{pool}], and for chloroplasts around 30 mV (*58*), the same as the estimate for Q_{pool}. Values for Q_{III} are some 50 mV more positive than for Q_{pool} in all three systems [see (*49, 59*); see also Matsuura *et al.*, Paper 3, Chapter V, this volume]. These values for E_{m7} show that Q_I cannot be in exchange with Q_{pool}, but is the same conclusion justified for Q_{II} and Q_{III}? Couldn't the differences reflect systematic experimental errors, for example when comparing the E_m value obtained from the redox level of extracted ubiquinone with that for Q_{III} derived from redox titrations of cytochrome c_2 rereduction after flash excitation?

The final proof of the existence of a chemical species is usually its isolation in a functional state. This was achieved for Q_I and Q_{II} bound to the reaction centers of Rhodospirillaceae [see (*11*); Okamura *et al.*, Paper 5, Chapter V, this volume], and also Q_{II} operating between succinate dehydrogenase and the cytochrome *b*-c_1 complex in mitochondria in the form of a quinone binding protein [see Yu and Yu (*54*) and Paper 7, Chapter V, this volume]. The nature of Q_{III} as a separate form, however, is still doubtful (*60*). On the other hand Q_{III}, measured via the rereduction of cytochrome c_2, is preferentially retained with Q_I and Q_{II} when membranes of photosynthetic bacteria are extracted with organic solvents, and is preferentially reconstituted by adding back small amounts of ubiquinone [see (*59*); see also Baccarini-Melandri *et al.* (*48*) and Paper 4, Chapter V, this volume]. Thus, under extraction conditions Q_{pool} cannot be in rapid equilibrium with Q_{III}. If the rate of rereduction of cytochrome c_2 reflects the turnover of cyclic electron transport under the conditions of Takamiya *et al.* (*59*), then their additional conclusion that 90% of the ubiquinone complement is not obligatorily involved in electron flow is justified. A similar conclusion was reached by Crofts and Bowyer (*61*). In contrast, Baccarini-Melandri *et al.* (*48*) established, again by extraction–reconstitution, that the excess Q_{pool} is necessary for photophosphorylation with chromatophores in continuous light. If this discrepancey is real, different pathways in continuous light (high membrane energization) and in flashing light (low average membrane energization) should be considered. An alternative would be that Q_{pool} is somehow involved in stimulating photophosphorylation, which is not unbelievable in view of evidence that regulatory phenomena via a membrane protein kinase in chloroplasts might be controlled by the redox level in the PQ pool (*62*). Bouges-Bocquet [see (*49*) and Paper 2, Chapter VI, this volume], reporting a kinetic incompetence of the Q_{pool} in flashing light for chloroplasts, favors the first possibility. She suggests that at low light input there is a pathway with a higher ATP yield, not involving Q_{pool}; and at high light intensities, a pathway involving Q pool, with a lower coupling ratio. This suggestion reminds one of Heber's postulate of flexible coupling for physiological reasons (*63*).

These interesting ideas need firmer experimental grounds to be generally accepted. More direct kinetic measurements of Q_{III}, and also of Q_{II}, would be desirable. Ultraviolet measurements of one special quinone among large quinone pools are difficult, especially in pigmented photosynthetic membranes. In the less absorbing mitochondrial membranes of higher organisms such measurements should be possible, because up to half of the ubiquinone seems to be specialized as Q_{II} and Q_{III} (*36*). But so far, no kinetic inhomogeneity of the quinone complement in mitochondria has been detected. Measurements should also be possible in chloroplasts, where plastoquinone reactions can be determined indirectly by chlorophyll fluorescence (*57*).

Kinetic Homogeneity—Pool Behavior

The idea that a homogeneous quinone pool links the dehydrogenases to the cytochrome chain in mitochondria (*64*) (Fig. 2A) by diffusion in the membrane has existed for a long time.

Kröger *et al.* (*35*, *65*) correlated the rates of succinate and NADH oxidation with the redox level of the ubiquinone pool. Using inhibitors, they covered a wide range of relative electron input and output rates into and from the ubiquinone pool, establishing what is known as "pool behavior" and (Fig. 2A). The key findings were

1. The rate of electron input depends on the amount of ubiquinone;

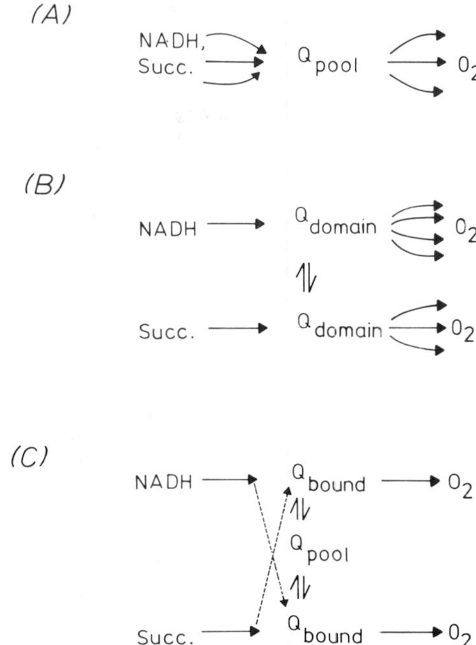

Fig. 2. Different possibilities for the interaction of the dehydrogenases and the cytochrome chain via UQ in mitochondria. Q stands for UQ, and succ. for succinate. The arrows on the left represent the dehydrogenases, the arrows on the right the cytochrome chains. (A), after Kröger and Klingenberg, 1973; (B), after Gutman, 1975; and (C), after Ragan *et al.*, 1978, all used with permission.

the rate of electron output depends on the amount of UQH_2 in the pool, with apparent first order kinetics.

2. The overall rate through the respiratory chain follows the equation

$$v = \frac{V_r \cdot V_o}{V_r + V_o} \tag{1}$$

where V_r is the maximal input capacity of the dehydrogenase, and V_o is the output capacity via the cytochromes.

3. The previously recognized "concentration lag" of inhibition by antimycin was variable, being larger with V_o/V_r. The lag was absent in reactions involving the cytochrome b-c_1 complex without the UQ pool [see however, (*66*, *67*)]. This lag is critically dependent on the high affinity of antimycin for its binding site, but the mathematical treatment can be extended for a test of pool behavior with weaker inhibitors (*68*).

These results suggest that interaction of the dehydrogenases and the cytochrome b-c_1 complex with ubiquinone occurs via bimolecular collisions, and that the individual dehydrogenases and cytochrome complexes are not linked stoichiometrically, but rather by a common redox pool. This redox pool might be physiologically important in linear electron transport, because it would always guarantee a maximal rate v at suboptimal V_o and V_r, (low substrate and/or O_2 concentration). In cyclic electron transport such an optimal matching of oxidation–reduction is not required, because electron input and output is always geared in the reaction centers. Nevertheless, cytochrome c_2 reduction by a special ubiquinone (Q_{III} in Fig. 1A) in cyclic electron transport of photosynthetic bacteria has been reported to involve bimolecular collisons (*67*, *69*). With the discovery of electron carriers between Q_{III} and cytochrome c_2, the Rieske Fe–S center (*47*) and cytochrome c_1 (*44*), this second-order reaction probably occurs between ubiquinone and one of these two redox centers. Plastoquinol oxidation in chloroplasts also follows second-order kinetics (*60*, *70*), and pool behavior was described for plastoquinone in chloroplasts, based on a "concentration lag" of DCMU in inhibiting electron transport, that varied with the relative activity of photosystems I and II (*71*).

Two alternative explanations are possible for the observation of the inhibition "lags" with antimycin and DCMU. One is that there is a diffusion-controlled collision between the dehydrogenase and cytochrome complexes (*36*) (see next sections). Another is that in spite of very high inhibitor affinities (low dissociation constants), the inhibitor molecules still might exchange rapidly between the individual binding sites [low activation energy (Lockau, personal communication)]. What are the alternatives to the two observations favoring a quinone pool by Kröger *et al.* (*35*, *65*)? First, an extension of their work by Gutman (*67*) should be mentioned. He studied simultaneous oxidation of succinate and NADH, with respect to equation (1) and found that

$$v > \frac{(V_n + V_s) \cdot V_o}{V_n + V_s + V_o} \tag{2}$$

where V_n and V_s are the input capacities of succinate and NADH dehydrogenase, respectively. In other words, because he found that the mutual inhibition of the two dehydrogenases via the reduction level of the UQ pool was less than expected from Eq. (1) Gutman concluded that each dehydrogenase is associated with a ubiquinone domain (Fig. 2B) and that these domains are in relatively slow exchange.

Another extension of the view of a homogeneous ubiquinone pool was provided by Ragan and co-workers (*36*, *39*), studying the interaction of isolated NADH dehydrogenase with isolated cytochrome b-c_1 complex.

This interaction was stoichiometric, although the preparations together retained about half of the lipid and UQ complement. Pool kinetics were observed if lipid was readded, and reached maximal rates only if UQ was also readded. In a system where endogenous lipids were replaced by defined lecithin molecules, pool behavior could be frozen out and stoichiometric interaction regained below the temperature of the gel/liquid crystalline phase transition (*72*).

The picture developed by Ragan and collaborators (Fig. 2C) as an alternative to the one by Kröger *et al.* (Fig. 2A) to explain pool behavior, emphasizes the diffusion and collision of the protein complexes, each carrying bound ubiquinone. It is interesting to conjecture why such differently bound quinones in mitochondria do not appear as different phases in measurements of redox changes of the ubiquinone complement (*35*). Does the "bound" UQ of Ragan reflect only the mixed, micellar state of the isolated complexes, with unique kinetic parameters (*73, 74*)? The report of Gutman and Silman (*38*) that no mutual inhibition of succinate- and NADH-dehydrogenase is observed when suboptimal amounts of UQ are reincorporated into submitochondrial particles, previously extracted with organic solvents, does favor the idea of independent ubiquinone domains. This contradicts the observation of Heron *et al.* (*36*) that readdition of lipid alone to isolated complexes reconstitutes pool behavior.

The diffusion-controlled rate limit between the dehydrogenase and the cytochromes in the respiratory chain of mitochondria was confirmed recently by Hackenbrock and co-workers (*75*). Submitochondrial particles were fused with lipid vesicles causing "two-dimensional" dilution of the electron-transfer complexes, seen as a decrease in particle density on freeze fracture faces. Electron transport from NADH or succinate to the cytochromes was inhibited, but electron transport starting from the cytochrome b-c_1 complex was not inhibited (suggesting that diffusion of cytochrome c (Fig. 1B) never became rate limiting!) In a later study, ubiquinone was also added during dilution with lipid, and preservation of NADH or succinate oxidation was observed (Schneider *et al.*, Paper 3, Chapter III, this volume). The preservation, however, was only partial, and was higher with UQ_3 than with UQ_{10} (see next sections).

In summary, it seems impossible at present to decide whether diffusion of protein complexes, or of ubiquinone is responsible for pool behavior, although one would guess that diffusion of UQ should be faster than of the much larger complexes.

Lateral Mobility, Physical State, and Location in the Membrane

Average values for lateral diffusion coefficients of lipid molecules and proteins in membranes are 10^{-8} and 10^{-9} cm^2sec^{-1}, respectively (*76*). The

MW of UQ_{10} is comparable to that of a lipid molecule, but its shape is much more stretched in its most stable conformation. Since the quinone ring is more hydrophilic than the isoprenoid side chain, especially when fully reduced, an orientation parallel to the lipid molecules in the membrane is imagined (*77–79*). Following Cherry's derivation (*76*) that the diffusion coefficient of a lipid molecule is inversely proportional to the square of the length of the side chains, one would conclude that UQ_{10}, with about double the length of a membrane lipid, diffuses more slowly than a lipid molecule. The assumption, however, that the quinone ring is located in the head group region of the lipids could not be substantiated (see next sections). Moreover, kinks of the side chain are feasible, the steric hinderance of the methyl groups being overestimated in our minds (*77*). If the quinone predominantly exists monomerically dispersed in the acyl chain region of the membrane, then the diffusion coefficients could be much larger than for a lipid molecule, probably around 10^{-6} cm^2sec^{-1} (*80*). As reasoned by Schneider *et al.* (*75*), this faster diffusion of quinone cannot be taken as an argument for quinone function (Fig. 2A) because the much slower diffusion of the protein complexes is still probably fast enough to account for the observed electron-transport rates.

What do we know about the physical state of polyprenyl quinones in membranes, and about their interaction with lipid molecules? The partition coefficients between phospholipids and water are 3.3×10^3 for UQ_0, about 10^5 for UQ_1, 2.7×10^3 for UQ_0H_2, and 4×10^4 for UQ_1H_2 [(*81*, *82*); see, however, Lenaz *et al.*, Paper 2, Chapter III, this volume, for somewhat lower values]; homologs of UQ and UQH_2 have a still higher preference for the lipid phase. This preference was also demonstrated for UQ_3, UQ_7, and UQ_{10} by studying the fluorescence quenching of a hydrophobic probe in mitochondrial membranes (*83*). [Reduced ubiquinones showed less quenching, reflecting their slightly increased tendency to leave the hydrophobic environment. A location in the lipid acyl chain region was also shown for UQ_6 by a spin label technique (*84*).

The effects of UQ_1 and UQ_9 on the phase transition of dipalmitoylphosphatidylcholine are presented in Fig. 3 in liposomes. Both cause a broadening of the T interval (UQ_9 to a larger extent than UQ_1, that is indicative of quinone–lipid interaction, as known for cholesterol (*85*). Interaction of isoprenoid quinones can also be studied in monolayer films on aqueous surfaces. Maggio *et al.* (*86*) concluded from such studies that UQ_3, UQ_7, and UQ_9, in contrast to α-tocopherols, have negligible interaction with saturated and unsaturated phospholipids, and form a separate phase. Later Quinn and Esfahani (*87*) found that with increased surface pressure UQ_{10} and $UQ_{10}H_2$ are progressively squeezed out from monolayers of lecithin and diphosphatidylglycerol into the side-chain region.

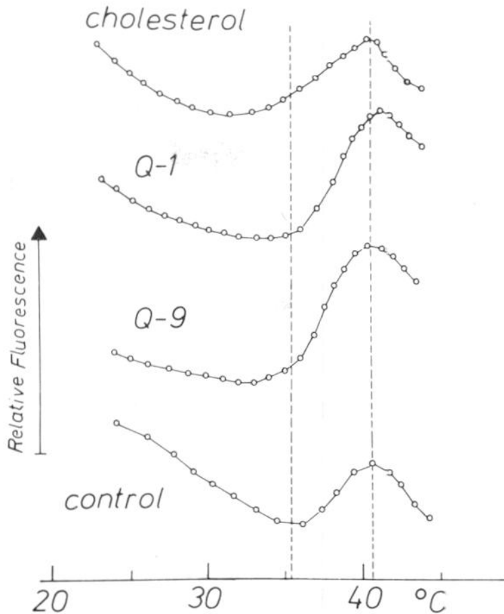

Fig. 3. Effect of UQ_1 and UQ_{10} on the T-interval of the phase transition of dipalmitoyl-phosphatidyl-choline in liposomes. Assay via fluorescence of 8-anilino-naphthalenesulfonate (*110*): The cuvette of a thermostated fluorometer contained in 2.5 ml, 100 m*M* KCl, 20 m*M* Tricine-NaOH, pH 8, 10 μg 8-anilino-naphthalenesulfonate and an aliquot of small liposomes corresponding to 1.6 mg dipalmitoyl-phosphatidyl-choline, containing 5 mole% of cholesterol, UQ_1 or UQ_9 where indicated. Fluorescence was excited and measured at 365 and 485 nm, respectively. Q stands for UQ.

Recently a similar study was reported for mixed monolayers of chloroplast galactolipids and plastoquinone or α-tocopherol. Again it was found that both were squeezed out from monolayers with saturated side chains, but were mixed if the side chains contained double bonds (*88*). Another pertinent study was carried out by McCloskey and Troy (*89*) with spin-labeled polyprenyl carboxy- and phosphoesters. Monomolecular dispersion was obtained with phosphoesters in saturated phospholipid, but neutral carboxyesters gave limited miscibility. Unsaturation of the lecithin acyl chains increased isoprenoid monomer solubility.

An extensive NMR study on the location and mobility of UQ homologs in liposomes was made by Kingsley and Feigenson (*81*). By using perdeuterated lecithin, the ^{1}H NMR signals of the quinone protons could be observed at quinone/lipid ratios down to 1/100 (corresponding to that in natural membranes). With paramagnetic shift agents it was found that the quinone ring, as indicated by the shift of the methoxy protons, is located in the hydrophobic region; the deeper, the longer the isoprenoid side chain. Ubiquinol rings reside closer to the membrane surface. The isoprenoid side chain is further apart from the surface. At the highly curved, inner surface of small lipid vesicles, the quinone rings, whether reduced or oxidized, seem much more exposed to shift agents. All ubiquinone homologs exhibit high mobility, as reflected by the relatively narrow NMR signals.

With UQ_{10}, and UQ_9, two signals attributed to the ring methoxy protons were observed. The relative height was concentration dependent, the upfield signal increasing with the UQ/lipid ratio (the lower limit for the observation was 1%). The authors conclude that this second signal comes from an aggregated form of UQ_{10}, that is not in equilibrium with the major part of the monomeric form. Indeed a UQ-rich fraction could be separated from the liposome suspension by centrifugation, and the lack of exchange probably reflects the slow exchange of ubiquinone between vesicles rather than within an individual vesicle (the UQ-rich fraction had a quinone/lipid ratio of one, and its vesicular nature is questionable). However, the splitting into two signals was the same in small and large liposomes, suggesting that the exchange between the two quinone states is also slow in larger membrane continua. In the aggregates the mobility of the quinone ring was higher than that of the monomer, as seen by the sharpness of the upfield signal. With UQ_3 and UQ_4 the upfield signal was absent, so these homologs did not form aggregates under the experimental conditions. The downfield signal, reflecting quinone monomer, was absent at temperatures below the phase transition with saturated phospholipids; in multilayered vesicles it was also absent above the transition temperature. This was explained by an immobilization of the monomeric quinone in the membranes under these conditions.

In summary, the ring of polyisoprenoid quinones in lipid membranes is highly mobile; it resides predominantly in the hydrophobic region, but segregation from phospho- and galactolipids, at high quinone/lipid ratios, has been established. Do quinone-rich phases occur in biomembranes, and if so, are they related to the quinone domains postulated on kinetic basis in the previous chapter? No clear answer is possible at present, but a few arguments suggest that segregation of a quinone-rich phase does *not* occur: (1) the ratio of quinone/lipid in biomembranes (1/100) is at the lower limit for segregation to occur, according to Kingsley and Feigenson; (2) the presence of unsaturated side chains should increase the solubility of isoprenoids (*86*, *88*, *89*); (3) the presence of protein could have an additional dispersing action, or even cause specific binding; (4) no hypochromic effect is seen in the UV spectrum of UQ in liposomes from soybean lipids, at quinone/lipid ratios of 1–5%, neither with UQ_{10} nor with UQ_1 (Hurt, unpublished); such an effect is expected in aggregates (*6*). The absorption peak of UQ_1 in ethanol (275 nm) was shifted to 279 nm in liposomes, which probably is an effect exerted by the head group region of the membrane. With UQ_{10} the shift is much less (275 to 275.7 nm), in accordance with its predominant residence in the hydrophobic region (see Lenaz *et al,* Paper 2, Chapter III, this volume, for similar, more extensive results on UV spectroscopy).

Another key result of Kingsley and Feigenson is that the rings of ubiquinone homologs, with long or short side chains, rapidly exchange through the membrane, confirming older data on electron and proton transport through membranes, by isoprenoid quinones (*90*).

Transverse Mobility

In the previous chapter evidence presented argues against the view of a stiffly anchored state, with very low tumbling rates, for polyprenol quinones in membranes (*77–79*). This view promotes a protein-catalyzed mechanism for electron flow-linked proton transport during energy conservation in respiration and photosynthesis (*91*, *92*) rather than the view of quinone loops that are dependent on quinone flip-flops through the membranes (*1*, *2*).

Artificial, quinoid compounds, lacking the long isoprenoid side chains, were first shown to catalyze vectorial redox reactions carrying protons into liposomes (*93*), and chloroplast thylakoids (*92*). Polyprenyl quinones have also been shown to react through lipid barriers (*94–96*).

In our own studies we additionally observed a striking differential behavior: quinones carrying a side chain longer than two isoprenoid units catalyzed the reduction of liposome-trapped ferricyanide, by external dithionite, with a completely different mechanism than quinones with no, or shorter side chains (*90*, *97–99*). The observed differences were

1. As shown in Table II, with polyprenyl benzoquinones, the rate was about 30 times faster if the side chain was longer than two isoprene units (except for the naphtoquinone series).
2. The activation energy for the reaction with long-chain homologs was about three times that for short-chain homologs (55 and 17 kJ $mole^{-1}$, respectively). The former value is similar to that for the transition of amphiphilic molecules from the micellar to the monomeric state in water (*100*).
3. With long-chain quinones the rate fell with pH. With short-chain homologs the rate rose with pH as expected for quinol oxidation reactions.
4. The reaction with long-chain homologs was dependent on external dithionite concentration, but not with short-chain homologs.
5. The reaction catalyzed by long-chain quinones was of higher order; this was particularly obvious at low quinone/lipid ratios. With short-chain quinones the rate was pseudo-first-order. The higher order with long-

chain quinones might be caused by the superposition of several pseudo-first-order reactions. This may imply that quinone clusters occur as the active, catalytic form of long-chain quinones, and are distributed in the lipid vesicles, with no exchange between cluster carrying- and -lacking vesicles.

6. During the reaction with short-chain quinones, semiquinone is formed transiently (peak of up to 30% of total quinones), but this is not observed with long-chain quinones.

From these observations we concluded that with long-chain quinones the rate of dithionite reduction is rate limiting; with short chain quinones oxidation by ferricyanide limits the overall rate. Based on the notion that *p*-phenylenediamines (that are much more water soluble than short-chain quinones), catalyzed the reaction even more efficiently than the long-chain benzoquinones, we concluded that the permeation of Q or QH_2 through the membrane is not the limiting step. The turnover numbers of the long-chain quinones in our system were calculated to be fast enough to account for the known electron-transport rates for respiration and photosynthesis (*98*). Lenaz *et al.* [see (*101*) and Paper 2, Chapter III, this volume] also found a differential efficiency of long and short-chain quinones in catalyzing redox reactions in mitochondrial and artificial membranes. Their conclusion, however, that UQ_3, in contrast to UQ_4 and UQ_5, does not cross the membrane is not supported by our experiments. In this volume they present results obtained with liposomes from egg lecithin that contradict our results obtained with soybean lipid, for reasons presently not understood.

The data from Kingsley and Feigenson discussed previously, complements our results. Our question—why does reduction of long-chain quinones outside liposomes become rate limiting, because of higher hydrophobicity, when oxidation inside does not (*90*)—finds an additional explanation in that the quinone rings are all exposed inside small lipid vesicles, but outside the rings of long-chain quinones are not. Most important in the context of this chapter is the finding that paramagnetic shift agents affect the whole quinone population in the liposomal system. This suggests that the flip-flop rate of the quinone rings through the membrane is faster than the resolution of NMR measurements. The minimal flip-flop rate is proportional to the observed shift in ppm (Hz), and was about ten times higher for UQ_1 than for UQ_{10}. The rate for UQ_{10} was about twice the minimal flip-flop rate derived from our measurements (*98*). The rate for UQ_1 was about 500 times faster, which confirms our conclusion that permeability, of either long or short-chain quinones, did not limit the vectorial redox reactions.

To explain the high order of rate, it is tempting to conclude that the quinone aggregates found by Kingsley and Feigenson (*81*) with UQ_9 and UQ_{10}, above 3% quinone/lipid, represent our postulated quinone clusters. Indeed, Kingsley and Feigenson suggested that the lack of transient semiquinone formation with the long-chain UQ homologs at 5% quinone/lipid, reflects this quinone-rich phase. Unfortunately, we are afraid it is not that simple. First the semiquinone transient is also not seen with UQ_3 and UQ_4, and they do not segregate according to Kingsley and Feigenson. Secondly, if (according to their data) a phase is formed with a quinone/lipid ratio of one, then 95% of the lipid should form vesicles carrying very little or no UQ. Accordingly, 95% of the entrapped ferricyanide should be reduced rather slowly; but this is not the case (*99*). Thirdly, we believe that soybean phospholipids, with their high degree of unsaturation, provide a better solvent for ubiquinone monomers than saturated lecithins; this is supported by the results given in Fig. 4. It shows the Arrhenius plots of the reaction with UQ_1 and UQ_9 in liposomes formed from either soybean phospholipids or dipalmitoyl-phosphatidylcholine, and it

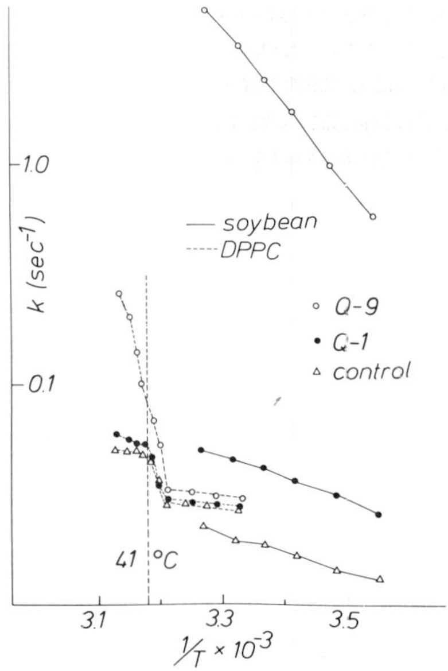

Fig. 4. Arrhenius plots of the vectorial redox reaction in liposomes catalyzed by UQ_1 and UQ_9. Liposomes were tested and prepared, either from soybean phospholipids or from dipalmitoyl-phosphatidylcholine (DPPC), containing UQ_1 or UQ_9 (1 mole%) where indicated, as previously described (*90*).

is obvious that the rate with UQ_9 in the former is faster than in the latter. More interestingly, the figure shows that the stimulating effect of the isoprenoid side chain is frozen out below the lipid phase transition of the pure lecithin. This is a clear indication that UQ_9 functions as a mobile carrier (*102*) that is immobilized in the gel state of the surrounding lipid. As mentioned previously, this immobilization has been demonstrated by Kingsley and Feigenson in their NMR study. Thinking of monomeric UQ_9, it finally occurred to us that the higher order of the rate could be simply explained by the fact that long-chain quinones cannot exchange rapidly between individual vesicles, whereas short-chain quinones can. In other words a statistical distribution of long-chain quinones onto the vesicles is almost static, and not averaged out with time. This automatically must lead to quinone-richer and -poorer vesicles, and to polyphasic reaction rates, without the assumption of quinone clusters.

If we nevertheless remain with the idea of quinone clusters, we should not think of them as ordered, possibly stacked, paracrystalline arrays across the membrane, because charge transfer from one ring to the next, and proton transfer via hydrogen bonds would be perpendicular to each other. Rather, a highly disordered aggregate similar to that reported by Kingsley and Feigenson, is feasible. If, on the other hand, we leave the idea of quinone clusters, we would again face the question of what enables the isoprenoid side chain to influence the vectorial redox process so profoundly. More specifically, since the formation of quinolate anions is the rate determining step for the oxidation of short-chain quinones (*90*), how does the presence of the side chain circumvent this, if not by some supermolecular structure of a quinhydrone type? We previously established that the double bonds in the side chain are not essential, and excluded the possibility that the side chain causes a leak of dithionite through the vesicles (*90*). There is no clear answer to the question, but we cannot exclude the possibility that, whatever the effect of the polyprenyl side chain it has some bearing on the mechanism of physiological redox reactions of quinones.

On the Physiological Role of the Isoprenoid Side Chain

The role of the isoprenoid side chain could be at least threefold: (1) anchoring of the quinone to the membrane, (2) determining the specificity for interaction of the quinone with redox protein complexes, and (3) affecting the redox reactions of the quinone directly. From the information given previously, the picture of the side chain as an "anchor" should be changed to one of a "sinker", since the side chain helps keep the quinone ring in the hydrophobic parts of the membrane. In contrast to earlier views (*77–79*), there is no stringent reason to assume that the side chain is stiff and always stretched.

The specificity of quinone reductases and dehydrogenases for the length and constitution of the side chain is complex and has been reviewed with an emphasis on mitochondria by Crane (*5*). Studying the differential specificity for long- and short-chain quinones, especially with isolated protein complexes, is complicated by the difficulty of applying the long-chain homologs to aqueous suspensions yet avoiding indirect effects from low water solubility. The presence of detergents is often required, and since kinetics of micelle interactions follow their own rules (*73*, *74*), an observed lower specificity for long-chain quinones might actually be caused by their low water solubility. An early observation during extraction–reconstitution studies is that succinate oxidase in mitochondria is not specific for the chain length, but NADH-oxidase is highly specific for the endogeneous UQ homolog [UQ_{10} in beef heart mitochondria (*37*); UQ_6 in yeast mitochondria (*103*)]. A similar specificity for respiratory membranes of *Rps. palustris* was reported (*14*). Lenaz *et al.* (*101*) investigated this differential effect in the mitochondrial membrane in greater detail. Their experimental evidence suggests that it is quinol oxidation which discriminates between long- and short-chain quinones during NADH-oxidation. Since this is not true for succinate oxidation, one had to postulate two different quinol dehydrogenases (*5*), an idea in disagreement with the current view of the respiratory chain. Lenaz *et al.* (*101*) avoided this discrepancy by suggesting that reduction of quinone by NADH- and succinate-dehydrogenase occured at different sides of the membrane, and that in the case of NADH oxidation, quinol had to cross the membrane to find the oxidation site. They also suggested that quinones with short chains, like UQ_3, are unable to cross the membrane at a fast rate. Unfortunately, recent evidence indicates that all quinones do cross membranes at fast rates.

An alternate explanation can be made from experiments by Schneider *et al.* (Paper 3, Chapter III, this volume). They report that UQ_3 was more effective than UQ_{10} in preserving NADH and succinate oxidation in mitochondria which were "diluted" by diffusion with phospholipids. Thus, the short chain ubiquinone seems to diffuse faster in the plane of the membrane. Does the lack of efficiency of UQ_{10} in this system reflect the aggregates of UQ_{10} in liposomes found by Kingsley and Feigenson (*81*)?

A difference on similar lines between UQ_1 and UQ_8 in *E. coli* was reported (*104*). Both quinones reconstituted lactate, as well as NADH oxidation and active transport of proline coupled to lactate oxidation in membrane vesicles from a UQ-deficient mutant. Active transport coupled to

NADH oxidation was restored by UQ_1 only. Our alternative to the rather complicated interpretation by the authors is the following: Lactate- and NADH-dehydrogenase are located on opposite membrane surfaces, and therefore are accessible to their hydrophilic substrates on different vesicles only [right-side-out, and inside-out, respectively (*105*)]. Active accumulation of an amino acid can occur in right-side-out vesicles only, and can either be driven by lactate or quinol oxidation. Since the water soluble UQ_1H_2 can mediate between different vesicles, but UQ_8H_2 cannot, proline uptake into right-side-out vesicles might be driven by UQ_1H_2, formed by NADH-oxidation in inside-out vesicles, but not by UQ_8H_2. This is another possible example of apparent differential specificity caused by differential solubility.

Reconstitution of electron transport in partially extracted chloroplasts does not seem to depend on the length of the isoprenoid side chain in plastoquinone, but shows specificity for the ring substituents (*106*, *107*), especially for the prenyl substitution. Accordingly, isolated plastoquinol-plastocyanin oxidoreductase (*108*) functions with PQ_1H_2 and PQ_9H_2 (*109*).

With chromatophores from Rhodospirillaceae, a double specificity for the length of the ubiquinone side chain was detected (Table II). Reconstitution of photophosphorylation in partially extracted preparations required at least three isoprene units in the side chain, while reconstitution in extensively extracted preparations required more than five units (compare second and third columns in Table II) (*48*). It was concluded that for reconstitution of Q_{III} (Fig. 1A) UQ_3 suffices, while for reconstitution of Q_{II} a higher homolog is necessary (Baccarini-Melandri *et al.*, Paper 4, Chapter V, this volume). For comparison, the dependence of rate constants on side chain length is given in Table II. Using a quinol-mediated ferricyanide reduction in liposomes, a striking parallelism between chromatophores and liposomes is revealed. The requirement for a minimum chain length of three isoprene units is as stringent for the reconstitution of quinol oxidation by the cytochrome complex (= Q_{III}) in chromatophores as it is for the liposomal model reaction. In light of this observation, one feels justified in believing that the isoprenoid side chain has an effect on the physiological mechanism of quinol oxidation, other than keeping the quinones in the membrane and providing for binding specificities.

CONCLUSIONS

From the preceding discussion we may conclude:

1. All quinones and hydroquinones of concern have high preference for the hydrophobic part of membranes. At physiological concentrations

they are monomerically dispersed in the lipid, although there is a tendency to segregate at higher concentration for quinones carrying long polyprenyl side chains, especially from saturated phospholipids. The role of protein on quinone solubility is not known and deserves attention.

2. All isoprenoid quinones are highly mobile in the membranes, laterally and transversly (see, however, Lenaz *et al.*, Paper 2, Chapter III, this volume).

3. The isoprenoid side chain might alter the redox mechanisms of quinones in membranes, in addition to the more feasible functions of sticking to hydrophobic regions and providing specificity for redox partners.

4. At present it is neither possible to exclude, nor to establish the role of the quinone pool as an obligate intermediate in the electron-transport system. Specially bound quinones exist, but the exchange rates with the quinone pool are not known, and might even depend on the state of the membrane (this holds especially for Q_{III} in Fig. 1). The pool varies in size from system to system. It is still the most likely explanation of "pool function behavior" in linear electron-transport chains.

5. The isoprenoid quinol–quinone systems have all the required features to function as hydrogen carriers through the hydrophobic region of membranes.

ACKNOWLEDGMENT

We are grateful to Dr. G. Feigenson for information prior to publication, and to the Deutsche Forschungsgemeinschaft for supporting our research.

REFERENCES

1. Mitchell, P. (1966). *Biol. Rev. Cambridge Philos. Soc.* **41,** 455–502.
2. Mitchell, P. (1976). *J. Theor. Biol.* **67,** 327–367.
3. Morton, R. A. (1965). "The Biochemistry of Quinones." Academic Press, New York.
4. Crane, F. L. (1968). *In* "Biological Oxidations" (T. P. Singer, ed.), pp. 533–580. Wiley (Interscience), New York.
5. Crane, F. L. (1977). *Annu. Rev. Biochem.* **46,** 439–469.
6. Trumpower, B. L. (1981). *J. Bioenerg. Biomembr.* **13,** 1–24.
7. Amesz, J. (1973). *Biochim. Biophys. Acta* **301,** 35–51.
8. Bendall, D. S. (1977). *Int. Rev. Biochem.* **13,** 41–78.
9. Witt. H. T. (1979). *Biochim. Biophys. Acta* **505,** 355–427.
10. Parson, W. W. (1978). *In* "The Photosynthetic Bacteria" (R. K. Clayton and W. F. Siström, eds.), pp. 455–469. Plenum, New York.

11. Wraight, C. A. (1979). *Photochem. Photobiol.* **30,** 767–776.
12. Crane, F. L. (1965). *In* "The Biochemistry of Quinones" (R. A. Morton, ed.), pp. 183–206. Academic Press, New York.
13. Carr, N. G., Exell, G., Flynn, V., Hallaway, M., and Talukdar, S. (1967). *Arch. Biochem. Biophys.* **120,** 503–507.
14. King, M. T., and Drews, G. (1973). *Biochim. Biophys. Acta* **305,** 230–248.
15. Zannoni, D., Melandri, B. A., and Baccarini-Melandri, A. (1976). *Biochim. Biophys. Acta* **423,** 413–430.
16. Kröger, A., and Klingenberg, M. (1967). *Curr. Top. Bioenerg.* **2,** 176–193.
17. Brodie, A. F. (1965). *In* "The Biochemistry of Quinones" (R. A. Morton, ed.), pp. 356–404. Academic Press, New York.
18. Kröger, A., Dadak, V., Klingenberg, M., and Diemer, F. (1971). *Eur. J. Biochem.* **21,** 322–333.
19. Scholes, P. B., and Smith, L. (1968). *Biochim. Biophys. Acta* **153,** 363–375.
20. Kröger, A., and Dadak, V. (1969). *Eur. J. Biochem.* **11,** 328–340.
21. Kröger, A., and Innerhofer, A. (1976). *Eur. J. Biochem.* **69,** 487–495.
22. Badziong, W., and Thauer, R. K. (1980). *Arch. Microbiol.* **125,** 167–174.
23. Kawamura, M., Mimuro, M., and Fujita, Y. (1979). *Plant Cell Physiol.* **20,** 697–705.
24. Douce, R., Christensen, E. L., and Bonner, W. D. (1972). *Biochim. Biophys. Acta* **292,** 148–160.
25. Peschek, G. A. (1980). *Biochem. J.* **186,** 515–523.
26. Kröger, A. (1978). *Biochim. Biophys. Acta* **505,** 129–145.
27. Thauer, R. K., Jungermann, K., and Decker, K. (1977). *Bacteriol. Rev.* **41,** 100–180.
28. Powls, R., and Redfearn, E. R. (1969). *Biochim. Biophys. Acta* **172,** 429–437.
29. Sato, M., Yamada, K., and Ozawa, H. (1972). *Biochem. Biophys. Res. Commun.* **46,** 578–582.
30. Redfearn, E. R. (1965). *In* "The Biochemistry of Quinones" (R. A. Morton, ed.), pp. 149–181. Academic Press, New York.
31. Stiehl, H. H., and Witt, H. T. (1969). *Z. Naturforsch., Teil B* **24,** 1588–1598.
32. Bailey, J. L., and Whyborn, A. G. (1963). *Biochim. Biophys. Acta* **78,** 163–174.
33. Lichtenthaler, H. V. (1969). *Protoplasma* **68,** 65–77.
34. Zannoni, D., Melandri, B. A., and Baccarini-Melandri, A. (1976). *Biochim. Biophys. Acta* **449,** 386–400.
35. Kröger, A., Klingenberg, M., and Schweidler, S. (1973). *Eur. J. Biochem.* **34,** 398–368.
36. Heron, C., Ragan, C. I., and Trumpower, B. L. (1978). *Biochem. J.* **174,** 791–800.
37. Lenaz, G., Daves, G. D., and Folkers, K. (1968). *Arch. Biochem. Biophys.* **123,** 539–550.
38. Gutman, M., and Silman, N. (1972). *FEBS Lett.* **26,** 207–210.
39. Ragan, C. I., and Heron, C. (1978). *Biochem. J.* **174,** 783–790.
40. Cox, G. B., and Gibson, F. (1974). *Biochim. Biophys. Acta* **346,** 1–25.
41. Mitchell, P. (1975). *FEBS Lett.* **59,** 137–139.
42. Crofts, A. R., Crowther, D., and Tierney, G. V. (1975). *In* "Electron Transfer Chains and Oxidative Phosphorylation" (E. Quagliarello, S. Papa, F. Palmieri, E. C. Slater, and N. Siliprandi, eds.), pp. 233–241. Elsevier/North-Holland, Amsterdam.
43. Baccarini-Melandri, A., and Melandri, B. A. (1977). *FEBS Lett.* **80,** 459–464.
44. Wood, P. (1981). *Proc. Int. Congr. Photosynth., 5th, Halkidiki, Greece, 1980,* Vol. II, pp. 591–598.
45. van Gorkom, H. J. (1974). *Biochim. Biophys. Acta* **347,** 439–442.
46. Evans, M. C. W., and Rutherford, A. W. (1980). *Eur. Bioenerg. Conf., 1st, Short Rep., Urbino* pp. 33–34.

47. Bowyer, J. R., Dutton, P. L., Prince, R. C., and Crofts, A. R. (1980). *Biochim. Biophys. Acta* **592,** 445–460.
48. Baccarini-Melandri, A., Gabellini, N., Melandri, B. A., Hurt, E., and Hauska, G. (1980). *J. Bioenerg. Biomembr.* **12,** 95–105.
49. Bouges-Bocquet, B. (1980). *FEBS Lett.* **117,** 54–58.
50. Prince, R. C., Baccarini-Melandri, A., Hauska, G., Melandri, B. A., and Crofts, A. R. (1975). *Biochim. Biophys. Acta* **387,** 212–227.
51. Hauska, G., McCarty, R. E., Berzborn, R. J., and Racker, E. (1971). *J. Biol. Chem.* **246,** 3524–3531.
52. Wildner, G., and Hauska, G. (1974). *Arch. Biochem. Biophys.* **164,** 136–144.
53. Ohnishi, T., and Trumpower, B. L. (1980). *J. Biol. Chem.* **255,** 3278–3284.
54. Yu, C.-A., and Yu, L. (1980). *Biochemistry* **19,** 3579–3585.
55. Urban, P. F., and Klingenberg, M. (1969). *Eur. J. Biochem.* **9,** 519–525.
56. Takamiya, K.-I., and Dutton, P. C. (1979). *Biochim. Biophys. Acta* **546,** 1–16.
57. Golbeck, J. H., and Kok, B. (1979). *Biochim. Biophys. Acta* **547,** 347–360.
58. Bouges-Bocquet, B. (1975). *Proc. Int. Congr. Photosynth. 3rd, Rehovoth, 1974* **1,** 579–588.
59. Takamiya, K.-I., Prince, R. C., and Dutton, P. C. (1979). *J. Biol. Chem.* **254,** 11307–11311.
60. Rich, P. R., and Bendall, D. S. (1980). *Eur. Bioenerg. Conf., 1st, Short Rep., Urbino* pp. 59–60.
61. Crofts, A. R., and Bowyer, J. (1977). *In* "The Proton and Calcium Pumps" (G. F. Azzone, M. Aaron, J. C. Metcalfe, E. Quagliarello, and N. Siliprandi, eds.), pp. 55–64. Elsevier/North-Holland, Amsterdam.
62. Horton, P., Allen, J. F., Black, M. T., and Bennett, J. (1981). *FEBS Lett.* **125,** 193–196.
63. Heber, U. (1976). *J. Bioenerg. Biomembr.* **8,** 157–172.
64. Green, D. E. (1962). *Comp. Biochem. Physiol.* **4,** 81–122.
65. Kröger, A., Klingenberg, M., and Schweidler, S. (1973). *Eur. J. Biochem.* **39,** 313–323.
66. von Jagow, G., and Bohrer, C. (1975). *Biochim. Biophys. Acta* **387,** 409–424.
67. Gutman, M. (1977). *In* "Bioenergetics of Membranes" (L. Packer, G. C. Papageorgiou, and A. Trebst, eds.), pp. 165–175. Elsevier/North-Holland Amsterdam.
68. Fernandes-Moreira, M. T., Rich, P. R., and Bendall, D. S. (1980). *Eur. Bioenerg. Conf., 1st, Short Rep., Urbino* pp. 61–62.
69. Prince, R. C., Bashford, C. L., Takamiya, K.-I., van den Berg, W. H., and Dutton, P. L. (1977). *J. Biol. Chem.* **253,** 4137–4142.
70. Haehnel, W. (1973). *Biochim. Biophys. Acta* **305,** 618–631.
71. Siggel, U., Renger, G., Stiehl, H. H., and Rumberg, B. (1972). *Biochim. Biophys. Acta* **256,** 328–335.
72. Heron, C., Gare, M. G., and Ragan, C. I. (1979). *Biochem. J.* **178,** 415–426.
73. Weiss, H., and Wingfield, P. (1979). *Eur. J. Biochem.* **9,** 151–160.
74. Engel, W. D., Schägger, H., and von Jagow, G. (1980). *Biochim. Biophys. Acta* **592,** 211–222.
75. Schneider, H., Lemasters, J. J., Höchli, M., and Hackenbrock, C. R. (1980). *J. Biol. Chem.* **255,** 3748–3756.
76. Cherry, R. J. (1979). *Biochim. Biophys. Acta* **559,** 289–327.
77. Trumpower, B. L., and Landeen, C. E. (1977). *Ealing Rev.* **1,** 4–7.
78. Robertson, R. N., and Boardman, N. K. (1975). *FEBS Lett.* **60,** 1–6.
79. De Pierre, J. W., and Ernster, L. (1977). *Annu. Rev. Biochem.* **46,** 201–262.
80. Marcus, M. F., and Hawley, M. D. (1970). *Biochim. Biophys. Acta* **201,** 1–8.

81. Kingsley, P. B., and Feigenson, G. W. (1981). *Biochim. Biophys. Acta* **635,** 602–618.
82. Ragan, C. I. (1978). *Biochem. J.* **172,** 539–547.
83. Chance, B. (1972). *In* "Biochemistry and Biophysics of Mitochondrial Membranes" (G. F. Azzone, E. Carafoli, A. L. Lehninger, E. Quagliarello, and N. Siliprandi, eds.), pp. 85–99. Academic Press, New York.
84. Oettmeier, W., Norris, J. R., and Katz, J. J. (1976). *Biochem. Biophys. Res. Commun.* **71,** 445–451.
85. Chapman, D., and Wallach, D. F. H. (1968). *In* "Biological Membranes" (D. Chapman, ed.), pp. 125–202. Academic Press, New York.
86. Maggio, B., Diplock, A. T., and Lucy, J. A. (1977). *Biochem. J.* **161,** 111–121.
87. Quinn, P. J., and Esfahani, M. D. (1980). *Biochem. J.* **185,** 715–722.
88. Liljenberg, C., Wachtmeister, G., and Oquist, G. (1981). *Proc. Int. Congr. Photosynth., 5th, Halkidiki, Greece, 1980,* Vol. I, pp. 235–242.
89. McCloskey, M. A., and Troy, F. A. (1980). *Biochemistry* **19,** 2061–2066.
90. Futami, A., Hurt, E., and Hauska, G. (1979). *Biochim. Biophys. Acta* **547,** 583–596.
91. Papa, S. (1976). *Biochim. Biophys. Acta* **456,** 39–84.
92. Hauska, G., and Trebst, A. (1977). *Curr. Top. Bioenerg.* **6,** 152–220.
93. Hinkle, P. (1973). *Fed. Proc., Fed. Am. Soc. Exp. Biol.* **32,** 1988–92.
94. Yaguzhinsky, L. S., Boguslawsky, L. I., and Ismailow, A. D. (1974). *Biochim. Biophys. Acta* **368,** 22–38.
95. Anderson, S. S., Lyle, I. G., and Paterson, R. (1976). *Nature* (*London*) **259,** 147–148.
96. Masters, B. R., and Mauzerall, D. (1978). *J. Membr. Biol.* **41,** 4377–4388.
97. Hauska, G. (1977). *FEBS Lett.* **79,** 345–347.
98. Hauska, G. (1977). *In* "Bioenergetics of Membranes" (L. Packer, G. C. Papageorgiou, and A. Trebst, eds.), pp. 177–187. Elsevier/North-Holland, Amsterdam.
99. Futami, A., and Hauska, G. (1979). *Biochim. Biophys. Acta* **547,** 597–608.
100. Assiansson, E. A. G., Wall, S. N., Alengren, M., Hoffmann, H., Kielmann, I., Ulbricht, W., Zana, R., Lang, J., and Tondre, C. (1976). *J. Phys. Chem.* **80,** 905–922.
101. Lenaz, G., Mascarello, S., Landi, L., Cabrini, L., Pasquali, P., Parenti-Castelli, G., Sechi, A. M., and Bertoli, E. (1977). *In* "Bioenergetics of Membranes" (L. Packer, G. C. Papageorgiou, and A. Trebst, eds.), pp. 189–198. Elsevier/North-Holland, Amsterdam.
102. Krasne, S., Eisenman, G., and Szabo, G. (1971). *Science* **174,** 412–414.
103. Castelli, A., Lenaz, G., and Folkers, K. (1969). *Biochem. Biophys. Res. Commun.* **34,** 200–204.
104. Stroobant, P., and Kaback, H. R. (1979). *Biochemistry* **18,** 226–231.
105. Futai, M. (1974). *J. Bacteriol.* **120,** 861–865.
106. Krogman, D. W., and Olivero, E. (1962). *J. Biol. Chem.* **237,** 3292–3295.
107. Trebst, A., Eck, H., and Wagner, S. (1963). *In* "Photosynthetic Mechanisms of Green Plants," pp. 174–194. Natl. Acad. Sci., Washington, D.C.
108. Wood, P., and Bendall, D. S. (1976). *Eur. J. Biochem.* **61,** 337–344.
109. Hurt, E., and Hauska, G. (1981). *Eur. J. Biochem.* **117,** 591–599.
110. Sackmann, E. (1974). *Ber. Bunsenges. Phys. Chem.* **78,** 929–941.

Studies on the Interactions and Mobility of Ubiquinone in Mitochondrial and Model Membranes

2

GIORGIO LENAZ
MAURO DEGLI ESPOSTI
ENRICO BERTOLI
GIOVANNA PARENTI-CASTELLI
SERGIO MASCARELLO
ROMANA FATO
CLAUDIO CASALI

INTRODUCTION

The role of ubiquinone (Q)[1] in the mitochondrial respiratory chain was investigated in our laboratory by testing different Q homologs as electron carriers (*1–4*). Evidence has been provided that the incapability of short-chain isoprenoid homologs to restore NADH oxidation, in contrast with succinate oxidation (*1*, *2*), may reflect a sidedness of the Q redox cycle in the membrane (*3*, *4*). The hypothesis has been advanced that short-chain ubiquinones, in contrast with long-chain ubiquinones, are unable to cross

[1] Abbreviations: BHM, beef heart mitochondria; EPR, electron paramagnetic resonance; P, partition coefficient; PL, phospholipids; Q, ubiquinone; SMP, submitochondrial particles.

Function of Quinones in Energy Conserving Systems

ISBN 0-12-701280-X

the lipid bilayer (*3*, *4*). Direct studies of the permeability of different Q homologs in artificial systems (*3*, *5*) support this suggestion. Several investigators currently employ exogenous Q homologs as electron donors or acceptors (*6–10*).

The use of exogenous ubiquinones, in particular short-chain homologs, may not be free of complications, since little is known about the interaction of such quinones with the electron-transfer chain and with the endogenous quinone pool (*6*); moreover, the intramembrane location and the effects on the lipid phase organization of ubiquinones having different isoprenoid side-chain lengths may be quite diverse (*11–13*).

For this reason we have undertaken a study of the interaction of the different Q homologs with lipid vesicles and mitochondrial membranes.

EXPERIMENTAL RESULTS

Physical Properties of Ubiquinones

It is tacitly assumed that exogenous ubiquinones, being very hydrophobic, are almost completely incorporated into the membrane phase; however no study is available to quantitatively evaluate such incorporation. Little is also known about the physical state of quinones in the aqueous phase, even if such compounds are often used as substrates for electron-transfer activities.

We investigated the physical state of ubiquinones in water by use of EPR spin labels which are sensitive to the mobility and order of their environment (*14*). Figure 1 shows the EPR spectra of the spin label 5-*N*-oxyl-4′,4′-dimethyloxazolidine stearate (5-NS) in water and in presence of 5 m*M* Q_4; the effect of a detergent (5 m*M* Triton-X-100) is also shown for comparison. The splitting of the high-field line induced by Q_4 indicates a mixed population of spin labels, with free probe present in water and probe immobilization in a medium of higher viscosity. These results indicate that Q_4 is present in water in some kind of organized arrangement; similar data obtained with other ubiquinones suggests they may exist in water in a micellar-like organization approaching that of a detergent.

The spectroscopic properties of some oxidized and reduced ubiquinones in solvents of different polarity were compared with ubiquinones incorporated into phospholipid vesicles. Table I shows that the absorbance maximum for oxidized ubiquinones having two or more isoprenoid chains is at 270 nm in isooctane, 275 nm in absolute ethanol, and

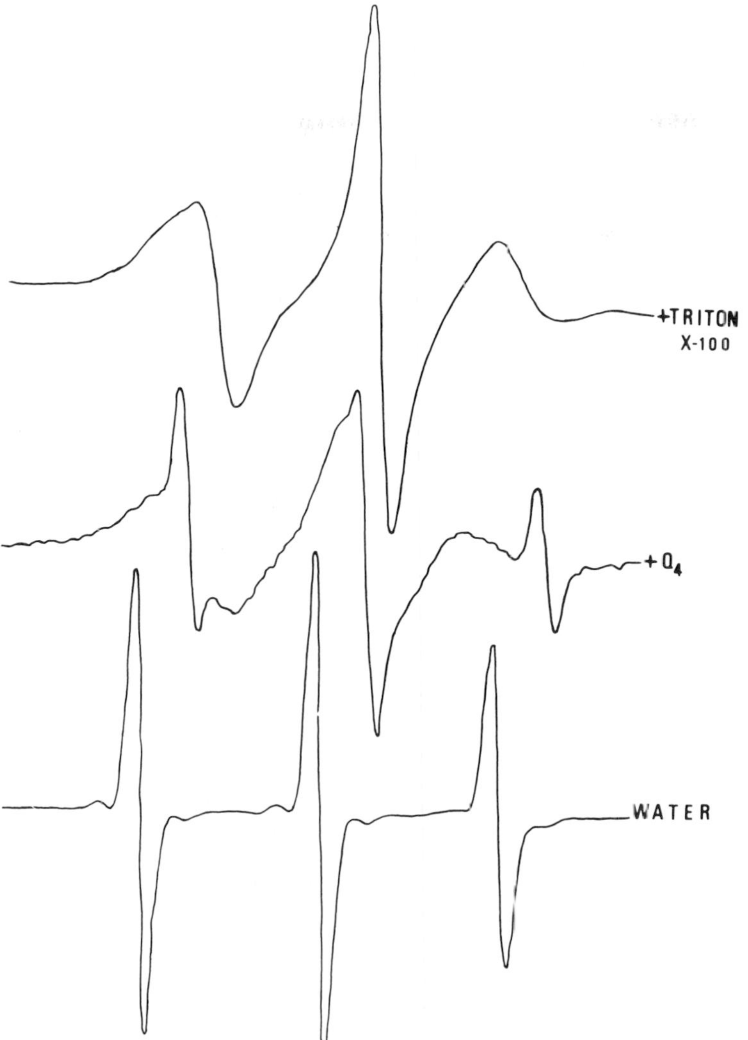

Fig. 1. EPR spectra of 5-NS in water and in presence of Q_4 and Triton-X-100. The probe concentration was 0.05 m*M*; Q_4 and Triton concentrations were 5 m*M*.

284 nm in water; for ubiquinols the maximum is 289 nm in both isooctane and ethanol, and 294 nm in water. The apparent extinction coefficients at λ_{max} are lower in water than in organic solvents; the red shift and absorption quenching in water confirm that ubiquinones exist in non-monomeric form.

TABLE I

Spectroscopic Characteristics[A] of Ubiquinones in Different Solvents and in Phospholipid Vesicles[B]

	Q_1		Q_1H_2		Q_3		Q_3H_2		Q_4		Q_4H_2		Q_7		Q_7H_2	
Solvents	λ_{max}	ϵ	λ_{max}	ϵ	λ_{max}	ϵ	λ_{max}	ϵ	λ_{max}	ϵ	λ_{max}	ϵ	λ_{max}	ϵ	λ_{max}	ϵ
H_2O	278	12.5	284	—	286	2.4	294	2.9	284	3.8	294	—	284	—	294	—
Ethanol	275	13.5	289	5.0	275	14.8	289	4.5	275	14.0	289	4.4	275	14.3	290	4.1
Isooctane	271	13.4	289	5.2	270	13.4	289	5	270	14.4	289	—	270	14.0	289	—
PL vesicles	278	12.5	—	—	271	17	289	5	270	17	289	5	270	—	289	—

[A] λ_{max} (nm) and ϵ coefficients (mM^{-1} cm^{-1}), calculated at λ_{max} = 310 nm.

[B] Ubiquinones and phospholipid were co-sonicated as described by Futami *et al.* (*5*).

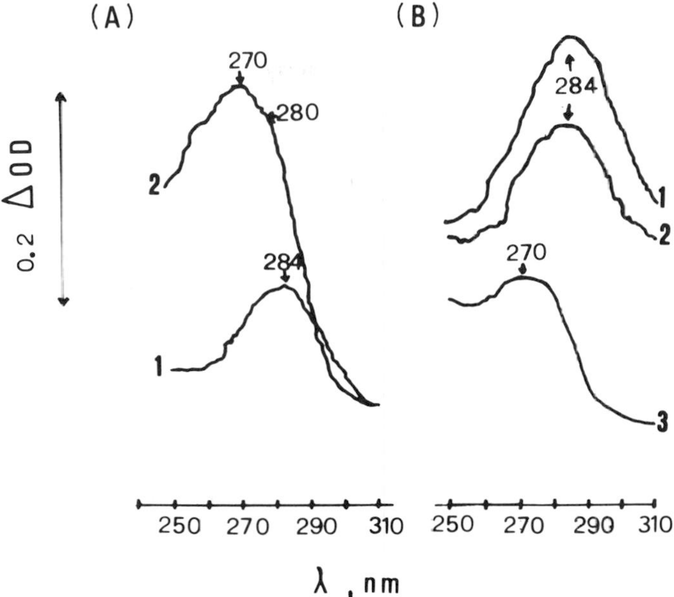

Fig. 2. UV spectra of ubiquinones in water and when incorporated into mixed PL. (A) Curve 1, spectrum of 18 μM Q_4 in water; Curve 2, spectrum of 18 μM Q_4 incorporated into PL by extensive sonication. (B) Curve 1, spectrum of 40 μM Q_5 in water; curve 2, spectrum of 40 μM Q_5 incorporated into PL without sonication; curve 3, same as curve 2 after a pentane wash. Concentration of Q_5 was calculated to be 9 μM.

Ubiquinone incorporation into phospholipid vesicles from water by sonication induces a blue shift of the maximum to 270 nm with no significant extinction coefficient differences in organic solvents (Fig. 2); a shoulder near 280 nm, however, shows the presence of conspicuous amounts of nonincorporated quinones (mainly lower homologs or ubiquinones incubated without sonication). The shoulder is completely removed by a mild pentane wash (see next section). The oxidized ubiquinones in phospholipid vesicles can be completely reduced by $NaBH_4$, with maximum absorbance at 289 nm. The behavior of Q_1 is unique, its spectroscopic properties, even after extensive sonication with phospholipid, appear similar to those in water.

Incorporation of Ubiquinones

Incorporation of ubiquinones was followed by equilibration with the membranes either by incubating at 30 °C for 10 to 60 min or by sonication. The unbound quinones were removed by a pentane wash, that extracts only nonincorporated quinone (λ_{max} 284 nm); controls showed that pentane does not extract any endogenous ubiquinone from mitochondria. The concentration of ubiquinone was measured either directly or after extraction according to Kröger and Klingenberg (*15*).

A typical curve for incorporation of Q_8 into phospholipid vesicles and beef heart mitochondria (BHM) is shown in Fig. 3A. The curves are nonlinear and appear sigmoidal for the higher ubiquinones, reaching saturation at 200–250 nmols of Q added/mg phospholipid. The extent of incorporation is much greater in BHM than in vesicles for all ubiquinones. Studies with Q-depleted BHM, prepared by pentane extraction after lyophilization (*16*), show an even greater incorporation.

The binding curves approach linearity for all ubiquinones in the range of 0.05–0.3 m*M* Q added; in this range it is possible to calculate an apparent "partition" coefficient (P) as described in Fig. 3. A double reciprocal plot of Q incorporated as a function of phospholipid concen-

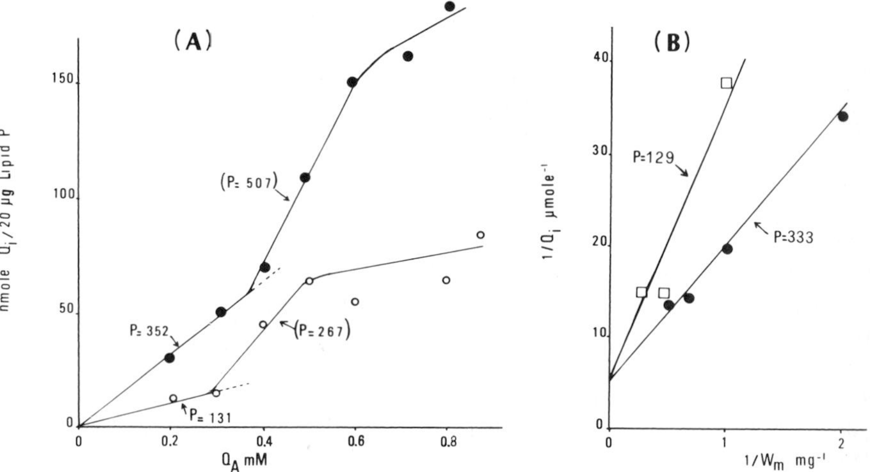

Fig. 3. (A) Incorporation of Q_8 into lipid vesicles (○) and BHM (●). The plots show apparent partition coefficients (P) calculated in the linear range by means of

$$P = \frac{Q_i}{W_m} \bigg/ \frac{Q_A - Q_i}{W_{H_2O}}$$

(where Q_i, μmole Q incorporated; Q_A, μmole Q added; W_m, mg of membrane lipids; W_{H_2O}, mg water). (B) Double reciprocal plots of incorporated Q_8, as a function of W_m in lipid vesicles (□) and BHM (●). The plot is linear according to

$$\frac{1}{Q_i} = \frac{W_{H_2O}}{P \cdot Q_A W_m} + \frac{1}{Q_A}.$$

Extrapolation of $1/Q_i$ to infinite PL concentration results experimentally in $Q_A = Q_i$. The concentration of Q_A was 0.2 m*M*.

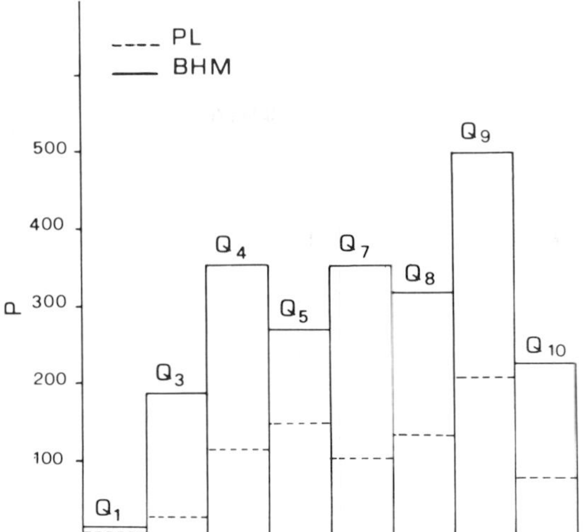

Fig. 4. Partition coefficients (P) of ubiquinones in egg lecithin vesicles (PL) and in BHM. The values were calculated from double reciprocal plots (see Fig. 3B).

tration is linear (Fig. 3B) and also allows calculations from the slope of a partition coefficient. The partition coefficients obtained by both methods are almost superimposable as shown in Fig. 4 for different homologs in lipid vesicles and BHM. There is an increase of P with increasing isoprenoid chain length, incorporation being about three-fold greater into BHM than into vesicles. The composition of the vesicles (either egg lecithin or mixed phospholipid) has no effect, suggesting that the increased incorporation into BHM may be due to the presence of proteins. Reduced ubiquinones have a lower P than the corresponding oxidized forms in both vesicles and BHM.

Effect of Ubiquinones on the Physical State of the Membrane

Ubiquinones induce striking effects on the physical state of membrane lipids, both in vesicles and mitochondrial membranes, as shown by the changes in fluorescence polarization of the hydrophobic probe perylene (*17*) incorporated into the membranes. Figure 5 shows that short-chain ubiquinones enhance polarization in both vesicles and membranes, although the changes are qualitatively and quantitatively different. Long-chain ubiquinones, on the contrary, have a tendency to decrease polarization. The changes reflect an increase in membrane viscosity induced by the lower homologs, and a slight disorganization of

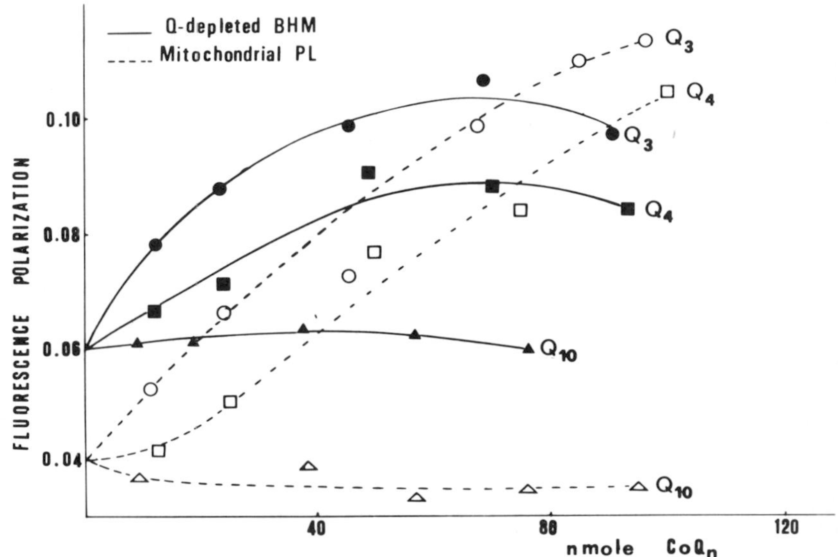

Fig. 5. Fluorescence polarization of perylene in vesicles of mitochondrial PL and in Q-depleted BHM after addition of different Q homologs. Polarization was calculated (*17*) after excitation of perylene (0.1 m*M*) at 412 nm and emission at 445 nm, in a Perkin-Elmer MPF-4 spectrophotofluorimeter equipped with polarization accessories.

the bilayer by the higher homologs. No appreciable differences have been found between oxidized and reduced quinones.

In mitochondrial membranes the increase of membrane order induced by lower ubiquinones is accompanied by functional changes, i.e., uncoupling at all three oxidative-phosphorylation sites and loss of oligomycin sensitivity of ATPase; these effects are reversed by longer quinones (*18*). The efficacy of the homologs in decreasing oligomycin sensitivity (Fig. 6) follows the pattern observed in enhancing polarization. The small effect of Q_1 may be due to its very low incorporation into the membrane.

Permeability of Ubiquinones across Phospholipid Membranes

We previously investigated the transmembrane mobility of ubiquinones in phospholipid vesicles by using exogenous ubiquinols to reduce cytochrome *c* trapped inside the vesicles (*3*). This was also investigated by Futami *et al.* (*5*) using ubiquinones incorporated into liposomes as

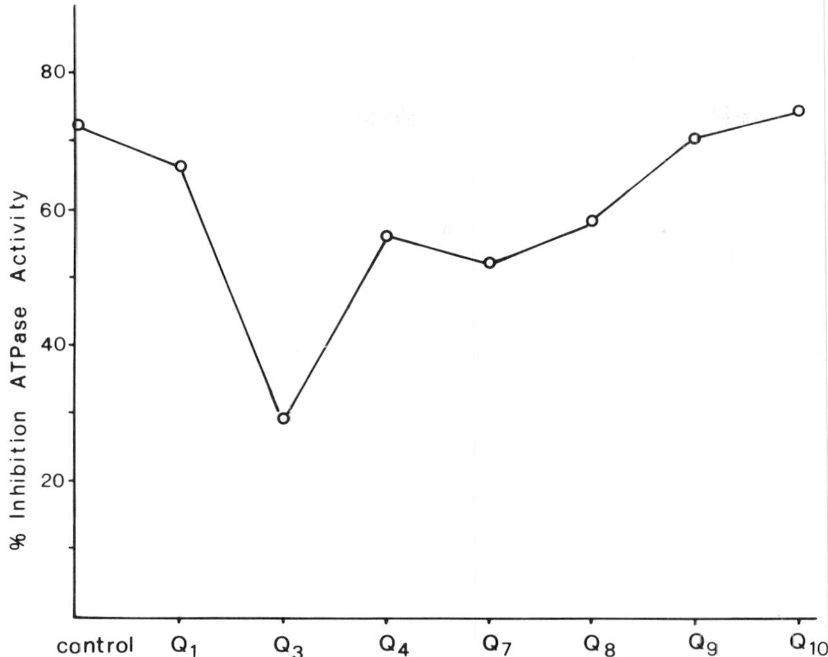

Fig. 6. Effects of different Q homologs on oligomycin inhibition of ATPase activity in SMP. Q concentration was 0.3 m*M*; oligomycin was 0.45 μg/mg protein.

electron shuttles between exogenous dithionite and internal ferricyanide. From both approaches a lower transmembrane mobility of the short-chain homologs in comparison with long-chain homologs was suggested. Such artificial systems, however, may be complicated by a number of factors, including the diverse incorporation of different homologs into the lipid core. We have confirmed an increased capability of external dithionite to reduce internal ferricyanide in presence of ubiquinones of increasing chain length. However, when comparing such results with the amounts of Q homologs incorporated by sonication into the lipid bilayer (by UV spectra as in Fig. 2), it appeared that the capability to transfer electrons across the membrane was related to the concentration of Q incorporated into the membrane core, regardless of the isoprenoid chain length (Fig. 7). Ubiquinones incorporated into liposomes prevent the reduction of internal ferricyanide by external ubiquinols; this effect is in accordance with the idea that ubiquinone partition into the membrane is the only determinant of transmembrane electron transfer.

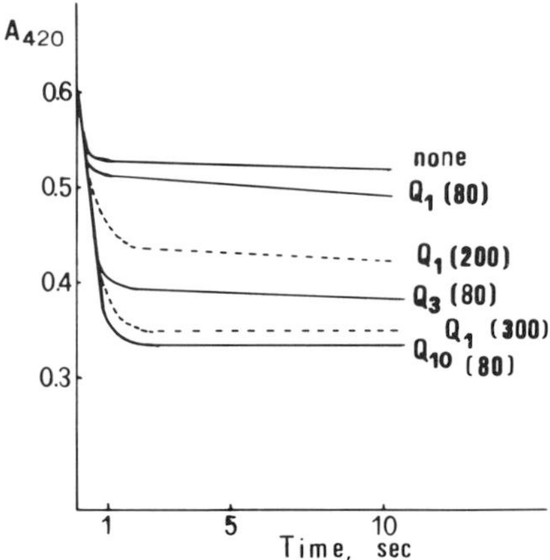

Fig. 7. Reduction by external sodium-dithionite of ferricyanide trapped inside vesicles of egg lecithin containing different Q homologs. PL (2 mg/ml) were sonicated as described by Futami *et al.* (*5*) in presence of 0.1 *M* K-ferricyanide and different ubiquinones. Numbers in parentheses refer to the initial concentration of ubiquinones in nmole/mg PL. After dialysis, aliquots were transferred to cuvettes and the reaction was followed in a Cary-15 spectrophotometer equipped with a rapid mixing apparatus, after addition of dithionite (10 m*M*) by recording the absorbance decrease at 420 nm due to reduction of ferricyanide. Vesicles sonicated with ubiquinones in the absence of ferricyanide were used to assay the amount of ubiquinone incorporated in the membrane phase as described in Fig. 2.

Sidedness of Ubiquinol Oxidation

Previous experiments (*4*) showed that short-chain ubiquinols are oxidized faster in inside-out submitochondrial particles than in BHM, suggesting that the site of ubiquinol oxidation in Complex III may be located near the matrix side of the membrane. Preliminary stopped-flow experiments performed in collaboration with Papa and co-workers in Bari, have clearly shown that endogenous cytochrome *c* reduction by exogenous ubiquinols is faster in SMP than in BHM; the effect is particularly striking with hydrophilic Q_0, but is found clearly also with ubiquinols up to four isoprenoid units; the initial rates and half-times are compatible with better accessibility of the matrix side of the membrane to exogenous ubiquinols (Table II).

DISCUSSION

The results of the present study caution any investigator using exogenous ubiquinone homologs to consider the partition of such molecules be-

TABLE II

Initial Oxidation Rates and Half-Times for Endogenous Cytochrome *c* Reduction by Exogenous Ubiquinols[A]

	BHM		SMP	
Q_nH_2	$t_{1/2}$ (msec)	V_i (nmole/min/mg protein)	$t_{1/2}$ (msec)	V_i (nmole/min/mg protein)
Q_0H_2 (125 μM)	8,300	1.8	1,100	8.1
DQH_2 (200 μM)	225	82	110	119
Q_1H_2 (28 μM)	211	111	93	476
Q_2H_2 (28 μM)	268	141	156	252
Q_3H_2 (40 μM)	600	35	260	211
Q_4H_2 (100 μM)	1,800	24	840	31

[A] The reduction of cytochrome *c* was followed at 550–540 nm in a dual wavelength spectrophotometer equipped with a stopped-flow apparatus at 25 °C using an extinction coefficient of 19.1 mM^{-1} cm^{-1}. The main chamber contained BHM (2 mg/ml) or SMP (1.5 mg/ml) in 200 mM sucrose and inhibited with rotenone and valinomycin (0.5 μg/mg protein), K-malonate (10 mM), oligomycin (2 μg/mg protein) and KCN (1 mM). Reaction was started by injection of the methanolic solution of quinol at a final quinol concentration in the range of the K_m values separately determined with isolated *b*-c_1 complex (Degli Esposti *et al.*, unpublished data used with permission). The final pH was 7.3. The values were corrected for antimycin-insensitive cytochrome *c* reductase (usually less than 5%). DQH_2 = duroquinol (Degli Esposti *et al.*, unpublished data, used with permission).

tween the water and membrane phases. Our method is suitable for distinguishing between incorporated and nonincorporated Q and for evaluating quantitatively the incorporated species only. The physical state of ubiquinones in water is nonmonomeric, shown by the restriction of the mobility of spin labels in presence of ubiquinones. A similar conclusion is supported by the red shift and decrease of absorbance of ubiquinones in water as compared with nonpolar media. When incorporated into lipid vesicles and mitochondria, ubiquinones have spectral features resembling those in isooctane (λ_{max} 270 nm) rather than in ethanol (λ_{max} 275 nm), indicating that the quinone ring is buried in the hydrophobic phase of the membrane and does not protrude towards the polar phase. Our results do not disagree with those of Szarkowska and Klingenberg (*19*) showing maximal absorption of endogenous Q_{10} in the 265–275 nm region; however, their spectra were built indirectly, not allowing resolution at 270 nm.

The deep location of ubiquinone in the hydrophobic region of the membrane is compatible with the apparent inaccessibility of mitochondrial endogenous ubiquinol to external ferricyanide (*20*) and the unreactivity of external ubiquinols with endogenous ubiquinol in an artificial system and in mitochondrial membranes (*21*).

Short-chain homologs, however, have a strong tendency to remain in the water phase: apparent partition coefficients may be calculated from incorporation studies, showing that the affinity of ubiquinones for the membrane increases with the length of the chain and is higher for mitochondrial membranes than for lipid vesicles. Even extensive sonication, although strongly enhancing the apparent partition, fails to induce total incorporation of short-chain ubiquinones. In the extreme case of Q_1 the large majority is found in a polar phase, that might be either bulk water or a polar compartment of the membrane

The greater incorporation of exogenous ubiquinones into mitochondrial membranes than into phospholipid vesicles might suggest an involvement of membrane proteins in the binding. This experimental evidence qualitatively agrees with the findings of "Q-binding proteins" (*22–24*), and indirectly supports the hypothesis that a fraction of protein-bound Q is involved during the ubiquinone redox processes (*25*).

The complex kinetics of incorporation (Fig. 3) may be related to the effects exerted directly by ubiquinones on the physical state of the membrane. Previous EPR studies in phospholipid vesicles (*13*) had shown that only reduced short-chain ubiquinones enhanced membrane order; all others induced disorganization. The present fluorescence polarization studies in both vesicles and membranes show that short-chain ubiquinones, both oxidized and reduced, enhance membrane viscosity, while long-chain ubiquinones fluidize the bilayer. These results suggest that short-chain ubiquinones insert their full length between the phospholipid molecules, possibly exposing the polar ring to the water phase (*11*, *13*), whereas this arrangement is forbidden to long-chain homologs due to excessive side-chain length. Such stacking could perhaps be partly responsible for the red-shifted peak found in the UV spectra of short ubiquinones in phospholipid vesicles.

The ordering effect induced by short homologs appears related to functional alterations in mitochondria, including loss of respiratory control (*18*) and oligomycin sensitivity of ATPase (*26*). It is conceivable that ubiquinones inserted between the lipids can affect, either by direct binding or indirectly through the lipids, the organization of multiprotein complexes such as ATPase.

A better knowledge of the state of different ubiquinone homologs in lipid membranes allows alternative interpretations of previous experiments (*3*, *5*) where short homologs were not effective in supporting transmembrane electron transfer in artificial systems. Such capability has now been found to be related to the concentration of ubiquinone incorporated into the hydrophobic milieu of the membrane (λ_{max} 270 nm), regardless of isoprenoid chain length. Therefore it is not easy to establish a correlation between rate of Q-mediated transmembrane electron movements and the lifetime of membrane-bound semiquinones (*27*)

Because of their high activities and relative impermeability through bilayers, lower homologs of ubiquinone are suitable for probing the location of ubiquinone oxidation and reduction sites. Moreover it has been shown in our laboratory that short-chain ubiquinones can interact directly with the redox sites without a need for endogenous Q_{10} (*21*). In collaboration with Papa and co-workers, we have shown (by a stopped-flow technique) that exogenous ubiquinols are oxidized with faster kinetics in inside–out SMP than in BHM. This finding supports our previous observations (*3*, *4*) and indicates that the oxidation site of ubiquinol is located near the matrix side of the membrane. The observed H^+ ejection into the external medium during reduction of Complex III (*28*) does not appear to be directly linked to ubiquinol oxidation. Two alternative hypotheses may be advanced: (1) ubiquinol oxidation at the matrix side is accompanied by H^+ ejection to the cytoplasmic side through a H^+ channel, or (2) H^+ ejection is the result of proton-translocating activity of the protein during electron flow (*28*). The above interpretations do not question the general views of the chemiosmotic hypothesis (*29*), but can indicate some details about the mechanism of reduction of Complex III. Although we do not have direct data bearing on the so-called Q cycle (*29*), we question the transmembrane mobility of ubiquinone as a direct determinant of the H^+ movements, as the function here of Q-binding proteins may be linked to scalar rather than vectorial effects.

ACKNOWLEDGMENTS

We are grateful to Hoffmann-La Roche, Basel, for the kind gifts of ubiquinones. This study has been supported in part by grants from the C.N.R., and from the Ministero della Pubblica Istruzione, Roma.

REFERENCES

1. Lenaz, G., Daves, D. G., and Folkers, K. (1968). *Arch. Biochem. Biophys.* **123,** 539–550.

2. Lenaz, G., Pasquali, P., Bertoli, E., Parenti-Castelli, G., and Folkers, K. (1975). *Arch. Biochem. Biophys.* **169,** 217–226.

3. Lenaz, G., Mascarello, S., Landi, L., Cabrini, L., Pasquali, P., Parenti-Castelli, G., Sechi, A. M., and Bertoli, E. (1977). *In* "Bioenergetics of Membranes" (L. Packer, G. C. Papageorgiou, and A. Trebst, eds.), pp. 189–198. Elsevier, North-Holland, Amsterdam..
4. Lenaz, G., Landi, L., Cabrini, L., Pasquali, P., Sechi, A. M., and Ozawa, T. (1978). *Biochem. Biophys. Res. Commun.* **85,** 1047–1053.
5. Futami, A., Hurt, E., and Hauska, G. (1979). *Biochim. Biophys. Acta* **547,** 583–596.
6. Crane, F. L. (1977). *Annu. Rev. Biochem.* **46,** 439–469.
7. Rieske, J. S. (1967). *In* "Oxidation and Phosphorylation" (R. W. Estabrook and M. E. Pullman, eds.), Methods in Enzymology, Vol. 10, pp. 239–245. Academic Press, New York.
8. Nelson, D., and Gellerfors, P. (1978). *In* "Biomembranes: Part D: Biological Oxidations, Mitochondrial and Microbial Systems" (S. Fleischer and L. Packer, eds.), Methods in Enzymology, Vol. 53, pp. 80–91. Academic Press, New York.
9. Malviya, A. N., Nicholls, P., and Elliott, W. B. (1980). *Biochim. Biophys. Acta* **589,** 137–149.
10. Lang, B., Burger, G., and Bandlow, W. (1974). *Biochim. Biophys. Acta* **368,** 71–85.
11. Cain, J., Santillan, G., and Blasie, J. K. (1972). *In* "Membrane Research" (C. F. Fox, ed.), pp. 3–14. Academic Press, New York.
12. Chance, B., Erecinska, M., and Radda, G. (1975). *Eur. J. Biochem.* **54,** 521–529.
13. Spisni, A., Masotti, L., Lenaz, G., Bertoli, E., Pedulli, G. F., and Zannoni, C. (1978). *Arch. Biochem. Biophys.* **190,** 454–458.
14. Seelig, J. (1976). *In* "Spin Labeling, Theory and Applications" (L. Berliner, ed.), pp. 373–409. Academic Press, New York.
15. Kröger, A., and Klingenberg, M. (1978). *In* "Biomembranes: Part D: Biological Oxidations, Mitochondrial and Microbial Systems" (S. Fleischer and L. Packer, eds.), Methods in Enzymology, Vol. 53, pp. 580–591. Academic Press, New York.
16. Szarkowska, L. (1966). *Arch. Biochem. Biophys.* **113,** 519–525.
17. Shinitsky, N., Dianow, A. G., Gitler, C., and Weber, G. (1971). *Biochemistry* **10,** 2106–2113.
18. Bertoli, E., Parenti-Castelli, G., Sechi, A. M., Trigari, G., and Lenaz, G. (1978). *Biochem. Biophys. Res. Commun.* **85,** 1–6.
19. Szarkowska, L., and Klingenberg, M. (1963). *Biochem. Z.* **338,** 674–684.
20. Harmon, H. J., Hall, J. D., and Crane, F. L. (1974). *Biochim. Biophys. Acta* **344,** 119–155.
21. Cabrini, L., Landi, L., Pasquali, P., and Lenaz, G. (1981). *Arch. Biochem. Biophys.* **209,** 11–29.
22. Yu, C.-A., Yu, L., and King, T. (1977). *Biochem. Biophys. Res. Commun.* **78,** 259–265.
23. Yu, C.-A., and Yu, L. (1980). *Biochim. Biophys. Acta* **591,** 409–420.
24. Nishi, N., Kataoka, M., Soe, G., Kakuno, T., Ueki, T., Yamashita, J., and Horio, T. (1979). *J. Biochem. (Tokyo)* **86,** 1211–1224.
25. Baccarini-Melandri, A., Gabellini, N., Melandri, B. A., Hurt, E., Hauska, G., Rutherford, W. A., and Crofts, A. K. (1980). *Eur. Bioenerg. Conf., 1st, Short Rep., Urbino* pp. 11–13.
26. Degli Esposti, M., Bertoli, E., Parenti-Castelli, G., and Lenaz, G. (1980). *J. Bioenerg. Biomembr.* **13,** 37–50.
27. Futami, A., and Hauska, G. (1979). *Biochim. Biophys. Acta* **547,** 583–596.
28. Papa, S. (1976). *Biochim. Biophys. Acta* **456,** 39–84.
29. Mitchell, P. (1975). *FEBS Letts.* **56,** 1–6.

Lateral Diffusion of Ubiquinone in Mitochondrial Electron Transfer

3

HEINZ SCHNEIDER
JOHN J. LEMASTERS
CHARLES R. HACKENBROCK

INTRODUCTION

Since its discovery some 20 years ago, ubiquinone has become recognized as a component required for mitochondrial electron transfer (*1–5*). However, whether ubiquinone operates as an independent, mobile electron-transfer component or as a prosthetic group of a redox protein remains an important, unsettled question.

The concept that ubiquinone is a mobile electron carrier was introduced by Green and Wharton (*6*). This concept was adopted by Kröger and Klingenberg (*7, 8*), who demonstrated that 80–90% of the total mitochondrial ubiquinone exists in a common pool in the membrane and reacts in a functionally homogeneous fashion. The mobile carrier concept implies lateral diffusion of the ubiquinone molecule for the purpose of shuttling electrons between spatially separated membrane dehydrogenase and cytochrome b-c_1 complexes. The mobile carrier concept, however, was not compatible with the generally held notion that the structural organization of the mitochondrial membrane approached a solid-state system, and that the specific proteins of the electron-transfer sequence were immobilized in a rigid protein–protein lattice (*9–11*). Not until a few years ago was it realized that the mitochondrial membrane is in a highly fluid state and that its inte-

Function of Quinones in Energy Conserving Systems

ISBN 0-12-701280-X

gral proteins can diffuse laterally and independently of one another (*12*, *13*). These recent observations (*12*, *13*) confer additional meaning to the mobile electron carrier concept, *i.e.*, that lateral diffusion and collisional interaction of redox protein complexes with ubiquinone and other redox protein complexes may be involved in electron transfer.

Recently, Ragan and co-workers (*14*, *15*) proposed that electron transfer from NADH-ubiquinone reductase (Complex I) to ubiquinol-cytochrome *c* reductase (Complex III) occurs through rapid association and dissociation of Complex I–Complex III units at rates corresponding to electron-transfer rates from NADH to cytochrome *c*. In such a system, ubiquinone is thought to function as a coenzyme, and independent lateral mobility is not necessarily required. Another challenge to the ubiquinone mobile carrier concept emerged from the recent discovery of a stable form of ubisemiquinone in succinate-linked electron-transfer reactions (*16*). It was argued that ubiquinone radicals can exist only when stabilized by a protein (*17*). Indeed, Yu *et al.* (*18*, *19*) isolated a hydrophobic, "ubiquinone binding protein," and suggested that this protein serves as the true electron carrier with ubiquinone functioning as its prosthetic group (*20*). Although considerable evidence exists for protein diffusion in the mitochondrial membrane (*12*, *13*, *21*, *22*), independent diffusion of ubiquinone has not been determined in any convincing manner.

In a recent report from this laboratory (*21*), we determined that a diffusion mediated step occurs in the electron-transfer segments from membrane dehydrogenases to b-c_1 cytochromes. To increase the surface area of the inner-membrane bilayer we employed a new method. By fusion of exogenous lipids with the native membrane (*22*), we demonstrated that decreases in electron-transfer rates from NADH (as well as succinate) to cytochrome *c* occur. These rate decreases are proportionally related to an increase in the average distance between integral proteins as the bilayer surface area is increased (*21*). The data indicated that membrane dehydrogenase complexes, cytochrome b-c_1 complexes, and ubiquinone are diffusible redox components. However, the data did not reveal whether ubiquinone diffuses independently to collide with diffusing redox proteins, or exists as a prosthetic group of diffusing proteins. To distinguish between these two possibilities we have utilized ubiquinone-enriched liposomes for fusion with mitochondrial inner membranes. In this report we describe the effects of ubiquinone incorporation on electron transfer from NADH and succinate to cytochrome *c*, demonstrating that ubiquinone is an independent and diffusible redox component of the mitochondrial inner membrane.

EXPERIMENTAL RESULTS

Enrichment of Mitochondrial Inner Membranes with Phospholipid and Ubiquinone

The mitochondrial inner membranes were obtained by removal of the outer membranes from isolated rat liver mitochondria using a controlled digitonin incubation (*23*). The topographically complex configuration of the inner membrane was converted to a simple spherical configuration by repeated washing in diluted isolation medium (*24*). Small, unilamellar liposomes were prepared by sonication as previously described (*22*). Ubiquinone-containing liposomes were obtained by dissolving phospholipid (azolectin) together with ubiquinone (Q_{10}, Q_6, or Q_3)[1] in chloroform, evaporating the solvent, hydrating the dried lipid mixture, and sonicating the mixture (*21*, *22*). Four buoyant-distinct membrane fractions were collected after sucrose density gradient centrifugation of the liposome–membrane fusion mixture (*21*, *22*). These four fractions were designated Band 1, Band 2, Band 3, and Pellet in order of increasing density.

Structural and Compositional Analysis of the Four Phospholipid and Ubiquinone Enriched Membrane Fractions

Ultrastructural analysis of the four buoyant-distinct membrane fractions revealed an increase in the membrane surface area as the buoyancy of the membranes decreased (*22*). Related to the increase in membrane surface area and decrease in buoyant density, the average distance between the integral proteins increased as shown in Fig. 1. These observations indicate that the membrane lipid bilayer became enriched with exogenous phospholipid during fusion with liposomes, causing an increase in the membrane surface area. As the integral proteins re-randomized by lateral diffusion into the newly expanded bilayer, the average distance between them increased in proportion to the increase in bilayer lipid. The lipid phosphorus-to-heme *a* ratio increased progressively from Pellet to Band 1, with a six-fold increase in membrane bilayer lipid in Band 1 as compared to Pellet (Table I). In phos-

[1] Abbreviations: Q_{10}, ubiquinone-10; Q_6, ubiquinone-6; Q_3, ubiquinone-3; Q_1, ubiquinone-1; DBH, ubiquinone having a decyl side chain; PMS, phenazine methosulphate; DCIP, dichlorophenolindophenol; TTFA, thenoyltrifluoroacetone.

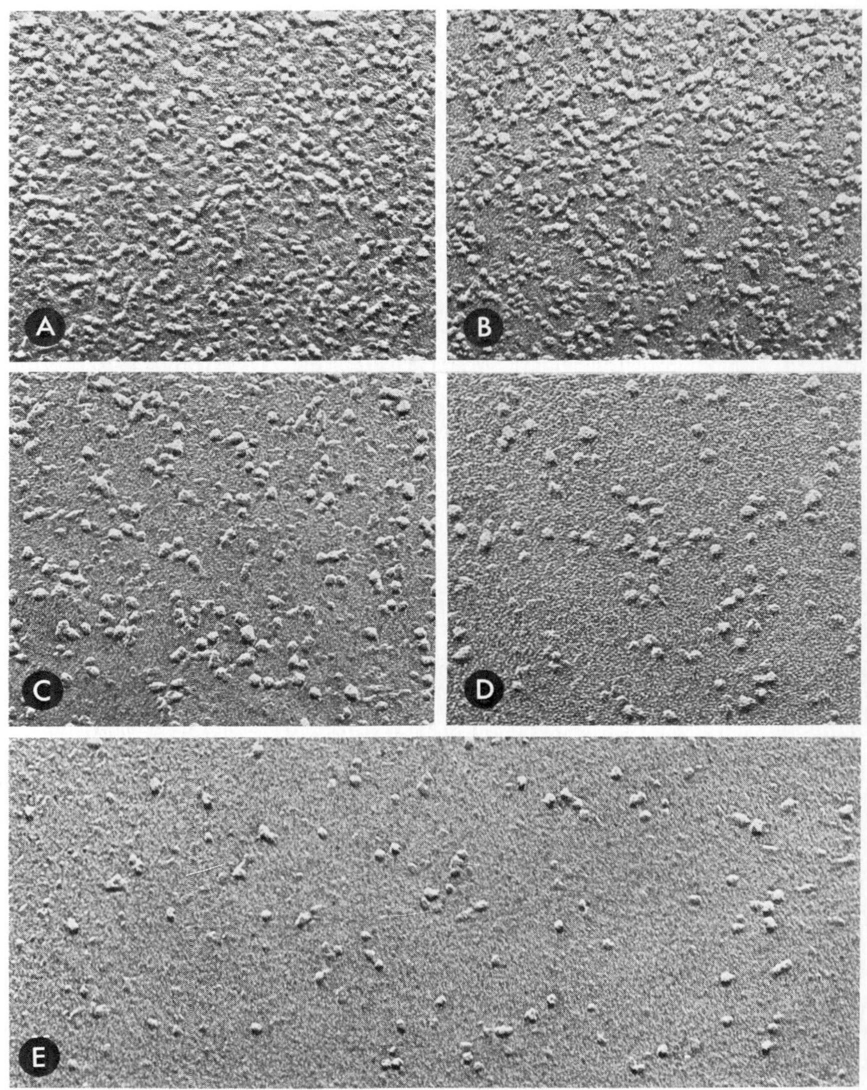

Fig. 1. Freeze fracture convex faces of mitochondrial inner membranes from an untreated inner-membrane fraction (control) and the four buoyant-distinct, lipid-enriched membrane fractions after fusion with liposomes.
(A) Untreated inner membrane fraction (control); (B) Pellet fraction; (C) Band 3 fraction; (D) Band 2 fraction; (E) Band 1 fraction.

TABLE I

Compositional Analysis of Phospholipid and Phospholipid plus Ubiquinone-Enriched Mitochondrial Inner Membranes[A]

	Molar ratio of lipid phosphorus to heme *a* × 10^3	Molar ratio of ubiquinone (Q) to heme *a*			
		Phospholipid only (control)	Phospholipid + Q_{10}	Phospholipid + Q_6	Phospholipid + Q_3
Pellet	1.1	7.5	13.3	12.8	23.7
Band 3	1.6	7.6	18.7	18.1	31.1
Band 2	3.3	8.1	35.0	41.6	43.1
Band 1	7.5	8.6	89.8	72.1	85.5

[A] Lipid phosphorus was determined according to Bartlett (*25*). Heme *a* values were measured as described in Schneider *et al.* (*21*) using an extinction coefficient of 13.1 mM^{-1}cm^{-1} (*26*). Ubiquinone content was measured after Kröger and Klingenberg (*27*).

pholipid-enriched membranes (control) the ubiquinone-to-heme *a* ratios remained constant; however, in phospholipid plus ubiquinone enriched membranes, the ubiquinone-to-heme *a* ratios increased in proportion to the increase in lipid phosphorus (Table I). In all ubiquinone-enriched membrane fractions, a ratio of approximately 10 ubiquinone molecules per 1000 phospholipid molecules was maintained (Table II); this corresponds closely to the ubiquinone concentration in native membranes (*21*). These results demonstrate that phospholipid and ubiquinone incorporation occur simultaneously and at a constant ratio. Only in Q_3 enriched Pellet and Band 3 was a higher ubiquinone-to-phospholipid ratio observed. Since the hydrophobicity of ubiquinone decreases as a function of side-chain length (*28*), it is possible that the lower hydrophobicity of Q_3 may facilitate its transfer between liposomes and membranes by means other than fusion.

TABLE II

Molar Ratios of Ubiquinone to Lipid Phosphorus in Phospholipid and Phospholipid plus Ubiquinone-Enriched Mitochondrial Inner Membranes[A]

	Phospholipid only (control)	Phospholipid + Q_{10}	Phospholipid + Q_6	Phospholipid + Q_3
Pellet	6.3	11.1	10.6	19.8
Band 3	4.7	11.5	11.3	19.4
Band 2	2.6	10.6	12.6	13.1
Band 1	1.2	11.9	9.6	11.4

[A] nmole ubiquinone (Q)/μmole lipid phosphorus.

Effects of Newly Incorporated Ubiquinone on Electron Transfer from Ubiquinol (DBH) to Cytochrome *c*

In control membranes enriched only with phospholipid, ubiquinol-cytochrome *c* reductase activity decreased in proportion to the increase in bilayer lipid. However, incorporation of ubiquinone along with phospholipid caused a complete restoration of activity in all lipid-enriched membrane fractions other than Band 1 (Fig. 2). Band 1 did, however, show a significant recovery in activity when compared with control Band 1. Ubiquinone incorporation did not alter antimycin inhibition of ubiquinol-cytochrome *c* reductase, which was greater than 95% in all membrane fractions. These results show that endogenous ubiquinone is required for electron transfer from DBH through the membrane-bound cytochrome b-c_1 complex to cytochrome *c*. Further, the results show that adequate enrichment of the membrane with ubiquinone restores the decrease in electron-transfer activity that accompanies the lateral dilution of endogenous ubiquinone and redox proteins by phospholipid enrichment alone. This finding provides good evidence that ubiquinone is a diffusible redox component.

Effects of Newly Incorporated Ubiquinone on NADH-Linked Electron Transfer

Ubiquinone-3 incorporation caused a substantial loss of rotenone sensitivity in NADH-ubiquinone reductase. Presentation and discussion of this and other effects of Q_3 on NADH-linked electron transfer exceed the purpose of this report and will be described elsewhere.

As shown in Fig. 3, NADH dehydrogenase activities in the four buoyant-distinct, phospholipid-enriched membranes, as well as phospholipid plus ubiquinone-enriched membranes, increased in proportion to the increase in bilayer lipid. However, either Q_{10} or Q_6 enrichment, caused an average increase in the turnover number in all fractions. Although the mechanisms underlying these increases in activity and turnover number are unknown, it appears that NADH dehydrogenase is influenced by both the local lipid environment and the concentration and type of ubiquinone in the membrane.

In control, phospholipid-enriched membranes, NADH-ubiquinone reductase activity decreased after incorporation of relatively large amounts of phospholipid, *i.e.*, in Band 1 and Band 2 membranes. As shown in Fig. 4, incorporation of Q_{10} and Q_6 along with phospholipid restored this activity in Band 2 but not in Band 1. This lack of restora-

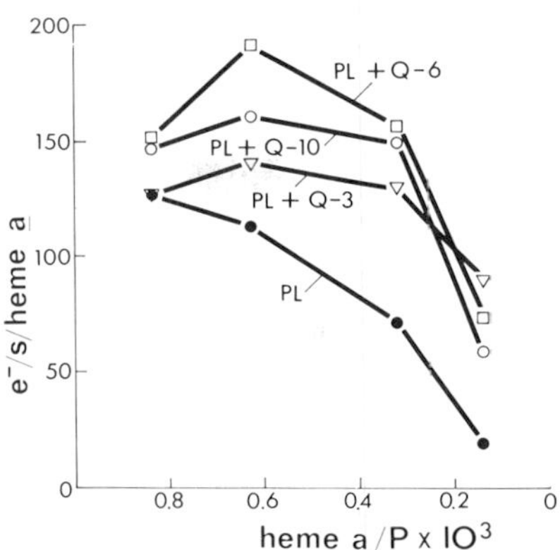

Fig. 2. Specific activity of ubiquinol-cytochrome *c* reductase in control, phospholipid enriched membranes (PL) and phospholipid plus ubiquinone enriched membranes (PL + Q_{10}, PL + Q_6, PL + Q_3) in relation to the heme *a*-to-lipid phosphorus (P) ratio. Ubiquinol-cytochrome *c* reductase activity was determined with reduced DBH as electron donor as described in Schneider *et al.* (*21*).

tion in Band 1, similar to the incomplete restoration of ubiquinol-cytochrome *c* reductase activity in ubiquinone-enriched Band 1 (Fig. 2), may represent a direct effect of the foreign bilayer lipid on the intrinsic electron-transfer capacity of the protein complexes due to the large amount of nonmitochondrial lipid present in Band 1 membranes.

NADH-cytochrome *c* reductase activity was the most sensitive of all NADH-linked activities to phospholipid enrichment of the membrane

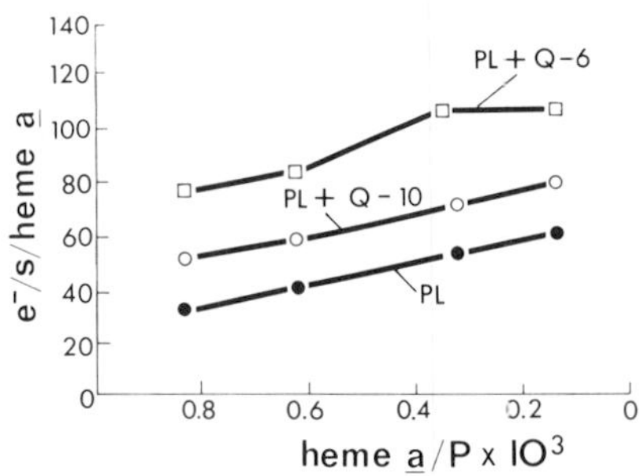

Fig. 3. Specific activity of NADH dehydrogenase in control, phospholipid-enriched membranes (PL), and phospholipid plus ubiquinone enriched membranes (PL + Q_{10}, PL + Q_6) in relation to the heme *a*-to-lipid phosphorus (P) ratio. Determination of NADH dehydrogenase activity was carried out with ferricyanide as electron acceptor as described in Schneider *et al.* (*21*).

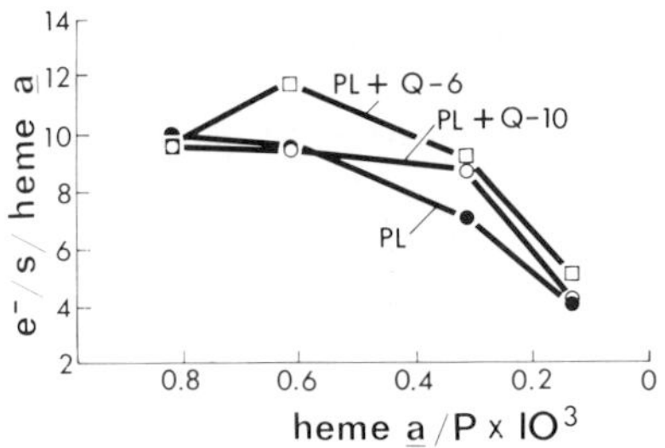

Fig. 4. Specific activity of NADH-ubiquinone reductase in control, phospholipid-enriched membranes (PL) and phospholipid plus ubiquinone enriched membranes (PL + Q_{10}, PL + Q_6) in relation to the heme *a*-to-lipid phosphorus (P) ratio. Ubiquinone-1 was used as electron acceptor for determination of NADH-ubiquinone reductase activity as described in Schneider *et al.* (*21*, used with permission).

and decreased sharply in proportion to the increase in bilayer lipid. Compared to the control phospholipid-enriched membranes, Q_{10} and Q_6 incorporation increased NADH-cytochrome *c* reductase activity in all fractions except Pellet (Fig. 5). This tendency of the newly incorporated ubiquinone to restore NADH-cytochrome *c* reductase activity that was previously lost upon phospholipid incorporation alone, strongly supports the view that ubiquinone diffusion mediates electron transfer between independently diffusing NADH dehydrogenase complexes and cytochrome b-c_1 complexes. However, ubiquinone enrichment failed to completely restore NADH-cytochrome *c* reductase activity. That the newly incorporated ubiquinone did not completely restore

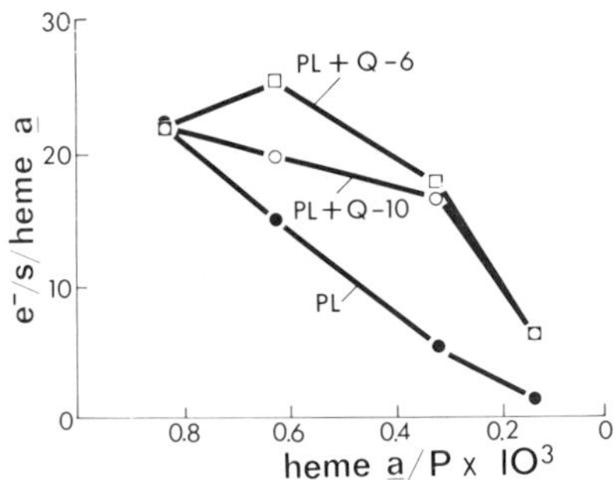

Fig. 5. Specific activity of NADH-cytochrome *c* reductase in control, phospholipid-enriched membranes (PL), and phospholipid plus ubiquinone enriched membranes (PL + Q_{10}, PL + Q_6) in relation to the heme *a*-to-lipid phosphorus (P) ratio. The assay for NADH-cytochrome *c* reductase activity was described in Schneider *et al.* (*21*, used with permission).

NADH-cytochrome *c* reductase activity in highly lipid-enriched Bands 2 and 1 membranes is most likely related to the increase in the average distance between NADH dehydrogenase complexes and cytochrome *b*-c_1 complexes as the membrane bilayer surface area increases upon phospholipid incorporation.

Effects of Newly Incorporated Ubiquinone on Succinate-Linked Electron Transfer

Succinate dehydrogenase can be activated by succinate as well as by ubiquinone, especially in its reduced from (*29*). Therefore, we routinely incubated membranes at room temperature in the presence of 20 m*M* succinate before determining succinate-linked electron-transfer reactions. Figure 6 shows that different ubiquinones incorporated with phospholipid affected the activity of succinate dehydrogenase (here determined as the TTFA insensitive portion of succinate-PMS reductase) in different ways. Compared to control membranes enriched with phospholipid only, in which the activity decreased in proportion to the lipid increase in the membrane bilayer, Q_{10} enriched membranes showed a further decrease in activity. Ubiquinone-6 incorporation, however, tended to increase activity, especially in Bands 3, 2, and 1. Ubiquinone-3 incorporation showed virtually no difference compared to control. These results suggest that different degrees of activation of succinate dehydrogenase occur in relation to the chain length of the incorporated ubiquinone molecule. Ubiquinone enrichment did not generally reverse the decrease in succinate dehydrogenase activity induced by the increase in membrane phospholipid content alone.

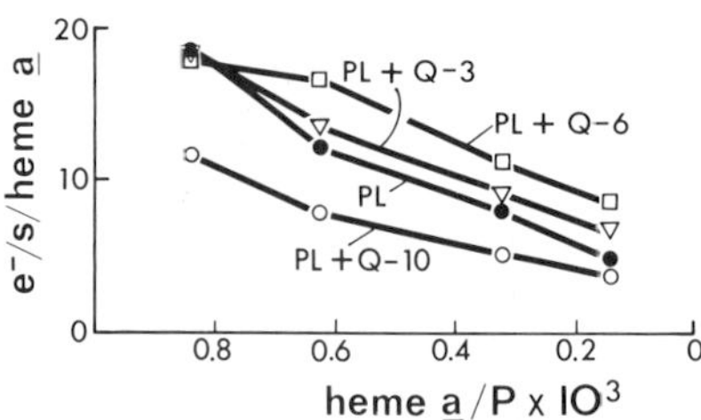

Fig. 6. Specific activity of succinate dehydrogenase in control, phospholipid-enriched membranes (PL), and phospholipid plus ubiquinone enriched membranes (PL + Q_{10}, PL + Q_6, PL + Q_3) in relation to the heme *a*-to-lipid phosphorus (P) ratio. Succinate dehydrogenase activity was determined as the TTFA-insensitive portion of succinate-PMS reductase activity employing DCIP as terminal electron acceptor, as described in Schneider *et al.* (*21*, used with permission).

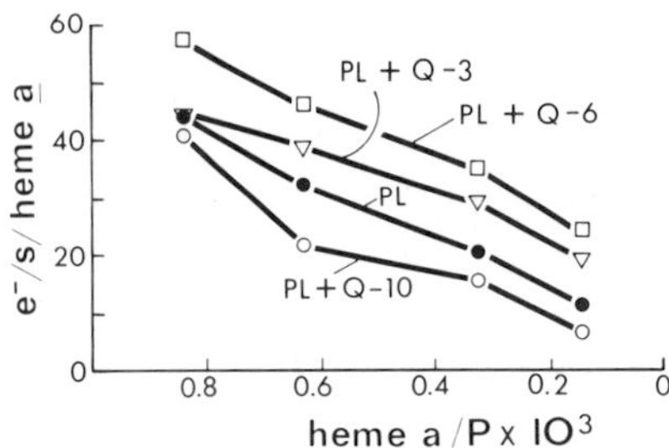

Fig. 7. Specific activity of succinate-ubiquinone reductase in control, phospholipid-enriched membranes (PL), and phospholipid plus ubiquinone enriched membranes (PL + Q_{10}, PL + Q_6, PL + Q_3) in relation to the heme *a*-to-lipid phosphorus (P) ratio. Ubiquinone was the mediator for determining succinate-ubiquinone reductase activity, employing DCIP as terminal electron acceptor, as described in Schneider *et al.* (*21*, used with permission).

Similar effects of ubiquinone incorporation were observed in electron transfer from succinate to Q_1 (Fig. 7). Ubiquinone-10 enrichment lowered the activity of succinate-ubiquinone reductase in all fractions, compared to control, while Q_3 and Q_6 enhanced activity somewhat. These observations on succinate-ubiquinone reductase activity again demonstrate the different modulatory effects of the various species of ubiquinone on succinate dehydrogenase activity. As with NADH-cytochrome *c* reductase activity, succinate-cytochrome *c* reductase activity was highly sensitive to incorporation of exogenous phospholipid

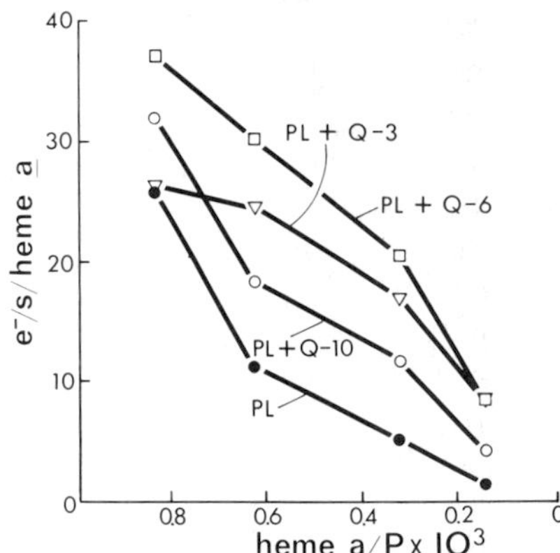

Fig. 8. Specific activity of succinate-cytochrome *c* reductase in control, phospholipid-enriched membranes (PL), and phospholipid plus ubiquinone enriched membranes (Q_{10}, Q_6, Q_3) in relation to the heme *a*-to-lipid phosphorus (P) ratio. Measurements of succinate-cytochrome *c* reductase activity were carried out as described in Schneider *et al.* (*21*, used with permission).

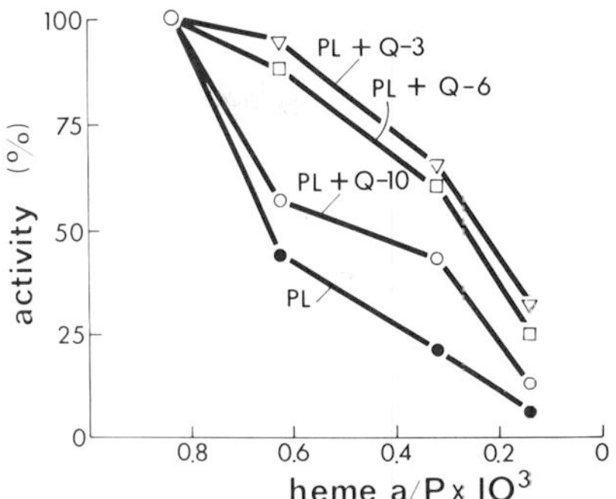

Fig. 9. Comparison of relative percent activities of succinate-cytochrome *c* reductase in control, phospholipid-enriched membranes (PL), and phospholipid plus ubiquinone enriched membranes (PL + Q_{10}, PL + Q_6, PL + Q_3) in relation to the heme *a*-to-lipid phosphorus (P) ratio.

into the membrane lipid bilayer. In control phospholipid-enriched membranes, succinate-cytochrome *c* reductase activity decreased dramatically as the lipid content of the membrane increased. As shown in Fig. 8, Q_3, Q_6, and Q_{10} enrichment resulted in substantial increases of succinate-cytochrome *c* reductase activity in all but one membrane fraction. To distinguish between modulatory effects of succinate dehydrogenase and a diffusion-linked restoration of activity, we compared the percent activity of succinate-cytochrome *c* reductase in ubiquinone-enriched and control membranes. Figure 9 reveals that all three classes of newly incorporated ubiquinone partly restore activity that is decreased upon phospholipid incorporation alone. It is noteworthy that Q_3, and to a slightly lesser degree Q_6, resulted in a greater restoration than Q_{10}, indicating that ubiquinone chain length is the factor determining the degree of restoration. To a somewhat lesser extent, there is also a chain-length dependence in the ubiquinone-mediated increase of NADH-cytochrome *c* reductase activity (Fig. 5). The increased ability of ubiquinones with shorter chain lengths to partly restore electron transfer from membrane dehydrogenase complexes to cytochrome b-c_1 complexes may be related to the greater lateral mobilities of these short-chain ubiquinones, since lateral mobility of a lipid molecule is inversely proportional to the square of its length (*30*).

DISCUSSION

The results presented here on the effects of ubiquinone and phospholipid incorporation into the mitochondrial inner membrane reveal a diffusional mechanism of electron transfer involving the membrane dehydrogenase complexes, cytochrome b-c_1 complexes, and ubiquinone. Incorporation of exogenous phospholipid into the mitochondrial inner membrane resulted in an increase in the membrane surface area. This was accompanied by an increase in the average distance between randomly distributed integral proteins, suggesting free diffusion of the proteins into the newly expanded membrane bilayer. Related to these increased distances between the integral proteins, the electron-transfer rates from NADH and succinate to cytochrome *c* decreased in proportion to the increase in membrane bilayer lipid. This data indicates that membrane dehydrogenase complexes, cytochrome b-c_1 complexes, and ubiquinone interact as diffusing electron-transfer components.

Incorporation of ubiquinone along with phospholipid resulted in a partial restoration of the decreases in NADH- and succinate-cytochrome *c* reductase activity that occurred upon incorporation of phospholipid alone. In addition, ubiquinone incorporation together with phospholipid restored the decrease in electron transfer from DBH, a reduced ubiquinone analog, to cytochrome *c* that occurred after enrichment of the inner membrane with phospholipid alone. The ability of newly incorporated ubiquinone to recover electron-transfer activities that were decreased by enriching the membrane bilayer exclusively with phospholipid establishes ubiquinone as an individual, diffusible electron-transfer component. The demonstration that the extent of recovery of NADH- and succinate-cytochrome *c* reductase activity was related to the side-chain length of the newly incorporated ubiquinone molecules, with the shorter side-chain homologs giving greater degrees of recovery, strongly supports the view that ubiquinone diffusion mediates electron transfer between independently diffusing membrane dehydrogenase complexes and cytochrome b-c_1 complexes. Our findings represent the first definitive evidence that ubiquinone is an independent, mobile electron carrier since the early proposals of Green and Wharton (*6*), and Kröger and Klingenberg (*7*, *8*), that suggested ubiquinone was a mobile electron carrier operating in a functionally homogeneous fashion in a common pool between the dehydrogenase and b-c_1 complexes.

Our results do not appear to support other proposals on the function of ubiquinone as a coenzyme (*14*, *15*) or as a prosthetic group of a protein that serves as the true electron carrier between the dehydrogenases and the b-c_1 cytochromes (*20*). In view of our observation that newly incorporated ubiquinone participates in electron-transfer reactions in addition to native ubiquinone, we expect that the majority of ubiquinone is unbound and freely mobile. Of course, we do not exclude the possibility of some protein-bound forms of ubiquinone (*16*, *17*). In fact, such forms would be consistent with the observation of Kröger and Klingenberg (7) of two ubiquinone populations in the mitochondrial inner membrane, a large pool (80–90%) that reacts in a kinetically homogeneous fashion and a small pool (10–20%) that is nonreducible by NADH or succinate. These two populations may represent free ubiquinone and protein-bound ubiquinone, respectively.

As illustrated in Fig. 10, our model of electron transfer from NADH and succinate to cytochrome *c* includes the following features

1. Individual electron-transfer components, such as the membrane dehydrogenase complexes, cytochrome b-c_1 complexes, and ubiquinone diffuse randomly and independently of one another in the membrane.
2. Ubiquinone transfers electrons from the dehydrogenase complexes to cytochrome b-c_1 complexes through diffusion and collisional interaction with these complexes.
3. Protein–protein interactions between the dehydrogenase complexes and cytochrome b-c_1 complexes are not required for electron transfer.

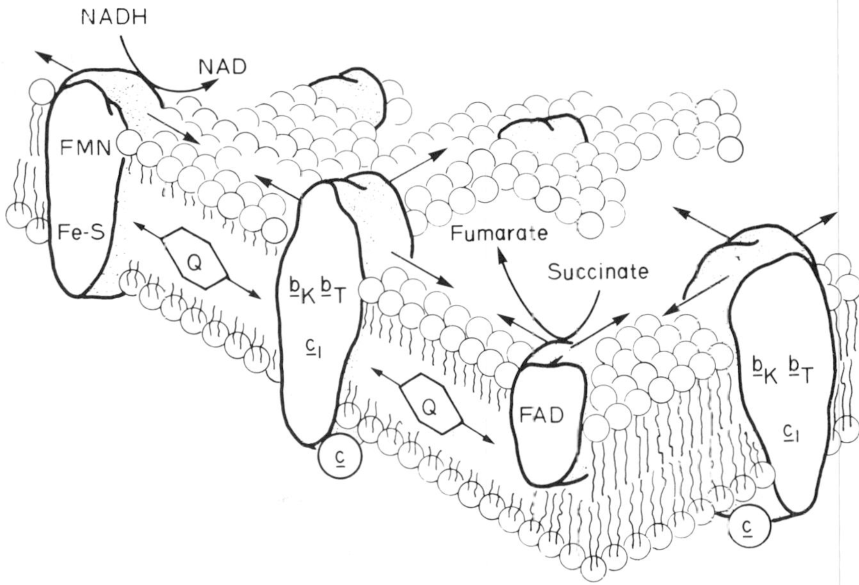

Fig. 10. Diffusional model of electron transfer between membrane dehydrogenases and b-c_1-cytochromes. Lateral diffusion of ubiquinone (Q) results in electron transfer between independently diffusing membrane dehydrogenase complexes (NADH – NAD; succinate-fumarate) and cytochrome b-c_1 (b_K, b_T, c_1) complexes.

Our model accounts for the nonstoichiometric relationship of the membrane dehydrogenases to the b-c_1-cytochromes (*31*), the abundance of ubiquinone in the membrane in relation to the other electron-transfer components (*6*, *31*), and the reaction of the ubiquinone pool in a kinetically homogeneous fashion (*7*, *8*). It explains the decreases in NADH- and succinate-cytochrome *c* reductase activities that result from increasing the average distance between interacting redox components upon phospholipid enrichment of the membrane, and the partial restoration of these activities by additional ubiquinone incorporation. The model also explains the influence of the ubiquinone chain length on these activities, since the diffusion rate of ubiquinone should be chain-length dependent. Finally, our model is consistent with studies of electron transfer between isolated complexes of succinate-ubiquinone reductase and ubiquinol-cytochrome *c* reductase in a nonionic detergent system, that revealed direct complex–complex interactions are not necessary for electron transfer to occur (*32*).

ACKNOWLEDGMENTS

We thank Linda Fuller and Phil Ives for their expert technical assistance, Dr. M. Höchli for freeze-fracture electron microscopy, Dr. B. L. Trumpower, Department of Biochemistry, Dartmouth Medical School, New Hampshire for the ubiquinone analog DBH, and Hoffman-La Roche, Basel, Switzerland for ubiquinone-1 and ubiquinone-3. This investigation was supported by Research Grants PCM 77-20689 and PCM 79-10968 from the National Science Foundation, and GM-28704 from the National Institutes of Health, to C.R.H.

REFERENCES

1. Lester, R. L., and Fleischer, S. (1959). *Arch. Biochem. Biophys.* **80,** 470–473.
2. Szarkowska, L. (1966). *Arch. Biochem. Biophys.* **113,** 519–525.
3. Ernster, L., Lee, I.-Y., Norling, B., and Persson, B. (1969). *Eur. J. Biochem.* **9,** 299–310.
4. Rossi, E., Nelson, B., Persson, B., and Ernster, L. (1970). *Eur. J. Biochem.* **16,** 508–513.
5. Yu, L., Yu, C.-A., and King, T. E. (1978). *J. Biol. Chem.* **8,** 2657–2663.
6. Green, D. E., and Wharton, D. C. (1963). *Biochem. Z.* **338,** 335–348.
7. Kröger, A., and Klingenberg, M. (1973). *Eur. J. Biochem.* **34,** 358–368.
8. Kröger, A., and Klingenberg, M. (1973). *Eur. J. Biochem.* **39,** 313–323.
9. Fleischer, S., Fleischer, B., and Stoeckenius, W. (1967). *J. Cell Biol.* **32,** 193–208.
10. Sjöstrand, F. S., and Barajas, L. (1970). *J. Ultrastruct. Res.* **32,** 293–306.
11. Capaldi, R. A., and Green, D. E. (1972). *FEBS Lett.* **25,** 205–209.

12. Höchli, M., and Hackenbrock, C. R. (1976). *Proc. Natl. Acad. Sci. U.S.A.* **73,** 1636–1640.
13. Höchli, M., and Hackenbrock, C. R. (1978). *Proc. Natl. Acad. Sci. U.S.A.* **76,** 1236–1240.
14. Ragan, C. I., and Heron, C. (1978). *Biochem. J.* **174,** 783–790.
15. Heron, C., Ragan, C. I., and Trumpower, B. L. (1978). *Biochem. J.* **174,** 791–800.
16. Ruzicka, F. J., Beinert, H., Schefler, K. L., Dunham, W. R., and Sands, R. H. (1975). *Proc. Natl. Acad. Sci. U.S.A.* **72,** 2886–2890.
17. Salerno, J. C., Harmon, H. J., Blum, H., Leigh, J. S., and Ohnishi, T. (1977). *FEBS Lett.* **82,** 179–182.
18. Yu, C.-A., Yu, L., and King, T. E. (1977). *Biochem. Biophys. Res. Commun.* **78,** 259–265.
19. Yu, C.-A., and Yu, L. (1980). *Biochemistry* **19,** 3579–3585.
20. Yu, C.-A., Nagaoka, S., Yu, L., and King, T. E. (1978). *Biochem. Biophys. Res. Commun.* **82,** 1070–1078.
21. Schneider, H., Lemasters, J. J., Höchli, M., and Hackenbrock, C. R. (1980). *J. Biol. Chem.* **255,** 3748–3756.
22. Schneider, H., Lemasters, J. J., Höchli, M., and Hackenbrock, C. R. (1980). *Proc. Natl. Acad. Sci. U.S.A.* **77,** 442–446.
23. Schnaitman, C., and Greenawalt, J. W. (1968). *J. Cell Biol.* **38,** 158–175.
24. Hackenbrock, C. R. (1972). *J. Cell Biol.* **53,** 450–465.
25. Bartlett, G. R. (1959). *J. Biol. Chem.* **234,** 466–468.
26. Vanneste, W. H. (1966). *Biochem. Biophys. Acta* **113,** 175–178.
27. Kröger, A., and Klingenberg, M. (1966). *Biochem. Z.* **344,** 317–336.
28. Quinn, P. J., and Esfahani, M. A. (1980). *Biochem. J.* **185,** 715–722.
29. Gutman, M. (1978). *Mol. Cell. Biochem.* **20,** 41–60.
30. Cherry, R. J. (1976). *In* "Biological Membranes" (D. Chapman and D. F. H. Wallach, eds.), Vol. 3, pp. 47–102. Academic Press, New York.
31. Kröger, A., and Klingenberg, M. (1967). *Curr. Top. Bioenerg.* **2,** 176–193.
32. Weiss, H., and Wingfield, P. (1979). *Eur. J. Biochem.* **99,** 151–160.

Effects of Lipid Concentration and Fluidity on Oxidoreduction of Ubiquinone by Isolated Complex I, Complex III, and Complex I–III

4

VERONICA M. POORE
C. IAN RAGAN

INTRODUCTION

Recent interest in the function of ubiquinone-10 in the respiratory chain has concentrated on the stabilized ubisemiquinones detected by electron spin resonance (ESR) spectroscopy (*1*, *2*). The relationship of the bound ubisemiquinones to the larger mobile pool of ubiquinone molecules in the membrane is not clear. In addition, work from this laboratory (*3–6*) and that of Hackenbrock (*7*) indicated that protein mobility and protein–protein interaction between the enzymes catalyzing ubiquinone oxidoreduction are important for electron transfer from substrates to cytochrome *c*.

Problems concerning molecular mobility in membranes may be studied by changing the physical state or the amount of phospholipid present, thus affecting the rates of reaction that depend on the diffusion of molecules (either proteins or ubiquinone) in the membrane. Such approaches were applied to intact rat liver mitoplasts (*7*) and to the interaction of Complex I (NADH-ubiquinone oxidoreductase) (*3–5*) and L-3-glycerolphosphate dehydrogenase (*6*) with Complex III (ubiquinol-cytochrome *c*

Function of Quinones in Energy Conserving Systems

ISBN 0-12-701280-X

oxidoreductase). In this paper we extend previous observations (*8*) of the effects of phospholipid on the interaction of ubiquinone analogs with isolated Complex I and Complex III, and of electron transfer from NADH to cytochrome *c* catalyzed by reconstituted Complex I–III. We have used the fluorescence depolarization of diphenylhexatriene (DPH) to monitor changes in the phospholipid fluidity of complexes whose lipid composition, concentration, and physical state were varied.

EXPERIMENTAL RESULTS

Isolated Complex I contains 0.20 to 0.25 μmole phospholipid P per mg protein; all except 0.04 to 0.05 μmole phospholipid per mg protein can be removed by treatment with cholate and ammonium sulphate (*9*). The residual lipid is cardiolipin (*9*). The endogenous phosphatidylethanolamine and phosphatidylcholine can also be replaced by DMPC[1] with retention of most of the ubiquinone-reductase activity of the enzyme (*9*). Similar results were found for Complex III (*5*). As previously reported (*8*), we found that variations in the lipid content of Complex I had no effect on the rate of reduction of electron acceptors other than ubiquinone analogs. Even for these acceptors, only the rotenone-sensitive component was affected. Figure 1 shows Arrhenius plots of rotenone-sensitive NADH-UQ_1 or UQ_2 reductase activities obtained with Complex I samples containing different phospholipid compositions and concentrations. Complex I containing natural phospholipids gave linear plots over the temperature range 4 °–38 °C. Nonlinear behavior was shown by Complex I whose phosphatidylethanolamine and phosphatidylcholine were replaced by DMPC, which in the pure state undergoes a phase transition at 25 °C (*10*). Thus we could clearly attribute changes in activation energy to changes in the physical state of the phospholipid. The increase in activation energy at lower temperatures was greater at high lipid-to-protein ratios, and with UQ_1 (Fig. 1A) rather than UQ_2 (Fig. 1B) as acceptor. At a lipid-to-protein ratio similar to that of native Complex I, the Arrhenius plots were linear.

The transition from liquid crystalline to gel state lipid could affect enzyme activity by changing the conformation of the protein or by preventing a conformational change from occurring during catalysis. If this transition did occur, we would expect the activity of the enzyme toward all

[1] Abbreviations: DMPC, dimyristoylphosphatidylcholine; UQ_1, ubiquinone-1; UQ_2, ubiquinone-2; DPH, diphenylhexatriene; P, degree of fluorescent polarization; UQ_{10}, ubiquinone-10.

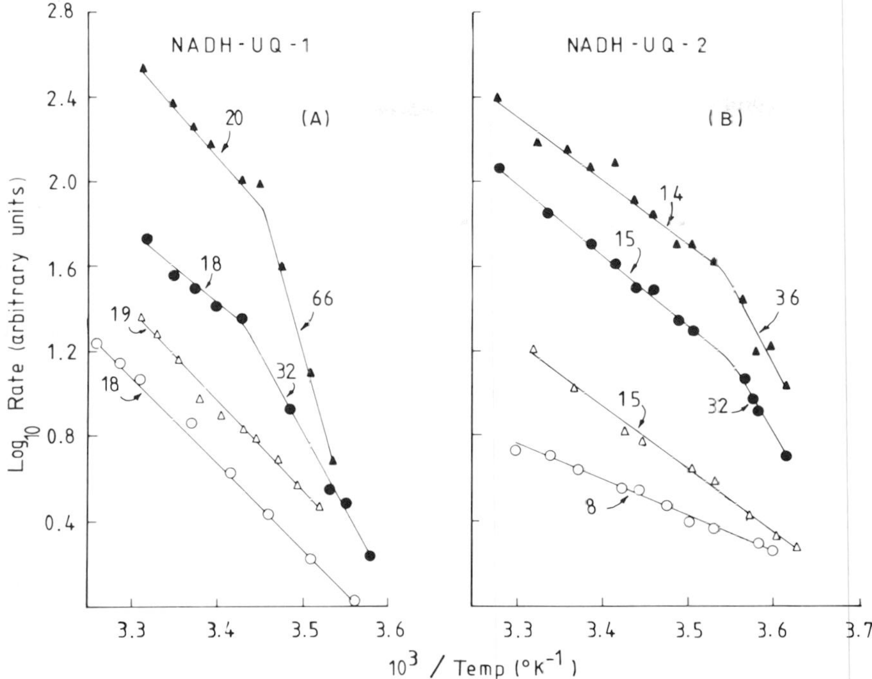

Fig. 1. Arrhenius plots of rotenone-sensitive NADH–UQ_1 or UQ_2 oxidoreductase activity. Complex I lipids were replaced with DMPC as described in Heron *et al.* (*9*). Samples were further supplemented with DMPC dispersed in 0.35% cholate to give the indicated lipid-to-protein ratios and then dialyzed extensively to remove cholate. Lipid/protein ratios in (A) were as follows: (○), 0.22 μmole/mg protein; (△), 0.28 μmole/mg protein; (●), 2.5 μmole/mg protein; (▲), 9.4 μmole/mg protein. Ratios in (B) were as follows: (○), 0.22 μmole/mg protein; (△), 0.28 μmole/mg protein; (●), 2.44 μmole/mg protein; (▲), 9.02 μmole/mg protein. Native Complex I is indicated by open circles, all other symbols refer to DMPC-replaced enzyme. Numbers on the lines are activation energies in kcal/mole.

acceptors to be altered in parallel, and the layer of lipid immediately adjacent to the protein to be the determining factor; therefore, the effects would be seen at lower as well as higher lipid-to-protein ratios. Since these effects were not found, we conclude that changes in the physical state of the lipid influence the accessibility of ubiquinone analogs to reduction sites that are known to be in the lipid phase (*8*). We would expect these effects to be greater with the less hydrophobic UQ_1, that partitions less favorably into the lipid phase (*8*), and greater with increasing size of the lipid phase relative to protein (as found in practice).

DMPC-replaced Complex III is extremely unstable above 15 °C and reproducible Arrhenius plots for temperatures above this could not be obtained. However, the activation energy for UQ_2H_2-cytochrome *c* oxidoreductase below this temperature (13 kcal/mole) was similar to that found for the native enzyme (11 kcal/mole) even at very high lipid-to-protein ratios. We therefore conclude that changes in the physical state of the lipid did not have appreciable effect on Complex III activity (at least under the conditions of assay used).

In the experiments of Fig. 2, Complex I samples were equilibrated with DPH and the degree of fluorescent polarization of the probe was measured as a function of temperature. Results were plotted as $\log_{10}P$ against reciprocal temperature. The degree of fluorescent polarization, P, is related to the apparent microviscosity of the probe environment (*11*), but in view of the assumptions inherent in the calculation of microviscosity we have chosen to use the experimental quantity P throughout. Using the formula of Shinitzky and Barenholz (*11*), the log of the apparent microviscosity is approximately linearly related to log P over a range of log P from -1.0 to -0.5 which covers most of the data presented here. Native Complex I or lipid-depleted Complex I (Fig. 2A) gave high values of P, similar

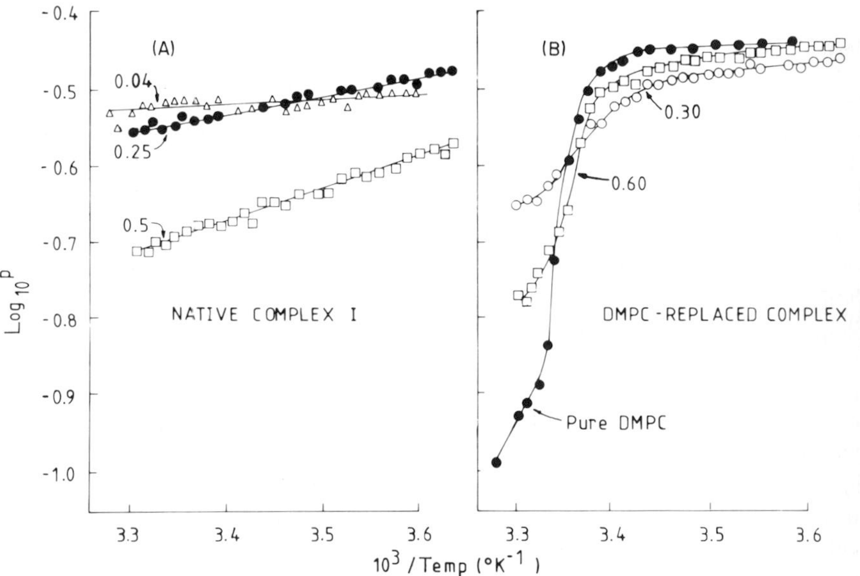

Fig. 2. DPH fluorescence in Complex I. Fluorescence depolarization of DPH was measured in Complex I (A) or DMPC-replaced Complex I (B) containing 1 nmole of DPH/mg protein. Lipid supplementation was with soyabean phosphatidylcholine (A) or DMPC (B). The former lipid is highly unsaturated and remains in the liquid crystalline state over the temperature range used. The numbers by each curve give the lipid content in μmole/mg protein.

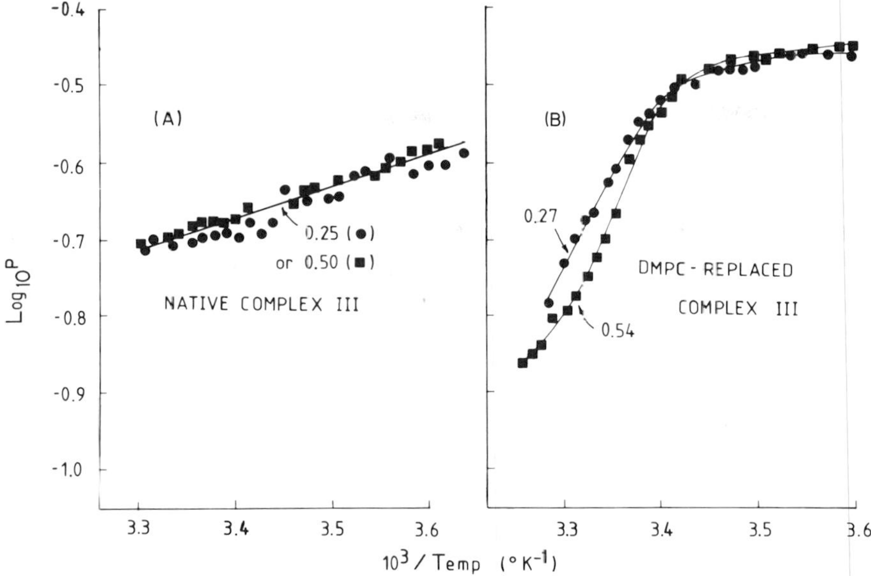

Fig. 3. DPH fluorescence in Complex III. Conditions and data presentation are as in Fig. 2.

to those obtained with pure DMPC in the gel state. Increasing the lipid-to-protein ratio gave lower values of P, indicating increased fluidity of the lipid phase. There were no abrupt changes of P with temperature.

DMPC-replaced Complex I showed greater changes in P with temperature, which approached that shown by DMPC alone at very high lipid-to-protein ratios (Fig. 2B). Above the transition temperature of the lipid, fluidity increased with increasing lipid-to-protein ratio and was higher than that of native Complex I at comparable lipid concentrations. At low temperatures, the fluidity was largely independent of lipid-to-protein ratio and the same as that of pure DMPC. However, the transition was less abrupt and occurred at lower temperatures at lower lipid concentrations. Thus, comparing Figs. 1 and 2, we see that fluidity alone does not cause the break in the Arrhenius plot, since both native Complex I and DMPC-replaced Complex I approach the same fluidity at low temperatures. Rather, it seems clear that the amount of lipid is the determining factor. Increasing the concentration of native lipid increases the overall fluidity and therefore has no effect on ubiquinone accessibility. Only with the DMPC-replaced enzyme is it possible to increase the lipid-to-protein ratio while maintaining low fluidity. In Fig. 3, comparable experiments with Complex III are shown. Native and DMPC-replaced Complex III show greater lipid

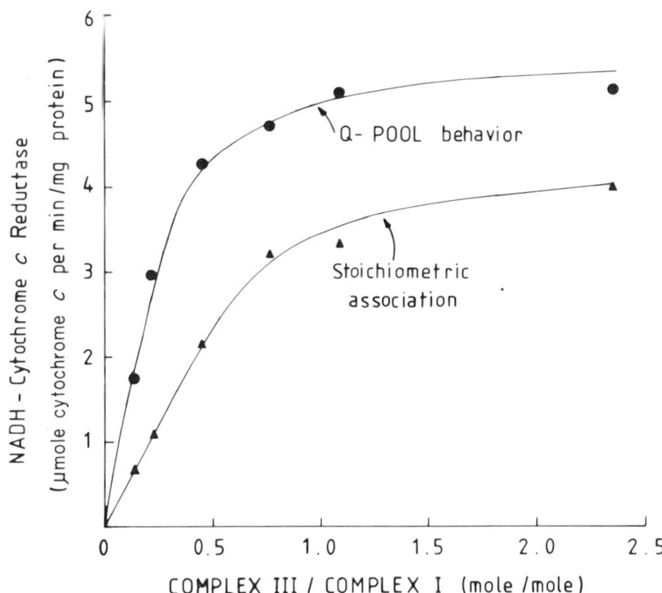

Fig. 4. Reconstitution of NADH-cytochrome *c* oxidoreductase from Complex I and Complex III. Complex I and Complex III were combined in the indicated molar proportions in the absence or presence of additional soyabean phospholipid and UQ_{10} (*4*) giving stoichiometric or Q-pool behavior, respectively. Stoichiometric association is characterized by a linear dependence of rate on Complex III concentration up to approximately 0.6 mole Complex III/mole Complex I. Q-pool behavior is characterized by a hyperbolic dependence of rate on Complex III concentration. The K_m for Complex III (0.23 mole/mole) can be calculated from the equations given by Kröger and Klingenberg (*13*).

fluidity than equivalent samples of Complex I at low lipid concentrations; otherwise the results are very similar to those of Fig. 2.

At low lipid-to–protein ratios (as in the isolated enzyme), Complex I and Complex III interact to form a stoichiometric I–III binary complex (*3*, *12*). The I–III unit is a fully active NADH-cytochrome *c* oxidoreductase. Electron transfer between such units is very slow and is not greatly stimulated by ubiquinone-10 (UQ_{10}). The same behavior occurs in the interaction between isolated L-3-glycerolphosphate dehydrogenase and Complex III (*6*). It is possible that electron transfer between the individual enzymes of the unit takes place through UQ_{10} immobilized within the binary complex. Certainly no more than two or three molecules of UQ_{10} per unit support maximal rates of electron transfer (*3*), which would be consistent with the two types of ubisemiquinone described in the b-c_1 region of the respiratory chain (*1*) and that described in Complex I (*2*).

In the mitochondrial membrane, the kinetics of ubiquinone oxidoreduction were explained by Kröger and Klingenberg in terms of independent enzymes operating in sequence through a mobile pool of ubiquinone molecules (*13*, *14*). Such Q-pool behavior can be restored to the I–III system or the L-3-glycerolphosphate dehydrogenase–III system by raising the lipid-to-protein ratio (*4*, *6*). This behavior is characterized, for example, by hyperbolic rather than linear dependence of reconstituted rate on the Complex I-to-III ratio and by antimycin-inhibition curves whose shapes are dependent on the Complex I-to-III ratio. An example of such behavior is shown in Fig. 4. The transition from stoichiometric to Q-pool behavior of the I–III system can be achieved not only by increasing the lipid-to-protein ratio but also by changing the physical state of the phospholipid. The experiment of Fig. 5 shows how this is done. Complexes I and III, whose endogenous lipids have been replaced by DMPC, associate stoi-

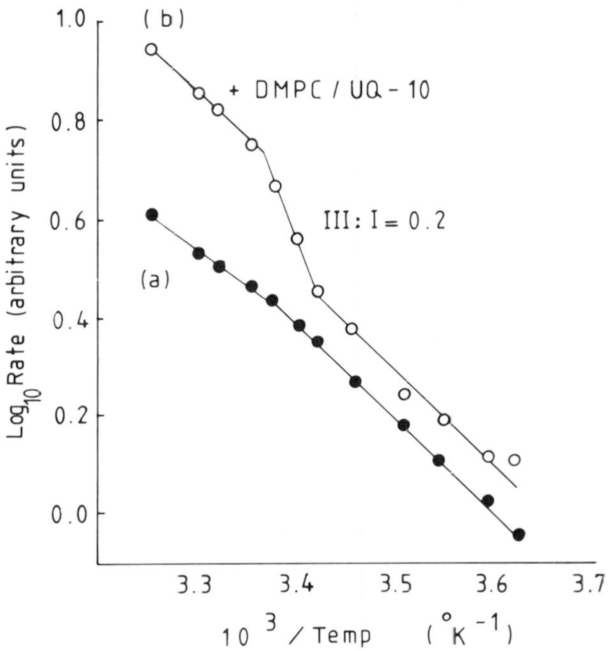

Fig. 5. Arrhenius plots for DMPC-replaced Complex I–III. The dependence of NADH-cytochrome *c* oxidoreductase on temperature was measured in curve (a), DMPC-replaced Complex I–III; or curve (b), DMPC-replaced Complex I–III whose lipid content had been doubled by further addition of DMPC. Ubiquinone-10 was also added to maintain high rates of electron transfer. The molar ratio of Complex III to Complex I was chosen as 0.2 to give an appreciable rate difference between Q-pool and stoichiometric behavior as in Fig. 4.

chiometrically as long as the lipid-to-protein ratio is the same as that of the original isolated enzymes. Stoichiometric behavior is maintained both above and below the transition temperature and, as shown in Fig. 5A, the Arrhenius plot for NADH-cytochrome oxidoreductase shows only a small increase in slope at lower temperatures. Since we know that the individual enzymes of the I–III unit do not undergo any intrinsic activity changes, the change in slope can be attributed to an effect of lipid structure on the transfer of ubiquinone and ubiquinol between the constituent complexes. The Arrhenius plot for Complex I–III whose lipid-to-protein ratio has been increased is strikingly different (Fig. 5B). In addition to differences in slope (activation energy) above and below the transition, the curve also shows a fairly abrupt discontinuity around 23 °C. Conditions in this experiment were chosen such that the rate per milligram of Complex I protein was 2.5-fold greater during Q-pool behavior than during stoichio-

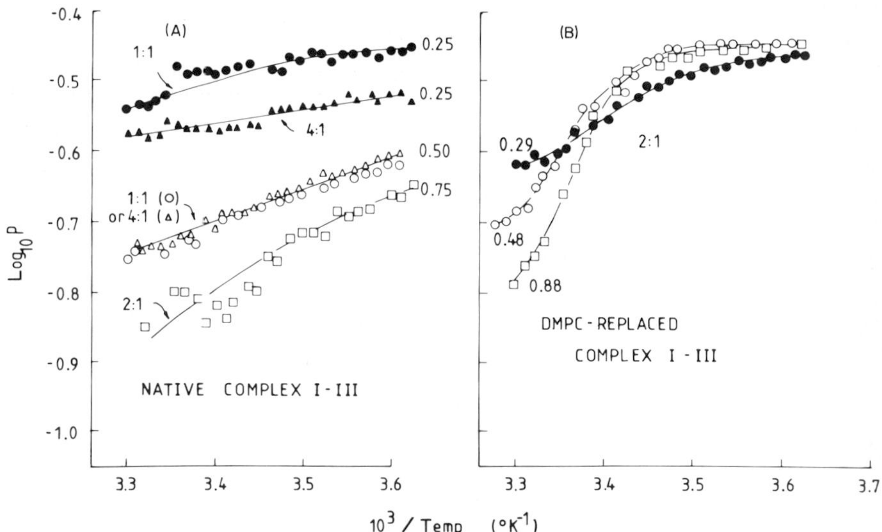

Fig. 6. DPH fluorescence in Complex I–III. Reconstitution of NADH-cytochrome *c* oxidoreductase with Complex I and Complex III was as described in Heron *et al.* (*5*). The lipid content of the native enzymes (A) was increased by addition of soyabean phospholipids (*4*) to give the amounts shown on the right hand ends of the curves in μmole/mg of total Complex I plus Complex III protein. Stoichiometric behavior was found at 0.25 μmole lipid/mg protein and Q-pool behavior at 0.50 and 0.75 μmole lipid/mg protein. The molar ratios of Complex III to Complex I are given at the left hand ends of the curves. DMPC-replaced Complex I–III (B) containing 2 mole of Complex III per mole Complex I was supplemented with DMPC to give the lipid contents shown at the left hand ends of the curves in μmole/mg protein. Stoichiometric behavior was shown by all samples at low temperatures and Q-pool behavior by samples containing 0.48 or 0.88 μmole lipid/mg protein at high temperatures.

metric behavior (see, e.g., Fig. 4). The change in position of the curve corresponds to a rather rapid activity change of this magnitude. Moreover, the slope of the Arrhenius plot at lower temperatures was identical to that of Fig. 5A (stoichiometric association) but greater than that of Fig. 5A at higher temperatures, a feature characteristic of Q-pool behavior (*5*). We conclude that at increased lipid-to-protein ratios, Q-pool behavior can be converted to stoichiometric association by transition of the lipid to the gel state.

DPH fluorescence measurements of samples similar to those of Fig. 5 are shown in Fig. 6. In Fig. 6A, Q-pool behavior induced by a higher lipid-to-protein ratio is characterized by increased overall fluidity of the lipid. Fluidity is also a function of the Complex I-to-Complex III ratio as expected from the results of Figs. 2 and 3. In Fig. 6B, the fluidity changes of DMPC-replaced Complex I–III are shown as a function of temperature. We see that the transition in activity (Fig. 5) takes place over the same temperature range as the changes in fluidity. Q-pool behavior is associated with a higher fluidity, and lowering the temperature causes an increase in P to values identical to those found for stoichiometric association. Thus, whichever way the transition from stoichiometric to Q-pool behavior is achieved, it is clear that stoichiometric behavior is associated with a less fluid lipid phase. Feinstein *et al.* (*15*) used fluorescence depolarization to measure the apparent microviscosity of mitochondrial membranes. Their value (0.9 poise at 30 °C) corresponds to $\log_{10}$ P of −0.85 at 10^3/temperature of 3.30. Thus we see that Q-pool behavior in isolated Complex I–III is associated with similar or slightly lower fluidity than the natural membrane which of course, only exhibits Q-pool behavior.

DISCUSSION

We have shown that the two types of behavior exhibited by the Complex I–III system are determined by the fluidity of the surrounding lipid phase. We have also shown that the intrinsic activity of the two constituent enzymes is not affected by lipid fluidity and therefore that nonlinearities in Arrhenius plots can be attributed to effects on whatever processes are involved in the transfer of reducing equivalents from one complex to the other. The accessibility of UQ_1 to Complex I can be affected by increased lipid fluidity, presumably through alteration in the partition of this rather hydrophilic analog between aqueous and lipid phases, or by restricting UQ_1 mobility in the lipid and making the reaction diffusion-limited. The magnitude of this effect was decreased by increasing the hydrophobicity of the ubiquinone analogs and was only observed at very high lipid-to-protein ratios. In contrast, the two types of Complex I–III behavior can be observed at relatively low lipid-to-protein ratios and activity depends on electron transfer mediated by the very hydrophobic natural ubiquinone UQ_{10}. The concentration of UQ_{10} does not affect the type of behavior shown, as was established for the L-3-glycerolphosphate dehydrogenase-Complex III system whose activity is very largely dependent on added UQ_{10}, unlike the Complex I–III system. Thus we think it unlikely that UQ_{10} mobility or accessibility is the determining factor for Q-pool rather than stoichiometric behavior.

The alternative view is that protein mobility is affected by changes in lipid fluidity and therefore protein–protein interactions between Complex I and III are important during electron transfer as suggested previously (*3–5*), and by Schneider *et al.* (*7*). This proposal allows cooperation between Complex I and Complex III on the reduction of UQ_{10} as suggested by the Q-cycle hypothesis of Mitchell (*16*). We see the permanant stoichiometric association of Complex I with Complex III resulting from low lipid fluidity. At higher fluidities, as in the mitochondrial membrane, the Complex I–III unit is still the catalytically active entity, but the association between the two enzymes is transient and Q-pool behavior results from rapid random dissociation and association of Complex I and Complex III molecules. In the dissociated state, the bound UQ_{10} may exchange or reach redox equilibrium with the pool of UQ_{10} molecules in the lipid phase. The same kind of explanation can be applied to the behavior of the L-3-glycerolphosphate dehydrogenase–Complex III system (*6*). The transition from Q-pool to stoichiometric behavior resulting from phase transition of the lipid to the gel state cannot be merely the result of immobilization of the complexes since electron-transfer activity between the complexes is preserved. Thus, during the transition, pairing of Complex I and Complex III must take place. The force driving this reaction may be the insolubility of the protein complexes in gel-state lipid leading to lipid–protein segregation as was seen by Hackenbrock and Höchli in mitochondrial membranes at low temperatures (*17*).

ACKNOWLEDGMENTS

This research was supported by the Science Research Council. UQ_1 and UQ_2 were gifts from Hoffman-La Roche, Basel, Switzerland.

REFERENCES

1. Ohnishi, T., and Trumpower, B. L. (1980). *J. Biol. Chem.* **255,** 3278–3284.
2. King, T. E. (1978). *FEBS Symp.* **45,** 17–31.
3. Ragan, C. I., and Heron, C. (1978). *Biochem. J.* **174,** 783–790.
4. Heron, C., Ragan, C. I., and Trumpower, B. L. (1978). *Biochem. J.* **174,** 791–800.
5. Heron, C., Gore, M. G., and Ragan, C. I. (1979). *Biochem. J.* **178,** 415–426.
6. Cottingham, I. R., and Ragan, C. I. (1982). *Biochem. J.* in press.
7. Schneider, H., Lemasters, J. J., Höchli, M., and Hackenbrock, C. R. (1980). *J. Biol. Chem.* **255,** 3748–3756.
8. Ragan, C. I., (1978). *Biochem. J.* **172,** 539–547.
9. Heron, C., Corina, D., and Ragan, C. I. (1977). *FEBS Lett.* **79,** 399–403.
10. Hinz, H. J., and Sturtevant, J. M. (1972). *J. Biol. Chem.* **247,** 6071–6074.
11. Shinitzky, M., and Barenholz, Y. (1978). *Biochim. Biophys. Acta* **515,** 367–394.
12. Hatefi, Y., Haavik, A. G., Fowler, L. R., and Griffiths, D. E. (1962). *J. Biol. Chem.* **237,** 2661–2669.
13. Kröger, A., and Klingenberg, M. (1973). *Eur. J. Biochem.* **34,** 358–368.
14. Kröger, A., and Klingenberg, M. (1973). *Eur. J. Biochem.* **39,** 313–323.
15. Feinstein, M. B., Fernandez, M. B., and Sha'afi, R. I. (1975). *Biochim. Biophys. Acta* **413,** 354–370.
16. Mitchell, P. (1980). *Ann. N.Y. Acad. Sci.* **341,** 564–584.
17. Hackenbrock, C. R., and Höchli, M. (1977). *FEBS Symp.* **42,** 10–36.

Ubisemiquinone Does Not Have a Pool Function in the QH_2: Ferricytochrome c Oxidoreductase

5

JAN A. BERDEN
SIMON DE VRIES
E. C. SLATER

INTRODUCTION

The widely held view that ubiquinone acts as a pool between the various dehydrogenases (Q reductases) and the QH_2 oxidase system of the respiratory chain is based on the findings that the rate of oxidation of QH_2 is faster than the rate of oxidation of NADH or succinate by oxygen, and that the oxidation of succinate or NADH is not linearly, but sigmoidally or hyperbolically inhibited by antimycin (*1*). Apparently the respiratory chain can be divided into two parts, separated by a diffusible component, ubiquinone, and only the second part is sensitive to antimycin. Assuming that the concentration of ubiquinone in the membrane is below the K_m of the Q reductases for Q and below the K_m of the QH_2 oxidase for QH_2, the rate of oxidation of substrate equals $V_1 \cdot V_2/(V_1 + V_2)$, where V_1 equals the rate of Q reduction when all Q present is in the oxidized state and V_2 equals the rate of QH_2 oxidation when all Q present is in the reduced state. A further assumption is that both the dehydrogenase substrate and oxygen are present in saturating concentrations.

Function of Quinones in Energy Conserving Systems

ISBN 0-12-701280-X

The use of antimycin, however, to quantitate the ratio of V_1 and V_2 is complicated by the finding that the effects of antimycin are cooperative (*2*), so that inhibition curves at least partly reflect the cooperative behavior of antimycin. Berden (*3*) suggested that ubisemiquinone, postulated by Wikström and Berden to be an intermediate in the oxidation of QH_2 (*4*), is an allosteric effector of the antimycin-binding protein, inducing the active state. However, another class of inhibitors of the QH_2 oxidase system, the hydroxyquinoline-*N*-oxides, do not show any cooperative behavior. Van Ark and Berden (*5*) found completely hyperbolic inhibition curves with HQNO,[1] reflecting the higher capacity of the QH_2 oxidase system relative to the Q reductase systems. Thus, using this inhibitor, the parameters defining the pool function of Q, V_1, and V_2 can be determined. The contribution of the allosteric effect of antimycin to the antimycin-inhibition curves may then be assessed by correcting for the contribution of the pool function.

Wikström and Berden (*4*) proposed a dual pathway for the two reduction equivalents from ubiquinol with formation of a semiquinone intermediate; the question arose whether this semiquinone can donate its electron to only one respiratory-chain assembly or to more or even all assemblies, in other words, whether it has a pool function. Two recent findings have made it possible to investigate this problem: First, Trumpower (*6*) established that extraction of Racker's OxF, later identified with the Rieske Fe–S protein (*7*), prevented reduction of cytochrome *b* in the presence of antimycin. In this respect, removal of the Fe–S protein has the same effect as inactivation of the respiratory chain by treatment with BAL (*8*). Second, Slater and de Vries (*9*) reported that this same Fe–S cluster is the site of irreversible inactivation by BAL. It should be possible, by comparing the inhibition of substrate oxidation by HQNO and BAL, both separately and together, to establish whether ubisemiquinone has a pool function. Another possibility is to measure the kinetics of cytochrome *b* reduction in the presence of antimycin after partial inactivation by BAL. If the semiquinone acts as a pool, all cytochrome *b* will be reduced with the same rate; if the semiquinone does not act as a pool the cytochrome *b* in the chains with an intact Fe–S cluster will be reduced rapidly and the rest slowly (or not at all). This is the approach used many years ago by Tsou (*10*) to establish that cytochrome *b* is not identical with succinate dehydrogenase.

[1] Abbreviations: HQNO, *n*-heptyl-4-hydroxyquinoline-*N*-oxide; BAL, British Anti-Lewisite (2,3-dimercaptopropanol); DCIP, 2,6-dichloroindophenol.

EXPERIMENTAL RESULTS

Comparison of the Inhibitory Effects of Antimycin and HQNO

In agreement with the data of van Ark and Berden (*5*), inhibition by HQNO of the succinate oxidase activity of EDTA particles (*11*) is hyperbolically related to the saturation of the binding site with HQNO. To calculate the saturation at the various concentrations of HQNO added, we determined the concentration of binding sites and the K_D fluorimetrically (*5*). The hyperbolic character of the inhibition curve (Fig. 1) is shown by the straight line in a reciprocal plot (Fig. 2). The

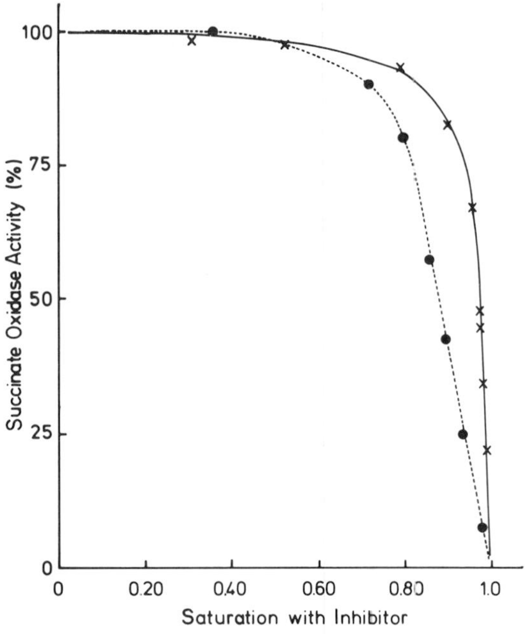

Fig. 1. Inhibition of the succinate oxidase activity of EDTA particles by HQNO and antimycin. The oxidation of 20 m*M* succinate by EDTA particles (0.5 mg/ml) was measured in a medium containing 0.25 *M* sucrose, 20 m*M* Tris-HCl buffer (pH 7.5), 0.2 m*M* EDTA, 4 μM cytochrome *c* and varying amounts of HQNO or antimycin. The oxygen uptake was measured with a Clark oxygen electrode. Temperature 30 °C. The concentration of binding sites, determined fluorimetrically (*5*), was 0.35 nmole/mg and the K_D for HQNO 70 n*M*. For antimycin a K_D of $3.2 \cdot 10^{-11}$ *M* was assumed at high saturation (*2*). The concentrations of the ethanolic solutions of the inhibitors were determined spectrophotometrically using an absorbance coefficient of 10.45 mM^{-1} cm^{-1} at 338 nm for HQNO (*5*), and a coefficient of 4.8 mM^{-1} cm^{-1} at 320 nm for antimycin (*22*). (x), HQNO; (●), antimycin.

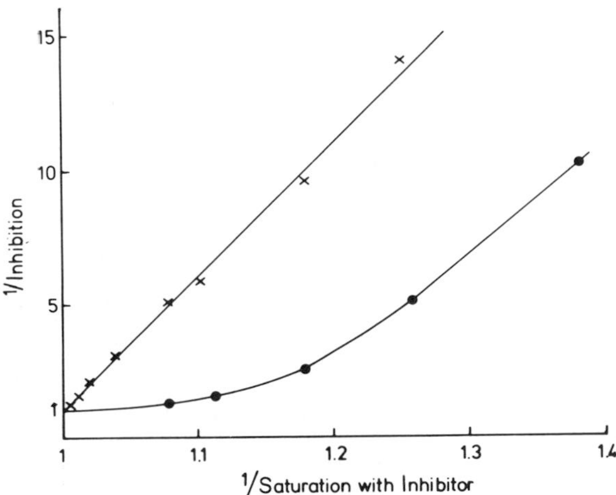

Fig. 2. Inhibition of the succinate oxidase activity of EDTA particles by HQNO and antimycin. The data of Fig. 1 are shown in a reciprocal plot. A hyperbolic inhibition results in a linear reciprocal plot with slope $V_2/V_1 + 1$ (5). (x), HQNO; (●), antimycin.

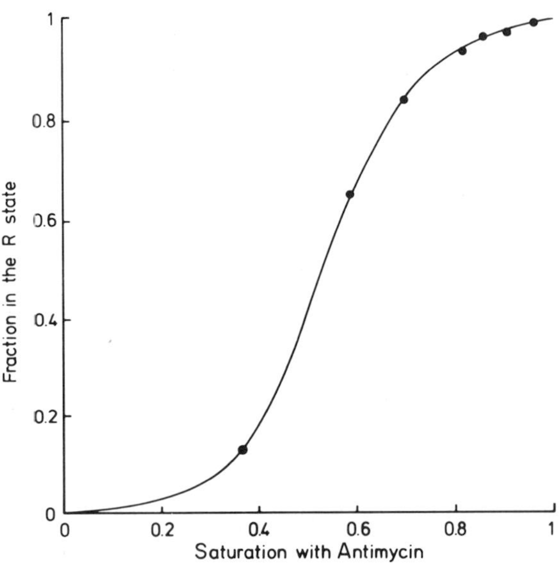

Fig. 3. Inhibition of the QH_2 oxidase activity of EDTA particles by antimycin. From Fig. 1 the percentage inhibition of the QH_2 oxidase activity at various levels of saturation with antimycin is calculated from the saturation with HQNO needed to obtain an equal inhibition of the succinate oxidase activity. The assumption is made that saturation with HQNO is proportional to inhibition of the QH_2 oxidase activity, as is concluded from Fig. 2 and Ref. 5. The inhibited state is called R state.

slope of this line equals $V_2/V_1 + 1$. From this slope and the measured velocity in the absence of HQNO, the values of V_1 and V_2 can be calculated from the equation $v = V_1 \cdot V_2/(V_1 + V_2)$.

Figure 2 shows that the inhibition by antimycin is not hyperbolic. This is also shown in Fig. 3, where the inhibition of V_2 is plotted as a function of the saturation with antimycin. While a straight line is obtained for the inhibition by HQNO, the curve for antimycin is sigmoidal, clearly demonstrating the allosteric nature of the effect of antimycin. As we have done before (*2*), the allosteric parameters according to the model of Monod *et al.* (*12*) were calculated. The active state (in the absence of antimycin) is called the T state and the inactive state, induced by binding of antimycin in a cooperative manner, is called the R state. In Fig. 4 we plotted $\log((1 - \bar{R})/\bar{R})$ against $\log(1 + \alpha) - \log(1 + \alpha c)$,

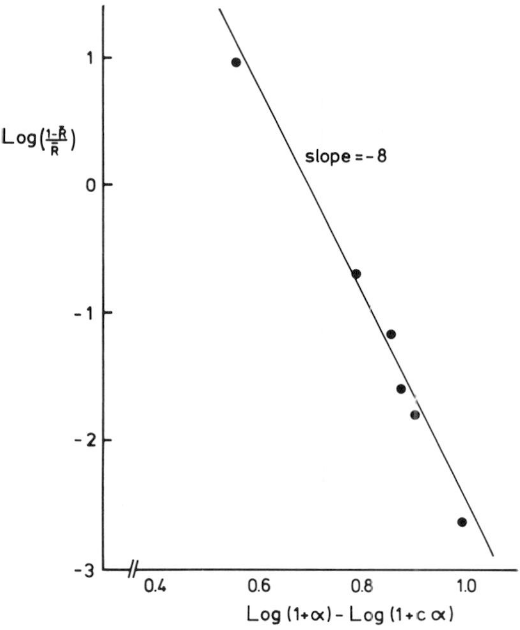

Fig. 4. Allosteric parameters of the inhibition of the QH_2 oxidase activity of EDTA particles by antimycin. The parameters n and L of the model of Monod *et al.* (*12*) are determined using the equation $\log((1 - \bar{R})/\bar{R}) = \log L - n[\log(1 + \alpha) - \log(1 + \alpha c)]$ (*2*). The values for $\bar{R}$ are taken from Fig. 3 and the values for c and the dissociation constants at various levels of saturation (needed to calculate α) are taken from Ref. 2. Extrapolation to the ordinate gives log L = 5.6; n = 8.

where $\overline{R}$ is the fraction in the R state, α is the concentration of free antimycin, divided by K_R, the intrinsic dissociation constant in the R state, and c is the ratio K_R/K_T. The binding data and the values for c, K_R and K_T given in Berden and Slater (*2*) were used for the construction of this figure. The resulting plot is very similar to that reported previously, where no correction was made for the pool function of ubiquinone. The value for the number of interacting sites, eight (the negative slope of the linear curve), is apparently not sensitive to the correction. The value for L, the ratio T state/R state in the absence of inhibitor, is given as the value of $(1 - \overline{R})/\overline{R}$ when α equals 0, i.e., the intersection with the ordinate. Extrapolation to the ordinate gives log L = 5.6.

Since under some conditions binding of antimycin does not induce the R state (*2*), one or more allosteric effectors must be involved. The possibility of an allosteric control by ubisemiquinone will be further discussed.

Comparison of the Effects of HQNO and BAL

The inhibitory effect of BAL on the respiratory chain can only be induced by aerobic incubation (*13*). Submitochondrial particles (EDTA particles, 10 mg/ml) were incubated (with shaking) for 20 min at 30 °C with varying amounts of BAL (0, 1, 2, 4, 8, and 15 m*M*) in a medium containing 250 m*M* sucrose, 20 m*M* Tris-HCl buffer (pH 7.5), 0.2 m*M* EDTA and 2 m*M* malonate to activate succinate dehydrogenase (*14*). After incubation, the particles were sedimented by centrifugation (30 min, 150,000 *g*), washed and resuspended in 250 m*M* sucrose, 20 m*M* Tris-HCl buffer (pH 7.5) and 0.2 m*M* EDTA. The succinate oxidase activity was measured in the same medium to which 4 μM cytochrome *c* was added, by measuring the oxygen uptake with a Clark electrode. In Fig. 5, the succinate oxidase activity of the sample (incubated in the absence of BAL and titrated with HQNO) is plotted against the degree of saturation with HQNO (open circles). Also shown is the activity measured in the absence of HQNO (after treatment with various concentrations of BAL), plotted against the degree of destruction of the Fe–S cluster of the Rieske protein, determined by measuring the EPR spectrum under appropriate conditions (*9*). A complicating factor, however, is that the BAL treatment also affects the activity of succinate dehydrogenase to some extent, especially at low concentrations of BAL (*13*). This was accounted for by measuring the succinate:DCIP oxidoreductase activity and correcting the measured rate of succinate oxidation for the inhibition of the succinate:Q reductase ac-

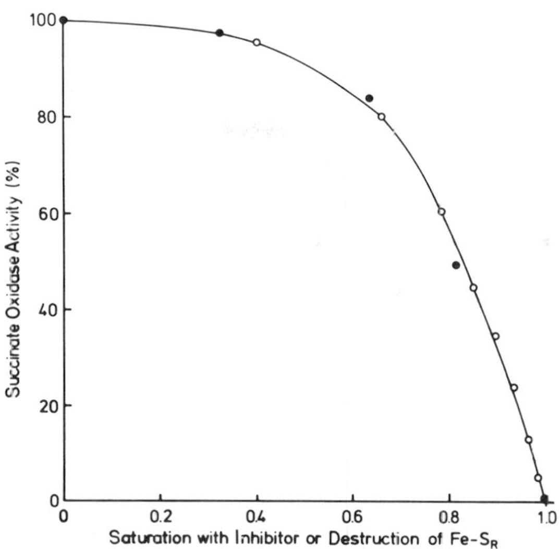

Fig. 5. Inhibition of the succinate oxidase activity of EDTA particles by HQNO and BAL. The oxidase activity was measured as in Fig. 1 and the saturation with HQNO was calculated as described in the legend to Fig. 1. BAL treatment was carried out as described in the text and the destruction of the Rieske Fe–S cluster was measured by EPR spectroscopy (*9*). The measured oxidase activity of BAL-treated preparations was corrected for the effect of the treatment on the succinate:Q reductase activity, measured as the DCIP reductase activity. (○), HQNO; (●), BAL.

tivity, which amounted to 10 to 20%. The corrected values, indicated by the closed circles in Fig. 5, fall close to the HQNO inhibition curve. The conclusion must then be that both the HQNO-sensitive step and the BAL-sensitive step have the same overcapacity relative to the Q-reduction step. This indicates that, although the site of action of the two inhibitors is different (*4*, *8*), both affect the same enzymatic unit, the QH_2 oxidase system.

This conclusion is confirmed by measurement of V_2 by titration of BAL-treated particles with HQNO. In Fig. 6 titrations of untreated particles, and of particles in which 82% of the Fe–S cluster is destroyed by BAL treatment, are shown in the reciprocal form. For the untreated particles $V_2/V_1 = 3.4$; for the treated particles $V_2/V_1 = 0.72$. Since the V_1 of the treated particles was diminished by 18%, the ratio of the two values of V_2 equals 3.4/0.72 times 1.18, that is 5.57. This means that the V_2 of the BAL-treated preparation is 18% of the V_2 of the untreated preparation, which is exactly the same as the percentage of residual intact Fe–S cluster. The results of similar determinations of V_2 by titration with HQNO for another series of BAL-treated particles are shown in Fig. 7. These data confirm the conclusion that HQNO-sensitive ac-

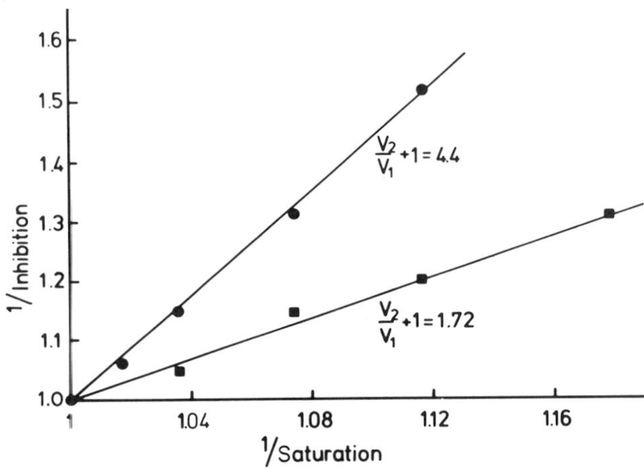

Fig. 6. Effect of BAL treatment on the HQNO-sensitive step in the succinate oxidase. The curve (●) represents the inhibition by HQNO of the succinate oxidase activity of particles not treated with BAL; $V_2/V_1 = 3.4$. The curve (■) represents the inhibition by HQNO of the succinate oxidase activity of BAL-treated particles of which 82% of the Rieske Fe–S cluster is destroyed; $V_2/V_1 = 0.72$. Since the V_1 was inhibited 18% by the BAL treatment, it can be calculated that the HQNO-sensitive step (V_2) was inhibited 82% by the BAL treatment (see text).

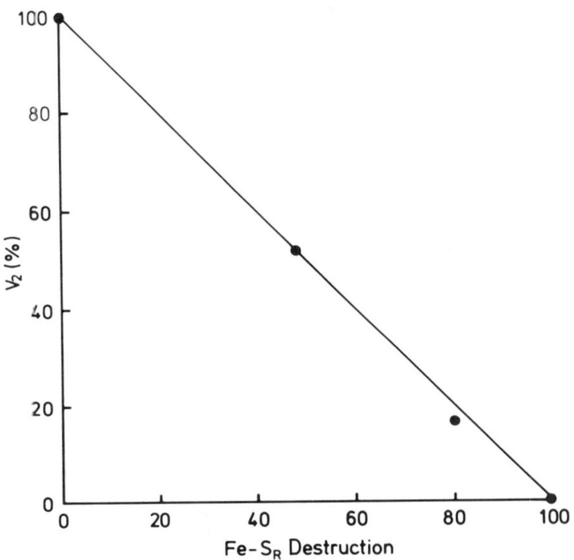

Fig. 7. Effect of BAL treatment on the HQNO-sensitive step in the succinate oxidase. The V_2/V_1 of BAL-treated particles was measured as in Fig. 6. From the V_2/V_1 and the rate of oxidation of succinate in the absence of HQNO the V_2 (and also the V_1) was determined using the equation $v = V_2V_1/(V_2 + V_2)$. The destruction of the Rieske Fe–S cluster by the BAL treatment was determined by EPR.

tivity is also inhibited by BAL treatment. Therefore, if ubisemiquinone is located between the sites of the two inhibitors, it does not act as a pool but reduces only the cytochrome *b* belonging to the same respiratory assembly as the Fe–S cluster that is reduced by ubiquinol.

Kinetics of Reduction of Cytochrome *b*

Van Ark has shown (*15*) that after partial inactivation by BAL of the succinate oxidase activity of submitochondrial particles, part of cytochrome *b* is reduced more slowly by succinate in the presence of antimycin. After complete inactivation by BAL treatment of the succinate oxidase the reduction of cytochrome *b* by succinate in the presence of antimycin is very slow and incomplete, in agreement with data in the literature (*8*, *16*). The two phases in the reduction of cytochrome *b* by succinate in the presence of antimycin, and after partial inactivation by BAL, are shown in Fig. 8. The relative amount of rapidly reduced cy-

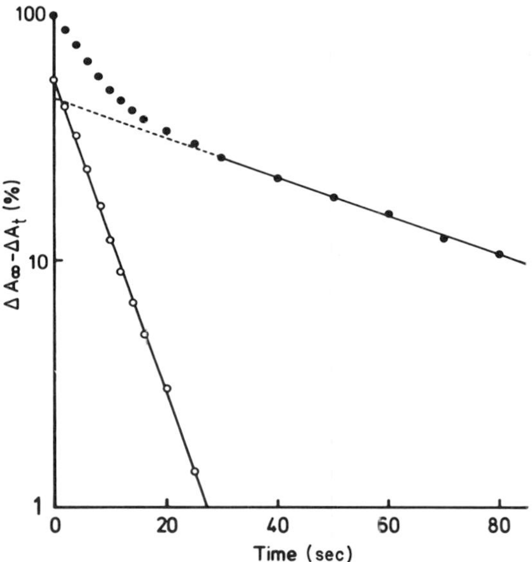

Fig. 8. Reduction of cytochrome *b* by succinate in the presence of antimycin after BAL treatment. To BAL-treated EDTA particles (2 mg/ml) suspended in the medium described in Fig. 1 but supplemented with 4 μM antimycin, 20 mM succinate was added and the absorbance change at 563 minus 577 nm followed. The temperature was 2 °C; 48% of the Rieske Fe–S cluster in the particles was destroyed by treatment with BAL. The absorbance change is plotted semilogarithmically (●). The curve (○) is obtained by correcting the total absorbance change for the contribution of the slow phase. The rate constant for the rapid phase equals 0.14 sec^{-1} and that for the slow phase 0.019 sec^{-1}.

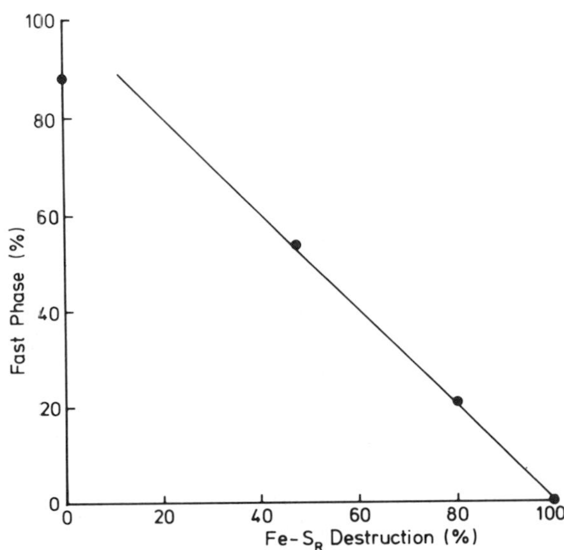

Fig. 9. Reduction of cytochrome *b* by succinate in the presence of antimycin after BAL treatment. The experiments were carried out as in Fig. 8 and the contribution of the fast phase to the total absorbance change is plotted as a function of the destruction of the Rieske Fe–S cluster by BAL. Without BAL treatment the fast phase never reaches 100% in EDTA particles, but it does in ATPMg particles (Berden *et al.*, unpublished observations used with permission).

tochrome *b* is a linear function of the destruction of Fe–S by BAL (Fig. 9). This confirms that only the cytochrome *b* belonging to respiratory-chain assemblies with intact Fe–S cluster are reduced rapidly, and that there is no pool function for the ubisemiquinone. Additionally, the rate of reduction of cytochrome *b* in the slow phase is not constant, but decreases with increasing destruction of the Fe–S cluster. This might be interpreted as an indication that ubisemiquinone can slowly reduce the cytochromes *b* of other respiratory-chain assemblies. On increasing destruction of the Fe–S cluster, less semiquinone can be formed and the slow reduction of cytochrome *b* belonging to assemblies with a destroyed Fe–S cluster is slowed down even further.

DISCUSSION

The allosteric character of the effect of antimycin on the oxidation of succinate is clearly demonstrated by the results shown in Fig. 3. The value of 8 for the number of interacting binding sites is unusually high; however, because we are dealing with enzyme complexes embedded in a membrane perhaps this is understandable. This finding is also consistent with the fact that in dimeric Complex III, no cooperative behavior of antimycin could be detected (*17*). Berden and Slater reported that the cooperative binding of antimycin is absent after extraction of cytochrome *c*; under these conditions no increased reduction of cytochrome *b* occurred upon addition of antimycin. In the present model of the respiratory chain shown in Fig. 10, two semiquinones can be distinguished, $Q_{in}^{\overline{\cdot}}$ and $Q_{out}^{\overline{\cdot}}$. The $Q_{in}^{\overline{\cdot}}$, is probably identical to the SQ_c of Ohnishi and Trumpower (*18*) and the QP_c radical of Yu *et al.* (*19*), and was also studied by de Vries *et al.* (*20*). Because this species of the semiquinone disappears on addition of antimycin, its presence might be related to the T state of the antimycin-binding site, but we have not yet established whether the disappearance of this semiquinone is linearly or sigmoidally related to saturation with antimycin. The other semiquinone ($Q_{out}^{\overline{\cdot}}$), not yet detected by EPR, might be related to the formation of R state, since its concentration is thought to increase under conditions of antimycin-induced reduction of cytochrome *b*.

The finding that treatment of submitochondrial particles with BAL under aerobic conditions specifically destroys the Rieske Fe–S cluster of Complex III was the essential information needed for investigating whether the semiquinone in Complex III ($Q_{out}^{\overline{\cdot}}$) has a pool function. The

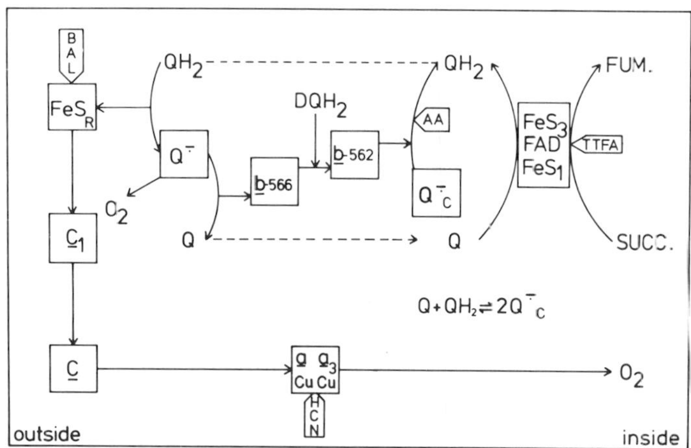

Fig. 10. Model for the electron transfer through the respiratory chain as favored by the authors. The arguments leading to this type of model are discussed extensively by van Ark (*15*).

data presented show very clearly that in an active respiratory chain no pool function is present between the Fe–S cluster and the antimycin- or HQNO-binding sites. However, the reduction kinetics of cytochrome *b* in the presence of antimycin suggest that a very slow reduction of cytochrome *b* of other respiratory-chain assemblies is possible. This implies that the semiquinone, produced by reduction of one oxidized Fe–S cluster, not only rapidly reduces the cytochrome *b* of the same respiratory-chain assembly, but also reduces, be it slowly, cytochrome *b* in other assemblies. The final level of *b* reduction will be determined by the equilibrium that is reached. In the presence of antimycin, cytochrome *b* cannot be oxidized, and if some or all of the Fe–S clusters are continuously oxidized by oxygen, the semiquinone formed will reduce more cytochrome *b* than under conditions of equilibrium in the absence of oxygen. The final level of reduction, then, is not related to the percentage of slowly reducible cytochrome *b*.

Although the data reported in this paper confirm the pool function of ubiquinone, we have not asked whether this pool function is complete. In many cases we have found that the V_2 measured via titrations of the succinate oxidase activity with HQNO differs from that measured via titrations of the NADH oxidation. The data (not shown) suggest that QH_2 formed by the NADH:Q oxidoreductase cannot react with all the QH_2 oxidase systems present. This could be due to a low mobility of the QH_2 (or a low concentration of total ubiquinone), or a low mobility of the NADH:Q reductase if we assume that the pool function behavior of ubiquinone is not due to the mobility of ubiquinone, but to the mobility of the complexes (*21*). This problem of the pool function of ubiquinone is at present under investigation.

ACKNOWLEDGMENTS

The authors thank Mr. J. van Dijk for the titrations of succinate oxidase activity with antimycin and Mr. Q. S. Zhu for the experiments on the kinetics of the reduction of cytochrome *b*. This work has been supported by grants from the Netherlands Organization for the Advancement of Pure Research (Z.W.O.) under the auspices of the Netherlands Foundation for Chemical Research (S.O.N.).

REFERENCES

1. Kröger, A., and Klingenberg, M. (1970). *Vitam. Horm.* (*N.Y.*) **28,** 533–574.
2. Berden, J. A., and Slater, E. C. (1972). *Biochim. Biophys. Acta* **256,** 199–215.
3. Berden, J. A. (1972). Ph.D. Thesis, Univ. of Amsterdam, Gerja, Waarland.

4. Wikström, M. K. F., and Berden, J. A. (1972). *Biochim. Biophys. Acta* **283,** 403–420.
5. van Ark, G., and Berden, J. A. (1977). *Biochim. Biophys. Acta* **459,** 119–137.
6. Trumpower, B. L. (1976). *Biochem. Biophys. Res. Commun.* **70,** 73–80.
7. Trumpower, B. L., and Edwards, C. A. (1979). *FEBS Lett.* **100,** 13–16.
8. Deul, D. H., and Thorn, M. B. (1962). *Biochim. Biophys. Acta* **59,** 426–436.
9. Slater, E. C., and de Vries, S. (1980). *Nature* (*London*) **288,** 717–718.
10. Tsou, C. L. (1951). *Biochem. J.* **49,** 512–520.
11. Lee, C. P., and Ernster, L. (1968). *Eur. J. Biochem.* **3,** 385–390.
12. Monod, J., Wyman, J., and Changeux, J. P. (1965). *J. Mol. Biol.* **12,** 88–118.
13. Slater, E. C. (1949). *Biochem. J.* **45,** 14–30.
14. Eisenbach, M., and Gutman, M. (1975). *Eur. J. Biochem.* **59,** 223–230.
15. van Ark, G. (1980). Ph.D. Thesis, Univ. of Amsterdam, Rodopi.
16. Ksenzenko, M. Y., and Konstantinov, A. A. (1980). *Biochemistry* (*Engl. Transl.*) **45,** 343–353.
17. Engel, W. D., and von Jagow, G. (1980). *Hoppe-Seyler's Z. Physiol. Chem.* **361,** 1279–1280.
18. Ohnishi, T., and Trumpower, B. L. (1980). *J. Biol. Chem.* **255,** 3278–3284.
19. Yu, C.-A., Nagaoka, S., Yu, L., and King, T. E. (1978). *Biochem. Biophys. Res. Commun.* **82,** 1070–1078.
20. de Vries, S., Berden, J. A., and Slater, E. C. (1980). *FEBS Lett.* **122,** 143–148.
21. Heron, C., Ragan, C. I., and Trumpower, B. L. (1978). *Biochem. J.* **174,** 791–800.
22. Strong, F. M., Dickie, J. P., Loomans, M. E., van Tamelen, E. E., and Dewey, R. S. (1960). *J. Am. Chem. Soc.* **82,** 1513–1514.

IV THERMODYNAMICALLY STABLE SEMIQUINONES IN MITOCHONDRIA AND PHOTOSYNTHETIC SYSTEMS

Electron Transfer between Primary and Secondary Electron Acceptors in Chromatophores and Reaction Centers of Photosynthetic Bacteria

1

ANDRÉ VERMÉGLIO

INTRODUCTION

Light-induced cyclic electron flow in photosynthetic bacteria is initiated in the so-called reaction center. After absorption of one photon, an electron is transferred from a special bacteriochlorophyll dimer $(BChl)_2$ to a bacteriopheophytin (Bpheo) molecule in a few picoseconds (*1*). This reduced Bpheo molecule then reduces a molecule of quinone (Q_1 or primary electron acceptor) within a few hundred picoseconds (*2*, *3*). The photooxidized primary donor $(Bchl)_2^{+}$ is rereduced by a *c*-type cytochrome (*4*) and the electron from $Q_1^{\overline{\cdot}}$ is transferred to the ferricytochrome *c* by passing through a quinone–cytochrome *b*-*c* complex (Q–*b*-*c*) to complete the cyclic electron transfer. The detailed mechanism of the electron transfer from $Q_1^{\overline{\cdot}}$ back to cytochrome c^+ is not well understood and has only been studied to some extent in the species *Rhodopseudomonas sphaeroides* and *Rps. capsulata* (*5*, *6*).

Of special interest is the mechanism of electron transfer between primary and secondary acceptors. Different lines of evidence suggest that this portion of the electron-carrier chain is similar to the acceptor side of photosystem II of green plants. In both cases, the primary and secondary acceptors are quinone molecules (*7–10*). These molecules act in series,

Function of Quinones in Energy Conserving Systems

ISBN 0-12-701280-X

and electrons are accumulated in pairs on the secondary acceptor quinone before transfer to subsequent electron carriers (*11–16*). This gating mechanism was clearly demonstrated for isolated reaction centers of *Rps. sphaeroides* R_{26} (*15*, *16*) and under certain conditions in chromatophores (*17*, *18*, *19*). Binary oscillations in the appearance and disappearance of semireduced ubiquinones in a series of saturating flashes were observed by both optical (*15*, *16*) and ESR techniques (*15*).

Measurements of the rate of electron transfer between primary (Q_1) and secondary (Q_2) quinones is, however, difficult. For example, ESR signals of the semireduced quinones are broad because of magnetic interaction with an iron atom, and only observable at low temperature (*20*, *21*). Therefore, in chromatophores, the electron transfer between primary and secondary quinones was studied only with indirect techniques, such as the double-flash experiments of Parson (*22*, *23*), or the kinetic analysis of the back-reaction between $(BChl)_2^{+\cdot}$ and reduced acceptors as proposed by Chamorovsky *et al.* (*24*). However, each of these two techniques has some limitations. In the double-flash experiment (*22*) the rereduction of $(BChl)_2^{+\cdot}$ by cytochrome *c* has to be much faster than the electron transfer from Q_1 to Q_2. For the species *Rhodospirillum rubrum, Rps. sphaeroides* and *Rps. capsulata* (the species most often studied when regarding electron-transfer mechanisms) this does not occur. (*4–6*). The method of Chamorovsky *et al.* (*24*) can only be applied under experimental conditions (i.e., high redox potential) where $(BChl)_2^{+\cdot}$ is not rereduced by cytochromes.

In a previous report we have shown (*25*) that in isolated reaction centers, the electron-transfer rate between primary and secondary ubiquinones can be followed by measurements of light-induced absorbance changes in the near infrared region between 730 and 780 nm. This is possible because the negatively charged primary ($Q_1^{-\cdot}$) and secondary ($Q_2^{-\cdot}$) ubiquinone molecules induce distinct absorbance changes, described as resulting from local electrostatic effects on the reaction center Bpheo and BChl molecules. This paper describes an extension of this method to chromatophores of *Rps. sphaeroides* R_{26} and *R. rubrum* G_9.

RESULTS AND DISCUSSION

Light-Induced Absorption Changes Related to $Q_1^{-\cdot}$ and $Q_2^{-\cdot}$ in Chromatophores of *Rps. sphaeroides* R_{26} and *R. rubrum* G_9

Figure 1A shows light-induced absorbance changes at a few selected wavelengths, under conditions expected to produce the state $Q_1Q_2^{-\cdot}$.

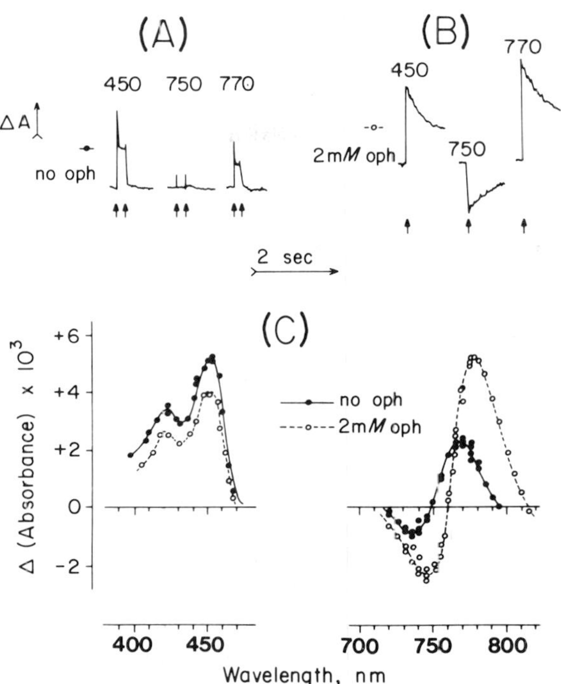

Fig. 1. Spectra of flash-induced absorbance changes in chromatophores of *Rps. sphaeroides* R_{26} in 10 m*M* Tris HCl, pH 7.5., 0.5 m*M* diaminodurene. (A) Kinetic responses to a pair of flashes. (B) Chromatophores as in A but with 2 m*M* orthophenanthroline (oph) added. (C) Spectra of the changes illustrated in (A) and (B). (●) related to state $Q_1Q_2^{\overline{\cdot}}$, (○) related to state $Q_1^{\overline{\cdot}}Q_2$.

Dark adapted chromatophores of *Rps. sphaeroides* R_{26} where subjected to saturating flashes in the presence of diaminodurene as the electron donor. The difference spectrum for state $Q_1Q_2^{\overline{\cdot}}$ was constructed by plotting the absorbance change that persists after the first flash and before the second (Fig. 1C, solid curve). To generate the state $Q_1^{\overline{\cdot}}Q_2$, chromatophores were subjected to single flash excitation in the presence of orthophenanthroline (2 m*M*) with diaminodurene as the electron donor (Fig. 1B). The spectrum of absorbance changes measured in this way is shown by the dashed curve in Fig. 1C. It is apparent that, as observed in isolated reaction centers (*25*), states $Q_1^{\overline{\cdot}}Q_2$ and $Q_1Q^{\overline{\cdot}}$ elicit different Bpheo absorption band shifts in the near-infrared part of the spectrum. Similar results were obtained for chromatophores of *R. rubrum* G_9 (Fig. 2).

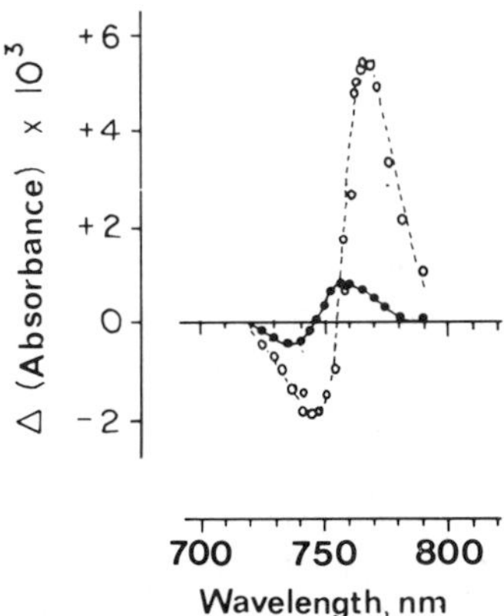

Fig. 2. Same as Fig. 1C but for *R. rubrum* G_9 chromatophores.

Absorption Transient in the Near Infrared Region: Evidence for its Relation to Electron Transfer Between Q_1 and Q_2.

Because states $Q_1^{-}Q_2$ and $Q_1Q_2^{-}$ induce different absorption changes in the near-infrared region, especially around 750 nm, an absorption transient reflecting electron transfer from Q_1 to Q_2 is expected. Absorption changes following laser excitation, shown in Fig. 3 for a few selected wavelengths, indeed exhibit a fast phase for chromatophore preparations of both R_{26} and G_9. Several arguments can be put forward to attribute the fast absorption changes observed (Fig. 3) to electron

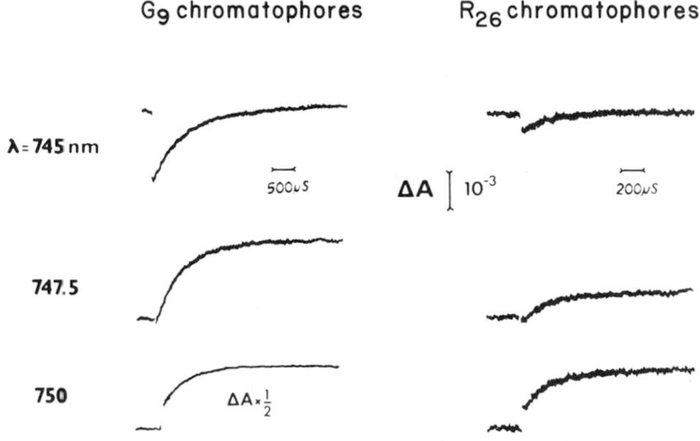

Fig. 3. Absorbance changes induced by a laser flash (15 nsec, 600 nm) at selected wavelengths for R_{26} and G_9 chromatophores. The chromatophore suspensions have an absorption of 0.4 at 750 nm in 10 m*M* tris, pH 8; no artifical donor added.

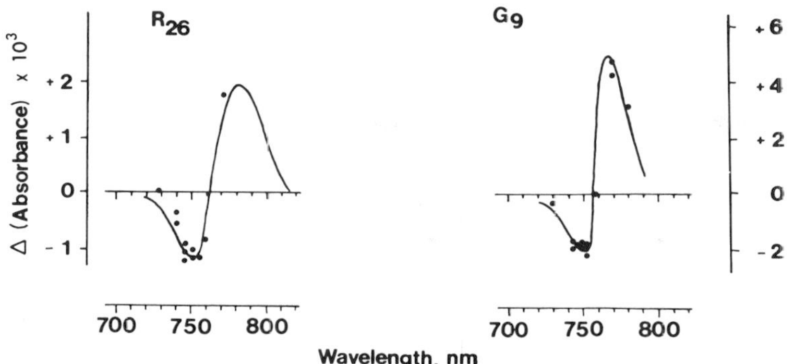

Fig. 4. Wavelength dependence of the amplitude (·) of the transient absorption changes illustrated in Fig. 3 for R_{26} and G_9 chromatophores. The solid curves were constructed from the spectra shown in Figs. 1 and 2. They correspond to the difference spectrum of states $(Q_1^{\bar{\cdot}}Q_2)$-$(Q_1Q_2^{\bar{\cdot}})$.

transfer between Q_1 and Q_2. Figure 4 shows the wavelength dependence of this fast phase. For both R_{26} and G_9 chromatophores, there is good agreement between the observed amplitude of the fast phase and the predicted amplitude from the difference spectra of $Q_1^{\bar{\cdot}}$ and $Q_2^{\bar{\cdot}}$ (Figs. 1 and 2). Additional evidence comes from the fact that 2.5 m*M* orthophenanthroline severely disminishes (Fig. 5) the fast phase under conditions (i.e., when Q_2 is fully oxidized prior to flash activation) where electron transfer between primary and secondary quinones is inhibited (*26*, *27*). This almost complete disappearance of the fast phase in the presence of orthophenanthroline is observed for R_{26} and G_9 chromatophores, as well as for isolated reaction centers from *Rps. sphaeroides* R_{26} (Fig. 5). Finally, in isolated reaction centers, one can show that the fast phase strictly depends on the amount of secondary acceptor

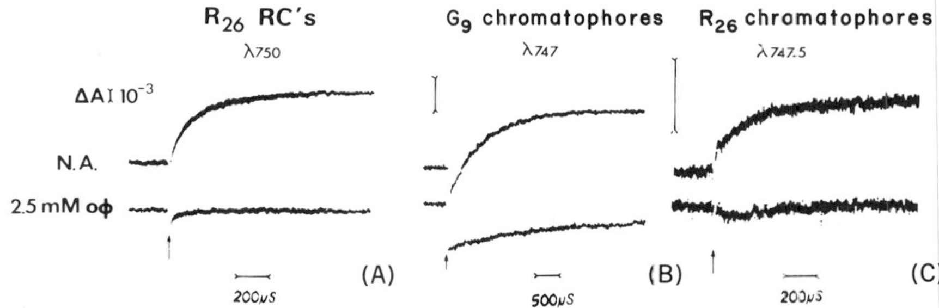

Fig. 5. Effect of 2.5 m*M* orthophenanthroline on the fast kinetics measured at the indicated wavelength. (A) Reaction centers from *Rps. sphaeroides* R_{26}. (B) G_9 chromatophores. (C) R_{26} chromatophores. No addition of artifical electron donor in all cases.

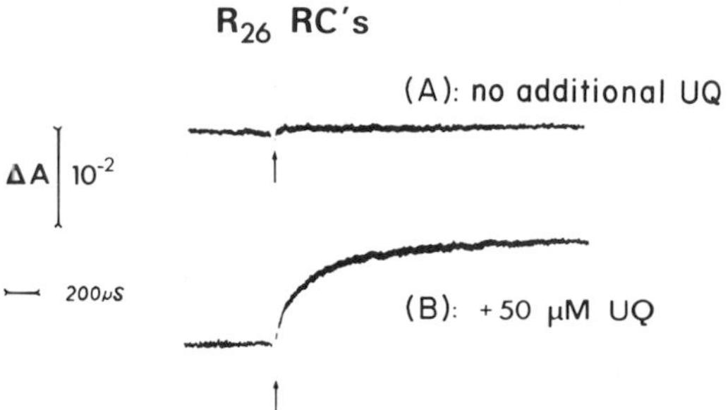

Fig. 6. Kinetics measured at 750 nm for isolated reaction centers from *Rps. sphaeroides* R_{26}. (A) Reaction centers (~2 μM) in 10 mM Tris HCl (pH 7.5) 0.05% LDAO. No ubiquinone added. These reaction centers contained only one molecule of ubiquinone as checked by the rate of the back-reaction after flash excitation ($t_{1/2}$ ~ 80 msec). (B) Same as (A) after addition of 50 μM ubiquinone-10.

bound. The fast phase is absent when reaction centers contain only acceptor Q_1 (Fig. 6A) and can only be observed by addition of exogenous ubiquinone (50 μM) (Fig. 6B).

These three types of experiments strongly suggest that the fast phase observed in the near-infrared region is related to electron transfer between primary (Q_1) and secondary (Q_2) acceptors.

Dependence of the Rate of Electron Transfer Between Q_1 and Q_2.

We have studied the dependence of the rate of electron transfer between Q_1 and Q_2, (as described in the preceding paragraph) upon different conditions such as the pH and temperature of the external medium. The kinetics of this electron transfer were also measured for two different states of Q_2 (fully oxidized and semireduced) prior to flash activation.

pH Figure 7 shows kinetics of absorbance changes measured around 750 nm for isolated reaction centers of *Rps. sphaeroides* R_{26} (Fig. 7A) and chromatophores (Fig. 7B) from the same species. The pH dependence of the rate of electron transfer is plotted in Fig. 8 for both reaction centers and chromatophores. At any given pH, the rate of electron transfer is three times faster for chromatophores than for isolated reaction centers, as already indicated (*19*), but the weak pH dependence is very similar in both cases. Moreover this pH dependence is similar to the one observed by Petty and Dutton (*5*) for the H^+ uptake by chromatophores from *Rps. sphaeroides* Ga. The pH dependence we observe for isolated *Rps. sphaeroides* R_{26} RC's (Fig. 8) is in good agree-

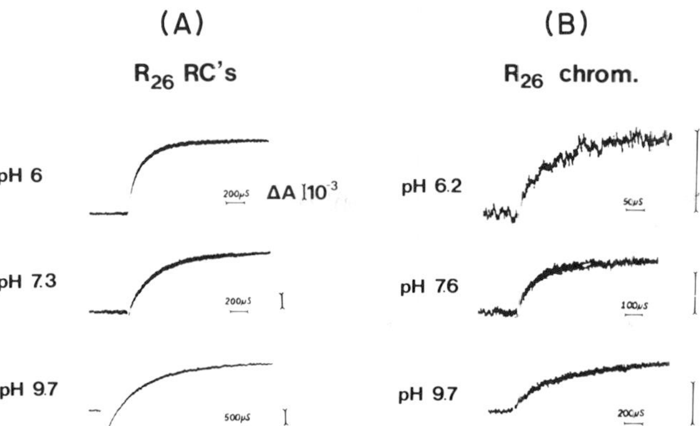

Fig. 7. Kinetics of absorbance changes at three different pH measured at the indicated wavelength. (A) isolated R_{26} reaction centers (2 μM). No artifical electron donor added (λ = 750 nm). (B) R_{26} chromatophores (absorption at 750 nm ~ 0.4). No artifical electron donor added (λ = 747.5 nm).

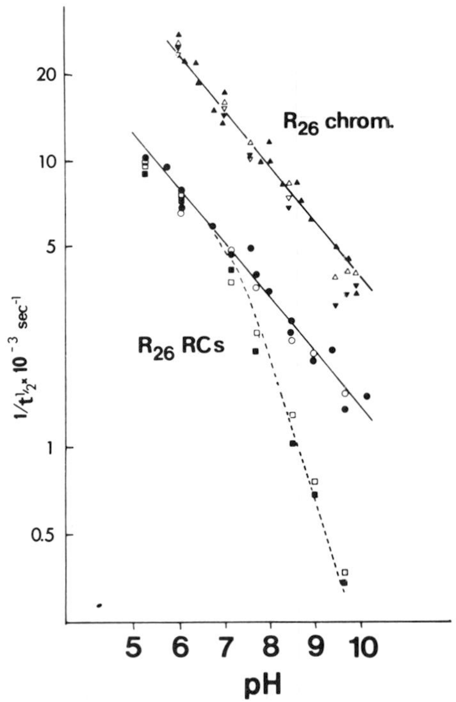

Fig. 8. The pH dependence of the electron transfer for chromatophores and isolated reaction centers from *Rps. sphaeroides* R_{26}. Reaction centers: (●) first electron transfer measured from experiments similar to those reported in Fig. 7A; (○) first and (□) second electron transfer from experiments similar to those shown in Fig. 11 (λ = 750 nm); and (■) second electron transfer from Fig. 11 (λ = 450 nm); Chromatophores: (▲) first electron transfer (λ = 750 nm) (Fig. 7B); (△) first and (▽) second electron transfer (λ = 750 nm); (▼) second electron transfer (λ = 450 nm).

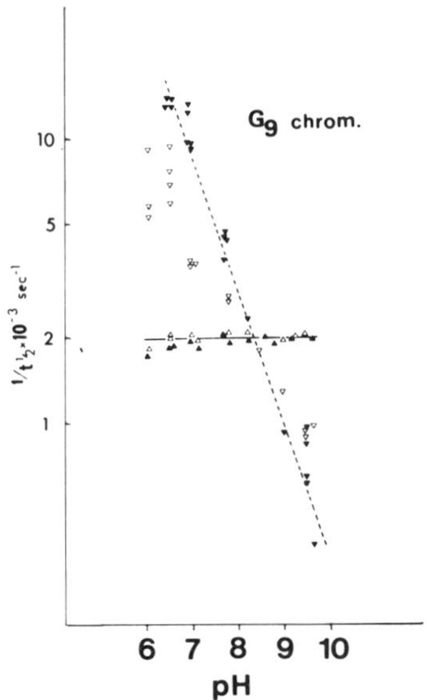

Fig. 9. Same as Fig. 8 but for *R. rubrum* G_9 chromatophores

ment with the dependence reported by Wraight (*28*). Chromatophores from *R. rubrum* G_9 exhibit no pH dependence for the rate of electron transfer from Q_1 to Q_2 between pH 5.5 and 10.5, the range studied (Fig. 9).

The non-integral slope (*Rps. sphaeroides* R_{26}) or the absence (*R. rubrum* G_9) of pH dependence of the rate of the first electron-transfer between primary and secondary acceptors ($Q_1^{\cdot-} Q_2 \rightarrow Q_1 Q_2^{\cdot-}$) indicates that this process is not limited by H^+ uptake.

Temperature An Arrhenius plot (Fig. 10) shows that the rate of electron transfer between Q_1 and Q_2 has an activation energy of about 10.5 kcal/mole between 3 and 24 °C for *R. rubrum* G_9 chromatophores. A similar temperature dependence was observed for *Rps. sphaeroides* R_{26} chromatophores (not shown). These activation energy estimates agree with the values reported by Parson (*22*) (8.3 kcal/mole for *Chromatium vinosum*) and by Chamorovsky *et al.* (*24*) (9.4–10.4 kcal/mole for *R. rubrum*). A value of 10.5 kcal/mole was also obtained for the activation energy of H^+ uptake by Petty and Dutton (*5*) for *Rps. sphaeroides* Ga chromatophores.

Redox states of Q_2 Kinetics measurements at 450 and around 750 nm of the electron transfer between $Q_1^{\cdot-}$ and Q_2 (first flash) and be-

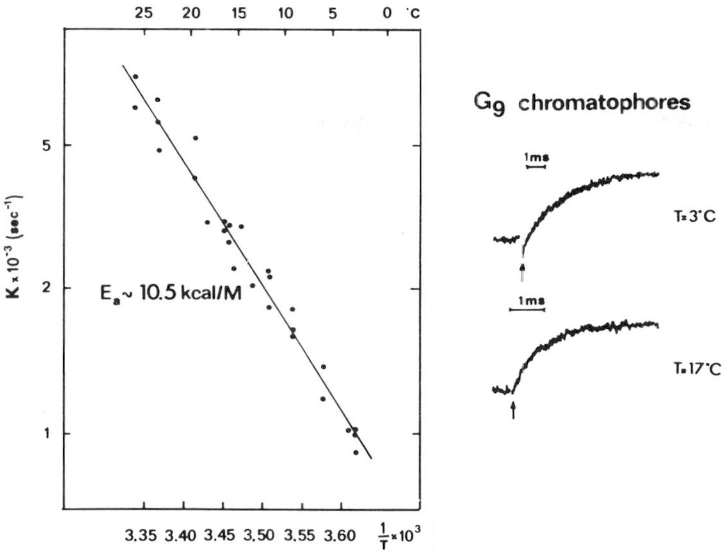

Fig. 10. Arrhenius plot of the rate of electron transfer between Q_1 and Q_2 for G_9 chromatophores suspended in 10 m*M* Tris HCl at pH 8. No electron donor added. The traces to the right show the kinetics measured at 747.5 nm at two different temperatures (T = 3 and 17 °C).

tween Q_1^- and Q_2^- (second flash) are shown in Figs. 11 and 12 for *R. rubrum* G_9 chromatophores and for *Rps. sphaeroides* R_{26} reaction centers. Plots of the reciprocal half-time of these electron transfers as a function of pH are shown in Figs. 8 and 9. Our measurements for isolated reaction centers agree with those reported by Wraight (*28*), although we obtain a better correlation between kinetic measurements at 450 and 750 nm on the second actinic flash (Fig. 11).

For chromatophores of *Rps. sphaeroides* R_{26}, the kinetics of electron transfer between Q_1 and Q_2 are not markedly dependent on the redox state of Q_2 (Fig. 8). More interesting is the situation in *R. rubrum* G_9 chromatophores. The kinetics of the first electron transfer between Q_1 and Q_2 do not depend on the medium pH, either in the presence or absence of an artificial electron donor (Fig. 9). However, the rate of the second electron transfer ($Q_1^- \cdot Q_2^- \rightarrow Q_1Q_2^{=}$) measured at 450 nm presents a pronounced pH dependence (~1 decade/two pH unit), indicating that this second transfer is limited by a proton uptake. No p*K* is observed in the pH range studied (6–9.5), in contradiction with the results obtained with *Rps. sphaeroides* reaction centers (Fig. 8) (*28*). The correlation between the kinetics measured at 450 nm and at 747.5 nm (Figs. 9 and 12) in the case of *R. rubrum* G_9 chromatophores is not good.

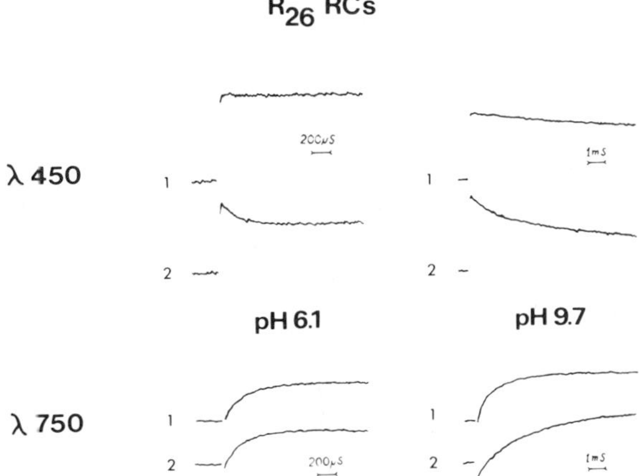

Fig. 11. Kinetics of absorbance changes measured at 450 and 750 nm on the first and second actinic flash for *Rps. sphaeroides* R_{26} isolated reaction centers in presence of 0.1 m*M* diaminodurene as electron donor at pH 6.1 and 9.7.

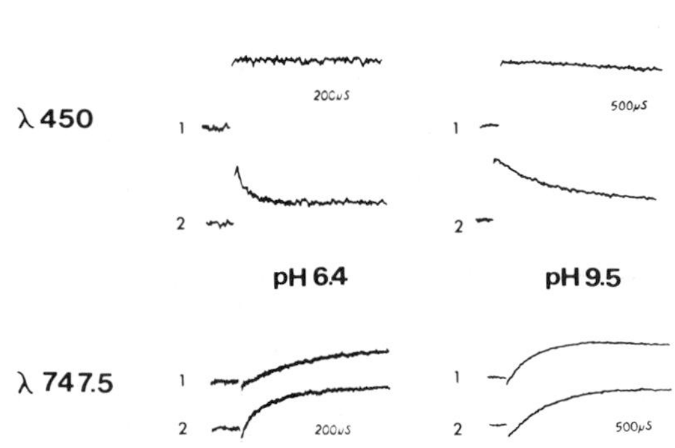

Fig. 12. Same as Fig. 11 but for *R. rubrum* G_9 chromatophores.

This discrepancy may be due to events (for example protonation) affecting the near-infrared kinetics or to some actinic effect of the near-infrared measuring light[1]. These possibilities are under investigation.

[1] The amplitude of the fast phase measured at 745 nm was not affected by diminishing the analyzing beam intensity by a factor of two.

CONCLUSIONS

We have presented in this paper several arguments suggesting that electron transfer between primary and secondary acceptors in chromatophores of *Rps. sphaeroides* R_{26} and *R. rubrum* G_9 can be kinetically resolved by measurements of light-induced absorption changes in the near-infrared region. This method is suitable for those species which do not possess a fast rereduction capability of $(Bchl)_2^{+}$ by cytochrome *c*, and can be applied regardless of the redox potential of the suspension.

The results we obtained for *Rps. sphaeroides* R_{26} chromatophores and reaction centers, and for *R. rubrum* G_9 are rather puzzling. No analogy is observed between the two species, or even between chromatophores and reaction centers of the same species. For example, the rate of electron transfer between Q_1 and Q_2 is three times slower in isolated reaction centers than in the intact chromatophores. This could reflect some alteration of the reaction center properties induced by extraction and purification procedures.

R_{26} chromatophores show a pH dependence for the first electron transfer between Q_1 and Q_2 (Fig. 8). This is not observed in the case of G_9 chromatophores (Fig. 9). However the second electron transfer for this species is limited by uptake of proton from pH 6 to 9.5. A pK is observed for R_{26} reaction centers, but not for G_9 chromatophores.

It is hoped that the combination of electron-transfer rate measurements with H^+ binding rate measurements in chromatophores of other species and under different experimental conditions will provide a better understanding of the mechanism of proton translocation in photosynthetic membranes.

REFERENCES

1. Holten, D., Hoganson, C., Windsor, M. W., Schenk, C. C., Parson, W. W., Migus, A., Fork, R. L., and Shank, C. V. (1980). *Biochim. Biophys. Acta* **592,** 461–477.
2. Rockley, M. G., Windsor, M. W., Cogdell, R. J., and Parson, W. W. (1975). *Proc. Natl. Acad. Sci. U.S.A.* **72,** 2251–2255.
3. Kaufmann, K. J., Dutton, P. L., Netzel, T. L., Leigh, J. S., and Rentzepis, P. M. (1975). *Science* **188,** 1301–1304.
4. Dutton, P. L., and Prince, R. C. (1978). *In* "The Photosynthetic Bacteria" (R. K. Clayton and W. R. Sistrom, eds.), pp. 525–570. Plenum, New York.
5. Petty, K. M., and Dutton, P. L. (1976). *Arch. Biochem. Biophys.* **172,** 335–345.
6. Crofts, A. R., and Wood, P. M. (1978). *Curr. Top. Bioenerg.* **7,** 175–244.
7. Okamura, M. Y., Isaacson, R. H., and Feher, G. (1975). *Proc. Natl. Acad. Sci. U.S.A.* **72,** 3491–3495.

8. Clayton, R. K., and Straley, S. C. (1972). *Biophys. J.* **12,** 1221–1234.
9. Halsey, Y. D., and Parson, W. W. (1974). *Biochim. Biophys. Acta* **347,** 404–416.
10. van Gorkom, H. J. (1974). *Biochim. Biophys. Acta* **347,** 439–442.
11. Mathis, P., and Haveman, J. (1977). *Biochim. Biophys. Acta* **461,** 167–181.
12. Pulles, M. P. J., van Gorkom, H. J., and Willemsen, J. G. (1976). *Biochim. Biophys. Acta* **449,** 536–540.
13. Velthuys, B. R., and Amesz, J. (1974). *Biochim. Biophys. Acta* **333,** 85–94.
14. Bouges-Bocquet, B. (1973). *Biochim. Biophys. Acta* **314,** 250–256.
15. Wraight, C. A. (1977). *Biochim. Biophys. Acta* **459,** 525–531.
16. Verméglio, A. (1977). *Biochim. Biophys. Acta* **459,** 516–524.
17. Barouch, Y., and Clayton, R. K. (1977). *Biochim. Biophys. Acta* **462,** 785–788.
18. de Grooth, B. G., van Grondelle, R., Romijn, J. C., and Pulles, M. P. J. (1978). *Biochim. Biophys. Acta* **503,** 480–490.
19. Bowyer, J. R., Tierney, G. V., and Crofts, A. R. (1979). *FEBS Lett.* **101,** 201–206.
20. Feher, G., Okamura, M. Y., and McElroy, J. D. (1972). *Biochim. Biophys. Acta* **267,** 222–226.
21. Wraight, C. A. (1978). *FEBS Lett.* **93,** 283–288.
22. Parson, W. W. (1969). *Biochim. Biophys. Acta* **189,** 384–396.
23. Case, G. D., and Parson, W. W. (1971). *Biochim. Biophys. Acta* **253,** 187–202.
24. Chamorovsky, S. K., Remennikov, S. M., Kononenko, A. A., Venedikto, P. S., and Rubin, A. B. (1976). *Biochim. Biophys. Acta* **430,** 62–70.
25. Verméglio, A., and Clayton, R. K. (1977). *Biochim. Biophys. Acta* **461,** 159–165.
26. Verméglio, A., Martinet, T. and Clayton, R. K. (1980). *Proc. Natl. Acad. Sci. U.S.A.* **7,** 1809–1813.
27. Wraight, C. A., and Stein, R. R. (1980). *FEBS Lett.* **113,** 73–77.
28. Wraight, C. A. (1979). *Biochim. Biophys. Acta* **548,** 309–327.

2

The Involvement of Stable Semiquinones in the Two-Electron Gates of Plant and Bacterial Photosystems

C. A. WRAIGHT

INTRODUCTION

Unlike the other major classes of biological redox components (flavoproteins, iron–sulfur proteins and cytochromes) quinones have frequently been considered to be free entities, more or less dissolved in the hydrocarbon matrix of the membrane. The notion of specific quinone-protein associations, although not entirely overlooked, lacked both experimental support and theoretical compulsion, partly because the redox properties of quinones *in vivo* seemed quite similar to those observed *in vitro*. *In vivo,* quinones generally exhibit a typical two-electron, two-proton oxidation–reduction with redox potentials in the same range as obtained in solution (*1–4*). It is now becoming clear, however, that although the bulk of the quinone in mitochondrial and photosynthetic membranes does behave "typically", active turnover of the redox chains involves only a few, specialized quinone molecules that are certainly protein bound.

The first demonstrated protein-bound quinone was the primary acceptor quinone of reaction centers from the photosynthetic bacterium, *Rhodopseudomonas sphaeroides* (*5–9*). The electron-acceptor system of these reaction centers is now recognized as a complex containing two ubiquinones acting in series as primary (Q_I) and secondary (Q_{II}) acceptor

Function of Quinones in Energy Conserving Systems

ISBN 0-12-701280-X

quinones (*9–11*). An iron atom (Fe^{2+}) is located close enough to interact magnetically with the semireduced forms of both acceptor quinones (*11–14*), but is not coordinated to them (*15*). Although iron may be involved in electron transfer between them (*17*), this function is not certain (*16*). Recently iron has also been demonstrated in the very similar plastoquinone-acceptor complex of photosystem II (PS II) in plants (*18*).

Many years ago Michaelis suggested that two-electron oxidation–reduction reactions might proceed via elemental one-electron steps (*19*). In spite of this early insight, one consequence of the "solutional" view of quinones *in vivo* was a general concern as to how two-electron redox components can be coupled to the one-electron reactions of iron–sulfur, heme proteins, and photochemistry. In the last seven years, some two-electron gating mechanisms ideally suited to the coupling of one- and two-electron couples were observed in the photosynthetic electron-transport systems of both plants (PS II) and purple photosynthetic bacteria. Appropriately, these processes were recognized at a time when strict two-electron reactivity of quinones *in vivo* had clearly been invalidated. These so-called two-electron "gating" systems involve highly atypical activities, achieved only by the specific association of quinones in a protein complex.

Manifestations of the Two-Electron Gates in Plants and Bacteria

The first observations of two-electron gating activities were made in plants. Bouges-Bocquet observed that under certain conditions electrons from PS II reached PS I only on alternate flashes (*20*). Velthuys and Amesz found a component of the PS II electron-acceptor system was stably reduced on odd flashes and reoxidized following even flashes (*21*). Pulles *et al.* (*22*) subsequently observed binary oscillations in the formation and disappearance of an absorbance signal in a series of flashes, attributed to anionic plastosemiquinone. A very similar activity, definitely attributed to ubisemiquinone, was observed in *Rps. sphaeroides,* first in isolated reaction centers (*10, 11*) and later in chromatophores (*23–25, 43*).

The operation of the gate and the origin of the binary oscillations can be approximately represented as (*10, 11*)

$$\text{1st flash:}\quad Q_IQ_{II} \xrightarrow{h\nu} Q_I^-Q_{II} \longleftarrow Q_IQ_{II}^- \text{ (stable)}$$

$$\text{2nd flash:}\quad Q_IQ_{II}^- \xrightarrow{h\nu} Q_I^-Q_{II}^- \xrightarrow{H^+} \xrightarrow{H^+} Q_IQ_{II}H_2 \longrightarrow Q_IQ_{II}\quad (A \rightarrow AH_2)$$

where Q_I and Q_{II} represent the primary and secondary acceptor quinone molecules (denoted Q and B in PS II) and A represents subsequent redox components including the quinone pool. The primary and secondary quinones in *Rps. sphaeroides* are both ubiquinone and in PS II they are both plastoquinone. However, oscillatory behavior is not dependent on this identity as it is also seen in subchromatophore particles from *Chromatium vinosum* (Wraight, unpublished observations), where the primary quinone is menaquinone (*26*) and the secondary quinone is ubiquinone (*27*).

The two-electron gate has many manifestations in the electron-transport chains of plants and photosynthetic bacteria, and activities reported to exhibit binary oscillations are summarized in Table I. Clearly, the gating processes in plants and bacteria are very similar and control identical

TABLE I

Manifestations of a Two-Electron Gate in Plants (PS II) and Bacteria

Observation/phenomenon	References: Bacteria	References: Plants
Availability of electrons to PS I on alternate flashes	N.A.	(*20*)
Enhanced reduction of Q_I(Q) on addition of herbicides, etc., or strong reductant. (Max. after odd flashes).	(*36*)	(*21*)
Anionic semiquinone formation (odd flashes) and disappearance (even flashes)	Optical (*10, 11, 57*) ESR (*11, 14*)	Optical (*22*)
Quinol formation (on even flashes)	(*10*)	—
H^+ uptake (on even flashes)	pH-dependent (*24, 28, 46*)	None detected to date
H^+ release (on even flashes)	(*47*)	(*37, 38*)
Electrochromic band shifts (max. on even flashes)	(*23*)	(*39*)
Rate of transfer from $Q_I \rightarrow Q_{II}$ (Q → B) (slow on even flashes)	pH-dependent (*28*)	pH-independent (*40, 41*)
Reversal of HCO_3^- depletion (suppressed after odd flashes)	No HCO_3^- effect known	(*42*)
Inhibition of $Q_I \rightarrow Q_{II}$ transfer by herbicides, etc., (suppressed after odd flashes)	(*36, 43*)	—
Oxidation–reduction of other identifiable electron transport components	Cytochrome *b* reduction (*25*)	Cytochrome *f* reduction (*44, 45*)

activities. However, there are some differences which may be of mechanistic significance. It is particularly noteworthy that binary oscillations in H^+ uptake (predicted from the preceeding simple scheme) were readily observed in bacterial reaction centers (*28*) and have been reported in chromatophores (*24*) but not detected in chloroplasts, despite considerable efforts. This is unfortunate, as studies on protonation events in bacterial reaction centers are most useful for probing the oxidation–reduction reactions of acceptor quinones (*29*). However, these studies do offer a possible explanation for the lack of oscillations in chloroplasts.

Figure 1 shows a detailed scheme of the electron and proton transfer reactions of the quinone complex on odd and even flashes (*28*). The pattern of H^+ uptake is pH-dependent. At low pH, strong binary oscillations are seen, with no protons bound on the first flash and two on the second. At high pH, the oscillations disappear and one proton is taken up on each flash. The oscillations in the formation and disappearance of the anionic semiquinone, however, are unaffected throughout the pH range. This is most simply accounted for by supposing that an unidentified acid–base group (N^-/NH) becomes deprotonated at high pH when the acceptor-quinone complex is fully oxidized (Q_IQ_{II}). Following a single flash, this group undergoes a p*K* shift in response to the appearance of anionic semiquinone (probably $Q_{II}^{\overline{\cdot}}$) and is reprotonated. This group could be functional in subsequent proton transfer to the fully reduced quinol ($Q_{II}H^- \rightarrow Q_{II}H_2$). Attempts to identify group N in bacterial reaction centers have been unsuccessful so far, but a modulating effect on p*K* may be exerted by phospholipid present in the reaction centers (Wraight, unpublished observations).

A similar sequence of events leading to a "referred" protonation, could be responsible for the lack of H^+ binding oscillations in chloroplasts. H^+ binding in chloroplasts is, in any case, kinetically obscured by a significant diffusion barrier—apparently proteinaceous in nature (*29*)—that retards H^+ uptake by as much as two orders of magnitude (*30*). As will be further discussed, referred protonation events could be of considerable importance in determining equilibrium redox properties of acceptor quinones.

Physicochemical Properties of the Acceptor Quinones in Bacteria

The operation of the two-electron gates clearly represents unusual properties of the two quinones involved. In particular, both Q_I and Q_{II} exhibit semiquinones with extraordinarily long life times—lasting 10 to 20 minutes under appropriate conditions (*10*, *11*, *31*). Furthermore, this stability is thermodynamic and not just kinetic (i.e., restricted access to other semiquinones for dismutation). Redox titrations of Q_I are clearly $n = 1$ (*32–34*) and estimates of the second reduction step place it at much lower potentials than the first (*35*, *49*). Q_{II} has also recently been titrated in chromatophores of *Rps. viridis* and *Rps. sphaeroides* (*51–53*), and exhibits two one-electron steps: in *Rps. sphaeroides* at pH 8.0, E_m (1) $\simeq +40$ mV for $Q_{II}/Q^{\bar{\cdot}}_{II}(H^+)$ and E_m (2) $\simeq -40$ mV for $Q^{\bar{\cdot}}_{II}(H^+)/Q_{II}H_2$.

These stable semiquinones have properties that contrast sharply with

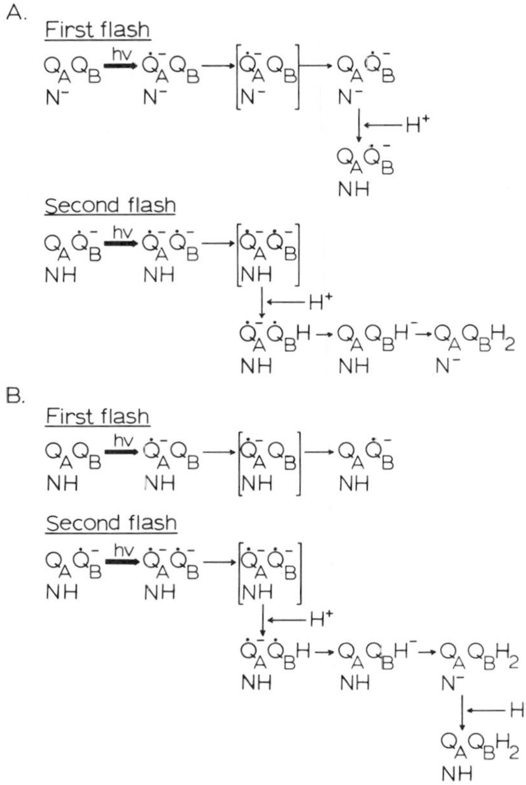

A.

First flash

$$\underset{N^-}{Q_A Q_B} \xrightarrow{h\nu} \underset{N^-}{\dot{Q}_A^- Q_B} \rightarrow \left[\underset{N^-}{\dot{Q}_A^- Q_B}\right] \rightarrow \underset{N^-}{Q_A \dot{Q}_B^-} \xrightarrow{+H^+} \underset{NH}{Q_A \dot{Q}_B^-}$$

Second flash

$$\underset{NH}{Q_A \dot{Q}_B^-} \xrightarrow{h\nu} \underset{NH}{\dot{Q}_A^- \dot{Q}_B^-} \rightarrow \left[\underset{NH}{\dot{Q}_A^- \dot{Q}_B^-}\right] \xrightarrow{+H^+} \underset{NH}{\dot{Q}_A^- \dot{Q}_B H} \rightarrow \underset{NH}{Q_A Q_B H^-} \rightarrow \underset{N^-}{Q_A Q_B H_2}$$

B.

First flash

$$\underset{NH}{Q_A Q_B} \xrightarrow{h\nu} \underset{NH}{\dot{Q}_A^- Q_B} \rightarrow \left[\underset{NH}{\dot{Q}_A^- Q_B}\right] \rightarrow \underset{NH}{Q_A \dot{Q}_B^-}$$

Second flash

$$\underset{NH}{Q_A \dot{Q}_B^-} \xrightarrow{h\nu} \underset{NH}{\dot{Q}_A^- \dot{Q}_B^-} \rightarrow \left[\underset{NH}{\dot{Q}_A^- \dot{Q}_B^-}\right] \xrightarrow{+H^+} \underset{NH}{\dot{Q}_A^- \dot{Q}_B H} \rightarrow \underset{NH}{Q_A Q_B H^-} \rightarrow \underset{N^-}{Q_A Q_B H_2} \xrightarrow{+H^+} \underset{NH}{Q_A Q_B H_2}$$

Fig. 1. A scheme for electron transfer and proton binding in the quinone acceptor complex of photosynthetic reaction centers. (Q_A and Q_B are equivalent to Q_I and Q_{II} in the text). (A) At high pH the proton taken up on the first flash is not initially bound to a semiquinone but to a non-chromophoric group (N^-/NH) in the reaction center complex. After the second flash the rate of disproportionation of the two semiquinones is limited by the uptake of a single H^+. This is the first proton onto the quinone (Q_B). The subsequent electron transfer produces fully reduced quinol which can obtain its second proton from the group NH, protonated on the first flash. (B) At low pH (<7) it has not been possible to kinetically distinguish the two protons taken up on the second flash, but the model derived at high pH can be readily extended to low pH. In this case, the second proton bound regenerates NH. The two protons would indeed be kinetically indistinguishable if the intervening steps, including electron transfer, were sufficiently fast. [After Wraight (*28*).]

the bulk quinones of bacteria (*3*) and with many quinones in solution (*4*), that exhibit strong two-electron redox behavior. Typical values would put E_m (*1*) several hundred millivolts more negative than E_m (*2*), corresponding to a very low stability constant for the semiquinone e.g., $K_{sq} < 10^{-7}$ in *Rps. sphaeroides* at pH 7 (*3*)

$$\boldsymbol{K}_{sq} = \frac{[Q^{\overline{\cdot}} \text{ or } Q{\cdot}H]^2}{[Q][QH_2]}$$

Insight into the significance of stable semiquinones can be obtained from a study of E_m/pH relationships. Some instructive but idealized examples of the many possible responses for a generalized, one-electron redox couple are shown in Fig. 2. In each case the redox behavior is modified by preferential stabilization (*e.g., binding*) of some of the components of the redox couple relative to others. In Fig. 2A, stabilization of all re-

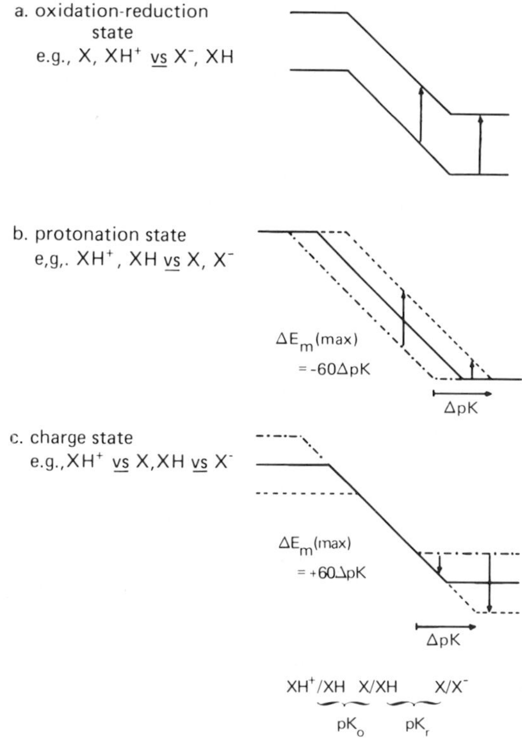

Fig. 2. Effects of preferential interactions with different states of a one-electron redox couple on E_m/pH curves. The specificity of the interaction is given on the left and the expected shifts in the E_m and pK values are on the right.

duced species, regardless of protonation state, causes a large shift upwards of the E_m/pH curve. Mechanistically, this might be achieved, in the case of the quinone/semiquinone couple, by π-electron donating ring substituents enhancing the aromaticity of the semiquinone.[1]

Many other stabilizing effects can arise from preferential interactions with the environment, and protonation (more generally, ligand binding) is one example. In Fig. 2B, a preferential interaction with the protonated forms, regardless of oxidation–reduction state, causes a p*K* shift to higher values and a corresponding increase in E_m in the pH-dependent region. Mechanisms for this type of stabilization could include hydrogen–bonding of the protonated forms to nucleophillic sites in the environment, such as a binding pocket of a protein. An opposite shift, or course, would arise if the deprotonated forms were hydrogen-bonded by electrophillic, proton-donor sites.

Figure 2C shows the effects of (de)stabilization processes active on the charged species. Destabilization of the negative species and stabilization of the positive species causes a p*K* shift to higher values and a corresponding decrease in E_m in the pH-independent regions. An obvious source of this type of effect would be an electric field provided by a local charge.

Some of these influences were considered previously (*28*, *48*), but it is important to realize that they are neither mutually exclusive nor necessarily independent. There are, in fact, many other possibilities, such as charged proton-donors or dipoles that could simultaneously contribute more than one form of interaction. The examples of Fig. 2 can be looked upon as elementary steps from which more complex effects can be built by appropriate combinations.

Figure 3 (solid lines) shows the E_m/pH curves for the component, one-electron reduction steps of a typical quinone. At low pH, the first reduction step is at lower potential than the second, the semiquinone is intrinsically unstable and quinone/quinol is a two-electron redox couple. At higher pH, the semiquinone passes through its p*K* and the first reduction step becomes pH-independent. Typical values for the p*K* of a benzosemiquinone in protic solvents would be 4–6, indicating a weak to moderate acid strength. The fully reduced quinol, however, is generally a strong base with two p*K* values in excess of 10–12. Thus, above the p*K* of the semiquinone there is usually a range of pH where the second reduction step is $Q^{\cdot-}/QH_2$, with a marked pH-dependence for the E_m (*2*) of −120

[1] Stabilization of a semiquinone by this means would be accompanied by a general destabilization of reduced species by electron-donating ring substituents. Thus, although the semiquinone may become relatively stabilized, the average potential for two-electron reduction of the quinone to the quinol could be lowered.

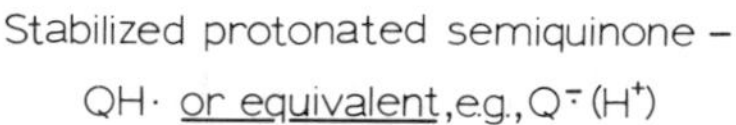

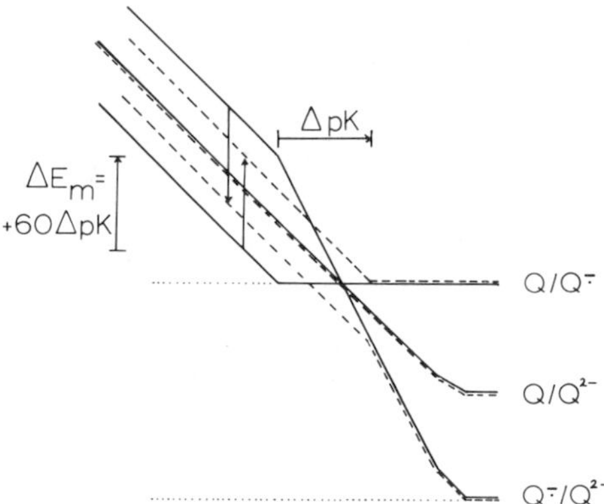

Fig. 3. E_m/pH relationships for the one-electron components of a two-electron (quinone-like) redox couple. The effect of specific stabilization of the unprotonated semiquinone is shown by the dashed lines. The dotted lines indicate that the $E_{m,pK}$ values may be operative even below the pK if protonation is slow or otherwise obstructed.

mV/pH. This will generally lead to a crossing of the midpoint potentials of the two one-electron steps and the emergence of a stable anionic semiquinone at high pH.

From this description of "typical" quinone properties it is clear that one way to stabilize the semiquinone is to lower the pK of the semiquinone. This could be achieved in one step by a simple charge stabilization or, *preferential binding,* of the anionic semiquinone relative to all other forms of the quinone, i.e., as in Fig. 2C. However, other combinations based on Fig. 2B can also achieve this end. In either case, a stable semiquinone is obtained by shifting the pH-dependent region far enough to the left that the stable, unprotonated semiquinone emerges in the pH range of interest.

The Stable Semiquinone of Q_{II}

The H^+ uptake by the secondary quinone of the acceptor-quinone complex of bacteria and plants generally occurs on each flash, accompanying the reduction of Q_{II} (B) to the semiquinone level (*28*, *37*, *38*). This is supported by dark redox titrations of Q_{II} in chromatophores from *Rps. sphaeroides* that show that the first and second reduction steps are pH-dependent, with a slope of −60 mV/pH, up to at least pH 9.5 where a pK on the semiquinone form may be evident (*53*). It is particularly noteworthy that throughout this pH region the absorption spectrum of the semiquinone shows it to be anionic in nature. It is clear, therefore, that association of the quinone with the protein of the acceptor-complex has extended the range of pH dependence for the quinone/semiquinone redox couple, through the species $Q^{\cdot}(H^+)$. Thus, it appears that stabilization of the semiquinone of Q_{II} is not achieved by depression of the p*K*.

Stabilization of the semiquinone in the pH-dependent region can be achieved by preferential interactions of the type shown in Fig. 2B. In Fig. 3, the dashed lines show the effect of stabilization of only the protonated semiquinone $Q^{\bar{\cdot}}H$ *or equivalent* $Q^{\bar{\cdot}}(H^+)$. The result, depending on the degree of stabilization, can be stability of the semiquinone throughout the entire pH range, as shown. A mechanistically attractive origin for this effect might be that the anionic semiquinone moiety is hydrogen bonded to a nearby proton-donor site. This interaction could be transmitted to an acid–base group (N^-/NH in Fig. 1), at the outside of the protein, which undergoes a p*K* shift. Because the semiquinone is acting as a hydrogen-bond acceptor, this interaction can be made specific for the anionic form. In this case, the average midpoint potential (i.e., Q/QH_2) is unaltered as the two one-electron steps move by equal but opposite amounts. Slight modifications of this outline can also lead to a stable $Q^{\bar{\cdot}}(H^+)$ species but with, additionally, a shift in the average E_m. For Q_{II} the average of the two reduction steps ($\frac{1}{2}[-40 + 40] = 0$ mV at pH 8.0) (*53*) is close to the $n = 2$ midpoint potential for the ubiquinone pool ($E_{m,8} = +30$ mV) (*3*).

On the basis of this analysis, I suggest that the involvement of the protein in the acceptor-quinone complex serves to extend the pH dependence of the semiquinone by referring the protonation to an external site on the complex. Thus, the anionic semiquinone is preferentially stabilized (bound) by a protonated form of the protein, giving the effective redox species $Q^{\bar{\cdot}}_{II}(H^+)$. The iron atom, which has no known function in electron transfer, could be involved in determining the redox behavior of the quinones. Removal of iron is also accompanied by loss of the H-subunit of the reaction centers (*55*), and causes Q_I to revert to $n = 2$ redox behavior (*56*). It was suggested that iron is involved in determining the redox properties of Q_I. Since the iron is probably almost equidistant from the two quinones (*14*), it may be involved in determining the redox properties of both Q_I and Q_{II}. An active participation, by iron, in propagating the referred protonation to the surface of the complex, is an attractive possibility.

The stability of the Q_{II} semiquinone that arises from the referred protonation is thermodynamic in the sense that the semiquinone is now at an intermediate free energy level between the quinone and quinol forms of

Q_{II}. It is also stabilized against dismutation via the pool quinones because the first reduction step of the pool quinones is probably at much lower potential ($E_{m,7}$ [1] ≤ -120 mV) [estimated from Takamiya and Dutton (*3*)], even though the average ($n = 2$) potential is slightly higher.

Transient Effects on the Redox Properties of Q_{II}

Immediately following the second flash (see Fig. 1) the acceptor-quinone complex is in the state $Q_I^- Q_{II}^-(H^+)$. The Q_{II} semiquinone probably now undergoes a true protonation to form $Q_{II}H\cdot(H^+)$, taking up a proton from the medium (*28*). This implies a large positive shift in the p*K* of this semiquinone in response to the anionic Q_I semiquinone. A likely corollary of such a p*K* shift, as indicated in Fig. 2B, is a downward shift of the E_m above the p*K*, prior to the H^+ uptake (*28*). This transient shift in the E_m of $Q_{II/Q_{II}^-H^+}$) may make the semiquinone reducing capacity sufficient for one-electron transfer to the pool quinone, or to some other low-potential component. This would, of course, be in competition with the protonation and subsequent electron transfer from Q_I^- to $Q_{II}H_2$. Nevertheless, it might provide a way to circumvent the oscillatory phenomena of the two-electron gates. Such behavior does occur occasionally and unpredictably in isolated reaction centers. I have previously suggested that a "ping-pong" process might account for this (*31*). The appearance of Q_I^- on the second flash could provide the "ping" for the "pong", the movement of an electron from Q_{II}^- to the pool. Oscillations in chromatophores are also frequently absent. The conditions necessary to obtain oscillations are uncertain (*23–25, 43*), and mechanisms may exist for circumventing them (*49, 50*).

The Semiquinone of Q_I

Dark titrations of the primary acceptor quinone have long been known to generate a stable semiquinone (*32–34*). Although the midpoint potential in such titrations is pH-dependent, the flash-induced reduction of Q_I is not accompanied by H^+ uptake and inhibition of electron transfer to Q_{II} inhibits H^+-uptake (*27, 33*). It was suggested, therefore, that under normal conditions of turnover Q_I operates at the midpoint potential above the p*K* of the reduced form (*32*). Furthermore, the optical spectrum of chemically reduced Q_I is clearly that of an anionic semiquinone (*5, 58*). Thus, as for Q_{II}, protonation appears to be a property of the acceptor complex and the semiquinone is $Q_I^-(H^+)$. It is thought that the protonation is a slow process, occurring on the time scale of a redox titration, not of a normal electron transfer (*32*). The E_m for Q_I reduction above the p*K* is -150 to -200 mV for a wide variety of bacterial species (*34*); a similar behavior can be ascribed to Q in PS II (*54*). In Fig. 3, the upper dotted line shows the "operating" midpoint potential for the $Q/Q^{\overline{\cdot}}$ couple.

Recently the second reduction step for Q_I has been reported (*35*, *51*–*53*). As yet there is no data on its protonation state but the very low E_m value estimated at pH 8.0 (*35*) is similar to that measured by titration at pH 10.0 (*53*), suggesting that it may be $Q^{\overline{\cdot}}/Q^{2-}$. If this is the case, then extreme protection against the aqueous environment would be sufficient to account for the stability of the semiquinone (see dotted lines in Fig. 3). However, an involvement of the iron atom, as previously discussed, is also a possibility.

Interactions between Q_I and Q_{II} and Inhibitors of Electron Transport from Q_I to Q_{II}

A general type of interaction responsible for E_m/p*K* shifts of the type shown in Fig. 2, is preferential binding of a "ligand" to certain species of a redox couple. Thus, preferential binding of the oxidized form of cytochrome *c* to negative membranes gives rise to a negative shift in the E_m (*59*). The magnitude of the E_m shift is a measure of the relative binding strengths of the two classes of redox forms; for example, oxidized cytochrome *c* binds seven times more strongly than the reduced form, giving rise to an E_m shift of -50 mV (*59*). Similarly, *o*-phenanthroline has been shown to raise the E_m of Q_I in the pH-dependent region but not above the p*K* (*32*–*34*). This was interpreted (*34*) as implying that *o*-phenanthroline binds preferentially to the protonated, reduced form of Q_I ($Q_I^{\overline{\cdot}}[H^+]$). We have recently shown that the strength of *o*-phenanthroline binding to the acceptor-quinone complex depends on the redox states of both Q_I and Q_{II}; this results in the one-electron redox equilibrium between $Q_I^{\overline{\cdot}}\,Q_{II}$ and $Q_IQ_{II}^{\overline{\cdot}}$ shifting towards Q Q_{II} (*36*). The redox potential span between the two quinones (ΔE_m) shifts from $+80$ mV to about -40 mV in the presence of saturating levels of *o*-phenanthroline.

We are now beginning to examine the interactions between Q_I, Q_{II} and *o*-phenanthroline in greater detail. Figure 4 shows the E_m/pH curves for Q_I in isolated reaction centers with variable levels of ubiquinone present. The reaction centers were incorporated into unilamellar phospholipid vesicles for stability and to provide a medium for the ubiquinone. It is clear that the presence of ubiquinone, in excess of one Q per reaction center, lowers the midpoint potential of Q_I. The full effect of the ubiquinone is

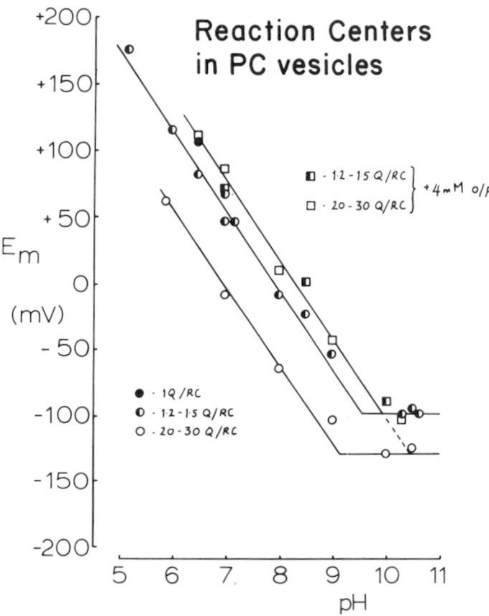

Fig. 4. The pH dependence of the E_m of the primary quinone (Q_I) in bacterial reaction centers. Reaction centers isolated from *Rps. sphaeroides*, R_{26}, were incorporated into egg phosphatidyl choline vesicles at a protein-to-lipid ratio of 1:10 (w/w) using a cholate, gel exclusion method. The reduction level of Q_I was determined by the flash-induced production of P^+ monitored at 433 nm. Redox mediators present were: 1,4-naphthoquinone, 1,2-naphthoquinone, 2-hydroxy-1,4-naphthoquinone, duroquinone, and diaminodurene (all 10 μM), phenazine methosulfate and phenazine ethosulfate (1 μM), and pyocyanine (3 μM). (◐), reaction centers (RC), as prepared, containing 1.2–1.5 Q/RC. The quinone level was reduced to 1 Q/RC by extraction by the method of Okamura *et al.* (*9*) (●). Reaction centers were also supplemented with quinone by drying down the phospholipid with ubiquinone-10 in chloroform-methanol prior to resuspension in cholate buffer and incorporation of reaction centers. The level of supplementation was 25–30 Q/RC, (○). The corresponding square symbols (◧, □) indicate titrations carried out in the presence of 4 m*M* *o*-phenanthroline.

probably not seen here as the protein-to-lipid ratio is very low (about 3 to 5 reaction centers per vesicle), and therefore the ubiquinone in the vesicle membrane is quite dilute. In chromatophores, relative lipid content is much lower and the ubiquinone is correspondingly more concentrated. The dependence of the E_m of Q_I on the level of ubiquinone present suggests that Q_{II} (or, more probably, $Q_{II}H_2$ at the potentials necessary for titration of Q_I) shifts the E_m of Q_I. The p*K* on $Q_I^-(H^+)$ may also be shifted. As a simple means of quantitating the level of restoration of Q_{II}, I have used the ratio of slow back reaction ($P^+Q_IQ_{II}^- \rightarrow P\ Q_I\ Q_{II}$; $t_{1/2} \simeq 2$ sec) to fast ($P^+Q_I^- \rightarrow P\ Q_I$; $t_{1/2} \simeq 100$ msec). Even at quite high levels of ubiquinone there is usually a significant amount (30–40%) of fast phase for reaction centers in phosphatidyl choline vesicles. This suggests that the binding of Q_{II} to reaction centers is not very strong in vesicles. Furthermore, the fact that the kinetics are biphasic (rather than monophasic with a variable half-time) indicates that, following a flash, Q_{II} cannot rapidly bind to reaction centers in the $P^+Q_I^-$ state of the complex. Thus, the ratio of the backreaction phases reflects the binding equilibrium established before the flash:

$$\text{fast} + Q \rightleftharpoons \text{slow}$$

$$K = \frac{(\text{slow})}{(\text{fast})(Q)}; \quad Q = \frac{1}{K} \cdot \frac{(\text{slow})}{(\text{fast})}$$

Figure 5 shows the dependence of the E_m of Q_I on the ratio of slow to fast back-reaction components, as an indicator of quinone level in the membranes. The data, although scattered, are remarkably consistent with expectation. The open circles represent points taken in the pH-dependent region of the Q_I midpoint potential (pH 5.5–9.0) and subsequently converted to an E_{m7} by subtracting 60(7.0 − pH) mV. The filled circles are values of the E_m taken above the pK (at pH 10.4–10.6). The two sets of data have been superimposed by an arbitrary adjustment of the vertical scale. It is apparent that the "Q-dependences" of the two E_m values are similar, approaching a value of −60 mV/decade [Q]. This is the expected

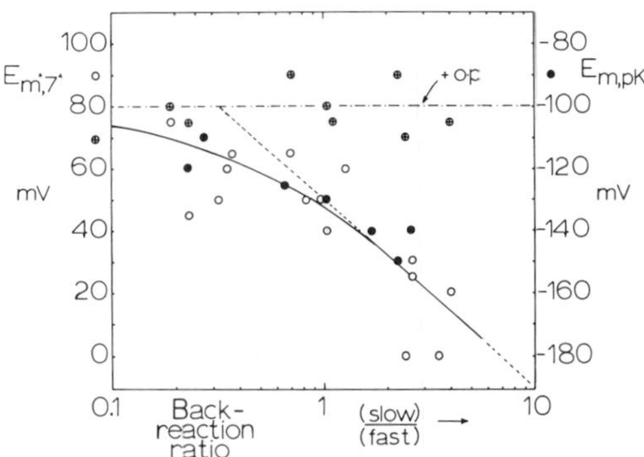

Fig. 5. Dependence of the E_m of the primary quinone (Q_I) on the presence of the secondary quinone (Q_{II}). Conditions as for Fig. 4. The secondary quinone level is indicated by the ratio of the slow back-reaction amplitude to the fast back-reaction as described in the text. The back-reaction was measured in separate samples in the absence of mediators or other donors to the reaction center. (●) $E_{m,pK}$ values determined at pH 10.4–10.6. (○), $E_{m,"7"}$ values determined in the pH-dependent region and adjusted to pH 7 by subtracting 60(7.0 − pH) mV. (⊕), $E_{m"7"}$ values, as for ○, but in the presence of 4 mM o-phenanthroline.

slope for a preferential binding interaction of a single $Q_{II}H_2$ with the oxidized form of Q_I.

Figures 4 and 5 also show that *o*-phenanthroline has little or no effect on the E_m of Q_I in the absence of $Q_{II}H_2$. In fact, the general effect of *o*-phenanthroline is simply to reverse the effect of $Q_{II}H_2$. This is indicative of a competitive displacement of $Q_{II}H_2$ by *o*-phenanthroline with *o*-phenanthroline exhibiting no preference for the redox states of Q_I, at least below the p*K* region. If the E_m is depressed above the p*K* (dotted line in Fig. 4) then *o*-phenanthroline must bind more strongly than $Q_{II}H_2$ to the protonated, reduced form of the complex ($Q_I^-[H^+]$). If the E_m is not lowered, then preferential binding of *o*-phenanthroline may require both the protonated form of the acceptor complex and $Q_{II}H_2$. In this case the interaction of *o*-phenanthroline with the acceptor complex would not be simply a competitive displacement of $Q_{II}H_2$. However, for reaction centers in vesicles with low levels of Q (1–2 Q per reaction center), 2–4 m*M* *o*-phenanthroline causes a significant (50–70%) loss of photochemical turnover, suggesting that it does displace Q_I (Wraight, unpublished observations). This is consistent with the proposed displacement of Q_{II}. Direct binding studies on the *o*-phenanthroline-reaction center system are currently underway to clarify these issues.

Electron Transfer through the Gate

The transfer of reducing equivalents from $Q_{II}H_2$ or B^{2-} (BH_2) to the electron-transport chains is little understood. Govindjee *et al.* (*60*) presented evidence indicating that this is the primary site of action of the bicarbonate effect in chloroplasts. In HCO_3^--depleted samples, the time for transfer is increased to 100–200 msec from a value less than 5 msec.

In bacteria, Okamura *et al.* (*9*) reported that reaction center quinones are more readily extracted under reducing conditions. This is consistent with a weaker binding of $Q_{II}H_2$ than of Q_{II}. Since the binding of Q_{II} does not appear to be very strong (see previous section) unbinding of $Q_{II}H_2$ and replacement by an oxidized Q from the pool may be a mechanism for transfer. However, it is probable that under conditions of rapid turnover, in bacteria (*61*, *62*) and plants (*64*), the pool quinones are bypassed and electron transfer occurs directly to the kinetically highly active components, including another specialized quinone (Q_z in bacteria (*61*), U in chloroplasts (*63*) and cytochromes. Such an activity need involve only rather small amplitude motions of the secondary quinone. Lack of any significant requirement for the pool is demonstrated by the kinetic competence of chromatophores after extraction of 90% of the quinone (*3*). It is also possible that the two-electron gating process is not functional under conditions of optimal turnover but only when communication to the pool quinone is forced on the reaction center by unfavorable redox status of the main chain.

CONCLUSIONS

Specific interactions of quinones of acceptor complexes of bacteria and PS II are responsible for the unusual stability of the semiquinone species. It is suggested that a hydrogen-bonding interaction to the anionic semiquinone of Q_{II} is transduced into a pK shift of an exposed acid–base group (N^-/NH) that becomes protonated. This serves to shift the pK of the protonated semiquinone complex to higher pH values and causes a net stabilization of the semiquinone throughout the pH range. In effect, binding of the anionic semiquinone is strong compared with the quinone and quinol, and requires a protonated state of the protein. The effective redox state is, therefore, $Q_{II}^{\bar{}}(H^+)$. A similar effect may be proposed for Q_I but this would be functionally irrelevant as the semiquinone is normally reoxidized before protonation can occur.

Complex interactions also occur between the two quinones of the acceptor complex. The dependence of these interactions on the redox state of the quinones can cause large transient perturbations in redox parameters of the quinones which would be responsible for unexpected kinetic effects. In view of the close proximity of many components in the redox complexes of biological electron-transport systems, such transient perturbations may be quite widespread and of considerable significance in determining the effective electron-transfer pathways.

ACKNOWLEDGMENTS

I thank Les Shipman for helpful discussions. This work was supported by grants from NSF (PCM 80-12032) and the USDA Competitive Research Grants Office (USDA/SEA 9-0308-0).

REFERENCES

1. Urban, P. F., and Klingenberg, M. (1969). *Eur. J. Biochem.* **9,** 519–525.
2. Parson, W. W. (1978). *In* "The Photosynthetic Bacteria" (R. K. Clayton and W. R. Siström, eds.), pp. 455–469. Plenum, New York.

3. Takamiya, K., and Dutton, P. L. (1979). *Biochim. Biophys. Acta* **546,** 1–6.
4. "The Chemistry of Quinoid Compounds" (1974). (S. Patai, ed.). Wiley (Interscience), New York.
5. Clayton, R. K., and Straley, S. C. (1972). *Biophys. J.* **12,** 1221–1234.
6. Clayton, R. K., and Yau, H. F. (1972). *Biophys. J.* **12,** 967–881.
7. Feher, G., Okamura, M. Y., and McElroy, J. D. (1972). *Biochim. Biophys. Acta* **267,** 222–226.
8. Cogdell, R. J., Brune, D. C., and Clayton, R. K. (1974). *FEBS Lett.* **45,** 344–347.
9. Okamura, M. Y., Isaacson, R. A., and Feher, G. (1975). *Proc. Natl. Acad. Sci. U.S.A.* **72,** 3491–3495.
10. Verméglio, A. (1977). *Biochim. Biophys. Acta* **459,** 516–524.
11. Wraight, C. A. (1977). *Biochim. Biophys. Acta* **459,** 525–531.
12. Feher, G. (1971). *Photochem. Photobiol.* **14,** 373–387.
13. Leigh, J. S., and Dutton, P. L. (1972). *Biochem. Biophys. Res. Commun.* **46,** 414–421.
14. Wraight, C. A. (1978). *FEBS Lett.* **93,** 283–288.
15. Eisenberger, P. M., Okamura, M. Y., and Feher, G. (1980). *Fed. Proc., Fed. Am. Soc. Exp. Biol.* **39,** 1802.
16. Okamura, M. Y., Feher, G., and Nelson, N. (1982). *In* "Photosynthesis" (Govindjee, ed.), Vol. 1. Academic Press, New York.
17. Blankenship, R. E., and Parson, W. W. (1979). *Biochim. Biophys. Acta* **545,** 429–444.
18. Klimov, V. V., Dolan, E., Shaw, E. R., and Ke, B. (1980). *Proc. Natl. Acad. Sci. U.S.A.* **77,** 7227–7231.
19. Michaelis, L. (1932). *J. Biol. Chem.* **96,** 703–715.
20. Bouges-Bocquet, B. (1973). *Biochim. Biophys. Acta* **314,** 250–256.
21. Velthuys, B. R., and Amesz, J. (1974). *Biochim. Biophys. Acta* **333,** 85–94.
22. Pulles, M. P. J., van Gorkom, H. J., and Willemsen, J. G. (1976). *Biochim. Biophys. Acta* **469,** 536–540.
23. de Grooth, B. G., van Grondelle, R., Romijn, J. C., and Pulles, M. P. J. (1978). *Biochim. Biophys. Acta* **503,** 480–490.
24. Barouch, Y., and Clayton, R. K. (1977). *Biochim. Biophys. Acta* **402,** 785–788.
25. Bowyer, J. R., Tierney, G. V., and Crofts, A. R. (1979). *FEBS Lett.* **101,** 201–206.
26. Feher, G., and Okamura, M. Y. (1977). *Brookhaven Symp. Biol.* No. 28, 183–194.
27. Halsey, Y. D., and Parson, W. W. (1974). *Biochim. Biophys. Acta* **347,** 404–416.
28. Wraight, C. A. (1979). *Biochim. Biophys. Acta* **548,** 309–327.
29. Renger, G., and Tiemann, R. (1979). *Biochim. Biophys. Acta* **545,** 316–324.
30. Auslander, W., and Junge, W. (1974). *Biochim. Biophys. Acta* **357,** 285–298.
31. Wraight, C. A., Cogdell, R. J., and Clayton, R. K. (1975). *Biochim. Biophys. Acta* **396,** 242–249.
32. Dutton, P. L., Leigh, J. S., and Wraight, C. A. (1973). *FEBS Lett.* **36,** 169–173.
33. Jackson, J. B., Cogdell, R. J., and Crofts, A. R. (1973). *Biochim. Biophys. Acta* **292,** 218–225.
34. Prince, R. C., and Dutton, P. L. (1976). *Arch. Biochem. Biophys.* **172,** 329–334.
35. Okamura, M. Y., Isaacson, R. A., and Feher, G. (1979). *Biochim. Biophys. Acta* **546,** 397–417.
36. Wraight, C. A., and Stein, R. R. (1980). *FEBS Lett.* **113,** 73–77.
37. Fowler, C. F., and Kok, B. (1974). *Biochim. Biophys. Acta* **357,** 299–307.
38. Saphon, S., and Crofts, A. R. (1977). *Z. Naturforsch,* **32,** 617–626.
39. Velthuys, B. R. (1978). *Proc. Natl. Acad. Sci. U.S.A.* **76,** 2765–2769.
40. Mathis, P., and Haveman, J. (1977). *Biochim. Biophys. Acta* **461,** 167–181.
41. Bowes, J. M., and Crofts, A. R. (1980). *Biochim. Biophys. Acta* **580,** 373–384.

42. Stemler, A. (1979). *Biochim. Biophys. Acta* **545,** 36–45.
43. Verméglio, A., Martinet, T., and Clayton, R. K. (1980). *Proc. Natl. Acad. Sci. U.S.A.* **77,** 1809–1813.
44. Velthuys, B. R. (1978). *Proc. Natl. Acad. Sci. U.S.A.* **75,** 6031–6034.
45. Velthuys, B. R. (1980). *Fed. Proc., Fed. Am. Soc. Exp. Biol.* **39,** 2125.
46. Wraight, C. A. (1977). *Bull. Am. Phys. Soc.* **22,** JI 10.
47. Fowler, C. F. (1976). *Int. Conf. Primary Electron Transp. Energy Transduction Photosynth. Bact., Brussels* Abstr. WB9.
48. Petty, K. M., Jackson, J. B., and Dutton, P. L. (1979). *Biochim. Biophys. Acta* **456,** 17–42.
49. Wraight, C. A. (1979). *Photochem. Photobiol.* **30,** 767–776.
50. Crofts, A. R. (1979). *In* "Light-Induced Change Separation in Biology and Chemistry" (H. Gerischer and J. J. Katz, eds.), pp. 389–410. Dahlem-Konferenzen, Berlin.
51. Rutherford, A. W., and Evans, M. C. W. (1979). *FEBS Lett.* **100,** 305–308.
52. Rutherford, A. W., Heathcote, P., and Evans, M. C. W. (1979). *Biochem. J.* **182,** 515–523.
53. Rutherford, A. W., and Evans, M. C. W. (1979). *FEBS Lett.* **110,** 257–261.
54. Knaff, D. (1975). *FEBS Lett.* **60,** 331–335.
55. Feher, G., and Okamura, M. Y. (1978). *In* "The Photosynthetic Bacteria" (R. K. Clayton and W. R. Sistrom, eds.), pp. 349–386. Plenum, New York.
56. Dutton, P. L., Prince, R. C., and Tiede, D. M. (1978). *Photochem. Photobiol.* **28,** 939–949.
57. Verméglio, A., and Clayton, R. K. (1977). *Biochim. Biophys. Acta* **461,** 159–165.
58. Slooten, L. (1972). *Biochim. Biophys. Acta* **275,** 208–218.
59. Dutton, P. L., Wilson, D. F., and Lee, C. P. (1970). *Biochemistry* **9,** 5077–5082.
60. Govindjee, Pulles, M. P. J., Govindjee, R., van Gorkom, H. J., and Duysens, L. N. M. (1976). *Biochim. Biophys. Acta* **449,** 602–605.
61. Prince, R. C., and Dutton, P. L. (1977). *Biochim. Biophys. Acta* **462,** 731–747.
62. Bashford, C. L., Prince, R. C., Takamiya, K., and Dutton, P. L. (1979). *Biochim. Biophys. Acta* **545,** 223–235.
63. Bouges-Bocquet, B. (1980). *Fed. Proc., Fed. Am. Soc. Exp. Biol.* **39,** 2125.
64. Bouges-Bocquet, B. (1980). *Int. Congr. Photosynth., 5th Halkidiki, Greece* Abstr. p. 82.

Enthalpy and Volume Changes Accompanying Electron Transfer from P_{870} to the Primary and Secondary Quinones in Photosynthetic Reaction Centers

3

HIROYUKI ARATA
WILLIAM W. PARSON

INTRODUCTION

The primary photochemical reaction in bacterial photosynthesis is electron transfer from a bacteriochlorophyll complex (P) to an initial acceptor complex, and then to quinone (Q_A). From Q_A^-, an electron passes to a second quinone (Q_B) (*1–3*). Free energy changes in the electron transfer processes have been calculated from the midpoint redox potentials (E_m) of the electron carriers (*1–4*). The E_m of P and Q_A at 295 °K are about +450 mV (*1*) and −180 mV (*5*), respectively. If the free energy of interaction between P^+ and Q_A^- is negligible, the standard free energy change for the formation of $P^+Q_A^-$ from the ground state (PQ_A) is about +0.63 eV. Recent studies of the decay kinetics of $P^+Q_A^-$ and $P^+Q_B^-$ (*6*) indicate that the standard free energy of $P^+Q_B^-$ is probably about 0.08 eV below that of $P^+Q_A^-$, or about 0.55 eV above the ground state. It is of interest to resolve these free energy changes into the underlying changes in enthalpy and entropy. Knowing the enthalpy changes is particularly pertinent to theoretical work on the rates of electron-transfer reactions, be-

Function of Quinones in Energy Conserving Systems

ISBN 0-12-701280-X

cause most theory has considered only differences in energy, not free energy (*3*).

From measurements of E_m temperature dependences, Case and Parson (*7*) estimated the enthalpy changes for reduction or oxidation of P, Q_A, Q_B and cytochrome *c* in *Chromatium vinosum* chromatophores. They concluded that most free energy stored in the $P^+Q_A^-$ and $P^+Q_B^-$ states takes the form of an entropy decrease. The enthalpies of these states appeared to be close to the ground state. Carithers and Parson (*8*) and Fleischman (*9*) measured the temperature dependence of delayed fluorescence from *Rhodopseudomonas viridis* chromatophores in the presence of *o*-phenanthroline, which inhibits electron transfer between Q_A^- and Q_B. From an Arrhenius plot of the intensity of delayed fluorescence, Carithers and Parson (*8*) concluded that $P^+Q_A^-$ was approximately 0.54 eV lower in enthalpy than the excited singlet state, P^*Q_A. Taking the excitation enthalpy of P^* in *Rps. viridis* as 1.26 eV, the enthalpy of $P^+Q_A^-$ would be approximately 0.7 eV above the ground state. This value does not agree with the potentiometric titrations that were done in *C. vinosum*.

Measurement of heat released or absorbed during a reaction is a direct way to determine the enthalpy change in the reaction. Callis *et al.* (*10*) used a capacitor microphone transducer to measure the volume change accompanying the primary photochemical process in chromatophores from *C. vinosum*. The amount of heat released or absorbed can be calculated from the volume change due to thermal expansion of the medium. Callis *et al.* (*10*) concluded that light-driven electron transfer from cytochrome *c*-555 to Q_B does not cause a significant enthalpy change. This agreed with the results from the study of temperature dependence of the E_m (*7*). Ort and Parson (*11–13*) made similar calorimetric measurements in purple membrane fragments of *Halobacterium halobium*. The present paper describes measurements of enthalpies of the $P^+Q_A^-$ and $P^+Q_B^-$ states in isolated reaction centers, using an improved model of the capacitor microphone. Some of the results are described in more detail elsewhere (*13a*).

EXPERIMENTAL RESULTS

Reaction centers of *Rhodopseudomonas sphaeroides* strain R-26 were prepared essentially as described by Clayton and Wang (*14*). For the experiments on reaction centers with two quinones (2-Q reaction centers), excess ubiquinone-10 was added in ethanolic solution to suspensions of reaction centers in 10 m*M* Tris HCl or phosphate, pH 8, and 0.1% lauryldimethylamine oxide. Unbound quinone then was removed by chromatography on DEAE-Sephacel. For preparation of reaction centers with one quinone (1-Q reaction centers), reaction centers (in 10 m*M* Tris HCl) were incubated with 1% lauryldimethylamine oxide and 1 m*M* *o*-phenanthroline for 3 to 6 hours at room temperature (*15*), and then chromatographed on DEAE-Sephacel.

Volume change measurements were made with the capacitor microphone described elsewhere (*11*, *16*). The apparatus was calibrated using an inert absorber (bromcresol purple) in very dilute buffer (less than 0.5 m*M* phosphate, pH about 8). The known thermal expansion coefficient, heat capacity, and density of pure water were used for the calculation [for details, see Callis *et al.* (*10*)]. Samples were illuminated by weak, 588 nm, 0.5 μsec flashes from a rhodamine-6-G dye laser. The flashes were repeated at about 15 sec intervals, and 16 signals were averaged.

A volume change due to a photochemical reaction can arise in two ways (*10–13*). First, there may be a volume difference, Δv_r, between reactants and products. Second, the solution may expand or contract through heating or cooling. The heat absorbed or released is equal to the enthalpy change of the reaction, ΔH_r. Thus, the volume change after flash excitation is

$$\Delta V = (n_e \mathrm{E}_e - n_r \, \Delta H_r) \frac{\alpha}{\rho C} + n_r \, \Delta v_r . \tag{1}$$

where n_e is the number of einsteins of photons absorbed, n_r is the number of moles of products generated, E_e is the energy of the photon (per einstein), ΔH_r is the enthalpy of the products (per mole) relative to the reactants, Δv_r is the difference in molal volume between reactants and products, α is the thermal expansion coefficient of the solution, C is the heat capacity of the solution at constant pressure, and ρ is the density of the solution.

If the photochemically active sample is replaced by a solution of an inert, non-fluorescent absorber, all of the energy of the absorbed photons is converted to heat. The volume change then becomes

$$\Delta V_{h\nu} = n_e \, \mathrm{E}_e \frac{\alpha}{\rho C} . \tag{2}$$

Figure 1 shows typical time courses of the volume change after flash excitation. Traces 1 and 2 are the volume changes of a bromcresol purple solution, which is photochemically inert. Trace 1 is a measurement at 22 °C; the expansion is due to the release of the energy of the absorbed photons as heat (Eq. 2). It relaxes slowly, depending on the capacitance of the microphone. Trace 2 is a measurement at 3.6 °C, where α of the

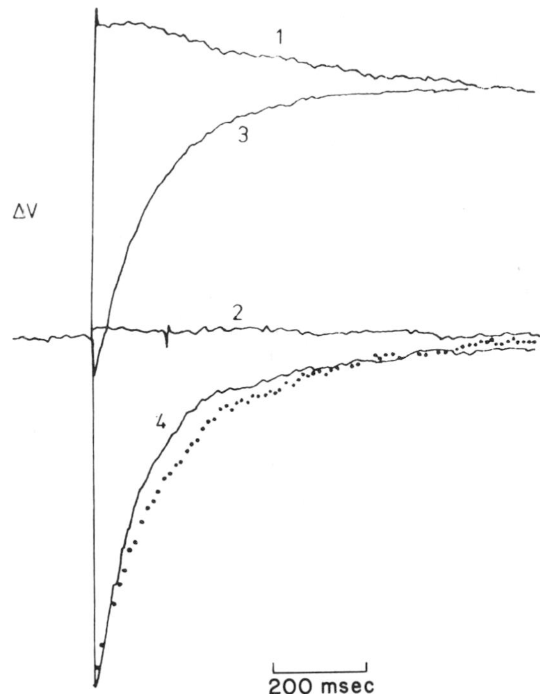

Fig. 1. Flash-induced volume changes of 1-Q reaction centers or bromcresol purple solution. Traces 1 and 2, bromcresol purple; traces 3 and 4, reaction centers. Temperature was 22 °C for traces 1 and 3, and 3.6 °C for traces 2 and 4. (···), Dotted line, trace 3 minus trace 1. Buffer solution was 10 m*M* phosphate, pH 8.0, with 0.05% Triton X-100. Flash intensity, 20 mJ; $A_{588} = 0.52\ \text{cm}^{-1}$ (the optical path length in the microphone cell is about 2 cm). The vertical scale is arbitrary.

buffer is close to zero. Little or no volume change is observed at this temperature. Traces 3 and 4 are the volume changes of 1-Q reaction centers. The reaction center solutions and the bromcresol purple had the same absorbance at the excitation wavelength. At 3.6 °C (trace 4), the reaction centers contract rapidly and then relax with a half-time of about 80 msec. The contraction must be due to the term $n_r \Delta v_r$ in Eq. 1. The decay kinetics correspond to those of $P^+Q_A^-$ (*17*, *18*). The volume change at 22 °C (trace 3) is the sum of the contraction $n_r \Delta v_r$ and an expansion due to release of heat (Eq. 1). The difference between traces 3 and 1 (dotted line) is:

$$\Delta V' \equiv \Delta V - \Delta V_{h\nu} = n_r \left(-\Delta H_r \frac{\alpha}{\rho C} + \Delta v_r \right). \tag{3}$$

This quantity becomes equal to ΔV near 3.6 °C, when $\Delta V_{h\nu}$ is zero. The initial value of $\Delta V'$ at 22 °C is approximately the same as ΔV measured at

3.6 °C (trace 4). $\Delta V'$ at 22 °C decays somewhat more slowly than the signal measured at 3.6 °C, in agreement with the known, inverse temperature dependence of the back-reaction between Q_A^- and P^+ (*3*).

Trace 1 in Fig. 2 shows the time course of the volume change with 2-Q reaction centers at 3.6 °C. 2-Q reaction centers complete the electron-transfer reaction from P to Q_B within one msec (*1, 6, 19–21*). The state $P^+Q_B^-$ then decays slowly with a half-time of about one sec (*6, 17, 18*). The decay half-time of the contraction (1.0 sec after correction for the relaxation time of the microphone) matches this value. In the presence of 2 m*M* *o*-phenanthroline (trace 2), the relaxation becomes as fast as it is with 1-Q reaction centers ($t_{1/2}$ = 70 msec), because *o*-phenanthroline blocks electron transfer between Q_A^- and Q_B (*22, 23*).

The contraction Δv_r is probably due to an increase in the number of electrically charged species in the system. Charged groups such as P^+, Q_A^-, and Q_B^- will cause an ordering (electrostriction) of molecules in their vicinity. Volume changes due to electrostriction in aqueous solutions usually do not depend strongly on temperature (*24*).

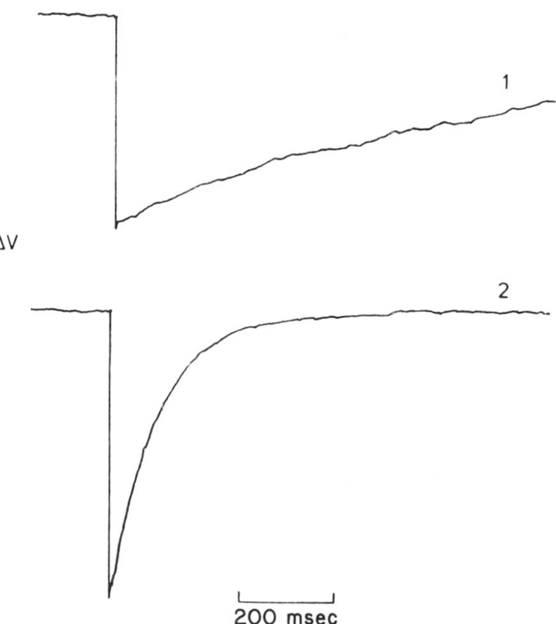

Fig. 2. Flash-induced volume change of 2-Q reaction centers. Reaction centers were suspended in 10 m*M* phosphate buffer, pH 7.7, with 0.1% lauryldimethylamine oxide and 100 m*M* NaCl. Trace 1, without *o*-phenathroline; Trace 2, with 2 m*M* *o*-phenenthroline. Temperature, 3.4 °C; flash intensity, 20 mJ; A_{588} = 0.49 cm^{-1}.

If n_r, ΔH_r and Δv_r do not depend on temperature, one can estimate ΔH_r by measuring the temperature dependence of ΔV and $\Delta V_{h\nu}$:

$$\frac{\dfrac{\partial \Delta v}{\partial T} - \dfrac{\partial \Delta v_{h\nu}}{\partial T}}{\dfrac{\partial \Delta v_{h\nu}}{\partial T}} = \frac{-\phi \dfrac{\Delta H_r}{\rho C} \dfrac{\partial \alpha}{\partial T}}{\dfrac{E_e}{\rho C} \dfrac{\partial \alpha}{\partial T}} = -\phi \frac{\Delta H_r}{E_e}, \tag{4}$$

where $\Delta v = \Delta V/n_e$, $\Delta v_{h\nu} = \Delta V_{h\nu}/n_e$, and $\phi = n_r/n_e$. ϕ (the quantum yield of the photochemical reaction) is essentially one in *Rps. sphaeroides* (*25*). If ΔH_r depends on temperature (i.e., if the reaction causes a change in heat capacity), it would be necessary to measure $\partial \Delta v/\partial T$ at the temperature where $\alpha = 0$.

One factor that could make Δv_r temperature dependent is the uptake of protons. If the photochemical reaction is accompanied by proton uptake, Δv_r becomes:

$$\Delta v_r = \Delta v_r^{o} + n_p \Delta v_r^{p}, \tag{5}$$

where Δv_r^{o} and Δv_r^{p} are the molal volume changes due to photochemical charge separation and to transfer of protons from the buffer to reaction centers, and n_p is the number of protons taken up per photochemical reaction. Although Δv_r^{p} may not depend strongly on temperature, n_p could depend on temperature. Δv_r^{p} depends on the nature of the buffer (*11*, *12*). Phosphate protonation by H_3O^+ causes loss of both a negative charge on the phosphate anion and of the positive charge of H_3O^+. This process is accompanied by an expansion of 24 ml/mole (*26*). Tris buffer protonation is accompanied by a much smaller volume change (−1 ml/mole) (*26*), since the number of charges in the solution remains constant.

In Fig. 3, faster time courses of the volume changes at 3.0 °C in the presence of different buffers are shown. The initial contractions that occur within 100 μsec are approximately the same in phosphate and Tris buffer (Fig. 3A). In Tris, the contraction is followed by a partial relaxation (expansion) with a time constant of about 0.5 msec; no corresponding expansion is observed in phosphate buffer. The expansion in Tris buffer is presumably caused by proton uptake, whereas the transfer of protons between phosphate and the reaction center apparently is not accompanied by a significant volume change. This means that contraction due to phosphate deprotonation is cancelled by expansion due to protonation of the reaction centers. The group on the reaction centers that takes up a proton is evidently an anionic species, such as a carboxylate or phenolate group, rather than an amino, histidine, or guanidino group. It could be Q_A^- itself. Similar observations were made previously in chromatophores of *C. vinosum* (*10*), but were misinterpreted as indicating that phosphate inhibited

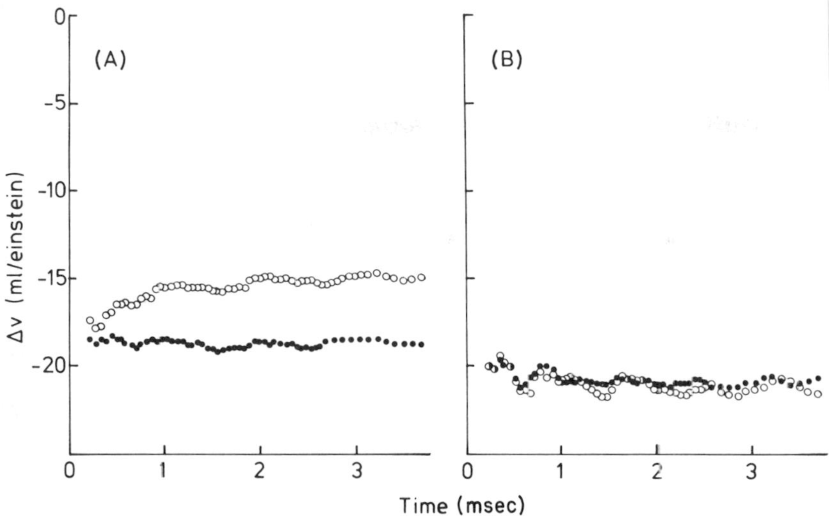

Fig. 3. Rapid time courses of the volume change of 1-Q reaction centers at 3.0 °C. (A) pH 8.0; (B) pH 6.5. ●, phosphate buffer; ○, Tris buffer (pH 8.0) or Aces buffer (pH 6.5). Flash strength, about 8 mJ. The buffer solutions (10 m*M*) contained 0.1% lauryldimethylamine oxide and 100 m*M* NaCl.

the proton uptake. From the differences between the volume changes measured with 1-Q reaction centers in Tris and phosphate buffers, we calculated $n_p = 0.26–0.27$ at pH 8.0.

Figure 3B shows time courses at pH 6.5 in phosphate and Aces (*N*-[2-acetamido]-2-amiomethane sulfonic acid) buffers. Aces is probably similar to Tris with respect to electrostrictive effect of proton transfer. The volume changes in the two buffers are similar at pH 6.5, indicating that light-induced proton binding does not occur at this pH. This agrees with results reported recently by Wraight (*27*).

Since volume changes due to proton uptake appear negligible at pH 6.5, or at higher pH in the presence of phosphate buffer, either of these conditions can be used to measure ΔH_r (Eq. 4). Phosphate buffer is particularly advantageous, because the enthalpy of phosphate protonation is small.

The temperature dependence of $\Delta v_{h\nu}$ and of Δv for 1-Q and 2-Q reaction centers is shown in Fig. 4. Initial volume changes are given in ml/einstein absorbed. $\Delta v_{h\nu}$ was measured with two different intensities (about 8 and 20 mJ). It increased almost linearly with increasing temperature, as expected from the known temperature dependence of α for water. The slope $\partial \Delta v_{h\nu}/\partial T$ of the line that gives the least-squares fit is 0.64 ± 0.005 ml einstein^{-1} deg^{-1}. The volume changes with reaction centers (in 10 m*M* phosphate, pH 6.4 or 6.5) were measured with a flash strength of about 8 mJ,

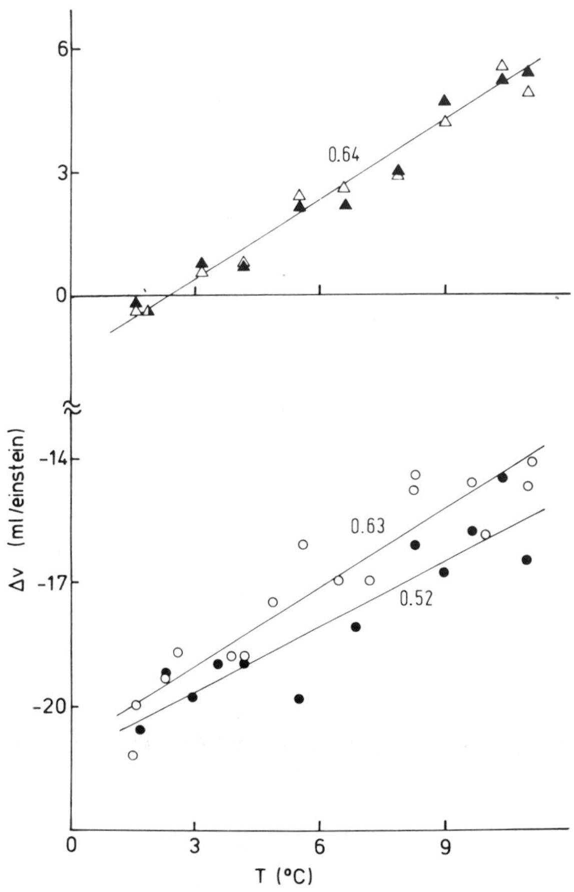

Fig. 4. Volume change of reaction centers as a function of temperature. △ and ▲, $\Delta v_{h\nu}$ measured with bromcresol purple solution; ○, 1-Q reaction centers; ●, 2-Q reaction centers. Flash strengths, approx. 20 mJ (△) or 8 mJ (▲, ○ and ●). buffer solutions (10 m*M* phosphate) contained 0.1% lauryldimethylamine oxide and 100 m*M* NaCl. pH 8.0 (△ and ▲), 6.5 (○) or 6.4 (●). Lines that give the least-squares fit are shown. The numbers indicate the slopes of the lines.

in the range where the volume changes are proportional to the flash strength (*i.e.*, ϕ is approximately one). For the 1-Q reaction centers, the slope $\partial \Delta v / \partial T$ of the least squares fit is 0.63 ± 0.012. Using Eq. 4, ΔH_r is calculated to be 0.03 ± 0.04 eV. Similar results were obtained with weaker flashes. The results were also similar with Aces buffer at pH 6.5 and phosphate buffer at pH 8.0. These results are collected in Table I. The mean value of all measurements with 1-Q reaction centers is ΔH_r = 0.05 ± 0.03 eV.

With 2-Q reaction centers, $\partial\Delta v/\partial T$ is significantly smaller (0.52 ± 0.02 ml einstein^{-1} deg^{-1}) than that with 1-Q reaction centers. The calculated ΔH_r is 0.40 ± 0.07 eV. Measurements were also made at pH 7.7, and with Aces or Tris buffer. The rapid time courses of 2-Q reaction centers were similar to those of 1-Q reaction centers: the initial contraction was followed by an expansion with a time constant of about 0.5 msec in Tris buffer at pH 7.7, but not in phosphate buffer. Aces buffer gave similar results to phosphate at pH 6.4. Table I shows the results with phosphate buffer at pH 6.4 and 7.7, and with Aces buffer at pH 6.4. The mean of all measurements with 2-Q reaction centers is $\Delta H_r = 0.42 \pm 0.02$ eV.

In the presence of *o*-phenanthroline, 2-Q reaction centers showed an enthalpy change of 0.13 ± 0.03 eV (Table I). This is similar to, but slightly more positive than that measured in 1-Q reaction centers.

DISCUSSION

The results obtained with both 1-Q reaction centers and 2-Q reaction centers in the presence of *o*-phenanthroline show that formation of the $P^+Q_A^-$ radical pair from the PQ_A ground state involves only a relatively small change in enthalpy. This agrees with the conclusion reached from the temperature dependence of the E_m of P and Q_A in *C. vinosum* chromatophores (*7*). The free energy of about 0.63 eV stored in the radical pair evidently takes the form of an entropy decrease. The standard entropy change for the formation of $P^+Q_A^-$ in 1-Q reaction centers is calculated to be about −50 cal deg^{-1} mole^{-1}.

Our results do not agree with the large, positive ΔH_r values obtained from the temperature dependence of delayed fluorescence from *Rps. viridis* chromatophores (*8*, *9*). This could reflect differences between the two species of bacteria, or between chromatophores and isolated reaction centers. The E_m of P decreases with increasing temperature in *C. vinosum* (*7*), but is less dependent on temperature in *Rps. viridis* (*8*). The E_m of Q_A is pH-dependent in chromatophores, but is independent of pH in isolated reaction centers (*28*, *29*). However, the discrepancy between the ΔH_r value calculated from delayed fluorescence and calorimetry probably does not reflect differences in the proton uptake that occurs during the two types of measurements. Both the delayed fluorescence measurements and our volume measurements at low pH were made under conditions that minimized proton binding. Another possibility is that the delayed fluorescence from chromatophores is influenced by an electrical potential difference between the positions of P^+ and Q_A^-. This difference could de-

TABLE I

Volume Changes and Enthalpy Changes Associated with Flash-Induced Charge Separation in Reaction Centers[A]

Sample	Buffer	pH	Intensity[B]	Δv_r[C] (ml/einst)	$\partial\Delta v/\partial T$ (ml/einst°C)	$\frac{\partial\Delta v'/\partial T}{\partial\Delta v_{h\nu}/\partial T}$[D]	ΔH_r[E] (eV)
1–Q	Pi	8.0	high	−20.2 ± 0.74	0.62 ± 0.018	−0.031 ± 0.029	0.07 ± 0.06
	Pi	8.0	low	−21.8 ± 1.26	0.68 ± 0.026	0.063 ± 0.041	−0.13 ± 0.09
	Pi	6.5	high	−19.3 ± 0.47	0.63 ± 0.012	−0.016 ± 0.020	0.03 ± 0.04
	Pi	6.5	low	−21.2 ± 0.53	0.61 ± 0.013	−0.047 ± 0.021	0.10 ± 0.04
	Aces	6.5	high	−22.0 ± 1.70	0.67 ± 0.042	0.047 ± 0.066	−0.10 ± 0.14
mean[F]							0.05 ± 0.03
2–Q	Pi	7.7	high	−18.8 ± 0.70	0.51 ± 0.017	−0.203 ± 0.028	0.43 ± 0.06
	Pi	7.7	low	−20.5 ± 0.79	0.51 ± 0.018	−0.203 ± 0.029	0.43 ± 0.06
	Pi	6.4	high	−19.9 ± 0.83	0.52 ± 0.020	−0.188 ± 0.032	0.40 ± 0.07
	Pi	6.4	low	−22.2 ± 0.99	0.56 ± 0.022	−0.125 ± 0.035	0.26 ± 0.07
	Aces	6.4	high	−18.6 ± 0.42	0.51 ± 0.011	−0.203 ± 0.019	0.43 ± 0.04
	Aces	6.4	low	−20.3 ± 0.50	0.50 ± 0.011	−0.219 ± 0.019	0.46 ± 0.04
mean[F]							0.42 ± 0.02

2–Q, 2m*M*							
o-phenanthroline	Pi	8.0	high	−22.5 ± 1.15	0.61 ± 0.026	−0.047 ± 0.041	0.10 ± 0.09
	Pi	8.0	low	−24.8 ± 1.26	0.59 ± 0.026	−0.078 ± 0.041	0.16 ± 0.09
	Pi	6.5	high	−20.4 ± 0.48	0.60 ± 0.011	−0.063 ± 0.019	0.13 ± 0.04
	Pi	6.5	low	−22.5 ± 0.72	0.60 ± 0.015	−0.063 ± 0.025	0.13 ± 0.05
mean[F]							0.13 ± 0.03

[A] The values of Δv_r and $\partial\Delta v/\partial T$ were obtained from least-squares fit to data sets similar to that shown in Fig. 4. The buffer solutions (10 m*M*) contained 0.1% lauryldimethylamine oxide and 100 m*M* NaCl.

[B] High and low intensities mean approximately 8 mJ and 3.7 mJ, respectively.

[C] Δv_r, Δv at the temperature where $\alpha = 0$.

[D] $\partial\Delta v'/\partial T = \partial\Delta v/\partial T - \partial\Delta v_{h\nu}/\partial T$; $\partial\Delta v_{h\nu}/\partial T = 0.64 \pm 0.005$, measured with bromcresol purple solution in 10 m*M* phosphate buffer (pH 8.0) with 1% lauryldimethylamine oxide and 100 m*M* NaCl.

[E] ΔH_r is calculated according to Eq. 4; $E_e = 2.11$ eV.

[F] In the calculation of the mean, each value of ΔH_r was weighted inversely by its own variance. This would not take into account systematic errors in ΔH_r.

pend on the integrity of the chromatophore membrane. The intensity of the delayed fluorescence also depends on the quantum yield of the photochemical reaction, which is only 40–50% in *Rps. viridis* (*8*, *30*). A temperature dependence of the quantum yield could influence the estimation of ΔH_r.

For the formation of $P^+Q_B^-$, we obtained an enthalpy change of 0.42 ± 0.02 eV. The entropy change associated with formation of $P^+Q_B^-$ is calculated to be about -10 cal deg^{-1} $mole^{-1}$. This contrasts markedly with the results on $P^+Q_A^-$, and the molecular basis for this difference is not clear. Our results with 2-Q reaction centers also disagree with the previous conclusions from measurements of the temperature dependence of E_m (*7*) and from calorimetry with *C. vinosum* chromatophores (*10*). In calorimetric measurements on chromatophores, the experimental conditions were such that cytochrome *c*-555 could rapidly donate electrons to P^+; electron transfer from cytochrome to P could be the origin of the difference in results. Another possibility is that an enthalpy decrease is caused by proton uptake in the chromatophores. At high pH, isolated reaction centers do bind protons after a single flash but the amount of proton uptake is relatively small. Also, the proton binding site in isolated reaction centers could differ from that in intact chromatophores, and differ from the site of proton uptake that follows two successive flashes (*27*).

Electron transfer from Q_A^- to Q_B involves an enthalpy increase of about 0.35 eV; so the activation enthalpy of this electron transfer is at least 0.35 eV. The activation enthalpy of this reaction has been measured in several different systems (*19*, *20*). It ranges from 0.33 to 0.55 eV, agreeing with our results.

Although we do not have any direct information on the origin of Δv_r, it is reasonable to ascribe the contraction to an electrostrictive effect of the positive charge of P^+ and the negative charge of Q_A^- or Q_B^-. In this respect, it is interesting that, although the enthalpy and entropy changes associated with the formation of $P^+Q_A^-$ and $P^+Q_B^-$ are very different, $\Delta \underline{v}_r$ is about the same. This is unexpected, because electrostrictive effects *in vitro* generally involve parallel changes in entropy and volume.

Note added August, 1981: Additional details on calorimetric measurements were published (*13a*), as well as a theory of electron transfer that distinguishes between enthalpy and free energy (*31*). We also obtained independent estimates of the free energy and enthalpy changes associated with the photoreduction of Q_A and Q_B in isolated reaction centers, from measurements of delayed fluorescence (*32*). The results for the free energies of $P^+Q_A^-$ and $P^+Q_B^-$ agree well with the values calculated from the midpoint redox potentials of P and the quinones. However, the standard enthalpy of $P^+Q_A^-$ is estimated to be about 0.75 eV below that of P^*Q_A, or about 0.63 eV above that of the PQ_A ground state. This disagrees with the enthalpy estimated from calorimetric measurements. The capacitor microphone measurements could be affected by a temperature dependence of $\Delta \underline{\nu}_r$.

ACKNOWLEDGMENTS

This work was supported by NSF grant PCM 77-13290. We thank C. C. Schenck for help with the preparation of the reaction centers.

REFERENCES

1. Parson, W. W., and Cogdell, R. J. (1975). *Biochim. Biophys. Acta* **416,** 105–149.
2. Dutton, P. L., and Prince, R. C. (1977). *In* "The Bacteria" (W. R. Sistrom and A. R. Crofts, eds.), Vol. 6, pp. 523–584. Academic Press, New York.
3. Blankenship, R. E., and Parson, W. W. (1979). *In* "Photosynthesis in Relation to Model Systems" (J. Barber, ed.), pp. 71–114. Elsevier, Amsterdam.
4. Prince, R. C., and Dutton, R. P. (1978). *In* "The Photosynthetic Bacteria" (R. K. Clayton and W. R. Sistrom, eds.), pp. 439–453. Plenum, New York.
5. Prince, R. C., and Dutton, P. L. (1976). *Arch. Biochem. Biophys.* **172,** 329–334.
6. Wraight, C. A., and Stein, R. R. (1980). *FEBS Lett.* **113,** 73–77.
7. Case, G. D., and Parson, W. W. (1971). *Biochim. Biophys. Acta* **292,** 677–684.
8. Carithers, R. P., and Parson, W. W. (1975). *Biochim. Biophys. Acta* **387,** 194–211.
9. Fleischman, D. (1974). *Photochem. Photobiol.* **19,** 59–68.
10. Callis, J. B., Parson, W. W., and Gouterman, M. (1972). *Biochim. Biophys. Acta* **267,** 348–362.
11. Ort, D. R., and Parson, W. W. (1978). *J. Biol. Chem.* **253,** 6158–6164.
12. Ort, D. R., and Parson, W. W. (1979). *Biophys. J.* **25,** 341–354.
13. Ort, D. R., and Parson, W. W. (1979). *Biophys. J.* **25,** 355–364.

13a. Arata, H., and Parson, W. W. (1981). *Biochim. Biophys. Acta* **636,** 70–81.

14. Clayton, R. K., and Wang, R. T. (1971). *In* "Photosynthesis and Nitrogen Fixation" (A. San Pietro, ed.), Methods in Enzymology, Vol. 23, pp. 696–704. Academic Press, New York.
15. Okamura, M. Y., Isaacson, R. A., and Feher, G. (1975). *Proc. Natl. Acad. Sci. U.S.A.* **72,** 3491–3495.
16. Parson, W. W. (1982). *In* "Visual Pigments and Purple Membranes" (L. Packer, ed.), Methods in Enzymology, Vol. 88. Academic Press, New York, in press.
17. Clayton, R. K., and Yau, H. F. (1972). *Biophys. J.* **12,** 867–881.
18. Blankenship, R. E., and Parson, W. W. (1979). *Biochim. Biophys. Acta* **545,** 429–444.
19. Parson, W. W. (1969). *Biochim. Biophys. Acta* **189,** 384–396.
20. Chamorovsky, S. K., Remennikov, S. M., Kononenko, A. A., Venediktov, P. S., and Rubin, A. B. (1976). *Biochim. Biophys. Acta* **430,** 62–70.

21. Vermeglio, A., and Clayton, R. K. (1977). *Biochim. Biophys. Acta* **461,** 159–165.
22. Parson, W. W., and Case, G. D. (1970). *Biochim. Biophys. Acta* **205,** 232–245.
23. Clayton, R. Y., Szuts, E. Z., and Fleming, H. (1972). *Biophys. J.* **12,** 64–79.
24. Millero, F. J. (1971). *Chem. Rev.* **71,** 147–176.
25. Wraight, C. A., and Clayton, R. K. (1973). *Biochim. Biophys. Acta* **333,** 246–260.
26. Neuman, R., Kauzmann, W., and Zipp, A. (1973). *J. Phys. Chem.* **77,** 2687–2691.
27. Wraight, C. A. (1979). *Biochim. Biophys. Acta* **548,** 309–327.
28. Reed, D. W., Zankel, K. L., and Clayton, R. K. (1969). *Proc. Natl. Acad. Sci. U.S.A.* **63,** 42–46.
29. Dutton, P. L., Leigh, J. S., and Wraight, C. A. (1973). *FEBS Lett.* **36,** 169–173.
30. Thornber, J. P., and Olson, J. M. (1971). *Photochem. Photobiol.* **14,** 329–341.
31. Kakitani, T., and Kakitani, H. (1981). *Biochim. Biophys. Acta* **635,** 498–514.
32. Arata, H., and Parson, W. W. (1981). *Biochim. Biophys. Acta* **638,** 201–209.

The Secondary Electron Acceptor of Photosystem II 4

H. J. VAN GORKOM
A. P. G. M. THIELEN
A. C. F. GORREN

INTRODUCTION

A remarkable aspect of the two-electron gate at the reducing side of the reaction center is that it was first discovered in photosystem II (*1*, *2*). For once, a property established in photosystem II served as a model for detection of its equivalent in reaction centers of the purple bacterium *Rhodopseudomonas sphaeroides* (*3*, *4*). Bouges-Bocquet discovered that electron donors available to photosystem I were produced by photosystem II once every two successive photoreactions (*1*). Independently, Velthuys and Amesz found that once every two photoreactions of system II, the electron leaving the primary acceptor ubiquinone (*Q*) remains on a secondary acceptor, and until the next photoreaction occurs, can be returned to Q by addition of DCMU[1] (*2*). Thus, a secondary electron acceptor of photosystem II, (denoted B in Paris and R in Leiden) was postulated, that is alternatively reduced and reoxidized by successive photoreactions of photosystem II. In isolated chloroplasts, and also in *Chlorella* treated with benzoquinone (*5*), R^- becomes largely oxidized in the dark; subsequent illumination by short, saturating flashes induces a net reduction on uneven flash numbers and oxidation on even flash num-

[1] Abbreviations: DCMU, 3-(3′,4′-dichlorophenyl)-1,1-dimethylurea; CCCP, carbonyl cyanide *m*-chlorophenylhydrazone.

Function of Quinones in Energy Conserving Systems

ISBN 0-12-701280-X

bers. The oscillation is damped as a result of misses in a statistical fraction of the centers that fail to produce a stable charge separation (observed also in oxygen evolution). We observed similar oscillations with a periodicity of two in absorbance changes around 320 nm, which could be accounted for if R^- were a plastosemiquinone anion (*6*), just like Q^- (*7*).

Later studies (*8–11*) confirmed these findings and added information on the kinetics and equilibria associated with the secondary acceptor. To some extent all these studies suffer because no reliable quantitative method of determining the concentration of R^- (or of R) has been developed. The reduction of Q by R^- upon addition of DCMU causes an increase of the chlorophyll fluorescence yield that can be measured with excellent signal-to-noise ratio. However, for this technique, the extent to which the reaction proceeds must be known and the quantitative relation between fluorescence and Q^- concentration is complicated. Absorbance changes, alternatively, should provide a direct, linear measure. The difficulty here is to avoid or subtract concomitant absorbance changes of other components. In fact, no difference spectrum of R^- minus R has been reported, and even a pure difference spectrum of Q reduction in whole chloroplasts is lacking. In this paper we report measurements of fluorescence and absorbance changes intended to provide a more solid quantitative basis for experiments on the two-electron gating mechanism. The reader should be cautioned that this mechanism does not occur in all photosystem II reaction centers.

RESULTS AND DISCUSSION

Tris-Washed Chloroplasts[2]

Figure 1 shows the results of fluorescence measurements, similar to those carried out by Velthuys and Amesz (*2*), on a typical preparation of Tris-washed chloroplasts. The fluorescence rise induced by DCMU addition, after illumination of a dark-adapted sample by a varying number of saturating flashes, constitutes a damped oscillation with a periodicity of two versus flash number. Oxygen evolution, which might introduce oscillations unrelated to R, was prevented by Tris washing, and the one-electron donor hydroxylamine was added to restore electron

[2] Tris-washing was carried out by incubating osmotically broken chloroplasts in 0.8 *M* Tris (*tris*[hydroxymethyl]aminomethane) at pH 8.0 for 30 min in darkness at 0 °C. The chlorophyll concentration was 0.25 m*M*.

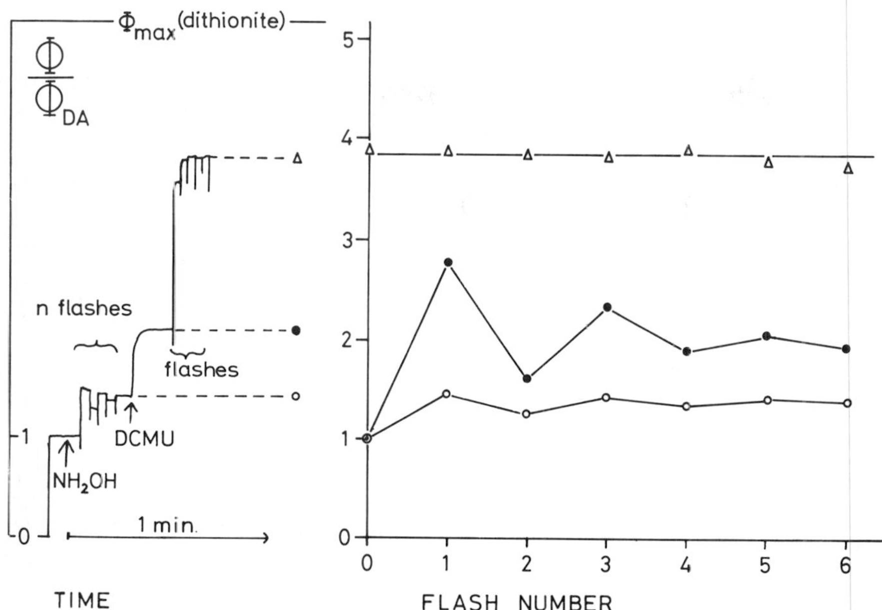

Fig. 1. Fluorescence oscillations in Tris-washed spinach chloroplasts. Fluorescence, measured at 675–695 nm, was excited by modulated green light which caused one hit per photosystem II per minute. The photomultiplier was shut off during each flash. Excitation and measurement were on the same side of the 1 mm cuvette, which contained 50 μM chlorophyll. Additions: 1 mM NH_2OH, 10 μM DCMU.

transport. Normally, after prolonged dark adaptation, no R^- is found in Tris-washed chloroplasts. Both the increase induced by DCMU and the fluorescence yield before DCMU addition oscillate with a periodicity of two. This suggests the presence of some Q^-R in equilibrium with Q R^-.

As reported earlier (*12*), the quenching of fluorescence by Q in these preparations followed a simple Stern–Volmer relation. The fraction of reaction centers in which Q was in the reduced state is given by:

$$\frac{[Q^-]}{[Q] + [Q^-]} = 1 - \frac{1/F - 1/F_M}{1/F_0 - 1/F_M} \qquad (1)$$

where F_0 and F_M correspond to the fluorescence levels after dark adaptation and after complete photoreduction of Q (triangles), respectively. In the experiments of Fig. 1, the concentration of Q^- after DCMU addition was always 2.0 times the concentration at the moment of DCMU addition. Also, the oscillation with flash number of these two concentrations was quantitatively explained when an equilibrium constant of

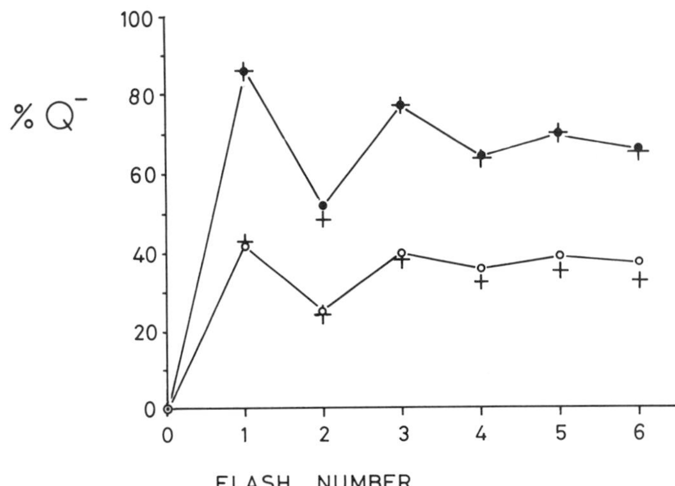

Fig. 2. Q^- concentrations calculated according to Eq. (1) from Fig. 1, corresponding symbols. (+), calculated oscillation pattern assuming $K(Q\,R^-/Q^-R)$ is one in the absence and zero in the presence of DCMU, and 15% misses on each flash.

1.0 for $[Q\,R^-]/[Q^-R]$ was assumed, and the observed 15% misses on the first flash were assumed to occur on each flash (Fig. 2). Therefore, in contrast to our earlier suggestion (*12*), we conclude that the remarkable stability of a large fraction of Q^- in our preparations of Tris-washed chloroplasts was due to an equilibrium constant close to unity between Q^-R and $Q\,R^-$. It should be noted, however, that for unknown reasons both this property and the absence of R^- after dark adaptation were sometimes less pronounced. Preparations, both from freshly picked leaves and from possibly old market spinach, in which a considerable fraction of the photosystem II centers appears truly blocked after Q, are not very exceptional, but this seems unrelated to Tris-washing.

There is good evidence that normally in undamaged photosystem II the equilibrium constant $[Q\,R^-]/[Q^-R]$ is much higher (*8*, *9*). The low value in Tris-washed chloroplasts may be due to damage at the oxidizing side. Similarly, in photosystem II particles prepared with deoxycholate we found the remarkable stability of the state Q^-R^- was dependent on the presence of an oxidized component at the donor side (*13*). Perhaps the lag observed in the reoxidation of Q^- in the state of Q^-R^- (*11*) can be similarly interpreted. These considerations restrict the usefulness of the apparently clear-cut information obtained with Tris-washed chloroplasts. Although in Tris-washed chloroplasts DCMU, orthophenanthroline, and even CCCP convert essentially all R^- into Q^-, this could not be assumed to occur a priori in untreated chloroplasts, and might even depend on the S-state. Moreover, in normal chloroplasts degree of Q reduction is not as simply related to the fluorescence yield as in Tris-washed chloroplasts.

Photosystems IIα and IIβ

In principle, the amount of oxidized Q can be determined in the presence of DCMU from the remaining ΔA_{320}, or more conveniently from the fluorescence induction curve. Since the distance $F_{max} - F(t)$ in such a curve is at any time proportional to the rate of Q reduction (*14*), the area over the fluorescence induction curve is proportional to the amount of reducible Q. Calibration of the area is done by comparison with an induction curve of dark-adapted chloroplasts and measurement of Q concentration by absorption change upon photoreduction. Kinetic analysis of the area growth shows the existence of two components (*15*), which represent two independent, permanently different types of photosystem II (*16*) (photosystem IIα and photosystem IIβ); the stoichiometric ratio of the two types given by the ratio of the two area components. The two types of photosystems II have different antenna properties (*17*, *18*), e.g., PS IIα occurs in a common pigment matrix, while PS IIβ occurs as separate units. Also, a difference in the midpoint potentials of the primary acceptors, Q_α and Q_β, has been found (*19*, *20*). In fact the Stern–Volmer type of relationship between fluorescence and Q^- in Tris-treated chloroplasts is due to sensitivity of photosystem IIβ towards Tris treatment (*21*).

When the experiments of Fig. 1 were repeated on untreated tobacco chloroplasts in the presence of NH_2OH, almost no oscillations were produced by the flashes (–○–, Fig. 3). Apparently the equilibrium $Q^-R \rightleftharpoons Q\ R^-$ is more to the right in untreated tobacco chloroplasts. The increased fluorescence level remaining after a flash is not completely explained by the amount of Q_β^- (see next sections) and this equilibrium. After the addition of 10 μM DCMU the fluorescence rise induced by continuous illumination oscillates with flash number, because the amount of Q^- produced from R^- depends on the number of flashes, as indicated by the DCMU-induced fluorescence rise (–●–).

We analyzed the growth of the area over the induction curves recorded when giving the continuous illumination. In Fig. 4 the total area and the areas of the α- and β-components (×, □, ●, respectively) when no flashes were given prior to the addition of DCMU were nor-

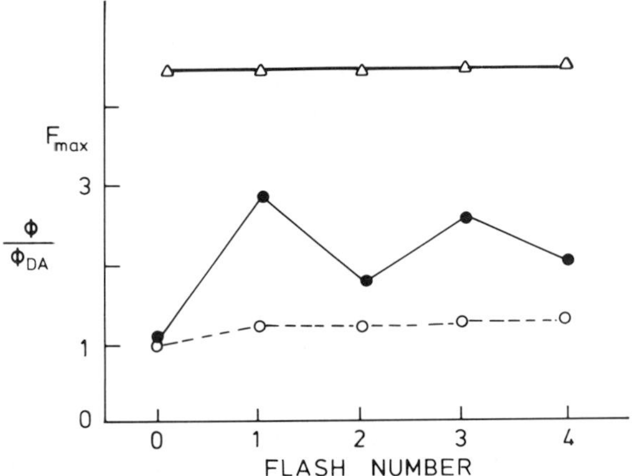

Fig. 3. Fluorescence measurements on chloroplasts from wild-type tobacco with NH_2OH as in Fig. 1, except that after the increase by DCMU, the maximum fluorescence was obtained by a short, continuous illumination rather than by flashes.

malized to 100%. (From the actual values of the areas of both components it followed that one-third of the PS II centers was of the β-type in these chloroplasts.) As expected, the total area oscillated with flash number. The area associated with the fast α-phase in the induction curve showed this pattern even more markedly and the initial rate constant of this phase (*17*) oscillated inversely. This was not surprising in view of the energy transfer between the units of PS IIα. As expected, for separate units the rate constant of the β-component did not change. The area was diminished about 20% after the first flash, with no change

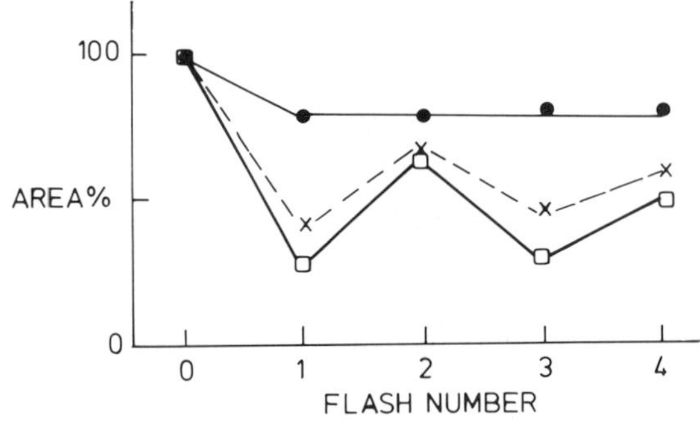

Fig. 4. Composition of the area over the fluorescence induction curves of tobacco chloroplasts (+ NH_2OH), recorded after a number of flashes and the addition of DCMU. The oscillations in the total area (×) are due only to oscillations in the area of the α-component (□); the β-component (●) does not show these variations.

after subsequent flashes. Figure 5 was calculated from the proportionality between the areas of both components and the amounts of non-reduced Q_α and Q_β. From these experiments it can be calculated that the oscillations of R occur only in photosystem IIα. Oscillations were also restricted to photosystem IIα in mutants of tobacco, with higher β-to-α ratios. So, unlike Tris-washed chloroplasts, where all functional photosystem II centers appear to have both Q and R, untreated chloroplasts can be predicted to have less R than Q.

Absorbance Difference Spectra

It is generally accepted that the secondary acceptor is a bound plastoquinone molecule, which, in the anionic semiquinone form, can store one electron between successive photoacts. This acceptance is probably inspired to some extent by analogy to the secondary acceptor in purple bacteria, because the experimental basis for the identification of R in photosystem II is far less convincing. Little more is known than that evidence of absorbance changes that oscillate with a periodicity of two have been observed near 320 nm and that this 320 nm band could be due to a semiquinone anion (*6*, *10*). To fill the gap, we have improved and extended these measurements and obtained a reasonably pure ultraviolet spectrum of the oscillating absorbance changes. Expectedly, it differs in detail from those estimated from data in Pulles *et al.* (*6*) and Mathis and Haveman (*10*), but it confirms the identification of R^- as a bound semiquinone anion.

To isolate the absorbance changes of R, the following procedure was used. Hydroxylamine (1 m*M*) was added to dark-adapted, uncoupled

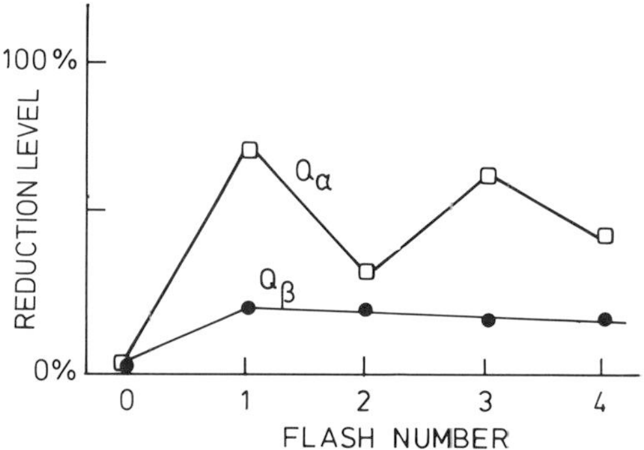

Fig. 5. Data of Fig. 4 translated into fractions of Q_α (□) and Q_β (●) found in the reduced state after a number of flashes and the addition of DCMU.

spinach chloroplasts. During illumination with a series of flashes spaced at 0.4 sec, absorbance changes that did not decay between flashes were measured. The intensity of the measuring light was adjusted to cause less than 0.1 photoreaction per center during the flash series, as checked by fluorescence induction measurements. The oscillating absorbance changes are superimposed on repetitive changes that were similar on all flashes of the series. The repetitive changes (+, Fig. 6) are rather unpredictable, especially above 300 nm, due to varying contributions by cytochrome *f* oxidation and cytochrome *b*-563 reduction. Also, one cannot simply subtract the observed value after oscillations have damped out, because of a gradual change of the amplitude during the flash series. This was clearly demonstrated by measurements in which the oscillations were prevented by preillumination with an exactly half-saturating flash.

An optimal discrimination against the repetitive absorbance changes was obtained by plotting on a log scale the differences between absorbance changes induced by each pair of successive flashes, $\Delta A_n - \Delta A_{n+1}$, positively for uneven and negatively for even values of n. A straight line was drawn to fit the points in the semilog plot versus n thus obtained, taking into account that the slope (i.e., the damping), should be wavelength independent and that the first two points are suspect.

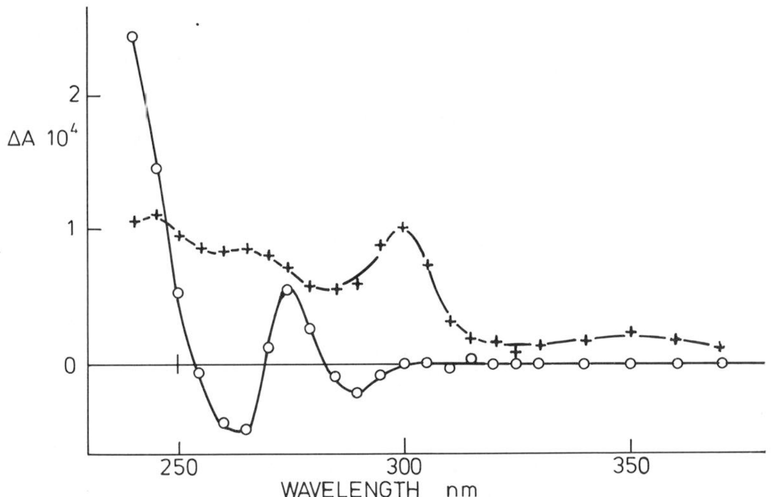

Fig. 6. Long-lived absorbance changes induced by each (+) or only the first (○) of a series of flashes, spaced at 0.4 sec, in dark-adapted, uncoupled spinach chloroplasts with 1 mM NH_2OH. Single beam measurements, averaged 16 to 64 times. Chlorophyll concentration 200 μM, optical pathlength 1.1 mm. Corrected for flattening.

Usually the second flash caused an additional reduction of cytochrome *b*-563 (*22*) (which, however, does not oscillate at the 0.4 sec flash spacing used). The first flash caused additional absorbance changes with the spectrum shown in Fig. 6, circles. As mentioned previously, in some preparations part of Q^- was stable between flashes; such preparations were discarded. The amplitude of the oscillating absorbance changes that would be obtained if there were no misses is equal to

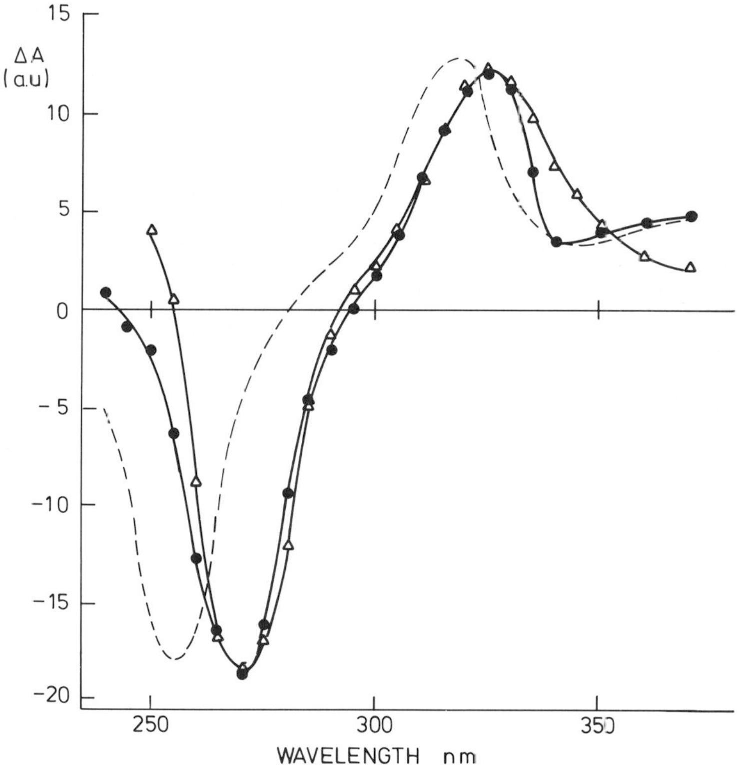

Fig. 7. (●), R^- spectrum of absorbance changes in spinach chloroplasts, which, in the measurements of Fig. 6 oscillated with a periodicity of two, determined from the measurements as explained in the text. Experimental error about one on this scale, reproducibility range about three. Average of two to five experiments; each measurement averaged 16 to 64 times. Plotted positive for uneven flash numbers. (△), Q^- in spinach chloroplasts; irreversible absorbance changes measured the same way but after addition of DCMU. Slowly decaying absorbance changes of cytochrome *f* were subtracted. All data were corrected for flattening. (- - -), PQ^- in vitro; difference spectrum of plastosemiquinone anion minus plastoquinone from Bensasson and Land (*23*), used with permission. The three spectra were normalized to the same amplitude and the ΔA units shown correspond to $\Delta\epsilon$ mM^{-1} cm^{-1} for the *in vitro* spectrum.

$$\frac{2(\Delta A_n - \Delta A_{n+1})}{(\rho - 1)^2 \cdot \rho^{n-1}}, \qquad \text{with } \rho = \frac{DA_{n+1} - \Delta A_n}{\Delta A_n - \Delta A_{n-1}}. \tag{2}$$

All absorbance changes were corrected for the particle flattening effect, according to the direct method described in Ref. *16*.

The spectrum of the oscillating absorbance changes thus obtained is shown in Fig. 7 (●), together with the spectrum of irreversible absorbance changes induced by a flash in the presence of DCMU and hydroxylamine (△). (Absorbance changes of cytochrome *f* decayed very slowly in these conditions and were subtracted). We ascribe these difference spectra to R^- minus R, and Q^- minus Q, respectively. The differences between the two spectra are significant and reproducible, but not necessarily due to Q and R; perhaps contributions of other absorbance changes were not all removed. Both spectra are very similar to the difference spectrum of plastosemiquinone anion minus plastoquinone [(---), from Bensasson and Land (*23*)], apart from a marked red shift, as observed earlier for the Q spectrum in deoxycholate particles (*7*). The spectrum in Fig. 6, (●), might possibly indicate that Q or R is somewhat less red-shifted in the dark-adapted state.

The absorbance units in Fig. 7 correspond to the differential extinction coefficients per m*M* per cm for the *in vitro* spectrum. Most likely, they also apply as such to the difference spectra of Q and R, which were arbitrarily normalized (*7*). The measured amplitudes then suggest that in dark-adapted spinach chloroplasts treated with hydroxylamine, the first flash produces 0.56 (±0.05) times the amount of stable semiquinones as produced by a flash in the presence of DCMU. For values of about 25 to 30% β centers, this suggests that R was about 15% reduced after dark adaptation. In similar experiments with chloroplasts from the Su/su mutant of tobacco, having about 55% β centers, we found about 0.34 R present per Q. This implies that in these chloroplasts, R was about 16% reduced. For both kinds of chloroplasts, this is in agreement with the value indicated by the fluorescence increase induced by DCMU in dark-adapted chloroplasts plus NH_2OH.

The findings described here should make it possible to use fluorescence measurements for quantitative studies on the behavior of R.

Correlation with the S-States

Figure 8 shows fluorescence measurements like those of Fig. 1, but on untreated chloroplasts without hydroxylamine (except triangles). Similar experiments were reported by Wollman (*5*). A periodicity of four, related to the turnover of the oxygen-evolving apparatus, is observed in the fluorescence level remaining after a flash (○), as found by

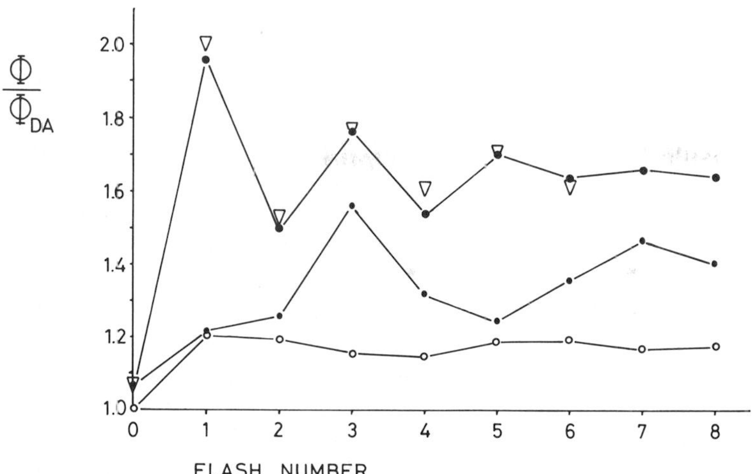

Fig. 8. Fluorescence measurements as in Fig. 1, but on normal oxygen-evolving spinach chloroplasts. (●), the initial maximum level reached upon DCMU addition. The lower dots indicate the irreversible part of the DCMU-induced increase. (▽), the (stable) fluorescence level obtained by adding NH_2OH and DCMU.

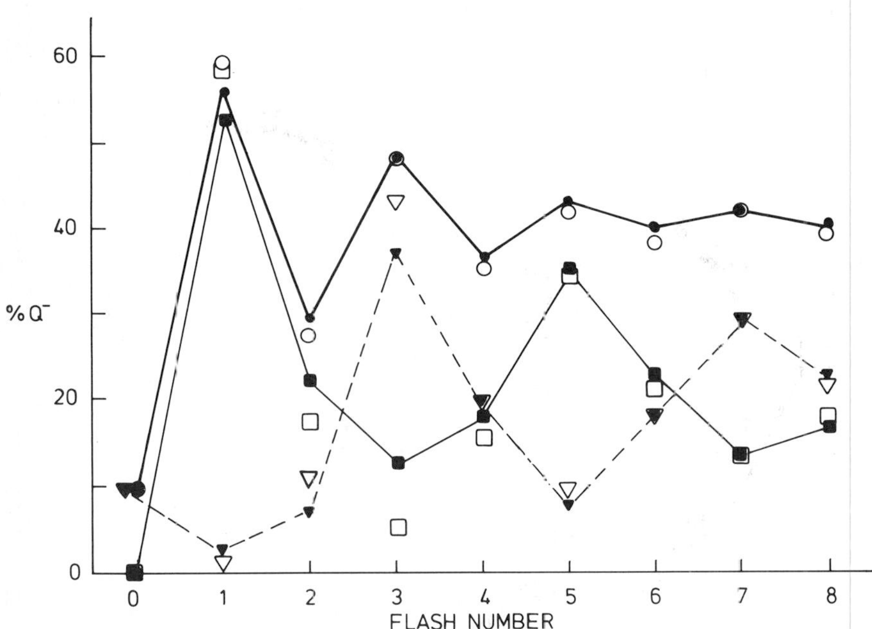

Fig. 9. Amounts of Q^- generated by DCMU addition, calculated from Fig. 8 as explained in the text: (●), total; (■), reversible; (▲) irreversible. Amounts of S-states from simulation: (○), $S_0 + S_2$; (□), S_2; (△), S_0. Initial distribution 0.1, 0.9, 0, 0; misses 0.1 and double hits 0.1 in all transitions. All values given as percent of total photosystem II, including β.

Delosme (*24*). Of the four successive oxidation levels of the oxygen-evolving enzyme, denoted S_0, S_1, S_2, and S_3 (S_4 being rapidly reduced to S_0), S_2 and S_3 seem to be associated with a higher fluorescence yield than S_0 and S_1. The DCMU-induced fluorescence increase, (●), is only partially stable (lower dots); a flash number-dependent fraction decays with a half-time of about 5 sec, as a result of back-reaction of Q^- with the higher S-states. Since the DCMU-induced fluorescence rise can occur in photosystem IIα only, the concentration of Q^- can be calculated from Eq. (1) if, for F_M, the maximum fluorescence minus the variable fluorescence of photosystem IIβ is taken. For F_0 we took the values indicated by the circles in Fig. 8, because we have no indication that any Q^- is associated with this fluorescence. A good fit of the data could only be obtained by assuming that throughout the flash sequence, R was associated with S_1 and S_3, and R^- was associated with S_0 and S_2, as illustrated in Fig. 9.

ACKNOWLEDGMENTS

This investigation was supported by the Netherlands Foundation for Chemical Research (SON), financed by the Netherlands Organization for the Advancement of Pure Research (ZWO). J. C. Lelie participated in part of the work.

REFERENCES

1. Bouges-Bocquet, B. (1973). *Biochim. Biophys. Acta* **314,** 250–256.
2. Velthuys, B. R., and Amesz, J. (1974). *Biochim. Biophys. Acta* **333,** 85–94.
3. Vermeglio, A. (1977). *Biochim. Biophys. Acta* **459,** 516–524.
4. Wraight, C. A. (1977). *Biochim. Biophys. Acta* **459,** 525–531.
5. Wollman, F.-A. (1978). *Biochim. Biophys. Acta* **503,** 263–273.
6. Pulles, M. P. J., van Gorkom, H. J., and Willemsen, J. G. (1976). *Biochim. Biophys. Acta* **449,** 536–540.
7. Van Gorkom, H. J. (1974). *Biochim. Biophys. Acta* **347,** 439–442.
8. Van Best, J. A., and Duysens, L. N. M. (1977). *Biochim. Biophys. Acta* **459,** 187–206.
9. Diner, B. A. (1977). *Biochim. Biophys. Acta* **460,** 247–258.
10. Mathis, P., and Haveman, J. (1977). *Biochim. Biophys. Acta* **461,** 167–181.
11. Bowes, J. M., and Crofts, A. R. (1980). *Biochim. Biophys. Acta* **590,** 373–384.
12. Van Gorkom, H. J., Pulles, M. P. J., and Etienne, A.-L. (1978). *In* "Photosynthetic Oxygen Evolution" (H. Metzner, ed.), pp. 135–145. Academic Press, New York.
13. Van Gorkom, H. J., Tamminga, J. J., Haveman, J., and van der Linden, I. K. (1974). *Biochim. Biophys. Acta* **347,** 417–438.

14. Bennoun, P., and Li, Y. S. (1973). *Biochim. Biophys. Acta* **292,** 162–168.
15. Melis, A., and Homann, P. H. (1975). *Photochem. Photobiol.* **21,** 431–437.
16. Thielen, A. P. G. M., and van Gorkom, H. J. (1981). *Biochim. Biophys. Acta* **635,** 111–120.
17. Melis, A., and Homann, P. H. (1978). *Arch. Biochem. Biophys.* **190,** 523–530.
18. Thielen, A. P. G. M., van Gorkom, H. J., and Rijgersberg, C. P. (1981). *Biochim. Biophys. Acta* **635,** 121–131.
19. Horton, P., and Croze, E. (1979). *Biochim. Biophys. Acta* **545,** 188–201.
20. Melis, A. (1978). *FEBS Lett.* **95,** 202–206.
21. Thielen, A. P. G. M., and van Gorkom, H. J. (1980). *Int. Congr. Photosynth., 5th, Halkidiki, Greece* Abstr., p. 568.
22. Velthuys, B. R. (1979). *Proc. Natl. Acad. Sci. U.S.A.* **76,** 2765–2769.
23. Bensasson, R., and Land, E. J. (1973). *Biochim. Biophys. Acta* **325,** 175–181.
24. Delosme, R. (1972). *Photosynth., Two Centuries Its Discovery Joseph Priestley, Proc. Int. Congr. Photosynth. Res., 2nd, Stresa, Italy, 1971* **1,** 187–195.

How Can We Reconcile the Notion of the "Transmembrane" Ubisemiquinone Pair with the Rapid Kinetics of its Formation Observed with Complex II?

5

HELMUT BEINERT
FRANK J. RUZICKA[1]

INTRODUCTION

Among the major EPR signals observed but not readily identified at <30 °K in whole tissues and mitochondria, was a signal with features at 322, 332, and 335 mT at 9.2 GHz. It apparently was superimposed on the signal originating from an Fe–S cluster of succinate dehydrogenase that occurs in the $+3$ state, as does the high potential iron-sulfur (Fe–S) protein of *Chromatium vinosum*. Examples of this signal, obtained from different materials under various conditions, are shown in Fig. 1.

The unidentified signal was absent in the fully oxidized state, reached a maximum during partial reduction, and disappeared again on full reduction. This indicates that it was due to a two-electron acceptor. Since the known two-electron acceptors in mitochondria are quinoid in nature (such as ubiquinone (UQ) and flavin), attention was drawn to these compounds as possible sources of the signal. Yet the evidence that the signal was ob-

[1] Present address: Department of Human Oncology, Clinical Sciences Center, University of Wisconsin, Madison.

Function of Quinones in Energy Conserving Systems

ISBN 0-12-701280-X

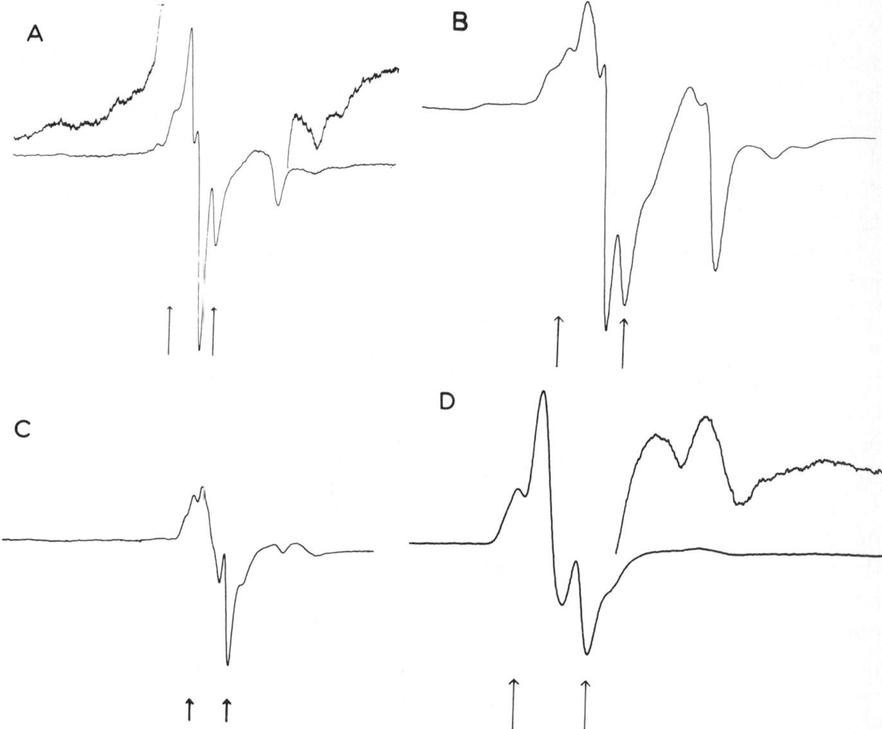

Fig. 1. EPR signal ascribed to a stabilized ubisemiquinone pair as seen with different materials under various conditions. The arrows point to the characteristic lines that appear at 322 and 332 mT at 9.2 GHz. (A) pigeon heart, rapidly freeze clamped (*1*) in a state of partial reduction; (B) beef heart mitochondria, approximately 60 mg protein per ml, 0.25 *M* sucrose, 0.01 *M* Tris chloride (pH 7.8), aerated and immediately frozen in isopentane at 133 °K; (C) submitochondrial particles from beef heart, 40 mg protein per ml, 0.25 *M* sucrose-Tris as above, treated with 4,5,6,7-tetrachloro-2-trifluoromethylbenzimidazole to prevent reduction by endogenous reductants, then partly reduced with dithionite; (D) submitochondrial particles from beef heart, lyophilized and extracted according to Norling *et al.* (*2*), followed by incorporation of UQ-10 (6 nmol/mg protein), then partly reduced by dithionite to maximal USQ formation. The conditions of EPR spectroscopy at 9.2 GHz for (A–D), respectively, were: microwave power, 0.27, 2.7, 0.27, and 9 mwatt; modulation amplitude, 0.8, 0.8, 0.8, and 0.4 mT; scanning rate, 20, 40, 20, and 40 mT per min; time constant, 0.5 sec throughout; and temperature, 12, 12, 13, and 10 °K. The wings in (A) and (D) were recorded at 2.7 and 0.27 mwatt and an additional 3.2- and 25-fold amplication, respectively; (D) was recorded at 13 °K.

served only at very low temperature and not readily saturated with microwave power, weighed against its origin from a semiquinone *per se*. We were able to show, however, that submitochondrial particles (SMP) treated by procedures designed to extract UQ (*2*) did not exhibit this unidentified signal at any oxidation state. The signal intensity at partial reduction depended on the degree of extraction; reincorporation of UQ into submitochondrial particles restored the original signal intensity observed on partial reduction (cf., Fig. 1D).

The signal was seen only in subcellular fractions that contained succinate dehydrogenase. It showed temperature dependence and saturation behavior similar to that exhibited by the +3 state Fe–S cluster of succinate dehydrogenase. Since the unidentified signal was maximal on partial reduction of the +3 state cluster of that enzyme, some relationship between this Fe–S cluster and ubisemiquinone (USQ) was suspected. This led to the idea of a Fe–S cluster–USQ interaction (*3*) that might produce the signal. EPR spectroscopy of submitochondrial particles at 35 GHz supported this notion in that the position in the magnetic field of the resonances observed at 9.2 GHz did not vary with frequency as expected for the true g values of a paramagnet. Rather, the separation of the lines appeared to remain constant. Computer synthesis of the 9.2 GHz spectra, based on a model of two interacting species using the g values of USQ and the +3-state Fe–S cluster of succinate dehydrogenase, indeed produced spectra approaching those of the unknown species (*4*). It soon became apparent, however, that better fits of simulated to experimental spectra were obtained if interaction was assumed to occur between two semiquinone molecules, with the +3 state Fe–S cluster perhaps secondarily involved.

Additional evidence for these proposals came from Ingledew *et al.*, who observed the signal during potentiometric titration of particles (*5*). These authors also noted that consideration of the known oxidation–reduction potentials eliminates flavin semiquinone as a possible partner of USQ (a possibility left open by us). The concentration of USQ represented in this signal was estimated to be in the range of succinate dehydrogenase concentration. This is remarkable, since from analogies drawn to similar chemical systems, it was thought that USQ was unstable and would readily disproportionate into UQ and UQH_2 [see (*6*, *7*); cf. (*8*)].

In the simulations of EPR spectra carried out by Schepler, Dunham, and Sands (cf. *4*), best fits were obtained when it was assumed that the USQ molecules in the pair were aligned along the direction of the largest g value. Since, for quinones in general, this g value [g_x, referred to as $g_\parallel$ in Ruzicka *et al.* (*4*)] is along the O–O axis of the quinone (*9*), it is likely that the USQ molecules of the pair are disposed edge-on relative to each other, and not face-to-face. An interesting further development ensued when Salerno *et al.* (*10*) investigated, in oriented multilayers of mitochondria, the response of the characteristic signal to the magnetic field direction. These authors came to the conclusion that the USQ pair (with its O–O axis aligned edge-on), is oriented such that the common O–O axis is normal to the plane of the membrane. This led to the proposal of a "transmembrane" USQ pair, as depicted in a number of recent articles and reviews [cf. (*10–12*)]. The report by Yu *et al.* (*13*) of a UQ-binding protein in succinate-cytochrome *c* reductase completes the picture, furnishing the presumptive anchoring point(s) for the members of the pair.

Simultaneous with these studies, aimed largely at structural aspects of the problem, investigations using titrations and kinetic approaches (*14–16*) made it appear likely that electron transfer in succinate dehydrogenase proceeds from substrate via flavin to the Fe–S clusters with the +3 state cluster (that of highest midpoint potential) being the exit port for electrons toward UQ. It was also shown that inhibitors of the succinate oxidase system, TTFA and the more recently discovered, more potent carboxins, blocked electron transfer at the +3 state cluster–UQ juncture. In the presence of these inhibitors, the USQ pair was never seen, and reduced succinate dehydrogenase could not be oxidized by UQ-type electron acceptors.

During studies (*16*) on the kinetics of reoxidation of reduced Complex II (succinate-UQ reductase) carried out with 2,3-dimethoxy-5-methyl-6-*n*-pentyl-1,4-benzoquinone (DPB), the signal of the USQ pair was readily observed (Fig. 2), although Complex II is practically devoid of UQ and the signal characteristic of the USQ pair has never been seen on reduction or reoxidation of Complex II with agents other than a UQ type compound. Thus, it is clear that the UQ pair can also be generated by the addition of UQ or even a relatively water-soluble analog such as DPB with only a saturated 5-carbon side chain. We were surprised to find that even at the earliest reaction times at 0 °C [which we were able to generate with our rapid freeze-quench apparatus (*17*)], maximal or close-to-maximal signal intensity was reached for the conditions used. These reaction times, calculated from the dimensions of the apparatus and the velocity of the syringe ram, were 1.5 msec. However, an unknown effective quenching time of the reaction will have to be added that is unlikely to exceed ~3 msec. Experiments were routinely done with ~6 μM Complex II and 230 μM DPB. The signal intensity was clearly dependent on the DPB concentration up to ~200 μM, but the characteristic signal was still seen with as little as 16 μM DPB at 0 °C.

In view of the preceeding notion about the "transmembrane" UQ pair (*10*), we have been pondering whether and how such a picture is compati-

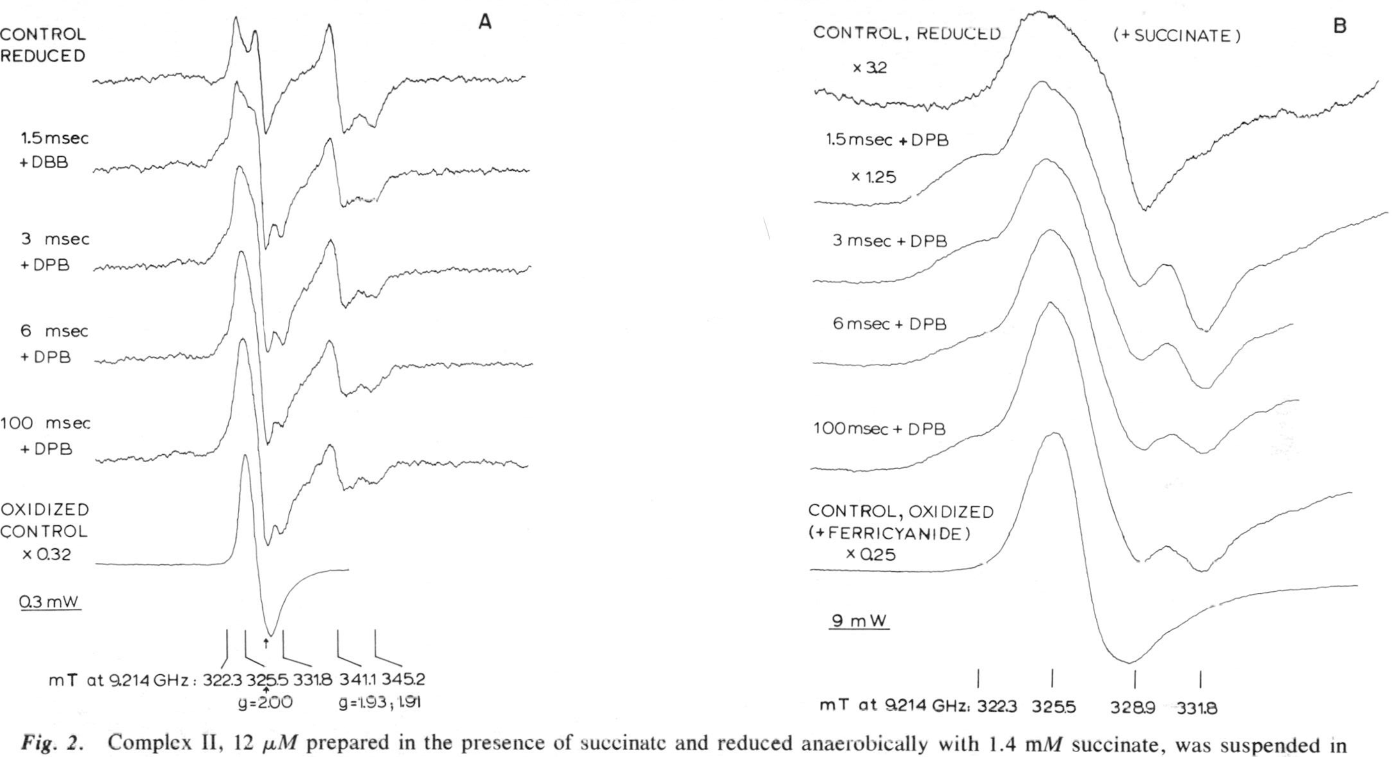

Fig. 2. Complex II, 12 μM prepared in the presence of succinate and reduced anaerobically with 1.4 m*M* succinate, was suspended in 50 m*M* Hepes buffer of pH 7.8 and was then rapidly mixed with an equal volume of 0.47 m*M* DPB in Hepes containing 12.5% ethanol (v/v) and collected by freeze quenching after the times indicated in the figure. The reduced control was produced by mixing Complex II with buffer alone and the oxidized control by mixing with 5 m*M* ferricyanide in buffer. The conditions of EPR spectroscopy at 9.2 GHz were for (A) and (B), respectively: microwave power, 0.27 and 9 mwatt; modulation amplitude, 0.8 mT throughout; scanning rate, 40 and 10 mT per min; time constant, 0.5 sec; and temperature, 13 °K for both.

ble with the observations from our (*16*) rapid freeze-quench studies. These questions are the principal object of this paper.

DISCUSSION

At the outset we should remember that the notion of the "transmembrane" USQ pair is based on the observation that the common O–O axis of the USQ pair is oriented normal to the plane of the mitochondrial membrane. However, our kinetic observations were made on Complex II, a fragment of the mitochondrial inner membrane; and the structure and disposition of Complex II in the membrane has only recently been studied (*18*, *19*). According to this work, Complex II spans the inner mitochondrial membrane, with the subunit bearing the flavoprotein exposed on the matrix side of the membrane, and a polypeptide C_{II-3} exposed on the cytoplasmic side. Two peptides of Complex II, C_{II-3} and C_{II-4} (MW 13,500 and 7000, respectively), are thought to "probably form the major part of the bilayer intercalated part of the Complex II and are thus available for interaction with the lipophilic ubiquinone molecules" (*18*). The orientation studies on the USQ pair of mitochondria [by Salerno *et al.* (*10*)] are not necessarily applicable to Complex II, although we would expect the USQ molecules to possibly occupy the same binding site or sites. However, since the EPR signals observed with mitochondria, Complex II and DPB are identical, it is very likely that the USQ molecules of the pair have essentially the same orientation with respect to each other. More recent simulations of EPR spectra by Salerno and Ohnishi (*20*), however, indicate that "the simulations are fairly insensitive to the orientations of the two quinone *g* tensors with respect to each other and to **R**." (**R** is the vector connecting the two interacting spins, assumed to be point dipoles); therefore, this last conclusion is not firm.

The picture conceived from the studies on mitochondria indicates that within the membrane—according to the schemes presented (*10–12*), rather deep within the membrane—there is a binding site occupied by two adjacent UQ molecules. During electron transfer from succinate via succinate dehydrogenase, these two adjacent UQ molecules receive two electrons and form a fairly stable USQ pair. Exchange of these UQ molecules with the UQ pool in the intermembrane space is likely, but such exchange is considered incidental to the electron-transfer function.

The picture obtained from the kinetic experiments with Complex II is that two molecules of DPB must, within ~5 msec at 0 °C, settle on a large fraction of the available Complex II assemblies (probably in the edge-to-edge alignment indicated by the EPR signal), and accept two electrons, one residing in each member of the pair. This would seem rather remarkable if the DPB molecules were to penetrate the membrane to any depth. Since the EPR signal of the USQ pair disappears within seconds, the bound DPB molecules are probably oxidized by accepting electrons from DPB in solution, and then dissociate (*20*).

From these observations, it seems that in Complex II preparations, the UQ binding peptide(s) C_{II-3} and (or) C_{II-4} are readily accessible from the aqueous phase and are free to react with DPB; that is, Complex II must be cleaved out of the membrane right at the face of C_{II-3} and (or) C_{II-4}. If the mutual orientation of the DPB molecules is edge-on, as observed for USQ in mitochondria, this orientation is probably conditioned by the binding site rather than by an intrinsic property of USQ and its analogs. To our knowledge no evidence for a pair formation with such a configuration has been seen by EPR with USQ in artificial membranes (*5*). If the orientation of the USQ molecules toward each other is indeed due to the binding peptide, we can conclude that the integrity of this site has been well preserved through the purification of Complex II, often considered a harsh procedure.

The results of Salerno *et al.* (*10*) on the orientation of the USQ pair with respect to the mitochondrial membrane do not necessarily imply that the pair lies within the membrane, but only indicate that it lies normal to it. One could, therefore, argue that the pair is only anchored at the membrane and protrudes out toward the intermembrane space, which would make the rapid reaction of Complex II with DPB more plausible. Such a view, however, is not supported by the results obtained by Ackrell *et al.* (*19*), who found that in the presence of succinate dehydrogenase, peptide C_{II-3} in Complex II was protected from extensive proteolytic digestion by chymotrypsin, so that full endogenous UQ reductase activity was preserved.

ACKNOWLEDGMENTS

The authors are indebted to Drs. B. A. C. Ackrell, T. Ohnishi, J. C. Salerno, and R. H. Sands for helpful discussions. Support of this work by a Research Grant (GM-12394) and a Research Career Award (5-KO6-GM-18442) from the Institute of General Medical Sciences, National Institutes of Health, USPHS, to H.B. is gratefully acknowledged.

REFERENCES

1. Orme-Johnson, N. R., Hansen, R. E., and Beinert, H. (1974). *J. Biol. Chem.* **249,** 1928–1939.

2. Norling, B., Glazek, E., Nelson, B. D., and Ernster, L. (1974). *Eur. J. Biochem.* **47,** 475–482.

3. Ruzicka, F. J., and Beinert, H. (1975). *Fed. Proc., Fed. Am. Soc. Exp. Biol.* **34,** 579.

4. Ruzicka, F. J., Beinert, H., Schepler, K. L., Dunham, W. R., and Sands, R. H. (1975). *Proc. Natl. Acad. Sci. U.S.A.* **72,** 2886–2890.

5. Ingledew, W. J., Salerno, J. C., and Ohnishi, T. (1976). *Arch. Biochem. Biophys.* **177,** 176–184.

6. Kröger, A. (1976). *FEBS Lett.* **65,** 278–280.

7. Land, E. J., and Swallow, A. J. (1970). *J. Biol. Chem.* **245,** 1890–1894.

8. Yu, C. A., Nagaoka, S., Yu, L., and King, T. E. (1980). *Arch. Biochem. Biophys.* **204,** 59–70.

9. Hales, B. J. (1975). *J. Am. Chem. Soc.* **97,** 5993–5999.

10. Salerno, J. C., Harmon, H. J., Blum, H., Leigh, J. S., and Ohnishi, T. (1977). *FEBS Lett.* **82,** 179–182.

11. Ohnishi, T. (1979). *In* "Membrane Proteins in Energy Transduction" (R. A. Capaldi, ed.), pp. 1–87. Dekker, New York.

12. Salerno, J. C., and Ohnishi, T. (1978). *In* "Tunneling in Biological Systems" (B. Chance, D. C. De Vault, H. Frauenfelder, R. A. Marcus, J. R. Schieffer, and N. Sutin, eds.), pp. 153–157. Academic Press, New York.

13. Yu, C. A., Yu, L., and King, T. E. (1977). *Biochem. Biophys. Res. Commun.* **78,** 259–265.

14. Beinert, H., Ackrell, B. A. C., Kearney, E. B., and Singer, T. P. (1975). *Eur. J. Biochem.* **54,** 185–194.

15. Ohnishi, T., Salerno, J. C., Winter, D. B., Lim, J., Yu, C. A., Yu, L., and King, T. E. (1976). *J. Biol. Chem.* **251,** 2094–2104.

16. Ackrell, B. A. C., Kearney, E. B., Coles, C. J., Singer, T. P., Beinert, H., Wan, Y.-P., and Folkers, K. (1977). *Arch. Biochem. Biophys.* **182,** 107–117.

17. Beinert, H., Hansen, R. E., and Hartzell, C. R. (1976). *Biochim. Biophys. Acta* **423,** 339–355.

18. Merli, A., Capaldi, R. A., Ackrell, B. A. C., and Kearney, E. B. (1979). *Biochemistry* **18,** 1393–1400.

19. Ackrell, B. A. C., Ball, M. B., and Kearney, E. B. (1980). *J. Biol. Chem.* **255,** 2761–2769.

20. Salerno, J. C., and Ohnishi, T. (1981). *Biochem. J.* **192,** 769–781.

Function and Properties of a Semiquinone Anion Located in the QH_2: Cytochrome c *Oxidoreductase Segment of the Respiratory Chain*

6

SIMON DE VRIES
J. A. BERDEN
E. C. SLATER

INTRODUCTION

The way in which electrons are transferred from ubiquinol to cytochrome *c* is still actively under discussion. The results of potentiometric titrations and measurements of pre-steady-state kinetics monitored optically, giving information mainly on the redox properties of the cytochromes, led to proposals (*1*, *2*) for electron transfer through QH_2: cytochrome *c* oxidoreductase, in which the formation of a semiquinone is a prerequisite for electron transfer. The presence of a semiquinone in preparations of the respiratory chain has been identified by EPR studies (*3–11*). Recently, Ohnishi and Trumpower detected two different populations of ubisemiquinone in isolated succinate : cytochrome *c* oxidoreductase, SQ_s and SQ_c, differing in relaxation time [see also (*12*)] and response to antimycin.

The results of the experiments presented in this paper indicate the existence of a very stable antimycin-sensitive semiquinone anion bound to

Function of Quinones in Energy Conserving Systems

ISBN 0-12-701280-X

QH_2 : cytochrome *c* oxidoreductase, presumably corresponding to SQ_c of Ohnishi and Trumpower (*10*). Quantitation of the EPR signal of the semiquinone anion showed that the maximal concentration is one-half that of the cytochrome c_1, at least in submitochondrial particles. To adequately describe the effect of pH on semiquinone anion concentration, one must consider that semiquinone can exist in significant amounts only when bound and that the binding site has a limited capacity. A modified Q-cycle scheme is presented showing the function of this semiquinone anion; part of this work has been published in de Vries *et al.* (*13*).

EXPERIMENTAL RESULTS

An intense signal at $g = 2.005$ is present in the EPR spectrum of submitochondrial particles, succinate : cytochrome *c* oxidoreductase and QH_2 : cytochrome *c* oxidoreductase preparations at pH values above 6.8,

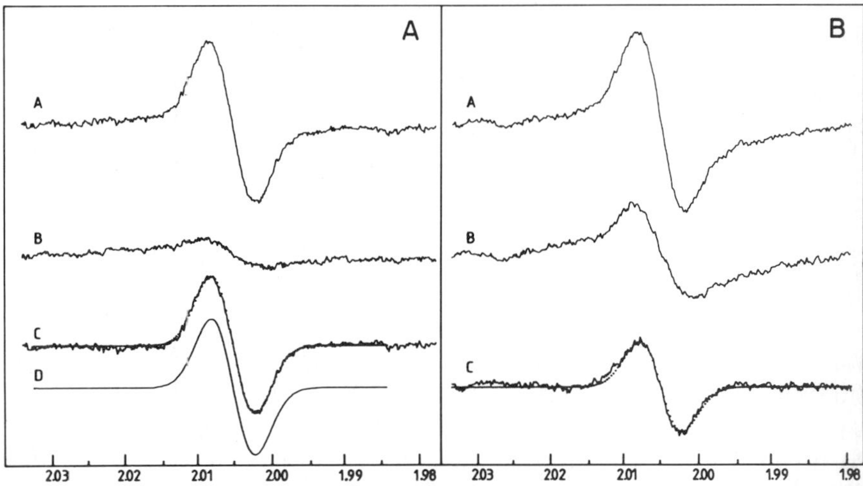

Fig. 1. EPR spectra of submitochondrial particles (18 μM cytochrome c_1) in the presence of 0.25 *M* sucrose, 50 m*M* Tris-HCl buffer (pH 8.4) and 4 m*M* KCN. (A) Curve A, after addition of 150 m*M* sodium fumarate and 1.5 m*M* sodium succinate ($E_h = 0$ mV) the particles were incubated for 5 min at room temperature; Curve B, same as Curve A, but 19 μM antimycin was also added; Curve C, difference A–B: (···), simulation (from Curve D); Curve D, simulation of the semiquinone using a Gaussian lineshape and $g = 2.005$, line width = 1 mT. (B) Curve A, after addition of 150 m*M* fumarate and 9.3 m*M* succinate ($E_h = -22$ mV); Curve B, Same as Curve A, but 19 μM antimycin was also added; Curve C, Difference A–B: (···), the same simulation (corrected for the difference in signal amplitude) as in (A). EPR conditions: Frequency (F), 9.13 GHz; Modulation Amplitude (MA), 1 mT; Power (P), 3 mW; Scanning Rate (SR), 2.5 mT/min; Temperature (*T*), 20 °C.

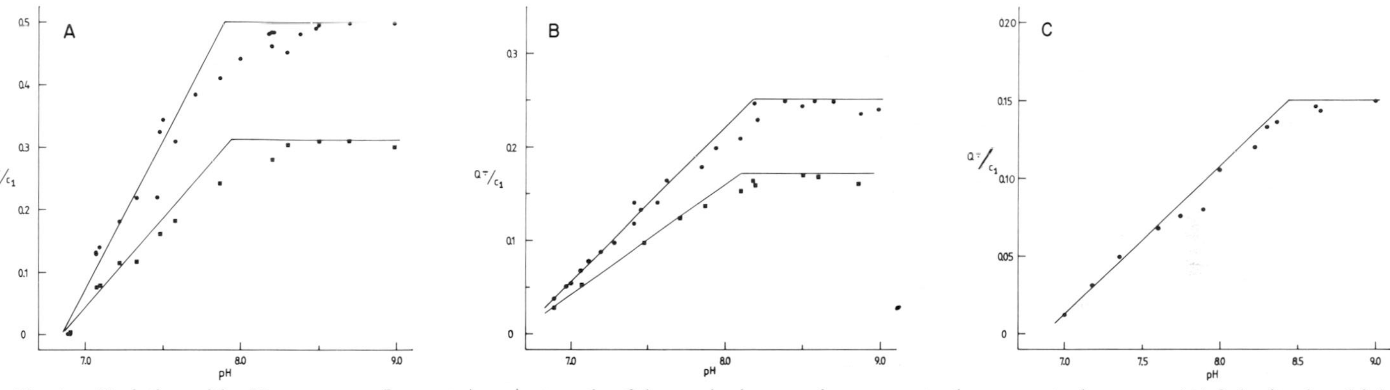

Fig. 2. Variation with pH at constant fumarate/succinate ratio of the semiquinone anion concentration per cytochrome c_1. (A) Submitochondrial particles in the presence of 4 m*M* KCN, 150 m*M* sodium fumarate. (●), after addition of 1.9 m*M* sodium succinate (E_h = 79 mV, pH 7); (■), after addition of 18.8 m*M* sodium succinate (E_h = 50 mV, pH 7). EPR conditions: F, 9.14 GHz; MA, 0.63 mT; *P*, 5 W; SR, 2.5 mT/min; *T*, 77 °K; (B) submitochondrial particles in the presence of 4 m*M* KCN, 150 m*M* sodium fumarate. (●), after addition of 1.5 m*M* sodium succinate (E_h = 82 mV, pH 7). (■), after addition of 6 m*M* sodium succinate (E_h = 65 mV, pH 7). EPR conditions as in Fig. 1. (C) succinate: cytochrome *c* oxidoreductase incubated with 1 m*M* KCN, 100 m*M* sodium fumarate, 1.4 m*M* sodium succinate (E_h = 78 mV, pH 7). EPR conditions as in Fig. 1.

temperatures between 4.2 °K and 37 °C, and at redox potentials between 0 and 100 mV maintained with a fumarate-succinate mixture (see Fig. 1). The signal is absent in pentane-extracted preparations, but present after reincorporation of ubiquinone (data not shown). Titration with antimycin showed that the signal intensity declined to a constant value after addition of one mole of antimycin per mole of cytochrome c_1 (*10*). The signal remaining at 20 °C (Fig. 1) and 77 °K (not shown) is that of a flavin semiquinone, probably that of succinate:Q oxidoreductase. It was concluded that the antimycin-sensitive signal originates from a semiquinone of ubiquinone bound to QH_2:cytochrome *c* oxidoreductase (*11*, *14*), and probably identical to SQ_c reported by Ohnishi and Trumpower (*10*).

In Figs. 2A, 2B, and 2C the effect of variation of pH at a constant fumarate to succinate ratio on the amount of semiquinone is shown. Since the semiquinone concentration strongly increases with increasing pH, the semiquinone anion, $Q^{\cdot-}$, is involved and not QH·, at least above pH 6.8. The maximal amount of semiquinone anion formed at high pH is dependent upon the fumarate to succinate ratio, temperature and preparation used (see Table I). The largest amount of semiquinone anion found in submitochondrial particles is equal to one-half the concentration of cytochrome c_1; smaller amounts are found in succinate:cytochrome *c* oxidoreductase [cf. (*10*) and Table I].

The results of potentiometric titrations at constant pH, obtained by measurement of semiquinone anion concentration by EPR at variable fumarate to succinate ratios, are summarized in Table II. The midpoint potential of the $E(Q)/E(QH_2)$ couple[1] is given, as a first approximation, by the potential at which the amount of semiquinone anion is maximal (*15*).

[1] Abbreviations: $E(Q)$, $E(Q^{\cdot-})$, $E(QH_2)$ denote bound ubiquinone, bound semiquinone anion, and bound ubiquinol, respectively; $E(Q^{\cdot-}_{max})$, maximal concentration of bound semiquinone anion.

TABLE I

Maximal Concentration of Semiquinone Anion under Different Conditions

Preparation	Q/c_1	Temperature	$(Q^{\cdot-}_{max})/c_1$	E_{m7}[A]
Submitochondrial particles	11	77 °K	0.5	77
		5 °C	0.35	82
		20 °C	0.26	84
		37 °C	0.13	84
Succinate or QH_2: cytochrome *c*		77 °K	0.3	75
oxidoreductase	1–2	20 °C	0.15	80

[A] $(E(Q)/E(QH_2))$ (mV)

TABLE II

Midpoint Potentials of Various Ubiquinone Couples[A]

Redox couple	Temperature	E_{m7} (mV)	Slope (mV/pH)
	Submitochondrial particles		
$E(Q)/E(QH_2)$	20 °C	84	−60
	77 °K	77	−60
$Q^{\overline{\cdot}}/QH_{2\,tot}(E_{m1})$	20 °C	204	−120[B]
	77 °K	184	−120[B]
$Q_{tot}/Q^{\overline{\cdot}}(E_{m2})$	20 °C	−36	0[B]
	77 °K	−30	0[B]
$E(Q^{\overline{\cdot}})/QH_{2\,tot}(E'_{m1})$	20 °C	233	−129
	77 °K	226	−130
$Q_{tot}/E(Q^{\overline{\cdot}})(E'_{m2})$	20 °C	−65	+9
	77 °K	−71	+10
	Succinate : cytochrome *c* oxidoreductase		
$E(Q)/E(QH_2)$	20 °C	80	−60
E_{m1}	20 °C	173	−120[C]
E_{m2}	20 °C	−13	0[C]
E'_{m1}	20 °C	230	−128
E'_{m2}	20 °C	−70	+8

[A] The midpoint potential of the $E(Q)/E(QH_2)$ couple was taken to be equal to the potential of the fumarate-succinate couple at which the amount of $Q^{\overline{\cdot}}$ is maximal. E_{m1} and E_{m2} were calculated from the amount of $Q^{\overline{\cdot}}$ found by EPR and the total amount of ubiquinone measured in the particles, assuming that the potential of $E(Q)/E(QH_2)$ is the same as that of $Q_{tot}/QH_{2\,tot}$. See text for calculation of E'_{m1} and E'_{m2}.

[B] (pH < 7.6); −60 (pH > 7.6)

[C] (pH < 7.9); −60 (pH > 7.9)

Since between pH 6.8 and 9.0 the maximum amount of semiquinone anion was always found at the same fumarate to succinate ratio, it follows that the midpoint potential of the $E(Q)/E(QH_2)$ couple changes by −60 mV/pH within this range of pH as for the free Q/QH_2 couple (*16*).

In Figs. 3A, 3B, and 3C E_m/pH plots constructed from the data in Fig. 2 are shown using the Nernst equations

$$E_h = 24 + 30\log(\text{fumarate/succinate}) - 60(\text{pH}-7) \quad (1)$$

$$E_h = E_{m7}(Q/QH_2) + 30\log(Q/QH_2) - 60(\text{pH}-7) \quad (2)$$

$$E_h = E_{m1} + 60\log(Q^{\overline{\cdot}}/QH_2) - 120(\text{pH}-7) \quad (3)$$

$$E_h = E_{m2} + 60\log(Q/Q^{\overline{\cdot}}) \quad (4)$$

Although below pH 7.6 in submitochondrial particles and pH 7.9 in succinate:cytochrome *c* oxidoreductase, the midpoint potentials of E_{m1} and E_{m2} vary with pH in the manner to be expected on the basis of Eqs. (1–4), above pH 7.6 and 7.9, respectively, the slope of the curves becomes −60 mV/pH. In the case of E_{m1} this could be explained by assuming that bound QH_2 has a p*K* of 7.6 or 7.9, thus

$$E(QH_2) \rightleftharpoons H^+ + E(QH^-) \rightleftharpoons E(Q^{\cdot -}) + H^+ + e^- \tag{5}$$

However, if this were true, the midpoint potential of the $E(Q)/E(QH_2)$ couple would vary −30 mV/pH above pH 7.6 and 7.9, respectively. This was not found [see Table II; see also (*10*), Fig. 6]. In addition, the p*K* of free QH_2, incorporated in phospholipid vesicles, was found to be greater than 12 (de Vries, unpublished observations). In any case, dissociation of a proton from QH_2 cannot explain the slope of −60 mV/pH in E_{m2}. A redox equilibrium with cytochrome *b*, e.g.,

$$Q^{\cdot -} + H^+ + b^{2+}H^+ \rightleftharpoons QH_2 + b^{3+} \tag{6}$$

could explain a slope of −60 mV/pH, but the p*K* is rather far from that reported to be involved in the redox reaction of cytochrome *b*, namely, 6.9 (*16*). Since the break in the slope of the curves in Fig. 3 cannot be explained by the assumption that there exists a redox-linked proton with a p*K* of 7.6, associated with a known redox carrier in the *b*-c_1 complex, we give the following interpretation.

The curves in Fig. 2 suggest that the amount of semiquinone anion that can be bound becomes limiting at high pH, when the binding site for the semiquinone anion is nearly saturated. To describe the experimental data including binding of the semiquinone anion, ubiquinone, and ubiquinol to a specific site, the following equations are needed

$$E(QH_2) \overset{K_r}{\rightleftharpoons} E + QH_2 \qquad K_r = \frac{[E]\cdot[QH_2]}{[E(QH_2)]} \tag{7}$$

$$E(QH_2) \rightleftharpoons E(Q^{\cdot -}) + 2H^+ + e^- \tag{8}$$

$$E(Q^{\cdot -}) \overset{K_s}{\rightleftharpoons} E + Q^{\cdot -} \qquad K_s = \frac{[E]\cdot[Q^{\cdot -}]}{[E(Q^{\cdot -})]} \tag{9}$$

$$E(Q^{\cdot -}) \rightleftharpoons E(Q) + e^- \tag{10}$$

$$E(Q) \overset{K_0}{\rightleftharpoons} E + Q \qquad K_0 = \frac{[E]\cdot[Q]}{[E(Q)]} \tag{11}$$

$$[E] = e_{tot} - [E(QH_2)] - [E(Q)] - [E(Q^{\cdot -})] \tag{12}$$

The derivation of the Nernst equation is straightforward (*17*).

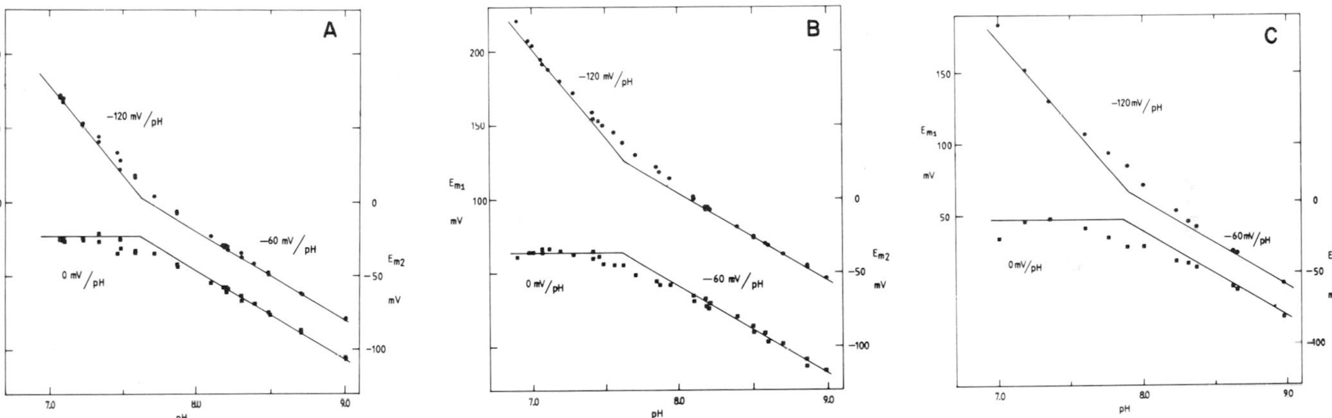

Fig. 3. E_m/pH plots (●, E_{m1}; ■, E_{m2}) constructed from the measured semiquinone anion concentration using Eqs. (1)–(4). (A), (B), and (C) data were taken from Figs. 2A, 2B, and 2C, respectively.

$$E_h = E'_{m1} + 60 \log E(Q^{\dot{-}})/QH_{2\,tot} - 60 \log\{E(Q^{\dot{-}}_{max}) - E(Q^{\dot{-}})\} - 120(pH-7) \quad (13)$$

where $E'_{m1} = E_m(E(Q^{\dot{-}})/E(QH_2)) + 60 \log K_r$

$$E_h = E'_{m2} + 60 \log Q_{tot}/E(Q^{\dot{-}}) + 60 \log\{E(Q^{\dot{-}}_{max}) - E(Q^{\dot{-}})\} \quad (14)$$

where $E'_{m2} = E_m(E(Q)/E(Q^{\dot{-}})) - 60 \log K_o$

In the derivation of Eqs. (13) and (14), it was assumed that all three forms of ubiquinone bind with a high affinity, but there is preferential binding of the semiquinone anion, and no measureable amount of free semiquinone anion exists. Consequently

$$K_o \approx K_r >> E >> K_s \approx 0;\ Q^{\dot{-}}_{tot} = E(Q^{\dot{-}})$$

$E(Q^{\dot{-}}_{max})$ was determined from the horizontal part of the curves in Fig. 2. Thus, $E(Q^{\dot{-}}_{max})$ equals $0.5/c_1$ for the upper trace in Fig. 2A, but $0.31/c_1$ for the lower trace. In Fig. 4 the values of E'_{m1} and E'_{m2} calculated from Eqs. 13 and 14 are shown. The Nernst plots now show no break at pH 7.6 and the slopes are close to the expected values of −120 mV/pH and zero for E'_{m1} and E'_{m2}, respectively. Indeed, as a consequence of normalizing to maximal $Q^{\dot{-}}$, the plots for E'_{m1} and E'_{m2} are now very similar for two different temperatures and preparations, in contrast to significant differences for the plots of E_{m1} and E_{m2} (see Fig. 3 and Table II).

DISCUSSION

The experiments show, in agreement with Ohnishi and Trumpower (*10*), that QH_2:cytochrome *c* oxidoreductase has a specific binding site for an antimycin-sensitive semiquinone anion. This site might be on a Q-binding protein (*18*). The maximal amount of semiquinone anion is half the cytochrome c_1 concentration in submitochondrial particles, which is more than in isolated succinate:cytochrome *c* oxidoreductase. The difference in stability is probably due to the isolation procedure and/or the lower ubiquinone content of purified preparations.

The anomalous redox behavior of the semiquinone anion can be described with a model that includes strong binding of all three forms of ubiquinone, but preferential binding of the semiquinone anion, and a limited capacity of the binding site. Since the midpoint potential of the $E(Q)/E(QH_2)$ couple varies by −60 mV/pH between pH 6.8 and 9.0, it follows that the quinol is bound. This is in disagreement with Ohnishi and Trumpower (*10*).

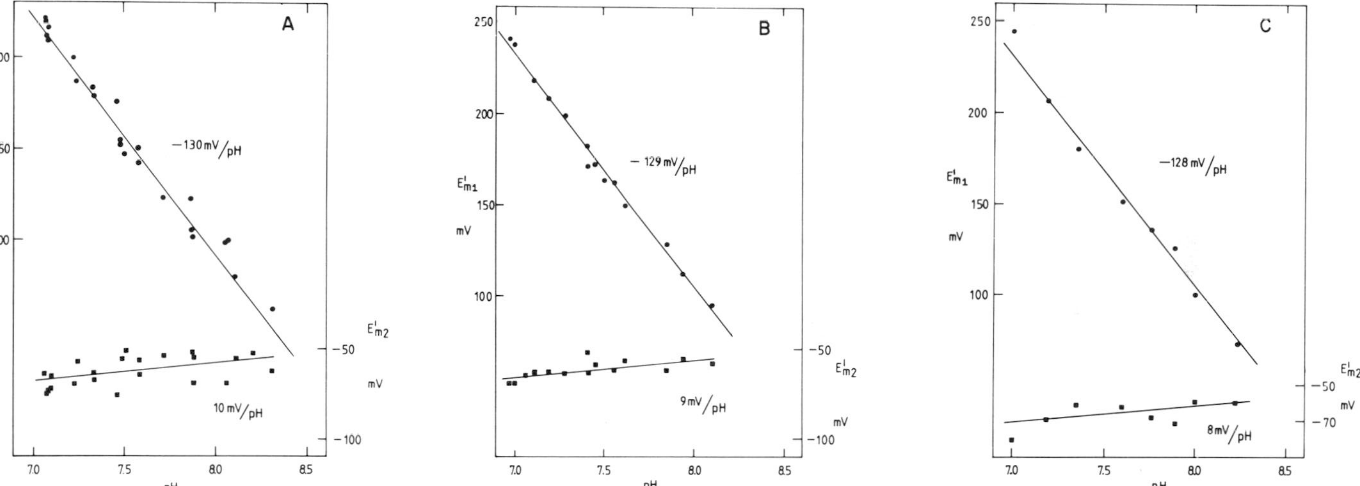

Fig. 4. E_m pH plots (●, E'_{m1}; ■, E'_{m2}) correcting for limited number of binding sites for the semiquinone anion using Eqs. 13 and 14. The lines drawn are least-square fits. (A), (B), and (C) data were taken from Figs. 2A, 2B, and 2C, respectively.

The finding that $E(Q^{-}_{max})$ is dependent upon the temperature and the potential is not understood. One would expect that at potentials different from the midpoint potential (cf. Fig. 2A, upper versus lower curves) or at different temperatures (cf. Fig. 2A versus Fig. 2B) the maximal amount of $Q^{\overline{\cdot}}$ would still be attained, but at a higher pH value. The fact that these findings are not described by the model may be due to the simplifying assumptions made in deriving Eqs. (13) and (14).

The calculated midpoint potential for bound ubiquinone is 84 mV (pH 7). Urban and Klingenberg (*16*) measured 65 mV (pH 7) for free ubiquinone. If this difference is not from experimental approach, the implication is that QH_2 binds about four times more firmly than Q.

What is the possible role for this semiquinone anion in electron transfer as catalyzed by QH_2:cytochrome *c* oxidoreductase? That the maximal amount of semiquinone anion is half the cytochrome c_1 concentration might be significant in connection with the evidence that the active enzymic unit in the membrane is a dimer (*19–21*), if it is also assumed that a stable semiquinone anion can be formed in only one of the two monomers

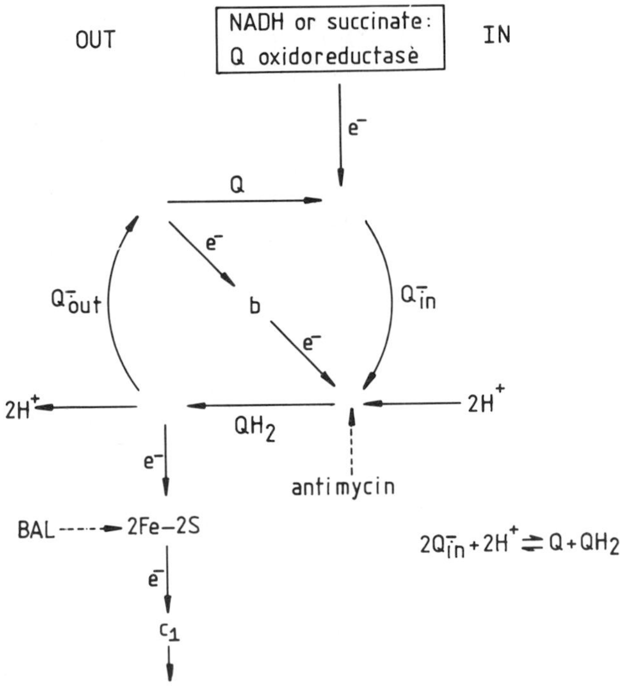

Fig. 5. Diagram representing a modified Q-cycle (*2*). See text for further explanation. The dotted lines indicate the sites of inhibition by BAL and antimycin.

(inside or outside). At the moment it seems unnecessarily complicated to construct a model that includes the dimeric nature of the enzyme. Therefore, the results will be discussed on the basis of the Q cycle as proposed by Mitchell (*2*).

Three arguments favor the proposal that the semiquinone investigated corresponds to $Q^{\dot{-}}_{in}$ in Fig. 5. First, the semiquinone anion is very stable. Second, it is abolished by antimycin. Third, after treatment of the submitochondrial particles with BAL (2,3-dimercaptopropanol), which specifically destroys the Rieske 2Fe–2S cluster (*22*) and so inhibits electron transfer from QH_2 to cytochrome *c* (*23*), the semiquinone anion is still present in the same amount. The scheme depicted in Fig. 5 explains [besides the points discussed by Mitchell (*2*)], that in the presence of antimycin, cytochrome *b* can be reduced via $Q^{\dot{-}}_{out}$, and in the presence of BAL, cytochrome *b* is reduced via $Q^{\dot{-}}_{in}$; but if both inhibitors are present cytochrome *b* remains oxidized (*24*). According to this scheme the "oxidant-induced reduction" of cytochrome *b* will proceed via $Q^{\dot{-}}_{out}$, since this is abolished in BAL-treated preparations. The antimycin-stimulated hydrogen peroxide formation (*25*) probably also proceeds via $Q^{\dot{-}}_{out}$.

We have not studied the mechanism of proton translocation. If this takes place via movement of ubiquinone across the membrane, as Mitchell (*2*) proposes, we suggest that this occurs in the equation

$$[2Fe{-}2S]^{2+} + Q(H_{in})_2 \rightleftharpoons [2Fe{-}2S]^{1+} + Q^{\dot{-}}_{out} + 2\,H^{+}_{out}$$

In this equation $\vec{H}^{+}/e^{-}$ equals two. The reaction in which the second electron cycles back through cytochrome *b* is also a possible source of proton translocation, as suggested by Papa (*26*). Further experiments on the role of the semiquinone anion in electron transfer are in progress.

ACKNOWLDEGMENTS

The authors thank Dr. S. P. J. Albracht for his valuable criticism and continuous interest. Part of this work has been supported by grants from the Netherlands Organization for the Advancement of Pure Research (Z.W.O.), under auspices of the Netherlands Foundation for Chemical Research (S.O.N.)

REFERENCES

1. Wikström, M. K. F., and Berden, J. A. (1972). *Biochim. Biophys. Acta* **283,** 403–420.
2. Mitchell, P. (1976). *J. Theor. Biol.* **62,** 327–367.

3. Backström, D., Norling, B., Ehrenberg, A., and Ernster, L. (1970). *Biochim. Biophys. Acta* **197,** 108–111.
4. Lee, I. Y., and Slater, E. C. (1974). *In* "Dynamics of Energy-Transducing Membranes" (L. Ernster, R. W. Estabrook, and E. C. Slater, eds.), pp. 61–75. Elsevier, Amsterdam
5. Ruzicka, F. J., Beinert, H., Schepler, H. L., Dunham, W. R., and Sands, R. H. (1975). *Proc. Natl. Acad. Sci. U.S.A.* **72,** 2886–2890.
6. Ingledew, W. J., Salerno, J. C., and Ohnishi, T. (1976). *Arch. Biochem. Biophys.* **177,** 176–184.
7. Konstantinov, A. A., and Ruuge, E. K. (1977). *FEBS Lett.* **81,** 137–141.
8. Salerno, J. C., Harmon, H. J., Blum, H., Leigh, J. S., and Ohnishi, T. (1977). *FEBS Lett.* **82,** 179–182.
9. Salerno, J. C., Blum, H., and Ohnishi, T. (1979). *Biochim. Biophys. Acta* **547,** 270–281.
10. Ohnishi, T., and Trumpower, B. L. (1980). *J. Biol. Chem.* **255,** 3278–3284..
11. Yu, C. A., Nagaoka, S., Yu, L., and King, T. E. (1980). *Arch. Biochem. Biophys.* **204,** 59–70.
12. Konstantinov, A. A., and Ruuge, E. K. (1977). *Bioorg. Chem.* **3,** 787–799.
13. de Vries, S., Berden, J. A., and Slater, E. C. (1980). *FEBS Lett.* **122,** 143–148.
14. Yu, C. A., Nagaoka, S., Yu, L., and King, T. E. (1978). *Biochem. Biophys. Res. Commun.* **82,** 1070–1078.
15. Clark, W. M. (1960). "Oxidation-Reduction Potentials of Organic Systems." Waverly Press, Baltimore, Maryland.
16. Urban, P. F., and Klingenberg, M. (1969). *Eur. J. Biochem.* **9,** 519–525.
17. Clark, W. M., Taylor, J. F., Davies, T. H., and Vestling, C. S. (1940). *J. Biol. Chem.* **135,** 543–568.
18. Yu, C. A., Yu, L., and King, T. E. (1977). *Biochem. Biophys. Res. Commun.* **79,** 939–946.
19. Weiss, H., and Kolb, H. J. (1979). *Eur. J. Biochem.* **99,** 139–149.
20. Von Jagow, G., Schägger, H., Riccio, P., Klingenberg, M., and Kolb, H. J. (1977). *Biochim. Biophys. Acta* **462,** 549–558.
21. de Vries, S., Albracht, S. P. J., and Leeuwerik, F. J. (1979). *Biochim. Biophys. Acta* **546,** 316–333.
22. Slater, E. C., and de Vries, S. (1980). *Nature* (*London*) **288,** 717–718.
23. Slater, E. C. (1949). *Biochem. J.* **45,** 14–30.
24. Deul, D. H., and Thorn, M. B. (1962). *Biochim. Biophys. Acta* **59,** 426–436.
25. Boveris, A., Cadenas, E., and Stoppani, A. O. M. (1976). *Biochem. J.* **156,** 435–444.
26. Papa, S. (1976). *Biochim. Biophys. Acta* **456,** 39–84.

Ubisemiquinones in Electron-Transfer Systems of Mitochondria

7

TOMOKO OHNISHI
JOHN C. SALERNO
HAYWOOD BLUM

INTRODUCTION

Extensive work by Klingenberg and Kröger on the role of ubiquinone (UQ) in the respiratory chain suggested that ubiquinone is a mobile hydrogen carrier that shuttles reducing equivalents between dehydrogenases and cytochrome b-c_1 complexes (*1*). Although Bäckström *et al.* (*2*) reported in 1970 that ubisemiquinone (USQ) EPR signals are detectable in respiring mitochondria, only recently has the possible importance of bound USQ attracted general attention. In 1972 Wikström and Berden (*3*) proposed a special mechanism of electron transfer in the ubiquinol-cytochrome c_1 segment of the respiratory chain to explain the peculiar phenomena of oxidant-induced reduction and reductant-induced oxidation of the *b*-cytochromes. In their model, electron transfer is split into two separate but tightly coupled electron-transfer pathways via the $QH_2/QH\cdot$ and $QH\cdot/Q$ couples. Subsequently, Mitchell (*4*) formulated the protonmotive Q cycle in which the vectorial movement of ubiquinone and ubiquinol across the membrane is coupled to proton transport for energy transduction at coupling sites 2 and 3. The Q-cycle mechanism predicts two distinct sites, namely "o" and "i" for the binding of USQ at cytoplasmic and matrix regions of the mitochondrial inner membrane, respectively. There have been several reports describing specifically bound, thermodynamically stable USQ in the mitochondrial respiratory chain (*5–10*). At least

Function of Quinones in Energy Conserving Systems

ISBN 0-12-701280-X

three different ubisemiquinone species have been identified, in the succinate-UQ reductase, NADH-UQ reductase and ubiquinol-cytochrome *c* reductase of the respiratory chain. These ubisemiquinones were designated SQ_s, SQ_n, and SQ_c, respectively. To date only limited information is available on SQ_n (*10*, *11*).

Detailed EPR studies have been conducted on the USQ species in the succinate-UQ reductase segment that gives unique signals at $g = 2.04$, 1.99, and 1.96 at low temperature and at intermediate redox states (*5*). The multiple signals arise from spin-spin interaction between two neighboring USQ moieties having equivalent oxidation–reduction properties (*5*, *6*). Computer simulation of the spectra from the spin-coupled USQ pair suggested an edge-to-edge orientation and a 7.7 Å distance between them (*5*). Using oriented multilayered preparations of beef heart mitochondria, Salerno *et al.* (*7*) demonstrated that the USQ pair is highly ordered in the mitochondrial membrane, with the intramolecular axis being perpendicular to the membrane plane. On the basis of the very short relaxation time of EPR signals from the spin-coupled USQ, Ingledew *et al.* (*6*) proposed a close proximity of iron-sulfur center S-3 to the USQ pair.

The power saturation analysis at higher temperature of the $g = 2$ USQ signal in submitochondrial particles or in succinate-cytochrome *c* reductase complex poised at intermediate oxidation–reduction potentials, shows the existence of two distinct USQ species that show different relaxation behavior and different responses to respiratory inhibitors such as TTFA and antimycin A (*8*, *12–14*). Existence of thermodynamically stable ubisemiquinone species was also reported in the isolated b-c_1 complex from plants (*15*) and yeast (*9*) mitochondria.

Konstantinov and Ruuge (*8*) reported that the SQ_c level (S_Q-1 in their notation) was enhanced by aeration, and in oxidized submitochondrial particles SQ_s (S_Q-2) was largely abolished by $K_3Fe(CN)_6$ addition. SQ_s was tentatively identified as the component responsible for the USQ spin coupling signals observed at lower temperature. They interpreted their results in terms of a Wikström-Berden-Mitchell model, assigning Q_s to center "i" and Q_c to center "o".

Quantitative information on the spatial organization of the redox components in the mitochondrial inner membrane is useful for understanding the mechanistic roles of these USQ species in electron and proton transfer processes in the Site II segment. We have developed quantitative procedures to determine the topographical distributions of redox components relative to the mitochondrial inner membrane utilizing the perturbation of the EPR spectra of membrane-bound redox centers by membrane-impermeable extrinsic paramagnetic dysprosium-complex probes. Dysprosium complexes elicit EPR spectra with a peak around $g = 15$, which exhibits extremely short intrinsic spin lattice relaxation times. The spectral perturbation is revealed as an enhanced saturation parameter ($P_{1/2}$) and/or broadened line width (ΔH) due to spin-spin interaction between intrinsic redox components and the dysprosium complex.

The saturation parameter ($P_{1/2}$) is the microwave power needed for half saturation of the EPR signal in a qualitative sense and has been defined quantitatively in Blum and Ohnishi (*16*) and estimated using simulations. To convert EPR parameters to distances, we have analyzed the effects of various dysprosium complexes on the $[4Fe-4S]^{+3(+2,+3)}$ cluster in the isolated *Pseudomonas gelatinosa* iron–sulfur protein and mammalian cytochrome *c*, where distance between the redox center and dysprosium ions can be estimated based on X-ray crystallographic information (*17*, *18*). In these systems, effective paramagnetic probes appeared to bind to the surface of the protein (*19*). Thus we used a model where $1/T_1$ is proportional to r^{-6} (r is the distance from the active iron to the dysprosium ion). In the temperature range of our EPR measurements, T_2 of the intrinsic redox components is unchanged; thus $\Delta P_{1/2}$ is proportional to $1/T_1$. We have used a parameter for distance measurements; $\Delta P_{1/2} = 4.12 \times 10^8 \times r^{-6}$ exp. $(-12.5/T)$ from the studies on *P. gelatinosa* iron-sulfur protein (*17*). $\Delta P_{1/2}$ is the enhancement of $P_{1/2}$ per m*M* of the paramagnetic probe and T is the sample temperature. To estimate distances from the protein surface to the paramagnetic cluster, we reduce r by 5 Å to account for the size of the Dy-EDTA complex, and 3 Å for the $Dy(NO_3)_3$.

The present paper summarizes EPR and thermodynamic parameters of two different species of stable USQ (SQ_c and SQ_s) in mammalian mitochondrial systems and the possible location of these two species is discussed in connection with neighboring redox components and the energy transduction mechanism at Site II.

EXPERIMENTAL RESULTS

Table I summarizes semiquinone stability constants (K) and midpoint potentials of the overall two hydrogen transfer step ($Q \rightarrow QH_2$) at pH 7.0 for SQ_c and SQ_s, determined in beef heart submitochondrial particles and resolved complexes. K values were calculated assuming that the pool size of Q_c and Q_s are one and two per cytochrome c_1 respectively, in both submitochondrial particles and succinate-cytochrome *c* reductase complex. For cytochrome b-c_1 Complex III, the total ubiquinone content in the preparations was used, which is lower than the content of cytochrome c_1

TABLE I

Stability Constants and $E_{m7.0}$ of Various Ubisemiquinone Species in Mitochondrial Systems

	System	K	$E_{m7.0}$	(mV)	Reference
Q_c[A]	Beef Heart SMP	5×10^{-2}	80	(−70 °C)	(*14*)
	Succinate-cyt *c* Reductase	8×10^{-3}	84	(50 °K)	(*13*)
	Cytochrome *b*-c_1 Complex III	3.7×10^{-2}	111	(room temp.)	(*21*)
	Cytochrome *b*-c_1 Complex III	4.5×10^{-2}	113	(60 °K)	(Ohnishi, unpublished data)
Q_s[B]	Beef Heart SMP	10	134	(12 °K)	(*14*)
	SCR	2×10^{-2}	94	(50 °K)	(*13*)

[A] UQ in ubiquinol-cytochrome *c* reductase segment of the respiratory chain
[B] UQ in succinate-UQ reductase segment

(*20*). The relatively high stability constant (5×10^{-2}) obtained for SQ_c in submitochondrial particles is not significantly altered in the resolved complexes. In submitrochondrial particles, the *K* value of the SQ_s species is three orders of magnitude greater than that of SQ_c, but is drastically lowered in the resolved complex. This further supports the notion that the microenvironment of these two USQ species is different.

Figure 1 presents the pH dependence of the maximum SQ_c concentration obtained from potentiometric titrations using succinate-cytochrome *c* reductase. It is clear that SQ_c is predominantly in the anionic form ($Q^{\overline{\cdot}}$) in the pH range of the experiment. A pK_S of 6.4 and $pK_R > 9$ gives a better fit to the experimental points than a pK_R of 8.0 as reported in an earlier paper (*13*). pK_S values below 6 were excluded because they require pK_R values as low as 8. E_m values obtained from the ambient redox potential (E_h) required to attain the maximum $g = 2.00$ signal show a pH dependence of −60 mV/pH up to the highest pH (8.5) so far examined (Ohnishi *et al.*, unpublished data). These results demonstrate that the semiquinone and quinol forms of the Q_c pool are $Q^{\overline{\cdot}}$ and QH_2 in the physiological pH range. Similar protonation states were found for the Q_s species (*14*). Power saturation behavior of the $g = 2.00$ EPR signal of succinate-cytochrome *c* reductase poised at an E_h value of 78 mV is shown in Fig. 2. It clearly reveals a biphasic curve that can be resolved into two components (shown by the dashed lines) attributable to two species of ubisemiquinone. One species (SQ_c) starts to show saturation below 0.1 mW power level, while the other signal (SQ_s) is not saturated even at 100 mW at 50 °K. Thus SQ_c and SQ_s can be monitored selectively at extremely low

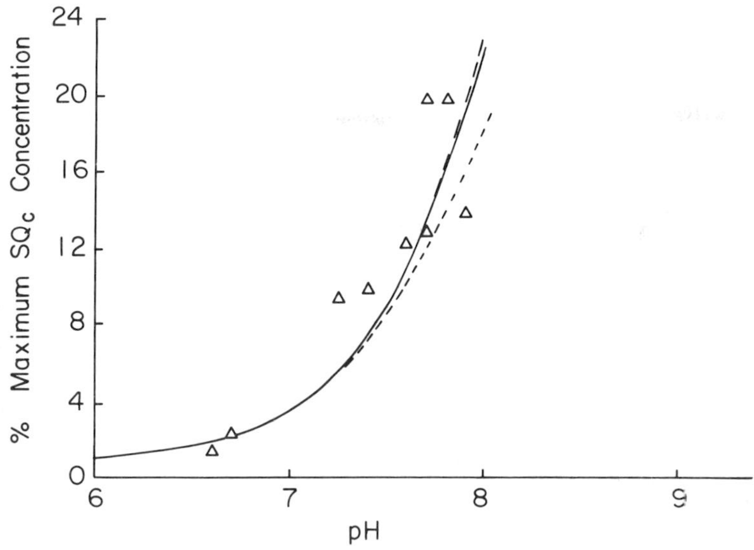

Fig. 1. The pH dependence of the maximum SQ_c (%) concentrations in potentiometric titrations with the succinate cytochrome *c* reductase complex. Theoretical curves: $pK_S = 6.4$; (— —) $pK_R = 14$, (——) $pK_R = 9$, (----) $pK_R = 8$.

and high microwave power levels, such as 10 μW and 100 mW, respectively. At 10 μW the relative contribution to the $g = 2.00$ signal from SQ_c and SQ_s is in a 3 to 1 ratio, while at 100 mW the ratio is 1 to 100. EPR spectra of the $g = 2.00$ signals arising from the two different SQ species in

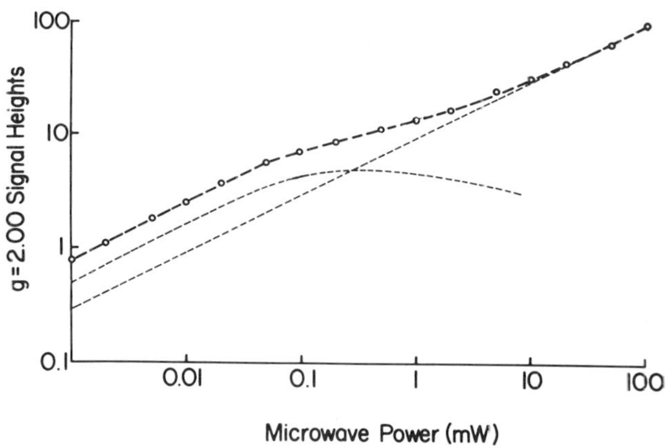

Fig. 2. Power saturation behavior of the $g = 2.00$ EPR signals in the isolated succinate-cytochrome *c* reductase complex poised at $E_h = 78$ mV monitored at 50 °K.

beef heart submitochondrial particles recorded at -70 °C are shown in Fig. 3. Spectrum A is attributable to SQ_c and spectrum B to SQ_s. The signal for SQ_c has a narrower peak-to-peak width (9.5 gauss); the signal for SQ_s has a peak-to-peak width of 12 gauss and elicits rapid spin relaxation. The spectrum of SQ_s exhibits a more Lorentzian line shape while the spectrum of SQ_c has a more Gaussian shape. The line shape and power saturation behavior of SQ_c was identified in the cytochrome *b*-c_1 Complex III, where no spectral overlap from the SQ_s species is present.

These two species of ubisemiquinone, SQ_c and SQ_s, can also be differentiated from their response to mitochondrial respiratory inhibitors, TTFA and antimycin A. Figure 4 presents potentiometric titrations of signals from spin-coupled $Q^{\overline{\cdot}}Q^{\overline{\cdot}}$, and the two different species of SQ in the presence and absence of TTFA at pH 7.2 using beef heart SMP and in the presence and absence of antimycin A, at pH 7.4 using isolated succinate-cytochrome *c* reductase complex.

The $Q^{\overline{\cdot}}Q^{\overline{\cdot}}$ interaction signal and the g = 2.00 signals monitored at a

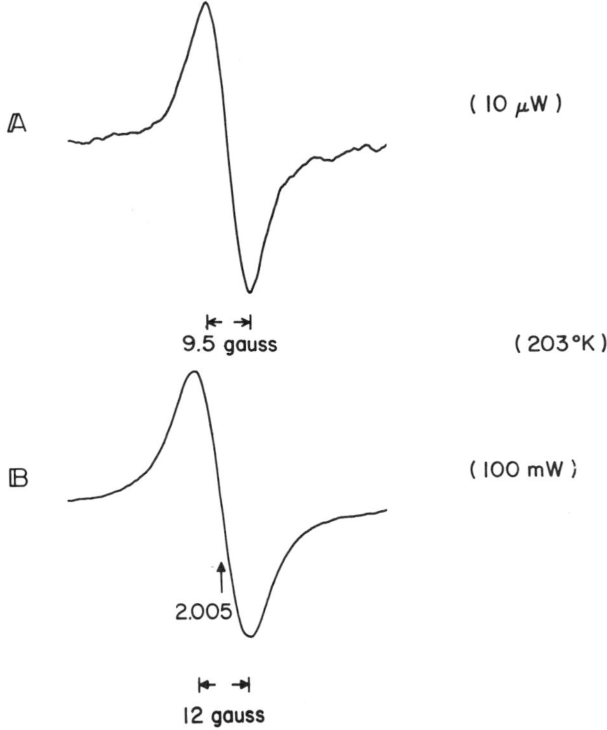

Fig. 3. EPR spectra of the g = 2.00 signals arising from USQ at two different microwave power levels in beef heart submitochondrial particles.

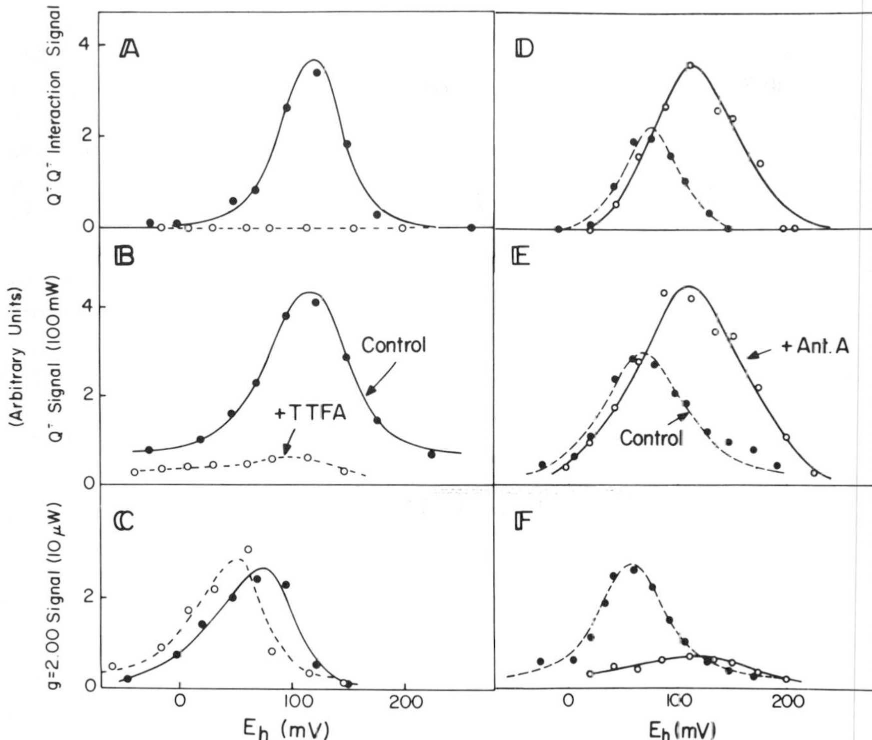

Fig. 4. Potentiometric titrations showing the effect of antimycin (D, E, F) and TTFA (A, B, C) on the g = 2.00 EPR signals of two different species of ubisemiquinone and on the low temperature EPR signals from the spin-coupled $Q^{\overline{\cdot}}Q^{\overline{\cdot}}$ pair in beef heart SMP (A), (B), (C) and in isolated succinate-cytochrome c reductase complex (D), (E), (F). Detailed conditions were described in Ohnishi and Trumpower (*13*) and Salerno and Ohnishi (*14*).

high microwave power (100 mW) show the same E_m value (120 mV) and are concommitantly quenched by 100 μW TTFA. The slowly relaxing g = 2.00 signal (SQ_c) monitored at 10 μW input microwave power peaks at lower redox potential (60 mV), and this SQ species is not significantly affected by TTFA. In contrast, antimycin A destabilizes the slowly relaxing SQ_c species (Fig. 4F), and significantly stabilizes SQ_s (Fig. 4E), giving an increase in maximal SQ_s spin concentration from 0.30 to 0.45 spins/c_1. Antimycin A also causes a positive shift in midpoint potentials of both the $Q^{\overline{\cdot}}Q^{\overline{\cdot}}$ interaction signals and the SQ_s g = 2.00 signal (Figs. 4D and E). The similar response of signals from spin-coupling of $Q^{\overline{\cdot}}Q^{\overline{\cdot}}$ pair and SQ_s to both TTFA and antimycin further strengthens the notion that these two EPR signals arise from the same pool of UQ, namely, Q_s. The slowly relaxing g = 2.00 signal (SQ_s) is not significantly affected by TTFA at concentrations which completely quench SQ_s signals. The residual $g = 2.00$ signal at 10 μW in the presence of antimycin arises mostly from the overlapping SQ_s species.

The proximity of the SQ_s pair to a transition metal ion, probably center S-3, was suggested previously based on their similar low temperature spin relaxation behavior. Center S-3 is located approximately 12 Å from the matrix surface of mitochondrial inner membrane using the Hipip-type cluster $[4Fe-4S]^{+3(+2,+3)}$ model from *P. gelatinosa* (*22*) for the distance estimate. Dysprosium probes exert no effect on center S-3 or on $Q^{\overline{\cdot}} Q^{\overline{\cdot}}$ interaction signals from the cytosolic surface of the mitochondrial inner membrane. Recent studies on the topographical distribution of two *b*-cytochromes, cytochrome c_1 and Rieske iron–sulfur cluster are presented below.

Figure 5 illustrates individual EPR signals of cytochromes *b*-562, *b*-566,

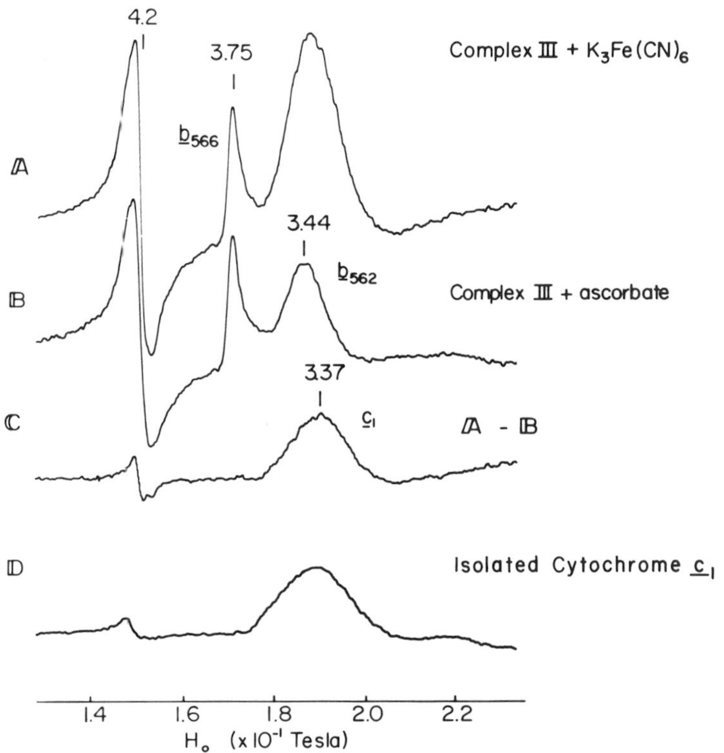

Fig. 5. EPR spectra of (A) cytochrome *b*-566, (B) *b*-562, (C) cytochrome c_1 obtained with cytochrome *b*-c_1 Complex III, and (D) isolated cytochrome c_1. EPR conditions: microwave input power, 5 mW; modulation amplitude, 2×10^{-3} tesla; sample temperature, 7 °K.

and c_1 in the magnetic field around 0.2 tesla using cytochrome b-c_1 Complex III and isolated cytochrome c_1. In spectrum A, all three cytochromes are oxidized (paramagnetic) while in spectrum B, cytochrome c_1 was reduced by the added ascorbate. The difference spectrum (A–B) gives the spectrum of cytochrome c_1 with the g_z peak at $g = 3.37$, which agrees well with the spectrum of isolated cytochrome c_1, as reported previously (*23, 24*). The distance of the cytochrome *b*-566 and *b*-562 heme irons from the barrier surface for dysprosium in ascorbate-reduced cytochrome b-c_1 Complex III was estimated from the analysis of $P_{1/2}$ enhancement of the $g = 3.75$ and 3.44 signals by dysprosium. The location of the cytochrome c_1 heme in isolated cytochrome c_1 was also analyzed.

As shown in Fig. 6, in the control sample which contains 10 m*M* La-EDTA, $P_{1/2}$ of the $g = 3.44$ and 3.75 signals at 5.5 *K* were 0.05 and

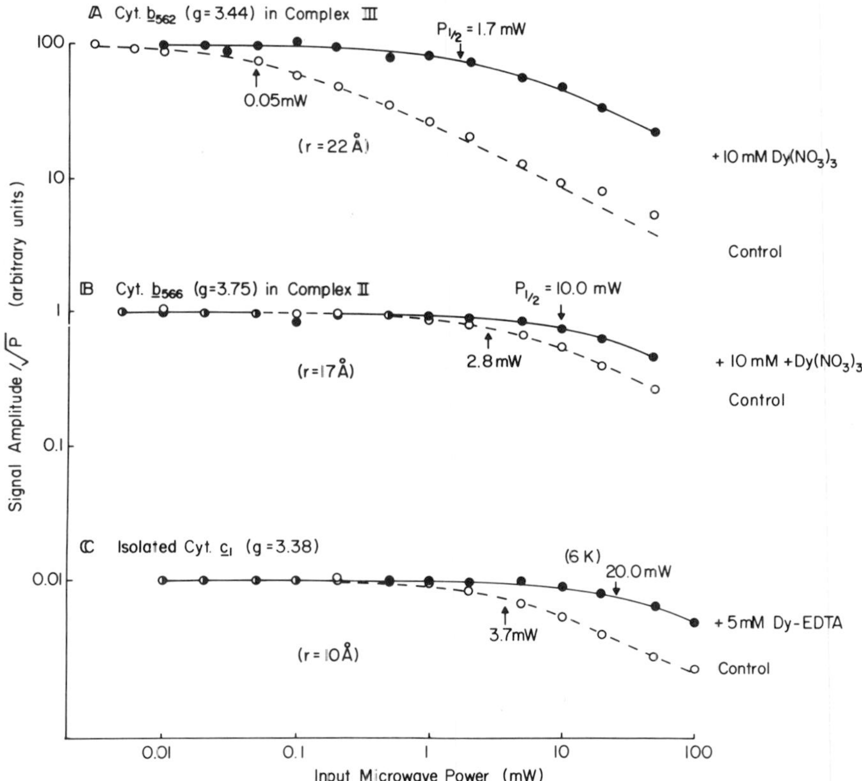

Fig. 6. Power saturation behavior of $g = 3.44$, 3.75 signals from (A) cytochrome *b*-562 in cytochrome b-c_1 Complex III, (B) *b*-566 in cytochrome b-c_1 Complex III and (C) $g = 3.37$ signal from cytochrome c_1 in the presence and absence of paramagnetic probes.

2.8 mW, respectively, and these were enhanced to 1.7 and 10 mW in the presence of 10 m*M* $Dy(NO_3)_3$. A distance of 22 Å to cytochrome *b*-562 heme and 17 Å to cytochrome *b*-566 heme from the barrier surface of *b*-c_1 Complex III was calculated. Negatively charged Dy-EDTA gave no significant spin relaxation enhancement. In the case of isolated cytochrome c_1, Dy-EDTA was more effective for the relaxation enhancement, while $Dy(NO_3)_3$ gave much less effect. An effective distance of 10 Å from the c_1 heme to the protein surface was obtained using negatively charged Dy-EDTA complex. The varying effects of spin probes with opposite charge should be tested in each system, because in most cases the surface charge distribution of mitochondrial membrane proteins is unknown.

Direct distance measurements of *b*-cytochromes from the cytosolic and matrix side surfaces of the mitochondrial inner membrane are very difficult because of the extremely small g_z EPR signals of cytochrome *b* and c_1, and are thus not available at present. However the results from the

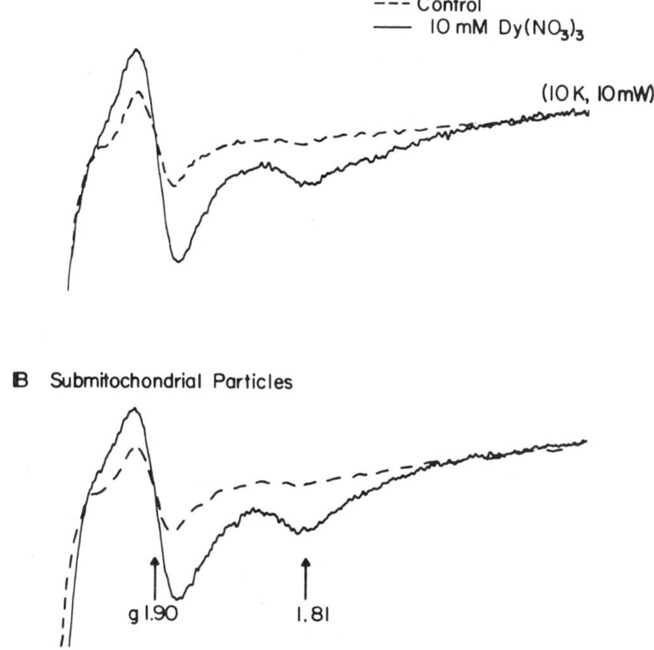

Fig. 7. EPR spectra of the Rieske iron–sulfur cluster examined (A) with and (B) without 10 m*M* $Dy(NO_3)_3$ in uncoupled pigeon heart mitochondria poised potentiometrically at E_h = 150 mV and in uncoupled pigeon heart submitochondrial particles reduced with 5 m*M* ascorbate.

b-c_1 Complex III studies already indicate that the distance of cytochrome *b*-562 and *b*-566 hemes from either surface (cytosolic or matrix side) of the membrane is greater than 22 and 17 Å, respectively.

Topographical assignment of the Rieske iron–sulfur cluster was also performed. Figure 7 very clearly illustrates that EPR signals of the Rieske iron–sulfur cluster exhibit both spectral broadening and relaxation enhancement in the presence of 10 m*M* $Dy(NO_3)_3$, when added on either the matrix or cytosolic surface of the mitochondrial inner membrane. Distance measurements conducted with the isolated Rieske iron–sulfur protein (*25*) showed the cluster is 19 Å from the barrier surface of the protein and therefore deeply buried within the molecule (*22*). In cytochrome *b*-c_1 Complex III, the Rieske iron–sulfur cluster is located about 20 Å from the barrier surface, as in the intact mitochondrial membrane preparations.

DISCUSSION

The most striking feature of the results of the potentiometric titrations is the high stability in the thermodynamic sense of the ubisemiquinone radicals. The stability constant of free ubisemiquinone in a hydrophobic milieu has been estimated to be 10^{-10} at pH 7.0 (*4*). The stability constants of the quinone in the dipolar coupled pair (Q_s) that were estimated from the spin-coupled signals, are greater than one (*K* approx. 10) whereas that of the "slowly" relaxing quinone (Q_c) are approximately 5×10^{-2}, at pH 7, assuming that the pool size of this Q species is equivalent to the cytochrome c_1 concentration. We refer to these stabilized ubisemiquinone species as SQ_s and SQ_c, because SQ_s appears to arise from the Q species in the succinate-ubiquinone reductase segment of the respiratory chain and SQ_c from the Q species in the ubiquinol-cytochrome *c* reductase segment (*10*, *13*).

The highly stabilized intermediate redox state and the previously reported fixed orientation of the Q pair in the membrane (*6*, *7*, *14*) suggest that the quinones are bound to a membrane protein. The semiquinone form must be many orders of magnitude more tightly bound than the quinol, which in turn must be at least a few times more tightly bound than the quinone. The exact values of the potentials reported here depend to some extent on two assumptions: (1) the number of binding sites of each type present and (2) no dependence of the properties of the binding site on the protonation and redox states of neighboring components. The error caused by the first assumption may be, at worst, 15–30 mV.

The presence of binding sites that stabilize ubisemiquinone is not surprising in view of the overall electron-transfer system in mitochondria. Obligatory two-electron reductants, such as NADH and succinate, reduce flavins in a two-electron step. In succinate dehydrogenase, the flavosemiquinone is stable enough ($K = 2.5 \times 10^{-2}$) to permit the reduction of the iron-sulfur centers of the dehydrogenase in sequential one-electron steps (*26*); both the upper and lower couples of the FAD in succinate dehydrogenase are lower in potential than that of the iron–sulfur center S-1 (*27*). A binding site that stabilizes ubisemiquinone would allow the reduction of Q by iron–sulfur centers in sequential one-electron transfer processes. The Q-binding proteins, QP_S, (*10,20*) or one of the two low molecular weight subunits of Complex II (*28–30*) are attractive candidates for SQ_s species. Without such a binding site the first one-electron reduction would be very unfavorable. This is probably a major cause of the inability of solubilized succinate dehydrogenases to reduce free ubiquinone.

Less is known about the Q_c binding site. A ubiquinone analogue, UHDBT, appears to inhibit electron transfer from the Rieske iron–sulfur cluster to cytochrome c_1 (*31*) (see also Bowyer, Paper 9, Chapter V, this volume). and oxidant-induced reduction of cytochrome *b* (*32*). This inhibitor shifts the g_y value of the Rieske iron–sulfur cluster from 1.90 to 1.89 and causes a positive shift of its midpoint redox potential by 70 mV, suggesting the Rieske iron–sulfur protein as a possible candidate for the SQ_c binding site (*31*). Recent photoaffinity labeling experiments using an arylazido-1-[^{14}C]-β-alanine ubiquinone derivative suggests a close proximity of the two *b* cytochromes to the SQ_c binding site (*33*).

However, even under conditions where both *b*-cytochromes and Rieske iron–sulfur cluster are paramagnetic, SQ_c spins have a saturation parameter $P_{1/2}$ (*16*) of 0.12 mW at 60 °K, the same as that of ubisemiquinone in alkaline liposomes ($P_{1/2}$ = 0.11 mW). This indicates that SQ_c is bound either to a specific Q binding protein or to the protein surface of neighboring redox components, keeping it away from these redox centers.

Of nearly equal importance to the unusual stability of the Q radicals is the fact that the equilibrium redox couples for both Q_s and Q_c are $Q/Q^{\overline{\cdot}}$ and $Q^{\overline{\cdot}}/QH_2$ in the physiological pH range. The second couple has an inherent $2H^+/e^-$ ratio, an important feature in view of the probable importance of bound quinone in proton translocation. By utilizing this couple, the site ratio can be preserved without recourse to a transmembrane electron carrying arm (*14*).

Attempts have been made to identify Q_s with "center i" and Q_c with "center o" (*8*). Several lines of evidence suggest that while Q_s is at the site of entry of electrons from SDH, the *b*-c_1 complex is unnecessary for quinone reduction by succinate. Therefore, it is likely that "center i" does not exist in its original sense. The paramagnetic probe data shows that both *b* hemes are deeply buried within the *b*-c_1 complex. Both considerations suggest that any Q cycle is likely to be local in nature, possibly involving a second Q binding site within the *b*-c_1 complex, and with access to the aqueous phases provided by channels.

To function as an efficient electron "pair splitter" any center "o" would have to be more complex than a simple binding site. A mechanism must exist to prevent the oxidation of bound QH_2 to Q by Rieske's iron–sulfur center in sequential one-electron steps, causing the H^+/e^- ratio to decrease to one. This could be accomplished if two or more species (eg., Rieske center, *b* cytochromes, and bound Q) exist in two states, one that sees only its oxidant and one that sees only its reductant. If the transitions between states in this "switching unit" are allowed only in specific redox and protonation states, a tightly coupled $2H^+/e^-$ system would result.

Loops in linear and Q-cycle schemes couple proton translocation to electron flow by placing electrons and protons together on a small mobile 'hydrogen carrier'. The coupling of protonation accessibility to the cytosolic (P) or matrix (N) aqueous phases with redox accessibility to the oxidizing or reducing sides of the coupling site is a common feature of tightly coupled, redox driven proton pumps. Proton pump schemes which ignore either redox or protonation accessibility are incompletely coupled, a fact that usually becomes apparent by analysis of their operation near equilibrium (or during reversed electron transfer). Coupled accessibility could be provided by protein conformational changes. A minimal model, based loosely on DeVault's transductase models (*27*), would involve a single electron carrier and associated ionizable group. This unit could exist in two states, a "reducing side/N side" state and an "oxidizing side/P side" state, with transitions between the two states allowed only in the oxidized/unprotonated and reduced/protonated states.

Proton translocation in the *b*-c_1 complex might occur by a similar mechanism instead of a Q cycle or other looplike scheme. All of the protein bound redox-active groups would be candidates for the role of the 'transductase' electron carrier except for cytochrome c_1; this leaves Rieske's center, the *b* cytochromes and bound quinone as possibilities.

A *b*-c_1 complex proton pump need not conform to the minimal model, although it would utilize the same general principles. The coupled accessibility changes could depend on the redox/protonation states of several groups within a "switching unit"; the electron-transfer accessibility considerations are not unlike the "center o" requirements mentioned earlier. A proton pump of this type could explain the oxidant-induced reduction of the *b* cytochromes as a consequence of the accessibility changes needed for proton translocation, if the accessibility of cytochrome *b* to its reductant was linked to the redox state of Rieske's center. The buried location of the *b* cytochromes is consistent with this kind of model.

It is unclear whether reducing equivalents enter the b-c_1 complex via hydrogen transfers (possibly through mobile ubiquinone) or by electron or hydrogen transfers within a transient dehydrogenase–b-c_1 supercomplex in which only bound quinones would be involved. Possibly the mode of reduction is different for dehydrogenases located on the P and N sides of the membrane. If reduction of the b-c_1 complex proceeds via electron transfer, a pump mechanism is implicated. On the other hand, a hydrogen-transfer mechanism could be accommodated by either a Q cycle or a pump mechanism.

Both proton pump and localized Q cycle models are at present viable alternatives for site II proton translocation. Both types of models are capable of rationalizing phenomena such as oxidant-induced reduction of the *b* cytochromes and the redox linked shifts in the spectra of Rieske's iron–sulfur center (*9*, *24*, *36*) in terms of features necessary for proton translocation. Both can tolerate the deeply buried location reported here for the *b* cytochromes, although this information favors pump models to some extent.

ACKNOWLEDGMENTS

The authors are grateful to Dr. John Bowyer for performing computer simulations of redox titration data, and for critically reading the manuscript. We thank Drs. C. A. Yu, L. Yu, T. E. King, B. L. Trumpower, and C. Edwards for providing purified enzyme preparations, and Mr. Tateo Hompo for skillful technical assistance. This work was supported by the National Institutes of Health Research Grant GM 12202 and the National Science Foundation Research Grant PCM 78-16779.

REFERENCES

1. Kröger, A., and Klingenberg, M. (1973). *Eur. J. Biochem.* **39,** 313–323.
2. Bäckström, D., Norling, B., Ehrenberg, A., and Ernster, L. (1970). *Biochim. Biophys. Acta* **197,** 108–111.
3. Wikström, M. K. F., and Berden, J. A. (1972). *Biochim. Biophys. Acta* **283,** 403–420.
4. Mitchell, P. (1976). *J. Theor. Biol.* **62,** 327–367.
5. Ruzicka, F. J., Beinert, H., Schepler, K. L., Dunham, W. K., and Sands, R. H. (1975). *Proc. Natl. Acad. Sci. U.S.A.* **72,** 2886–2890.
6. Ingledew, W. J., Salerno, J. C., and Ohnishi, T. (1976). *Arch. Biochem. Biophys.* **177,** 176–184.

7. Salerno, J. C., Blum, H., and Ohnishi, T. (1979). *Biochim. Biophys. Acta* **547,** 270–281.
8. Konstantinov, A. A., and Ruuge, E. K. (1977). *FEBS Lett.* **81,** 137–141.
9. Siedow, J. M., Power, S., de la Rosa, F. F., and Palmer, G. (1978). *J. Biol. Chem.* **253,** 2392–2399.
10. Yu, C. A., Nagaoka, S., Yu, L., and King, T. E. (1978). *Biochem. Biophys. Res. Commun.* **82,** 1070–1078.
11. Widger, W. L. (1979). Ph.D. Thesis, Dep. Chem., State Univ. of New York at Albany.
12. Salerno, J. C., and Ohnishi, T. (1978). *In* "Frontiers of Biological Energetics" (P. L. Dutton, J. S. Leigh, and A. Scarpa, eds.), Vol. 1, pp. 191–200. Academic Press, New York.
13. Ohnishi, T., and Trumpower, B. L. (1980). *J. Biol. Chem.* **255,** 3278–3284.
14. Salerno, J. C., and Ohnishi, T. (1980). *Biochem. J.* **192,** 769–781.
15. Rich, P. R., and Bonner, W. D. (1977). *In* "Function of Alternative Terminal Oxidases" (D. Lloyd, H. Dein, and G. C. Hill, eds.), pp. 149–158. Pergamon, Oxford.
16. Blum, H., and Ohnishi, T. (1980). *Biochim. Biophys. Acta* **621,** 9–18.
17. Blum, H., Cusanovich, M. A., Sweeney, W. V., and Ohnishi, T. (1981). *J. Biol. Chem.* **256,** 2199–2206.
18. Blum, H., Leigh, J. S., and Ohnishi, T. (1980). *Biochim. Biophys. Acta* **626,** 31–40.
19. Antholine, W. A., Hyde, J. S., and Swartz, H. M. (1978). *J. Magn. Reson.* **29,** 517–522.
20. Yu, C. A., and Yu, L. (1980). *Biochim. Biophys. Acta* **591,** 409–420.
21. Nagaoka, S., Yu, C. A., Yu, L., and King, T. E. (1980). *Arch. Biochem. Biophys.* **204,** 59–70.
22. Ohnishi, T., Harmon, H. J., Blum, H., and Hompo, T. (1980). *In* "Interaction Between Iron and Proteins in Oxygen and Electron Transport" (C. Ho, ed.) Elsevier/North-Holland, New York, in press.
23. DerVartanian, D. V., Albracht, S. P. J., Berden, J. A., van Gelder, B. F., and Slater, E. C. (1973). *Biochim. Biophys. Acta* **292,** 496–501.
24. Orme-Johnson, N. R., Hansen, R. E., and Beinert, H. (1974). *J. Biol. Chem.* **249,** 1928–1939.
25. Trumpower, B. L., and Edwards, C. (1979). *J. Biol. Chem.* **254,** 8697–8706.
26. Ohnishi, T., King, T. E., Salerno, J. C., Blum, H., Bowyer, J. R., and Maida, T. (1981). *J. Biol. Chem.* **256,** 5577–5582.
27. Ohnishi, T. (1979). *In* "Membrane Proteins in Energy Transduction" (R. A.Capaldi, ed.), pp. 1–87. Dekker, New York.
28. Vinogradov, A. D., Gavrikov, V. G., and Gavrikova, E. V. (1980). *Biochim. Biophys. Acta* **592,** 13–27.
29. Ackrell, B. A., Ball, M. B., and Kearney, E. B. (1980). *J. Biol. Chem.* **255,** 2761–2769.
30. Hatefi, Y., and Galante, Y. M. (1980). *J. Biol. Chem.* **255,** 5530–5537.
31. Bowyer, J. R., Dutton, P. L., Prince, R. C., and Crofts, A. R. (1980). *Biochim. Biophys. Acta* **592,** 445–460.
32. Bowyer, J. R., and Trumpower, B. L. (1980). *FEBS Lett.* **115,** 171–174.
33. Yu, C. A., and Yu, L. (1980). *Biochem. Biophys. Res. Commun.* **96,** 286–292.
34. Heron, C., Ragan, C. I., and Trumpower, B. L. (1978). *Biochem. J.* **174,** 791–800.
35. Schneider, H., Lemasters, J. J., Höchli, M., and Hackenbrock, C. R. (1980). *J. Biol. Chem.* **255,** 3748–3756.
36. DeVries, S., Albracht, S. P. J., and Leeuwerik, S. P. J. (1979). *Biochim. Biophys. Acta* **546,** 316–333.

V QUINONE BINDING PROTEINS IN THE MITOCHONDRIAL RESPIRATORY CHAIN AND PHOTOSYNTHETIC SYSTEMS

Quinones as Prosthetic Groups in Membrane Electron-Transfer Proteins I: Systematic Replacement of the Primary Ubiquinone of Photochemical Reaction Centers with Other Quinones

1

M. R. GUNNER
D. M. TIEDE
ROGER C. PRINCE
P. LESLIE DUTTON

INTRODUCTION

The photochemical reaction center protein from the photosynthetic bacterium *Rhodopseudomonas sphaeroides* offers an opportunity to study not only the mechanisms of electron transfer in proteins, but also the role played by the protein in binding, orienting and governing the physicochemical nature of its constituent ubiquinones (Q). This brief report confines itself to the tightly associated primary ubiquinone (Q_I) of the reaction center (RC) (*1*, *2*). Figure 1 summarizes the well-known reactions in isolated reaction centers when the secondary Q is absent (see O'Keefe *et al.*, Paper 2, Chapter V, this volume); k_2 is the rate of $(BChl)_2^{+}$ reduction

Function of Quinones in Energy Conserving Systems

ISBN 0-12-701280-X

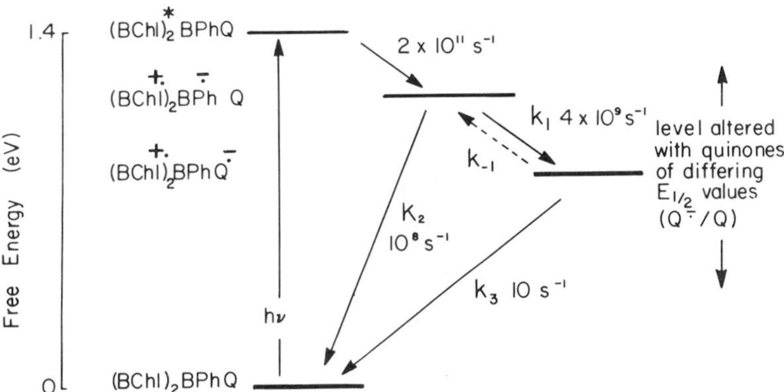

Fig. 1. A schematic of the reactions in isolated reaction centers from *Rps. sphaeroides*. The values for k_1 and k_3 are those seen in native reaction centers; here we present evidence that these values change when ubiquinone is replaced by other quinones. The value of k_2 seems constant under a wide range of conditions, including Q-extraction (W. W. Parson, personal communication). In native reaction centers the rates of the reactions are apparently unaltered between 4 and 300 °K.

by the flash reduced bacteriopheophytin ($BPh^{\bar{\cdot}}$). This is a nonproductive route, but in native reaction centers, k_2 is greatly exceeded by k_1, leading to the virtual quantitative reduction of Q_I. The $Q_I^{\bar{\cdot}}$ thus formed slowly and directly reduces the $(BChl)_2^{\dot{+}}(k_3)$, returning the system to the ground state; k_3, like k_2 is physiologically nonproductive, and while k_3 is dominant in the isolated reaction center, it is insignificant when electrons can move on to Q_{II}. The possibility of the electron on $Q_I^{\bar{\cdot}}$ returning "uphill" via k_{-1} to BPh has generally been considered unlikely in native reaction centers.

Despite the tight binding of Q_I, it can be removed (*1*, *2*); tetramethyl benzoquinone (*1*), naphthaquinone (*2*), and anthraquinone (*1*) have all been shown to function in place of the physiological ubiquinone. Here we present a progress report on our preliminary work, exploring the interaction of quinones with the reaction center protein.

EXPERIMENTAL RESULTS

The time resolution of the spectrophotometer used here precludes the measurement of $(BChl)_2^{\dot{+}}$ unless the electron gets as far as Q_I so that the system assumes the relatively stable (> 1 μsec) $(BChl)_2^{\dot{+}}$ BPh $Q^{\bar{\cdot}}$ state. Figure 2 shows the rate of reduction of $(BChl)_2^{\dot{+}}$ from this state, while in Fig. 3A the proportion of reaction centers that reached this state is given. The data are presented as a function of the $E_{1/2}$ of the replacement qui-

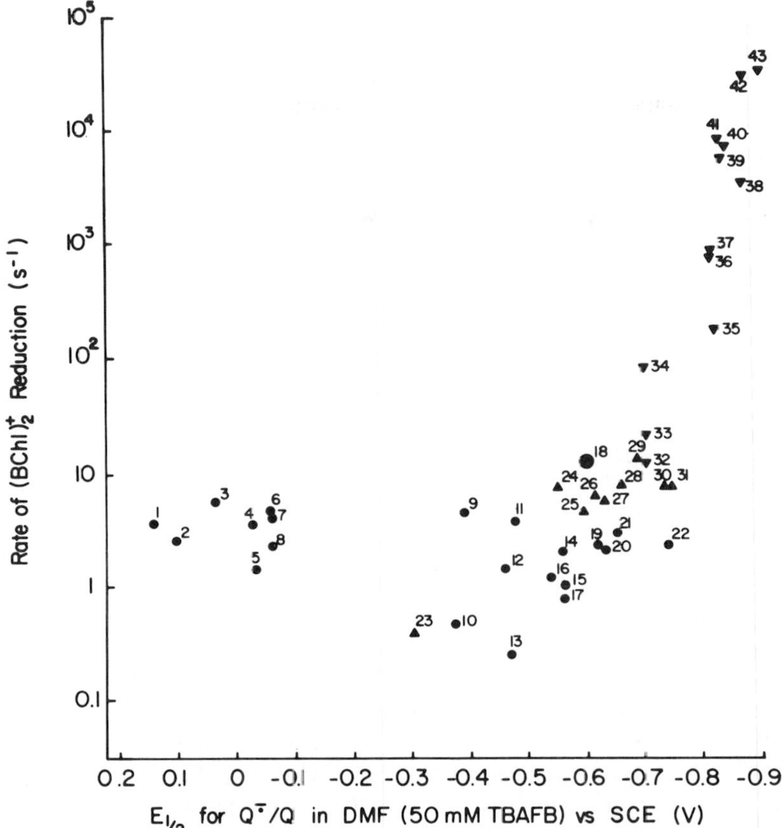

Fig. 2. The rate of $(BChl)_2^{\dot{+}}$ reduction in reaction centers with different quinones functioning as Q_I. Quinone-extracted reaction centers (0.5 μM) and replacement quinone (enough to saturate the reaction, varying from 0.5 to 25 μM) Tris-Cl 0.02% lauryldimethylamine-*N*-oxide pH 8, at 300 °K. Actinic light was provided by a 20 nsec laser flash for the anthraquinones, and a 6 μsec xenon flash for the other quinones. The key to the different quinones is found in Prince, *et al.* (Paper 1, Chapter II, this volume).

nones, the numbers by each point referring to the tables of Prince *et al.*, Paper 1, Chapter II, this volume.

DISCUSSION

The following points are emerging from this study:

1. Benzoquinones (BQ), napththaquinones (NQ), and anthraquinones (AQ) with amino, halogen, or alkyl substituents can be incorporated into

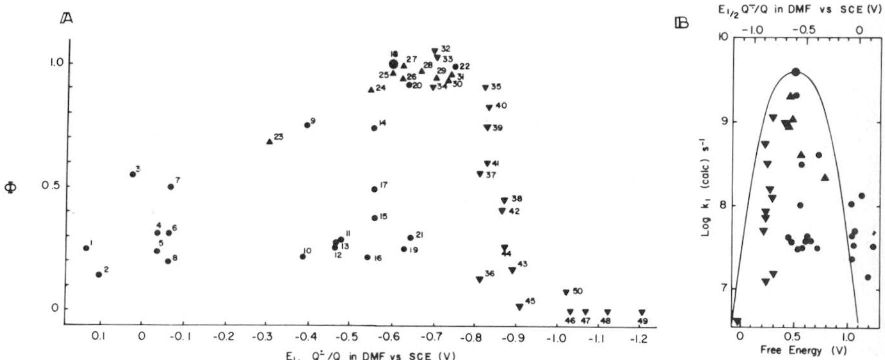

Fig. 3. (A) The proportion of the reaction centers attaining the stable $(BChl)_2^{+}$ BPH $Q^{\overline{\cdot}}$ state (ϕ); (B) the calculated value of k_1 as a function of the reaction exothermicity. Reaction conditions as in Fig. 2.

the reaction center to take the place of the native ubiquinone. Such modified reaction centers display $(BChl)_2^{+}$ reduction kinetics and ϕ values that correlate with $E_{1/2}$ of $Q/Q^{\overline{\cdot}}$ measured in DMF (see Prince *et al.*, Paper 1, Chapter II, this volume).

2. The $(BChl)_2^{+}$ reduction kinetics (Fig. 2) arise from two pathways: (a) with Q of high $E_{1/2}$, the e^- on $Q^{\overline{\cdot}}$ returns directly with rate k_3; (b) we propose that with Q of low $E_{1/2}$, the gap between $(BChl)_2^{+}$ BPh $Q^{\overline{\cdot}}$ and $(BChl)_2^{+}$ $BPh^{\overline{\cdot}}$ Q is sufficiently small to allow k_{-1} to compete with k_3. Appropriate to this, the rate of $(BChl)_2^{+}$ reduction with Q of low $E_{1/2}$ is temperature dependent, and proportional to the $E_{1/2}$ when $E_{1/2}$ varies from -0.7 to 0.85 V (*3*).

3. There are probably several influences on ϕ, the proportion of reaction centers that attain the stable $(BChl)_2^{+}$ BPh $Q^{\overline{\cdot}}$ state. For small BQs where $K_d > 100\ \mu M$ low values are not due to lack of binding of the Q. In some cases, low values of ϕ can be correlated with the size, position, or degree of substitution, providing information on the size and shape of the Q binding site, e.g., the position of substituents of dimethyl BQ has a marked effect on ϕ. For AQ, where $K_d \ll 0.1\ \mu M$ and the RCs are fully saturated with the AQ ϕ decreases with lowering $E_{1/2}$, perhaps as a result of a slowing of k_1. If this is so, k_1 may be obtained from $\phi = k_1/(k_1 + k_2)$. Support for this comes from the finding that the k_1 values obtained in this way can explain the low $E_{1/2}$ data of Fig. 2, where the observed rate of $(BChl)_2^{+}$ reduction should be $(k_{-1}\, k_2)(k_1 + k_2)$; a close fit is obtained with $k_{-1} = 10^4\ sec^{-1}$.

4. The operational $E_{1/2}$ of BPh can be calculated, relative to AQ, using $E_{1/2}AQ - E_{1/2}\,BPh = RT/nF\,(\ln k_1/k_{-1})$, giving a value of -1.1 V on

the scale of Figs. 2 and 3. Thus, the $\Delta E_{1/2}$ between ubiquinone and BPh is 0.5 V.

5. The rate of a reaction is predicted by $k = A \exp -Ea/kt$. The Ea can be expressed as $(\Delta E - \lambda)^2/4\lambda$ where λ is a measure of the energy necessary to effect the change in organization that accompanies the electron transfer (*5*) When $\lambda = \Delta E$, Ea = 0 and $k = A$. In the reaction center system, $BPh^{\overline{\cdot}}$ to ubiquinone displays an Ea = 0 (*6*); therefore, since ΔE = 0.5 V, λ = 0.5 V. Furthermore, since Ea = 0, $A = k$, which in this case is 4×10^9 sec^{-1} (*6*). If all replacement Q adopted the same values of λ and A as ubiquinone, the k_1 values calculated from ϕ plotted as a function of free energy, ΔE, should follow the expression $k_1 = 4 \times 10^9 \exp - (\Delta E - \lambda)^2/4 \lambda kT$, shown in Fig. 3B. As the exothermicity of the reaction increases, k_1 should increase, reach a maximum when E = λ and then decrease as the reaction becomes "overexothermic". The calculated values of k_1 shown in Fig. 3B show an encouraging adherence to the relationship (see also *7*).

6. Several ubiquinones seem to display low values of ϕ for reasons other than intrinsic restrictions on electron transfer, and more work is needed. Direct measurements of k_1 are needed to confirm the relationship with ϕ, and these have been commenced by R. M. Hochstrasser and colleagues. Nevertheless, our results add support to the application of electron tunneling theory (*7*) to the reaction between BPh and Q, and provide a very necessary latitude to the study of biological electron transfer that has been confined for too long to temperature/rate profiles in systems.

ACKNOWLEDGMENTS

This work was supported by grants from the National Science Foundation (PCM 79-09042) and Department of Energy (DE-AC02-80-ER 10590).

REFERENCES

1. Okamura, M. Y., Isaacson, R. A., and Feher, G. (1975). *Proc. Natl. Acad. Sci. U.S.A.* **72,** 3491–3496.
2. Cogdell, R. J., Brune, D. C., and Clayton, R. K. (1974). *FEBS Lett.* **45,** 344–349.
3. Dutton, P. L., Gunner, M. R., and Prince, R. C. (1982). *In* "Trends in Photobiology" C. Helene, M. Charlier, T. Montenay-Garestier, and G. Laustriat, eds.), pp. 561–570. Plenum, New York.
4. Gunner, M. R., Prince, R. C., and Dutton, P. L. (1981). In press.
5. Marcus, R. A. (1964). *Annu. Rev. Phys. Chem.* **15,** 155–196.
6. Peters, K., Avouris, P., and Rentzepis, P. M. (1978). *Biophys. J.* **23,** 207–217.
7. Miller, J. R. (1979). *J. Chem. Phys.* **71,** 4579–4595.

Quinones as Prosthetic Groups in Membrane Electron-Transfer Proteins II: The Interaction of the Reaction Center Secondary Quinone with the Q-b-c_2 Oxidoreductase

2

DANIEL P. O'KEEFE
ROGER C. PRINCE
P. LESLIE DUTTON

INTRODUCTION

The secondary quinone (Q_{II}) of reaction centers from *Rhodopseudomonas sphaeroides* occupies a position in the photosynthetic membrane that enables it to interact with both the primary quinone (Q_I) of the reaction center (RC) and the cytochrome *b* of the ubiquinone-cytochrome b-c_2 (Q-b-c_2) oxidoreductase. Previous studies have focused on the role of Q_{II} in semiquinone oscillations ($Q_{II}^{\overline{\cdot}}$ generation on odd numbered flashes, and disappearance on even flashes), and in proton binding to isolated reaction centers (*1–4*). We have examined these parameters in intact chromatophores and have found that the results are dramatically varied by binding of the reaction center to the Q-b-c_2 oxidoreductase.

The redox midpoint potentials of several species of partially reduced Q_{II} in chromatophores have been determined: E_{m8} = 40 mV for the $Q_{II}/Q_{II}H\cdot$

Function of Quinones in Energy Conserving Systems

ISBN 0-12-701280-X

couple and −40 mV for the $Q_{II}H\cdot/Q_{II}H_2$ couple (*5*). $Q_{II}H\cdot$ has a p*K* at 9.5 and hence the $Q_{II}/Q_{II}^{\overline{\cdot}}$ couple has a pH independent $E_m = -50$ mV. These $n = 1$ midpoints, and the $K_{Q\cdot}$ of 20, distinguish Q_{II} from the quinone pool ($E_{m8} = 30$ mV, $n = 1$, $K_{Q\cdot} < 10^{-7}$; see *6*).

It was assumed that reduction of Q_{II} by $Q_I^{\overline{\cdot}}$ was responsible for the binding of H_I^+ directly to Q_{II} (*7*). However, Wraight (*4*) surmised from discrepancies between proton binding and the appearance of $Q_{II}^{\overline{\cdot}}$ that H^+ binding must be somewhat remote from Q_{II}. That is, the generation of $Q_{II}^{\overline{\cdot}}$ induces a p*K* shift in a nonchromophoric protein group, which in turn, binds a proton. This model has proved useful in explaining the semiquinone oscillations observed in whole chromatophores.

EXPERIMENTAL RESULTS

Two Populations of Reaction Centers

The amount of cytochrome *b* reduction, H^+ binding, and the change in ΔA_{450} (the peak of the $Q^{\overline{\cdot}}$ absorption band) are plotted as a function of flash number at three different redox potentials in Fig. 1. Undamped

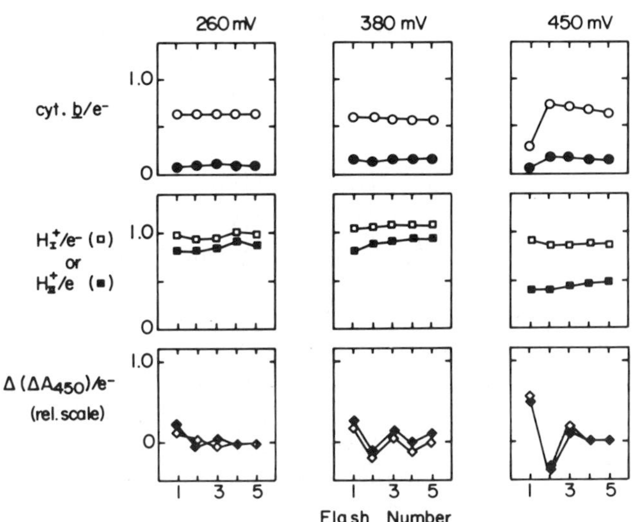

Fig. 1. Flash number dependence of cytochrome *b* reduction, proton binding, and ΔA_{450}; pH 6.0; (open symbols), +2 μM antimycin. $\Delta(\Delta A_{450}) = \Delta A_{450(n)}/\Delta A_{605(n)} - \Delta A_{450(n-1)}/\Delta A_{605(n-1)}$, where n = flash number.

binary oscillations in the semiquinone absorption band are only observed between 300 and 400 mV; below 300 mV no oscillations are observed (*9*, *13*), while above 400 mV they are large but highly damped. It is only at E_h >400 mV that cytochrome *b* reduction oscillates out of phase with, and as highly damped as, the 450 nm change, similar to the results of other investigators (*8*, *9*). The binary oscillations of ΔA_{450} seen between 300 and 400 mV appear to occur in only a fraction of the reaction center population, since concurrently, following each flash a reduction of as much as 0.6 cytochrome *b* per electron can be observed. There are only about 0.6 to 0.7 Q-*b*-c_2 oxidoreductase and cytochrome *b*-560 for each reaction center in the chromatophore (*10–12*), so this amount of cytochrome *b* reduction represents the transfer of about one reducing equivalent to each Q-*b*-c_2 oxidoreductase. The amount of change in ΔA_{450} is consistent with the other 0.3 reducing equivalents being transferred to Q_{II} after each flash, forming $Q_{II}^{\overline{\cdot}}$ on odd numbered flashes that disappear after even numbered flashes (*13*). Thus, the reaction centers can be divided into two populations: one where the Q_{II} is reduced and with $t_{1/2}$ = 15 msec reduces cytochrome *b*-560 of the Q-*b*-c_2 oxidoreductase (~70% of the reaction centers), and one where the Q_{II} is reduced to $Q_{II}^{\overline{\cdot}}$ and remains so until the next flash (~30% of the reaction centers). The recognition of two populations implies that the reaction center and its Q_{II} moiety exist in two noninteracting populations (RC:Q-*b*-c_2 pairs, and lone reaction centers), for if these populations were rapidly interchangeable on the timescale of this experiment, the oscillations of ΔA_{450} would be obliterated by the mixing of the Q-*b*-c_2 among the reaction centers between flashes.

At all E_h values examined, the binding of both H_I^+ and the antimycin sensitive H_{II}^+ were completely independent of the species of Q_{II} being generated. This finding suggests that a membrane Bohr-effect mechanism, such as that proposed by Wraight (*4*) for isolated reaction centers, is also operative in chromatophores; the binding of H_I^+ is probably not directly to $Q_{II}^{\overline{\cdot}}$ yielding $Q_{II}H\cdot$.

Interaction of Q_{II} with the Q-*b*-c_2 Oxidoreductase

Below E_h = 400 mV, the interaction of the Q-*b*-c_2 oxidoreductase with the reaction center causes the most pronounced differences between the lone reaction centers and the RC:Q-*b*-c_2 pairs. Figure 2 illustrates the reduction kinetics of cytochrome *b* (in the presence of valinomycin) at several E_h values, both in the presence and absence of antimycin. Antimycin stimulates the amount of cytochrome *b* reduction

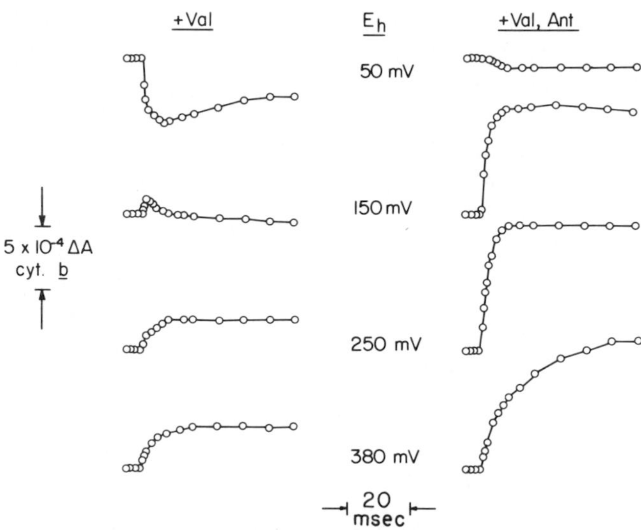

Fig. 2. Corrected cytochrome *b* kinetics. pH 6.0. Details of correction given in Ref. *12*.

after the flash ($E_h \geq$ 150 mV), and prevents the oxidation of cytochrome *b* ($E_h \leq$ 150 mV). At E_h = 50 mV cytochrome *b*-560 (E_{m6} = 110 mV, n = 1) is ~90% reduced, and flash activation results only in oxidation of the cytochrome *b*, whereas at 150 mV, both a transient reduction and subsequent oxidation occurs. At $E_h \geq$ 150 mV, flash activation in the presence of antimycin results in nearly complete reduction of the cytochrome *b*-560 of the Q-*b*-c_2 oxidoreductase; yet without antimycin present, only about one-third of the cytochrome *b*-560 becomes reduced. The latter is consistent with the cytochrome attaining redox equilibrium with the $Q_{II}/Q_{II}H\cdot$ couple (E_{m6} = 160 mV presumably; (*5*, *12*)) even though in the preceding section we concluded that Q_{II} is reduced to $Q_{II}^{\overline{\cdot}}$. A model consistent with both of these findings is shown,

$$Q_{II}^{\overline{\cdot}} + \text{ferri } b \rightleftharpoons \text{ferro } b + Q_{II} \quad (1)$$

$$\text{ferro } b + H^+ \rightleftharpoons \text{ferro } bH^+ \quad (2)$$

$$\text{ferro } bH^+ + Q_{II} \rightleftharpoons \text{ferri } b + Q_{II}H\cdot \quad (3)$$

If steps (2) and (3) were rapid compared to step (1), then in the absence of antimycin the kinetics of cytochrome *b* reduction would appear as a single reduction phase, with $Q_{II}H\cdot$ and ferrocytochrome bH^+ attaining redox equilibrium. Antimycin stimulates the amount of reduction either by slowing the protonation of ferrocytochrome *b* (step 2)), although this seems unlikely (*7*, *14*), or by slowing the oxidation of ferrocytochrome

bH$^+$ by Q_{II} (step (3)), effectively raising the functional E_m of cytochrome b-560. Slowing the oxidation by Q_{II} is consistent with the known effects of antimycin, and if this occurs, only steps (1) and (2) would be operational, and the amount of cytochrome b reduction would be greater because $Q_{II}^{\overline{\cdot}}$ could almost completely reduce cytochrome b-560. Therefore, in this model, ferrocytochrome b-560 functions as an intermediate in the delivery of H_I^+ to reduced Q_{II} unless antimycin is present.

Interaction of Q_{II} with the Quinone Pool

Continuous illumination results in only a small (<10%) reduction of the Q-pool (~18 Q/RC; E_{m6} = 150 mV, n = 2) unless antimycin is present, when up to 30% reduction is seen (6). One explanation for this effect is that only doubly-reduced Q_{II} in the lone reaction centers (generated on even numbered turnovers) can reduce or exchange with the Q pool in the absence of antimycin. The addition of the inhibitor, in slowing electron transfer through the cycle in the RC:Q-b-c_2 pairs, adds a second contributor of electrons to the Q-pool.

DISCUSSION

This work has focused on Q_{II} of the reaction center, the two populations of reaction centers, the interactions of Q_{II} with other redox components, and the effect of reaction center heterogeneity on these interactions, with the following conclusions:

1. There are two populations of reaction centers: the RC:Q-b-c_2 pairs (70%) and lone reaction centers (30%). There does not seem to be significant exchange between these populations although there may be exchange among the RC:Q-b-c_2 pairs. Whether there are two populations in intact cells is not yet clear.
2. H_I^+ is apparently not bound directly to Q_{II}.
3. $Q_{II}^{\overline{\cdot}}$ is initially generated in all reaction centers.
4. The Q_{II} in the lone reaction centers can only interact with members of the Q pool.
5. The Q_{II} in the RC:Q-b-c_2 pairs usually interacts with cytochrome b-560 of the Q-b-c_2 oxidoreductase, but can also interact with members of the Q-pool.

ACKNOWLEDGMENT

This work was supported by the National Institutes of Health Public Health Service Grant GM 27309.

REFERENCES

1. Wraight, C. A. (1977). *Biochim. Biophys. Acta* **459,** 525–531.
2. Vermeglio, A. (1977). *Biochim. Biophys. Acta* **459,** 516–524.
3. Vermeglio, A., and Clayton, R. K. (1977). *Biochim. Biophys. Acta* **461,** 159–165.
4. Wraight, C. A. (1979). *Biochim. Biophys. Acta* **548,** 309–327.
5. Rutherford, A. W., and Evans, M. C. W. (1980). *FEBS Lett.* **110,** 257–261.
6. Takamiya, K.-I., and Dutton, P. L. (1979). *Biochim. Biophys. Acta* **546,** 1–16.
7. Petty, K. M., and Dutton, P. L. (1976). *Arch. Biochem. Biophys.* **172,** 335–345.
8. DeGrooth, B. G., van Grondelle, R., Romijn, J. C., and Pulles, M. J. P. (1978). *Biochim. Biophys. Acta* **503,** 480–490.
9. Bowyer, J. R., Tierney, G. V., and Crofts, A. R. (1979). *FEBS Lett.* **110**, 201–206.
10. van den Berg, W. H., Prince, R. C., Bashford, C. L., Takamiya, K.-I., Bonner, W. D., and Dutton, P. L. (1979). *J. Biol. Chem.* **254,** 8594–8604.
11. Bowyer, J. R., Dutton, P. L., Prince, R. C., and Crofts, A. R. (1980). *Biochim. Biophys. Acta* **592,** 445–460.
12. O'Keefe, D. P., and Dutton, P. L. (1980). *Biochim. Biophys. Acta* **635,** 149–166.
13. O'Keefe, D. P., Prince, R. C., and Dutton, P. L. (1981). *Biochim. Biophys. Acta* **637,** 512–522.
14. Petty, K. M., and Dutton, P. L. (1976). *Arch. Biochem. Biophys.* **172,** 346–353.

Quinones as Prosthetic Groups in Membrane Electron-Transfer Proteins III: Functional Recognition of a Special Quinone Associated with the Ubiquinone–Cytochrome b-c_1 Oxidoreductase of Beef Heart Mitochondria

3

K. MATSUURA
N. K. PACKHAM
D. M. TIEDE
P. MUELLER
P. LESLIE DUTTON

INTRODUCTION

The mechanisms of proton translocation remain largely hypothetical in the multi-redox centered proteins of mitochondria and photosynthetic systems. Some progress has been made in recognizing a redox component that is functionally central to electron and proton transfer in the ubiquinone-cytochrome *b*-c_2 (Q-*b*-c_2) oxidoreductase of the photosynthetic bac-

Function of Quinones in Energy Conserving Systems

ISBN 0-12-701280-X

terium *Rhodopseudomonas sphaeroides*. The component in question was identified as a single (*1*, *2*) ubiquinone (*3*) firmly associated with the Q-*b*-c_2 oxidoreductase and tentatively designated Q_z.

Recently we constructed *in vitro* a functional, antimycin-sensitive electron-transfer cycle using isolated reaction centers (RC), horse heart cytochrome *c*, and isolated Q-*b*-*c* oxidoreductase (Complex III) from beef heart mitochondria (*4*). Remarkably, the flash-induced oxidation and reduction behavior displayed by mitochondrial cytochromes under a variety of conditions *in vitro* is very similar to the photosynthetic analog *in vivo*. Thus, advantages provided by single turnover flash activation of the photosynthetic system can be applied to this mitochondrial redox protein. In this paper we describe the properties of a Q_z-like component that is a key part of the mitochondrial Q-*b*-*c* oxidoreductase.

EXPERIMENTAL RESULTS

Figure 1 shows flash activated oxidation–reduction kinetics of cytochromes *b* and *c* under varying levels of reduction of Q-*b*-*c* oxidoreductase. The reduction level was altered by succinate. Traces A, B, and C were measured in sequence ~10 minutes after the succinate was added or increased in concentration. In (A), with cytochrome *c* reduced and cytochrome *b* oxidized before activation, the flash elicits prompt cytochrome *b* reduction (via the primary and secondary quinones of the reaction center) and cytochrome *c* oxidation (by $(BChl)_2^{+}$ of the reaction center). The subsequent reoxidation of the cytochrome *b* and rereduction of cytochrome *c* is slow, implying that passage of an electron from cytochrome *b* to cytochrome *c* is impeded. In (B), the system has become more reduced prior to activation and the reoxidation of flash-generated ferrocytochrome *b* is much faster. Parallel behavior is seen in the rereduction of flash generated ferricytochrome *c*. In (C) the succinate has reduced a significant portion of the cytochrome *b* complement. Under these conditions flash activation leads to net oxidation of the ferrocytochrome *b*; again the rate of oxidation matches that of ferricytochrome *c* rereduction. In (D) the dramatic inhibitory effect of antimycin is shown under conditions the same as described (C). Four tentative conclusions can be drawn from this semiquantitative work. (1) The similar rates of cytochrome *b* oxidation and cytochrome *c* reduction suggest a common rate-limiting step exists for the electron transfer from cytochrome *b* to cytochrome *c*. (2) The source of a common rate-limiting step appears to be a redox compo-

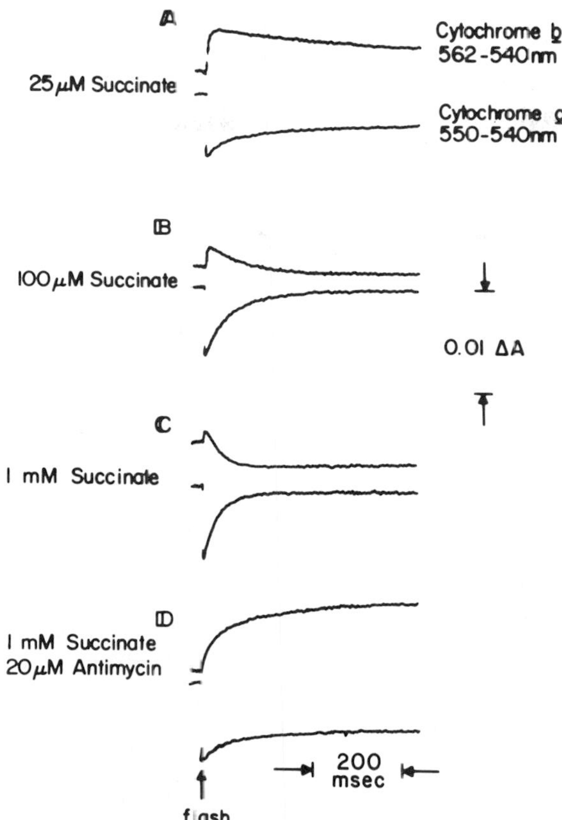

Fig. 1. Flash-induced oxidation–reduction kinetics of cytochromes b and c in a mixture of mitochondrial Q-b-c oxidoreductase (4 μM in c_1), cytochrome c (6 μM), and reaction centers (0.6 μM) in 10 mM Tris-HCl, pH 7.8.

nent that becomes reduced after cytochrome c but before cytochrome b. (3) The electron transfer from cytochrome b to cytochrome c is impeded unless this component is in a reduced state prior to flash activation. (4) The effect of antimycin demonstrates that electrons are flowing through the Q-b-c oxidoreductase in this constructed cyclic system.

The work with succinate provided a basic set of kinetics obtained in the absence of supplementary redox dyes or mediators. To obtain quantitative, redox potentiometric information about the agent involved between cytochrome b and cytochrome c, we repeated the experiment of Fig. 1, replacing succinate with redox mediators. Figure 2 shows plots of the initial rate of ferrocytochrome b oxidation (▲) and ferricytochrome c reduction (●) as a function of redox potential (E_h) established before the flash.

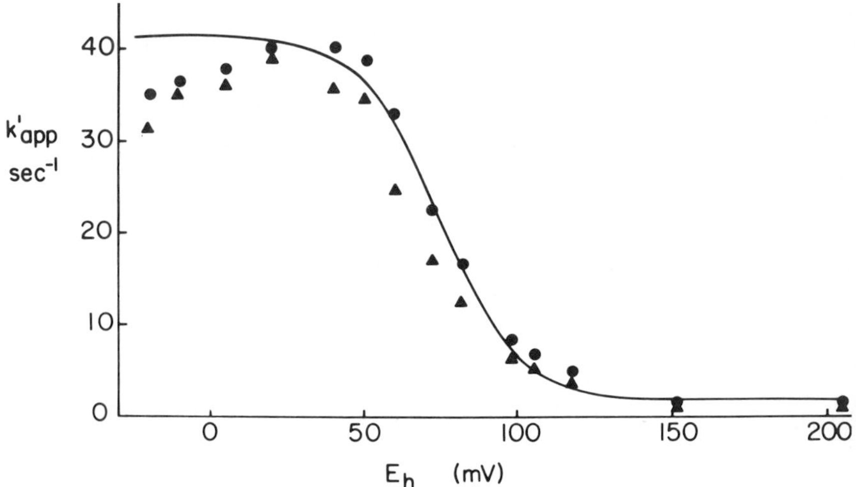

Fig. 2. E_h dependence of electron-transfer rate through Q-*b*-*c* oxidoreductase at pH 7.8. Redox mediators used were 0.1 m*M* Fe–EDTA (3.3 m*M*), 10 μM DAD and 10 μM pyocyanine. Other conditions were as in Fig. 1. The ratio of initial rate and extent of ferrocytochrome *b* oxidation (▲) and ferricytochrome *c* reduction (●) are plotted as $k'_{apparent}$ from similar kinetic traces to Fig. 1.

The E_h dependence of the two rates is similar, changing in $k_{apparent}$ from ~1 sec^{-1} to ~40 sec^{-1} and is described by a Nernst curve with a midpoint potential (E_m) of 75 mV at pH 7.8. The *n*-value of the curve is close to 2.0. Repetition at pH values from 6.3 to 8.5 revealed the E_m to have a −60 mV/pH unit dependency over this range. The E_m at pH 7.0 is approximately 115 mV for this redox component. The E_m value (pH 7.0) of cytochrome *b*-560 is 30 mV and that of cytochrome c_1 is 220 mV. The quantitative results obtained using redox potentiometry are consistent with the tentative conclusions drawn from the succinate experiment of Fig. 1.

DISCUSSION

The observtions reported for isolated mitochondrial Q-*b*-*c* oxidoreductase (Complex III) obtained *in vitro* in association with cytochrome *c* and for photoactivating reaction centers isolated from *Rps. sphaeroides* are entirely analogous to results obtained *in vivo* with chromatophore membranes that contain the reaction center in association with cytochrome c_2 and a Q-*b*-c_2 oxidoreductase. There is much to be learned from both systems.

Over the past five years evidence has accumulated suggesting that there is one molecule of Q_z per Q-*b*-c_2 oxidoreductase of *Rps. sphaeroides* (*1–3,5,6*); Q_z is probably a ubiquinone-10, since this is the only quinone species detected in *Rps. sphaeroides;* on extraction of native Q_z the reactions of Q_z are fully reconstituted by ubiquinone-10 (*3,6*). In the pH range 5 to 11, Q_z displays the equilibrium redox reaction equation

$$Q_z - 2e^- + 2H^+ = Q_zH_2, \quad (1)$$

with an E_m (pH 7.0) of 155 mV (n = 2) (see *1*). This compares with 115 mV (n = 2) obtained for the mitochondrial analog measured *in vitro,* which conforms to equation (1), at least from pH 6.3 to 8.5. The similarity of these equilibrium properties and the extraordinary similarity of kinetics of flash-induced cytochrome *b* and cytochrome *c* oxidation and reduction in bacterial and mitochondrial systems, strongly suggests that there is a Q_z in the mitochondrial system.

The E_m value obtained for the Q_z analog in mitochondria coincides with observations made by others. In particular, Rieske *et al.* (*7*) not only indicated the presence of a redox component (designated X) in "Complex III" that is indispensible for stability and function of the protein, but also estimated the E_m value to be 100 mV at pH 7.0. More recently, an important redox free radical was detected by several groups (*8*) using EPR on mitochondrial preparations. Redox potentiometric work (*8*) from EPR spectra revealed the redox component responsible displays a signal maximum at ~100 mV (pH 7.0) and hence an E_m at this value. The low stability constant (5×10^{-3}, pH 7.0) and pH dependency of the signal maximum indicates the component also conforms to the redox reaction equation (1), at least in the range studied, pH 6.3–8.0.

An additional piece of information emerged from mitochondrial studies (*8*). The two principal reactions that constitute the two-electron reduction of the mitochondrial Q_z analog were

$$Q_z + e^- = Q_z^{\bar{\cdot}};\ E_{m7} + 30 \text{ mV} \quad (2)$$

and

$$Q_z^{\bar{\cdot}} + e^- + 2H^+ = Q_zH_2;\ E_{m7} + 170 \text{ mV}. \quad (3)$$

In *Rps. sphaeroides,* in addition to ferricytochrome c_2 reduction and ferrocytochrome *b* oxidation stimulated by the Q_zH_2 (as shown in Figs. 1 and 2 for the mitochondrial system), there is also an apparent transmembrane electrogenic reaction and proton (H_{II}^+) binding; all four events are inhibited by antimycin (*5*). Combining the above information, we propose the general scheme for Q-*b*-*c* oxidoreductase shown in Fig. 3. The main points to be made are:

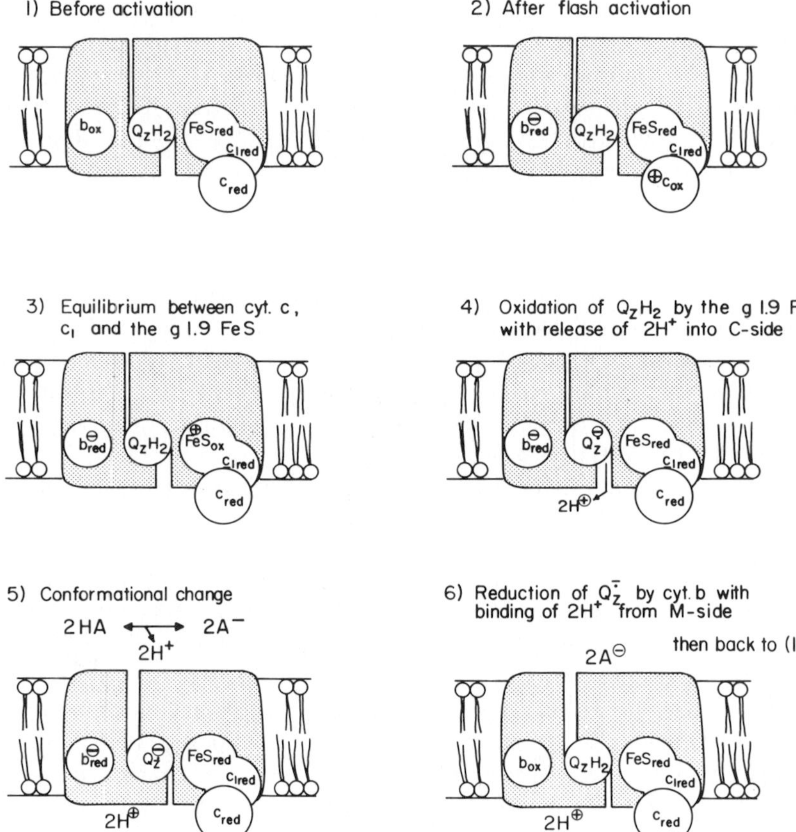

Fig. 3. A scheme for electron and proton transport in Q-*b*-*c* oxidoreductase.

1. An electron transfers from cytochrome *b* to the *g* 1.9 Fe–S center through a non-mobile (on a large scale), relatively stable protein-associated quinone, predominantly using a $Q_zH_2/Q_z^{\cdot -}$ redox couple.
2. Two protons per electron are translocated electrogenically through a proton channel in the oxidoreductase to $Q_z^{\cdot -}$ and from Q_zH_2 across the membrane.
3. A conformational change of the oxidoreductase accompanying the redox change of the Q_z (*7*) allows vectorial translocation of the proton.

The concept of this scheme may be similar to the redox-linked electrogenic proton pump proposed by Wikström and Krab (*9*) for cytochrome *c* oxidase.

ACKNOWLEDGMENT

This work was supported by the National Institutes of Health Public Health Service Grant GM 27309.

REFERENCES

1. Prince, R. C., and Dutton, P. L. (1977). *Biochim. Biophys. Acta* **462,** 731–747.
2. Prince, R. C., Bashford, C. L., Takamiya, K., van den Berg, W. H., and Dutton, P. L. (1978). *J. Biol. Chem.* **253,** 4137–4142.
3. Takamiya, K., Prince, R. C., and Dutton, P. L. (1979). *J. Biol. Chem.* **254,** 11307–11311.
4. Packham, N. K., Tiede, D. M., Mueller, P., and Dutton, P. L. (1980). *Proc. Natl. Acad. Sci. U.S.A.* **77,** 6339–6343.
5. van den Berg, W. H., Prince, R. C., Bashford, C. L., Takamiya, K., Bonner, W. D., and Dutton, P. L. (1979). *J. Biol. Chem.* **254,** 8594–8604.
6. Bowyer, J. R., Baccarini-Melandri, A., Melandri, B. A., and Crofts, A. R. (1978). *Z. Naturforsch., Teil C* **33,** 704–711.
7. Rieske, J. S. (1976). *Biochim. Biophys. Acta* **456,** 195–247.
8. Ohnishi, T., and Trumpower, B. L. (1980). *J. Biol. Chem.* **255,** 3278–3284.
9. Wikström, M., and Krab, K. (1979). *Biochim. Biophys. Acta* **549,** 177–222.

4

The Multifarious Role of Ubiquinone in Bacterial Chromatophores

A. BACCARINI-MELANDRI
N. GABELLINI
B. A. MELANDRI

INTRODUCTION

The non-sulfur purple bacteria (Rhodospirillaceae) can utilize energy from light under anaerobic conditions or from substrate oxidation under aerobic conditions in the dark. Growth conditions (especially light and oxygen concentration) induce a differentiation of the intracytoplasmic membranes: light-anaerobic conditions lead to a massive biosynthesis of photopigments and dark-aerobic conditions lead to an increase in dehydrogenases and terminal oxidases (*1*). Several lines of evidence strongly suggest the operation of a dual functional electron-transfer chain that shares the ubiquinone–cytochrome *c* oxidoreductase complex and competes for both respiration and photosynthesis (*1–4*). The same coupling factor is also active in oxidative and photosynthetic phosphorylation (*5*).

When Rhodospirillaceae are grown anaerobically in light, their cyclic electron-transfer system is formed essentially by a photosynthetic reaction center (RC) and by a UQ–cytochrome *c* oxidoreductase complex that shows striking similarities to the analogous mitochondrial complex. Kinetic and thermodynamic analyses of components involved in photosynthetic cyclic electron flow of chromatophores are greatly simplified (as compared to mitochondria) by the possibility that the system can be acti-

Function of Quinones in Energy Conserving Systems

ISBN 0-12-701280-X

vated photochemically, in a single turnover, by μsec flashes of actinic light. The activated reaction center delivers one reducing equivalent to the primary acceptor and one oxidizing equivalent to cytochrome c_2, all within a millisecond.

In the two organisms of choice in these studies, *Rhodopseudomonas capsulata,* Ala pho$^+$ and *Rps. sphaeroides,* Ga, ubiquinone-10 (UQ-10) is the only quinone species. Also, UQ-10 is present in large excess (25 to 30 UQ per RC in *Rps. sphaeroides* (*6*) and 30 to 40 UQ per RC in *Rps. capsulata* (*7*)). The chromatophore membrane shows a remarkable stability in its energy coupling properties after lyophilization and extraction with apolar solvents (*7*). By exploiting this experimental advantage, together with the possibility of rapid kinetic and thermodynamic analysis, evidence is appearing that different populations of UQ-10 are competing in cyclic electron flow and are required for particular redox events, as well as for coupled electron flow. In addition to a large, thermodynamically homogenous pool of UQ-10 (*6*), two ubiquinone molecules (UQ_I and UQ_{II}) act as primary and secondary acceptors from the reaction center in a two-electron gate mechanism (*8*, *9*); a third ubiquinone molecule (UQ_Z) appears to play a central role in the *b*-c_2 segment (*10*; for review, see *11*).

In this paper we report evidence obtained by extraction–reconstitution studies that strongly supports the role of specialized ubiquinone molecules in chromatophores. Studies utilizing UQ-10 and different UQ-homologs for the reconstitution of specific redox events and photophosphorylation in continuous light, document a close structural interaction between ubiquinone molecules and other membrane components (possibly proteins) and demonstrate the importance of the thermodynamically homogenous and easily extractable pool of UQ-10.

EXPERIMENTAL RESULTS

Kinetic Approach to Estimate the Amount of the Secondary Ubiquinone (UQ_{II})

The presence of the secondary acceptor in isolated reaction centers and in bacterial chromatophores can be monitored by several methods. The half-time needed for recovery in primary photochemistry of double flash experiments gives information on the rate of electron transfer between the primary and secondary acceptor and on its availability for reduction (*12*).

An indirect method utilized in isolated reaction centers consists of measuring the rate of $(BChl)_2^+$ rereduction[1] after a single turnover flash in the absence of a fast exogenous donor to the reaction center (*13*). Under these conditions $(BChl)_2^+$ rereduction can only be achieved by charge recombination. This reaction (back-reaction) is faster ($t_{1/2}$ = 60 msec) if electron flow from the primary to secondary acceptor is blocked (as for example by addition of *o*-phenanthroline) and is slowed ($t_{1/2}$ = 800 msec) when the electron is stabilized on the secondary acceptor. This last experimental approach can be applied also to bacterial chromatophores if the ambient redox potential of the assay is kept relatively high (+430 mV at pH 7.0), so that cytochrome c_2, the direct donor to $(BChl)_2^+$ is fully oxidized, and secondary electron transport is blocked by antimycin.

As shown in Fig. 1A, when bacterial chromatophores are poised at ambient E_h of 430 mV at pH 7.0, in the presence of antimycin, the decay of $(BChl)_2^+$ after a flash is monophasic with a half-time of about 800 msec; this compares well with values reported for isolated reaction centers possessing two ubiquinones. Addition of *o*-phenanthroline markedly increases the rate of $(BChl)_2^+$ rereduction ($t_{1/2}$ = 40 msec), as was shown for isolated reaction centers having only the primary acceptor (Q_I), or when the iron atom between the two quinones was extracted (*13*).

If the amount of the secondary quinone in bacterial chromatophores is reduced below two UQ per RC (Fig. 1B), the back-reaction shows biphasic kinetics. The decay pattern can be fitted accurately by the sum of two exponential rate functions, having half-times corresponding to the back-reaction from the primary or secondary Q. Therefore the percentage of slow back-reaction can be taken as an estimate of the reaction centers possessing both primary and secondary ubiquinone.

Kinetic Approach to Estimate the Amount of UQ_Z

On the basis of studies on the redox changes of cytochromes *b* and c_2 in UQ extracted and reconstituted chromatophores from *Rps. capsulata,* Ala pho$^+$, we previously suggested that UQ-10 participates in electron transfer between cytochrome *b* and cytochrome c_2 (*7*, *14*) and that a carrier Z, (postulated previously by Evans and Crofts (*15*) as mediating electron flow between cytochromes *b* and *c*), could be identified

[1] Abbreviations: E_h, oxidation–reduction potential; DAD, 2,3,5,6-tetramethyl-*p*-phenylendiamine; MOPS, 3-(*N*-morpholino)propanesulfonic acid; 1,2-NQ, 1,2-naphthoquinone; 1,4-NG 1,4-naphoquinone; (BChl), bacteriochlorophyll.

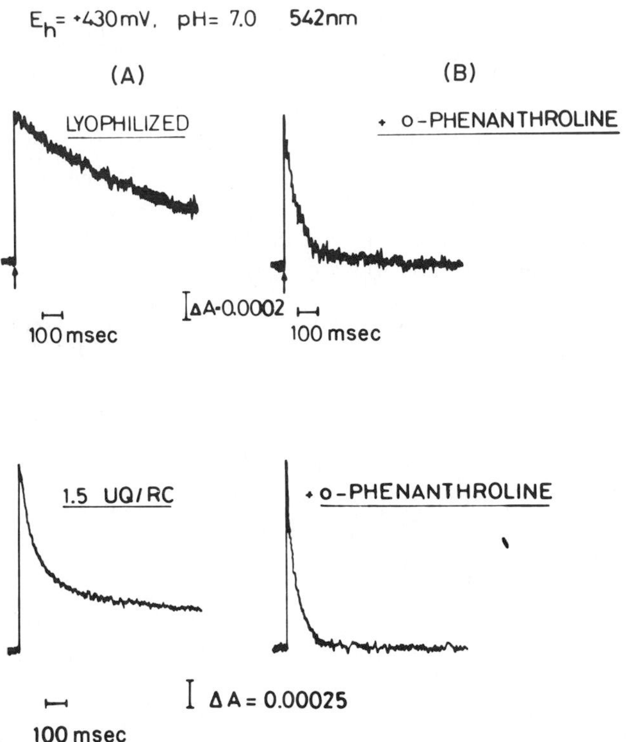

Fig. 1. Redox changes of $(BChl)_2$ upon flash excitation in lyophilized (A) and extracted (B) particles from *Rps. sphaeroides,* Ga. The assay contained: MOPS, 50 m*M*, pH 7.0; KCl 100 m*M*; valinomycin, 2 μM; antimycin, 4 μM and chromatophores corresponding to about 30 μM (BChl). The redox potential (430 mV) was increased by addition of small amounts of potassium ferricyanide; 2 m*M* *o*-phenantroline was added where indicated. The reaction was carried out in a gassed and stirred anaerobic cuvette. Near-infrared actinic flashes (15–20 μsec pulse width) were fired through the bottom of the cuvette. The preparation of lyophilized and extracted chromatophores is described in Baccarini-Melandri *et al.* (*18*).

as a UQ-10 molecule. This carrier has a midpoint potential of 155 mV in *Rps. sphaeroides,* Ga, and reduction of the carrier involves $2H^+$ and two electrons. At ambient E_h below the $E_{m,7}$ of this component, the rate of cytochrome c_2 rereduction after a flash is markedly accelerated, indicating a role for this carrier in cytochrome c_2 reduction (*16*).

Studies by Takamiya *et al.* (*10*) showed quite elegantly that extraction of UQ-10 affects the rate of rereduction of cytochrome c_2 after a single turnover flash, only when the amount of the remaining ubiquinone is isooctane extracted particles was below 3 to 4 UQ/RC. This finding was interpreted as evidence that Z is a specialized ubiquinone molecule (UQ_Z) distinct from the large pool and possibly bound to proteins. Extraction of ubiquinone below 3 to 4 UQ per RC also abolishes the III Phase of the carotenoid band shift (indicator of an electrogenic event within the b-c_2 segment. This event was shown to be correlated with the reactions leading to cytochrome rereduction (*17*).

Redox changes of cytochrome c_2 and the kinetics of the carotenoid band shift in lyophilized and extracted chromatophores from *Rps. sphaeroides,* GA, are shown in Fig. 2. Both measurements were performed at E_h below the $E_{m,7}$ of UQ_z, i.e., at 90 mV pH 7.0. In the left part of the Figure the kinetics of cytochrome c_2 are shown: the antimy-

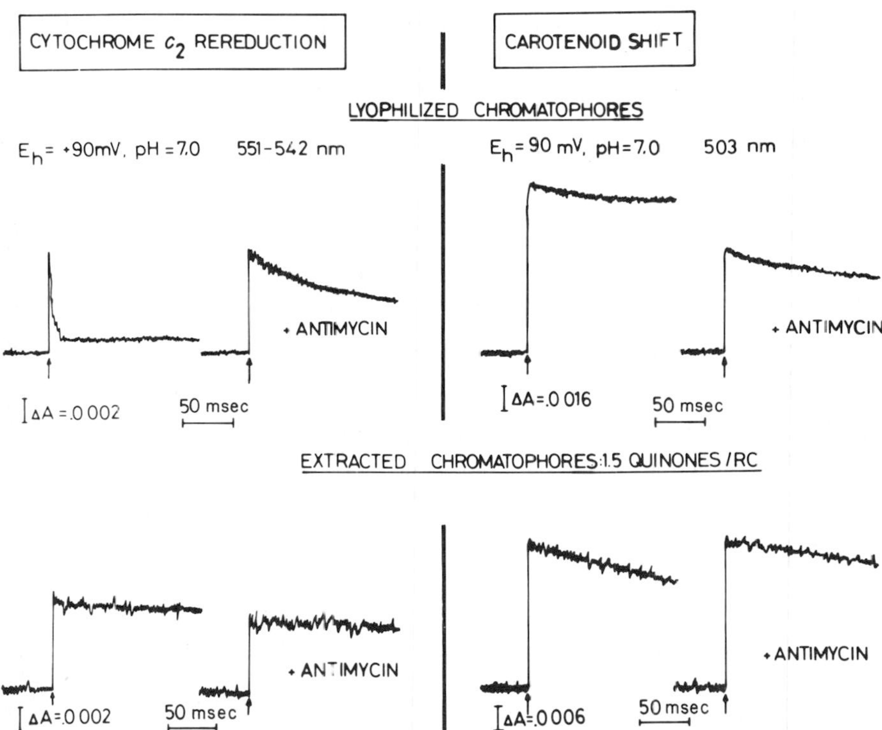

Fig. 2. Flash-induced cytochrome c redox change and carotenoid band shift in lyophilized and extracted chromatophores from *Rps. sphaeroides,* Ga. Lyophilized or extracted chromatophores were suspended in MOPS, 50 m*M*, pH 7.0; KCl 100 m*M* together with 1 μ*M* each of DAD, 1,2-NQ and 1,4-NQ in a gassed and stirred anaerobic redox cuvette. The E_h was adjusted to 90 mV. The assay conditions were similar in the two types of measurements except that valinomycin, 2 μ*M*, was present when cytochrome c redox changes were monitored. Antimycin, when added, was 4 μ*M*. Chromatophore concentrations were 18 μ*M* (BChl) and 7 μ*M* (BChl) for measurements of cytochrome c and of carotenoid shift, respectively. All other conditions as in Fig. 1.

cin A-sensitive fast rereduction of cytochrome c_2 disappears upon isooctane extraction in particles having 1.5 UQ per RC, as determined chemically. On the right side the signal of the carotenoid band shift, monitored at 503 nm in the presence and absence of antimycin, is shown in lyophilized and extracted (1.5 UQ/RC) membranes. Although the total extent of the signal is somewhat decreased upon isooctane extraction due to a loss of carotenoid pigments, the antimycin-sensitive part (III phase) is completely absent. Using these two parameters (i.e., the rate of flash-oxidized cytochrome c_2 reduction and the percentage of the antimycin-sensitive part of the carotenoid shift) one can establish the presence or absence of UQ_Z. We have utilized the carotenoid shift parameter to estimate the percentage of UQ_Z remaining following extraction with isooctane.

Different Extractibilities of UQ_{III} and UQ_Z

The two experimental approaches outlined in the previous paragraphs were utilized for a quantitative estimation of the content of UQ_{II} and UQ_Z. Lyophilized chromatophores from *Rps. sphaeroides,* Ga, were first depleted of a large part of the original UQ-10 present, so that the residual content was reduced below 4 to 5 UQ/per RC. In these particles 100% of UQ_{II} and 100% of UQ_Z were present.

Additional extractions with isooctane for variable times, utilizing a variable ratio of isooctane per BChl were then performed on 4 to 5 UQ per RC particles. Figure 3 shows a plot of the residual amounts of UQ_Z versus UQ_{II} in these preparations. It is evident that the content of UQ_Z decreases before UQ_{II} starts to be extracted; in several preparations

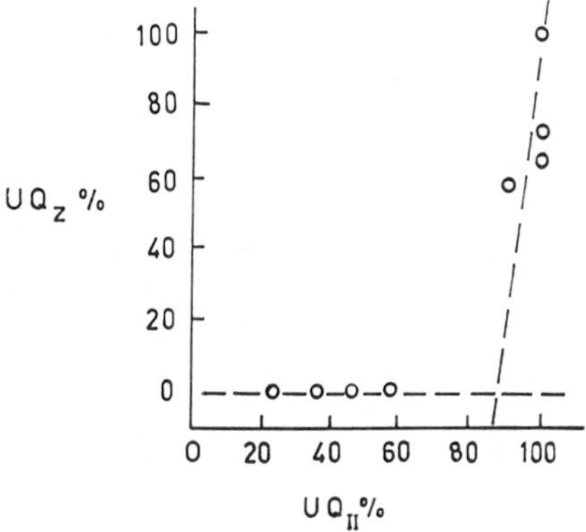

Fig. 3. Correlation plot of the percentage amounts of UQ_{II} and UQ_Z following progressive depletion of ubiquinone by isooctane extraction. The assay conditions for the evaluation of UQ_{II} and UQ_Z are described in the text and in the legends of Figs. 1 and 2.

where no UQ_Z was present, a large proportion (up to 60%) of UQ_{II} was retained. Therefore these two specialized ubiquinone molecules show a differential extractability in apolar solvents, probably reflecting different binding constants of the two quinone molecules to specific membrane components.

Structural Requirements of Quinone Coenzymes for Reconstitution of Electron Transfer and ATP Synthesis

In a previous paper (*18*) we examined the ability of different types of quinone homologs and analogs to restore phosphorylation. We found, in essence, that none of the quinone analogs tested were very effective for reconstitution, indicating a strict structural requirement of the ubiquinone ring for this function. A differential response was elicited, however, when UQ-10 homologs were utilized. UQ-1 and UQ-2 were always ineffective, while a variable response was found with homologs having a side chain equal to or longer than 15 C atoms (3 isoprenoid units). This difference in response appears to be related to the amount of UQ remaining after extraction. Typical data of this kind of behavior are shown in Table I. When extracted particles retaining two or more

TABLE I

Restoration of Photophosphorylation by Ubiquinone Homologs in UQ-10 Depleted Chromatophores[A]

	Photophosphorylation (μmole ATP h^{-1} mg $(BChl)^{-1}$)	
	Type A (2–4 UQ/RC)	Type B (1.2–1.5 UQ/RC)
Extracted	22	4
Reconstituted with:		
UQ-1 (40 UQ/RC)	29	5
UQ-2 (40 UQ/RC)	35	8
UQ-3 (40 UQ/RC)	181	39
UQ-5 (40 UQ/RC)	185	40
UQ-7 (40 UQ/RC)	186	180
UQ-10 (40 UQ/RC)	208	212
UQ-10 (40 UQ/RC) + antimycin A	0	0

[A] The rate of photophosphorylation was measured as described by Baccarini-Melandri and Melandri (*39*). The procedure for UQ extraction and reconstitution of chromatophores is described in detail in Baccarini-Melandri and Melandri (*7*) and Baccarini-Melandri *et al.* (*18*).

UQ-10 molecules (type A particles) were reconstituted with UQ homologs at a ratio of 40 UQ per RC, there was a sudden difference in restoration from UQ-2 to UQ-3; UQ-3 was as effective as UQ-10. Interestingly, a similar observation was also reported in an artificial system where a vectorial non-enzymatic redox reaction was measured across liposomes in which UQ homologs with variable chain length were incorporated (*19*, *20*).

In membranes with a lower content of residual ubiquinone, from 1.2 to 1.5 UQ/RC, UQ homologs (type B particles) with longer side chains were needed to reconstitute full activities of ATP synthesis. This differential response suggested that the structural requirement for UQ-10 to reconstitute phosphorylation could not be attributed to the restoration of the easily extractable Q pool, that was obviously absent in both A- and B-type preparations. Instead, it was attributed to the restoration of some specialized UQ molecules having a stricter interaction or a tighter binding constant than the pool with some membrane components. Since UQ_I is not extracted by apolar solvent, this specificity should be looked for in the other two specialized UQ-10 molecules (UQ_{II} and UQ_Z).

In previous investigations we found that all UQ homologs can fulfill the role of secondary acceptor from the reaction center (measured either by the capacity to bind scalar protons in continuous light (*18*), or by the slow rate of the back-reaction after a single turnover flash (*21*)). Alternatively, when the rate of rereduction of cytochrome c_2 after a flash (at E_h = 100 mV, pH 7.0) was analyzed in the same extracted particles (1.2 UQ/RC) reconstituted with UQ-1, UQ-3 and UQ-10, the results gave a clear explanation of the inefficiency of UQ-1; the function of UQ_Z as reductant of cytochrome c_2 could not be reconstituted by this short-chain homolog (*21*). However UQ-3 and UQ-10 were both capable of reconstituting a fast rate of cytochrome c_2 rereduction.

The structural specificity, therefore, should depend upon the reaction of UQ_{II} or UQ_Z with the oxidant or reductant (possibly cytochrome *b*-50). If so, the lack of active reconstitution of phosphorylation by UQ-3 in type B particles should be reflected in a slower turnover rate of the overall electron flow. A new experimental approach was recently developed in our laboratory to measure the rate of cyclic electron flow under multiple turnovers (*22*), based on the relaxation kinetics of $(BChl)_2$ during a train of flashes. Figure 4 shows a plot of the amount of reaction centers left oxidized before the flash in pseudosteady-state conditions (at the 5th to 7th flash) versus the dark time allowed between flashes. The curve fitting the points (○) is clearly biphasic. By computer analysis, this curve can be resolved into two first-order pro-

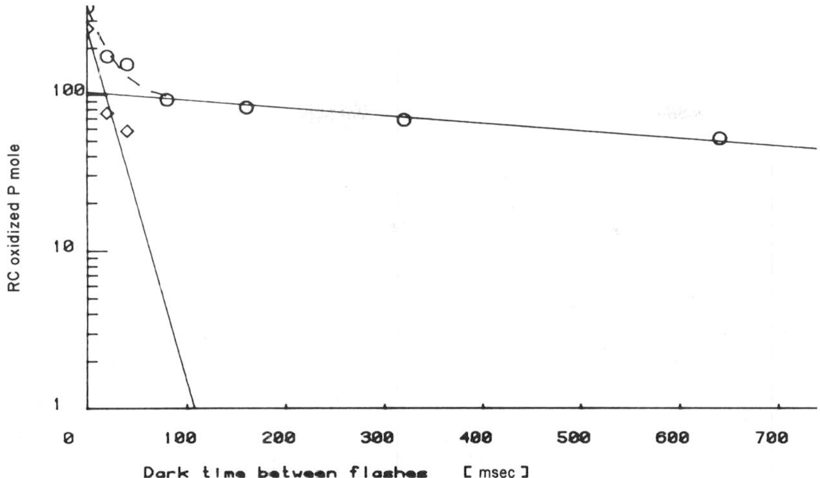

Fig. 4. Relaxation kinetics of $(BChl)_2^{+}$ rereduction upon multipulse excitation: lyophilized unextracted chromatophores. A semilogarithmic plot of the amounts of $(BChl)_2$ left oxidized before the 7th flash during a train of flashes fired at different frequencies is shown as a function of the dark time between flashes. The best fit of the kinetic resolution of $(BChl)_2^{+}$ rereduction with two additive first-order processes was obtained by computer analysis of the experimental points utilizing a Tektronic 4051 Graphic System. The assay conditions were similar to those of Fig. 2 except that no mediator dye was present and succinate (0.2 m*M*) and fumarate (2 m*M*) were added to poise redox conditions for optimal turnover of electron flow (*14*).

cesses, a fast one, reflecting the turnover of the reaction center through the cyclic chain and a slower one reflecting a nonphysiological reduction of the reaction center, possibly by the back-reaction (from UQ_{II}) in the chromatophore population lacking cytochrome c_2. In lyophilized particles the half-time of the fast phase (◇), i.e., of the cyclic electron transport system, is around 13 msec, a value very similar to that found in native chromatophores of *Rps. sphaeroides*, Ga.

If extracted particles retaining 1.2 UQ/RC (particles deprived of UQ_Z and to a large extent also of UQ_{II}) were reconstituted with UQ-3 or UQ-10, a differential behavior was observed (Figs. 5, 6). The turnover time of the cycle is markedly slower in UQ-3 reconstituted particles ($t_{1/2}$ = 57 msec) than in UQ-10 reconstituted particles ($t_{1/2}$ = 13 msec). The percentage of fast phase, as compared to slow, does not change appreciably indicating the lack of reconstitution of fast electron flow was not due to a different coupling between cytochrome c_2 and the reaction centers.

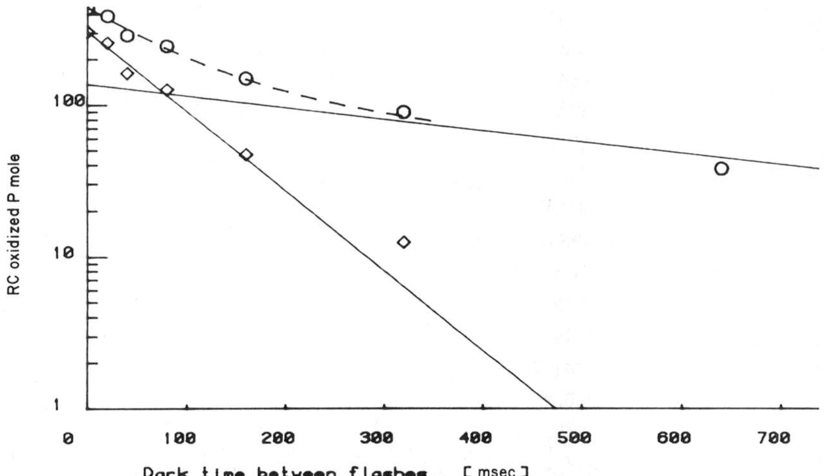

Fig. 5. Relaxation kinetics of $(BChl)_2^+$ rereduction upon multipulse excitation: isooctane extracted chromatophores reconstituted with UQ-3. Conditions as in Fig. 4.

DISCUSSION

The data presented in this paper and previously (*10*, *18*) strengthen the concept of heterogeneity of the ubiquinone complement present in membranes from facultative photosynthetic bacteria. The heterogeneity of the various populations of ubiquinone can be deduced from: (a) differential extractibility; (b) differential roles in specific redox events of the cyclic

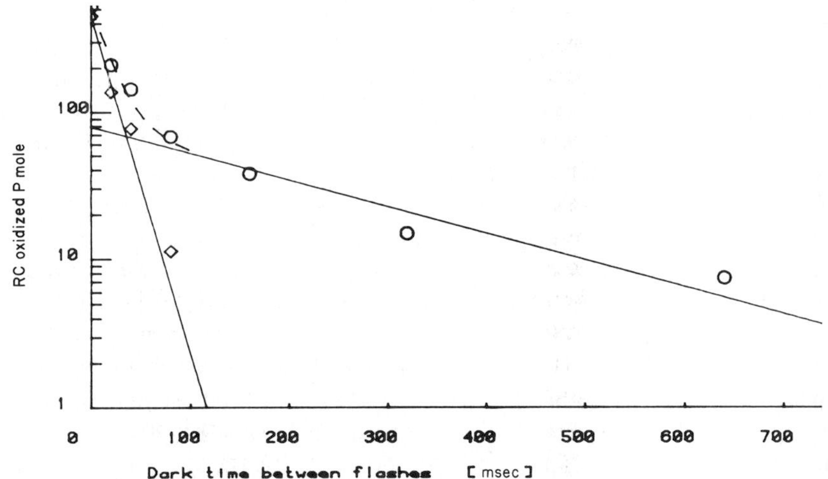

Fig. 6. Relaxation kinetics of $(BChl)_2^+$ rereduction upon multipulse excitation: isooctane extracted chromatophores reconstituted with UQ-10. Conditions as in Fig. 4.

electron-transfer chain, and (c) differential structural requirements for reconstitution of their functions. Experimental monitoring of the presence of UQ_{II} and UQ_Z and estimation of their content in extracted particles allowed us to show quite clearly that, after the detachment of the large homogenous quinone pool, a selective extraction of UQ_Z and UQ_{II} can also be successfully achieved.

Particles with a different ratio of these two quinone molecules could therefore be prepared from lyophilized chromatophores. These preparations were utilized to study the involvement of these specialized ubiquinone molecules in specific redox events. While the presence of UQ_Z is necessary for a fast reduction of flash-oxidized cytochrome c_2 and for an electrogenic event in the b-c_2 segment of the electron-transport chain (Fig. 2), it is not required for cytochrome b reduction. However, the redox event in the b-c_2 segment is dependent upon the presence of the secondary acceptor (UQ_{II}) (*21*). These observations support a linear mechanism of electron transfer in chromatophores, as earlier suggested by Crofts *et al.* (*23*).

In addition, we have shown (*18, 21*) that the function of UQ_{II} as acceptor from the reaction center can be reconstituted by all UQ-homologs; the reconstitution of the pool and of UQ_Z as reductant for cytochrome c_2 requires a quinone homolog with a side chain of at least 3 isoprenoid residues. In contrast, data on reconstitution of electron transfer (following trains of flashes) and ATP synthesis in particles lacking a large amount of UQ_{II} indicate that a quinone with a longer side chain ($\geq$UQ-7) is needed. This suggests a site-specific interaction either at the level of reoxidation of UQ_{II} by cytochrome b or at the level of reduction of UQ_Z by cytochrome b.

These data and other experimental work on the different redox properties of UQ_I, UQ_{II} (*24*), and UQ_Z (*16*) bring us to the conclusion that these quinones are distinct from the rest of the quinone complement. However, their interaction with specific membrane components or presence in a membrane environment that confers a distinctive kinetic and thermodynamic behavior remains to be demonstrated experimentally. Studies in this direction in other membrane systems (particularly in mitochondria), are quite suggestive of the possible existence of quinone-protein moieties catalyzing specific redox events (*25–28*). Whether quinones also function as specific prosthetic groups in energy transducing membranes is still a matter of speculation. The need for protein-quinone association, suggested earlier (*29*) as a means to confer stability to semiquinone redox intermediates was recently questioned (*30*). It should be recalled, however, that the binding of UQ_I to one of the subunits of chromatophore reaction centers (M subunit) was clearly demonstrated utilizing a photoaffinity technique by Marinetti *et al.* (*31*).

While the kinetics of cytochrome *b* and cytochrome c_2 reduction, following a single turnover flash, appear normal after extraction of a large amount (up to 80%) of UQ-10 (*10*), earlier observations indicate that the rate of photophosphorylation decreases in parallel with the amount of UQ-10 extracted by apolar solvents (*7, 32*). It appears, therefore, that a large excess of UQ-10 is required for phosphorylation in continuous light, but not for single turnover electron transfer. In a previous paper (*18*) we examined the dependence of light-induced phosphorylation upon the amount of UQ-10 reincorporated in isooctane extracted vesicles of *Rps. capsulata*, Ala pho^+, and found saturation is reached when the amount of UQ-10 is similar to that present in unextracted chromatophores.

We also found the pattern of binding scalar protons in continuous light, in the presence of antimycin A and an exogenous donor to the reaction center, (reflecting the photoreduction of the endogenous ubiquinone pool) matches the saturation curve observed for photophosphorylation. Both activities are almost saturated at a ratio of UQ-10 per reaction center of about 40 (*18*). This should indicate that the reconstitution of the secondary protonatable pool is necessary for photophosphorylation. Different views, emerging from several experimental approaches, on the role of the quinone pool in energy transducing membranes have appeared in the literature (*33–36*). However, the interpretations suggesting (a) fixed stoichiometries between complexes, (b) no requirement for free ubiquinone for redox activities (*37*), or (c) the lack of interaction between complexes in the absence of a freely diffusible pool of ubiquinone (see Schneider *et al.*, Paper 3, Chapter III, this volume) do not help in reconciling the data obtained in single turnover and in steady state in chromatophores.

It could be tentatively suggested that the pool is needed as an endogenous lipophylic redox buffer at some critical rate-limiting step in coupled electron flow. It should be remembered that a strict dependence of the rate of phosphorylation upon the ambient redox potential has been found in chromatophores (*38*). The possibility of measuring ATP synthesis in trains of differentially spaced flashes by the luciferine–luciferase technique and of simultaneously monitoring the rate of electron transport in pseudosteady state might help clarify this intriguing problem. Experiments along this line are in progress in our laboratory.

ACKNOWLEDGMENTS

Part of this work was supported by Consiglio Nazionale delle Ricerche (Italy), grant nr. 79.01931.04. We would like to thank Dr. G. Venturoli (University of Bologna) for performing the numerical analysis of the relaxation kinetics.

REFERENCES

1. Zannoni, D., and Baccarini-Melandri, A. (1980). *In* "Diversity of Bacterial Respiratory Systems" (D. Knowles, ed.), Vol. II, pp. 183–202. CRC Press, Cleveland, Ohio.
2. Baccarini-Melandri, A., Jones, O. T. G., and Hauska, G. (1978). *FEBS Lett.* **86,** 151–154.
3. Zannoni, D., Prince, R. C., Dutton, P. L., and Marrs, B. L. (1980). *FEBS Lett.* **113,** 289–293.
4. Baccarini-Melandri, A., and Zannoni, D. (1978). *J. Bioenerg. Biomembr.* **10,** 109–139.
5. Melandri, B. A., Baccarini-Melandri, A., San Pietro, A., and Gest, H. (1971). *Science* **174,** 514–516.
6. Takamiya, K., and Dutton, P. L. (1979). *Biochim. Biophys. Acta* **546,** 1–16.
7. Baccarini-Melandri, A., and Melandri, B. A. (1977). *FEBS Lett.* **80,** 459–464.
8. Wraight, C. A. (1977). *Biochim. Biophys. Acta* **459,** 525–531.
9. Vermeglio, A., and Clayton, R. K. (1977). *Biochim. Biophys. Acta* **461,** 159–165.
10. Takamiya, K., Prince, R. C., and Dutton, P. L. (1979). *J. Biol. Chem.* **254,** 11307–11311.
11. Wraight, C. A. (1979). *Photochem. Photobiol.* **30,** 767–776.
12. Halsey, H. D., and Parson, W. W. (1974). *Biochim. Biophys. Acta* **347,** 404–416.
13. Blankenship, R. E., and Parson, W. W. (1979). *Biochim. Biophys. Acta* **545,** 429–444.
14. Bowyer, J. R., Baccarini-Melandri, A., Melandri, B. A., and Crofts, A. R. (1978). *Z. Naturforsch. Teil C* **33,** 704–711.
15. Evans, E. H., and Crofts, A. R. (1974). *Biochim. Biophys. Acta* **357,** 89–102.
16. Prince, R. C., and Dutton, P. L. (1977). *Biochim. Biophys. Acta* **462,** 731–747.
17. Bashford, C. L., Prince, R. C., Takamiya, K., and Dutton, P. L. (1979). *Biochim. Biophys. Acta* **545,** 223–235.
18. Baccarini-Melandri, A., Gabellini, N., Melandri, B. A., Hurt, E., and Hauska, G. (1980). *J. Bioenerg. Biomembr.* **12,** 95–110.
19. Hauska, G. (1977). *FEBS Lett.* **79,** 345–347.
20. Futami, A., Hurt, E., and Hauska, G. (1979). *Biochim. Biophys. Acta* **547,** 583–596.
21. Baccarini-Melandri, A., Gabellini, N., Melandri, B. A., Jones, K. R., Rutherford, A. W., Crofts, A. R., and Hurt, E. (1982). *Arch. Biochem. Biophys.* (in press).
22. Jackson, J. B., Venturoli, G., Baccarini-Melandri, A., Melandri, B. A. (1981). *Biochem. Biophys. Acta* **636,** 1–8.
23. Crofts, A. R., Crowther, D., Bowyer, J., and Tierney, G. V. (1977). *In* "Structure and Function of Energy. Transducing Membranes" (K. Van Dam and B. F. Van Gelder, eds.), pp. 135–139. Elsevier, Amsterdam.
24. Rutherford, A. W., and Evans, M. C. W. (1979). *FEBS Lett.* **104,** 227–229.
25. Yu, C. A., Yu, L., and King, T. (1977). *Biochem. Biophys. Res. Commun.* **78,** 259–265.
26. Yu, C. A., and Yu, L. (1980). *Biochim. Biophys. Acta* **591,** 409–420.
27. Yu, C. A., and Yu, L. (1980). *Biochemistry* **19,** 3579–3585.
28. Ohnishi, T., and Trumpower, B. L. (1980). *J. Biol. Chem.* **255,** 3278–3284.
29. Mitchell, P. (1976). *J. Theor. Biol.* **62,** 327–367.
30. Rich, P. R., and Bendall, D. S. (1980). *Biochim. Biophys. Acta* **592,** 506–518.
31. Marinetti, T. D., Okamura, M. Y., and Feher, G. (1979). *Biochemistry* **18,** 3126–3133.
32. Yamamoto, N., Hatake-yama, K., Nishikawa, K., and Horio, T. (1970). *J. Biochem. (Tokyo)* **67,** 587–595.
33. Kröger, A., and Klingenberg, M. (1973). *Eur. J. Biochem.* **34,** 358–368.
34. Kröger, A., and Klingenberg, M. (1973). *Eur. J. Biochem.* **39,** 313–323.

35. Heron, C., Ragan, C. J., and Trumpower, B. L. (1978). *Biochem. J.* **174,** 791–800.
36. Schneider, H., Lemasters, J. J., Höchli, H., and Hackenbrock, C. R. (1980). *J. Biol. Chem.* **255,** 3748–3756.
37. Ragan, C. I., and Heron, C. (1978). *Biochem. J.* **174,** 783–790.
38. Baccarini-Melandri, A., Melandri, B. A., and Hauska, G. (1979). *J. Bioenerg. Biomembr.* **11,** 1–16.
39. Baccarini-Melandri, A., and Melandri, B. A. (1971). *In* "Photosynthesis and Nitrogen Fixation" (A. San Pietro, ed.), Methods in Enzymology, Vol. 23, pp. 556–561. Academic Press, New York.

Quinone Binding Sites in Reaction Centers from Photosynthetic Bacteria

5

M. Y. OKAMURA
R. J. DEBUS
D. KLEINFELD
G. FEHER

INTRODUCTION

Quinones have long been known to be involved in photosynthetic electron-transfer reactions (*1*). More recently, quinones were found to form an integral part of the reaction center (RC) of photosynthetic bacteria. The reaction center functions as an energy transducer, converting electromagnetic energy (the absorbed photon) into chemical free energy (the production of a charged donor–acceptor pair). The reaction center protein isolated from bacterial membranes with detergent consists of three subunits (L, M, H) with molecular weights (determined by SDS gel electrophoresis) of 21, 24, and 28 kdalton. It contains four bacteriochlorophylls, two bacteriopheophytins, one Fe^{2+}, and two quinones. For recent reviews on reaction centers, see Loach (*2*), Feher and Okamura (*3*), Dutton *et al.* (*4*), and Okamura *et al.* (*5*). For recent reviews on the role of quinones in reaction centers, see Parson (*6*), Bolton (*7*), and Wraight (*8*).

The primary photochemistry leading to charge separation occurs in several steps, shown schematically in Fig. 1 (*9*, 10). The process begins with absorption of a photon to form the excited state of a tetrapyrrole-pigment complex in which electron transfer occurs from a donor species,

Function of Quinones in Energy Conserving Systems

ISBN 0-12-701280-X

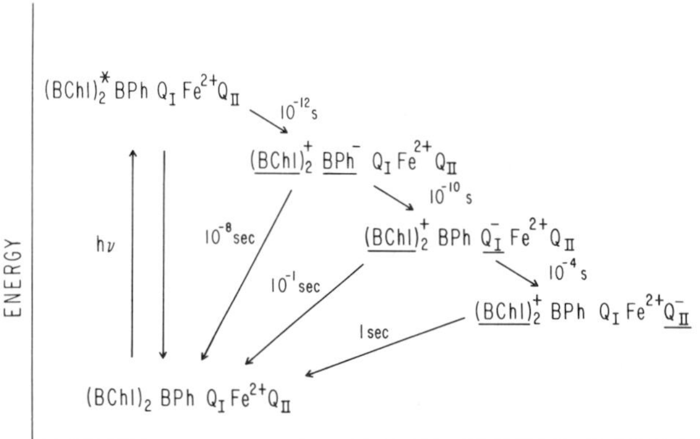

Fig. 1. Schematic representation of the electron-transfer reactions in reaction centers from photosynthetic bacteria. After photon absorption, the electron transfers through a series of reactants that are stabilized against charge recombination for progressively longer periods of time. Charged donor–acceptor species are underlined. Transfer times are given for room temperature and are rounded to the nearest power of 10.

identified to be a bacteriochlorophyll dimer $(BChl)_2$, to an acceptor species, thought to be a bacteriopheophytin molecule (BPh). At this stage, if no other acceptor species is present, decay to the ground state occurs in $\sim 10^{-8}$ sec leading to a loss of photochemical energy. To prevent this loss, another species must accept the electron in a time of less than 10^{-8} sec. The two quinones, Q_I and Q_{II}, perform this task in the bacterial reaction center. The primary quinone, Q_I, reacts with the reduced (BPh) in 10^{-10} sec. Subsequent electron transfer to the second quinone, Q_{II} occurs in 10^{-4} sec. The charge recombinations of $(BChl)_2{}^+$ with $Q_I{}^-$ and $Q_{II}{}^-$ are many orders of magnitude longer, i.e., 10^{-1} sec and 1 sec, respectively.

In reaction centers from *Rhodopseudomonas sphaeroides* R-26, both Q_I and Q_{II} are ubiquinone (UQ-); in *Chromatium vinosum,* Q_I is menaquinone and Q_{II} is UQ-50. In *Rps. sphaderoides,* both Q_I and Q_{II} are magnetically coupled to the Fe^{2+} in what has been called the ferroquinone complex. Under physiological conditions, Q_I accepts only one electron while Q_{II} can accept two electrons. The reactions of Q_I and Q_{II} have been investigated in detail and are discussed elsewhere in this volume.

One important approach in the investigation of quinones in reaction centers was based on the ability to extract them from reaction centers and subsequently reconstitute the reaction centers with the native quinone or a variety of other quinones. The procedure to remove quinones consists either of organic solvent extraction (*11*) or of a gentler method involving incubation with a high concentration of detergent in the presence of the electron-transfer inhibitor, orthophenanthroline (*12*). We shall review some of the properties of quinones bound to reaction centers isolated from *Rps. sphaeroides* R-26 and discuss some of the results from our laboratory. Many of the results are based on removal and reconstitution studies. The main questions we have posed are: Where on the reaction centers are the quinones bound? What is the nature of the quinone binding sites? How do the quinones interact with their neighbors, particularly (BPh) and Fe^{2+}? Some of the results discussed here have been presented previously, and others are preliminary in nature.

EXPERIMENTAL RESULTS

Quinone Assays by Photochemical Activity

The most straightforward but tedious procedure to determine the number of quinones bound to reaction centers is by chemical analysis (*12*). Several faster and simpler spectrophotometric procedures have been developed and are now routinely used.

PHOTOACTIVITY AT CRYOGENIC TEMPERATURES At temperatures less than 80 °K, electron transfer to Q_{II} is not observed to occur. Consequently, photochemical activity of reaction centers at these temperatures can be used to assay for electron transfer to Q_I. Most conveniently, the extent of bleaching of the spectral band at 890 nm, associated with the oxidation of $(BChl)_2$, can be used to monitor the electron transfer.

PHOTOACTIVITY AND CHARGE RECOMBINATION KINETICS AT ROOM TEMPERATURE Another quinone assay utilizes the flash-induced bleaching and recovery of $(BChl)_2$ monitored at 865 nm at room temperature (Fig. 2A). The amplitude of the bleaching is proportional to the amount of Q_I present (Fig. 2B). At room temperature, electron transfer from Q_I to Q_{II} occurs and forms the basis of an assay sensitive to the presence of Q_{II}. It utilizes the difference in charge recombination times between $(BChl)_2^+$ and Q_I^- (~0.1 sec) and between $(BChl)_2^+$ and Q_{II}^- (~1 sec; *13*, *14*). By decomposing the charge recombination kinetics into fast and slow phases, the amounts of photochemically active Q_I and Q_{II} can be determined (Fig. 2C). However, only 80%, rather than 100%, slow phase was observed in reaction centers having an average of 2.0 quinones per reaction center, determined by chemical analysis. In these reaction centers, ~0.2 quinones were not

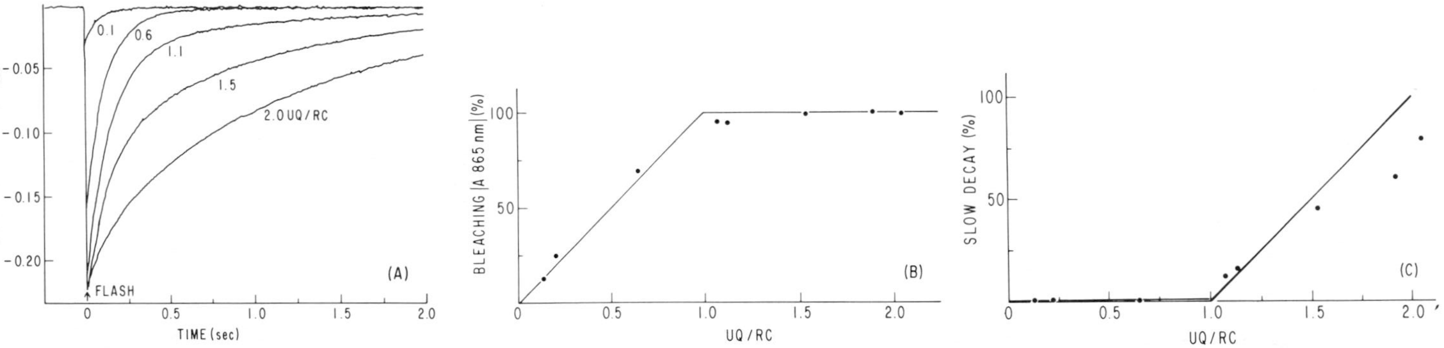

Fig. 2. Room temperature kinetic assay. (A) Formation and decay of $(BChl)_2^+$, monitored at 865 nm, in reaction centers containing different mole fractions of UQ-50. The reaction centers (2.0 μM in 0.025% LDAO, 10 mM Tris-Cl, pH = 8, at T = 23 °C) were given a 0.3 μsec laser flash (λ = 584 nm). The quinone content was determined chemically. (B) Amplitude of bleaching (expressed as percent of the maximum) as a function of mole fraction of bound quinones. The solid curve is the expected result if one quinone binds much more strongly than the second. (C) Percent slow phase as a function of mole fraction quinone. The slow component ($\tau_D \simeq 1.2$ sec) was obtained by plotting the logarithm of the amplitude of the curves in (A).

reduced by Q_I^- after a single flash. These quinones, however, appear to be reduced under continuous illumination of the reaction centers (see Fig. 3). Thus, they may be bound to an altered Q_{II} binding site or to another site nearby.

PHOTOOXIDATION OF CYTOCHROME c An alternate assay, sensitive to the presence of both quinones, utilizes the rapid oxidation of cytochrome *c* by reaction centers under continuous illumination (*13*, *15*, *16*). The number of cytochromes oxidized during the rapid phase of the absorption change at 550 nm equals the number of electrons used in the reduction of the bound quinones. The rise time of the fast phase of cy-

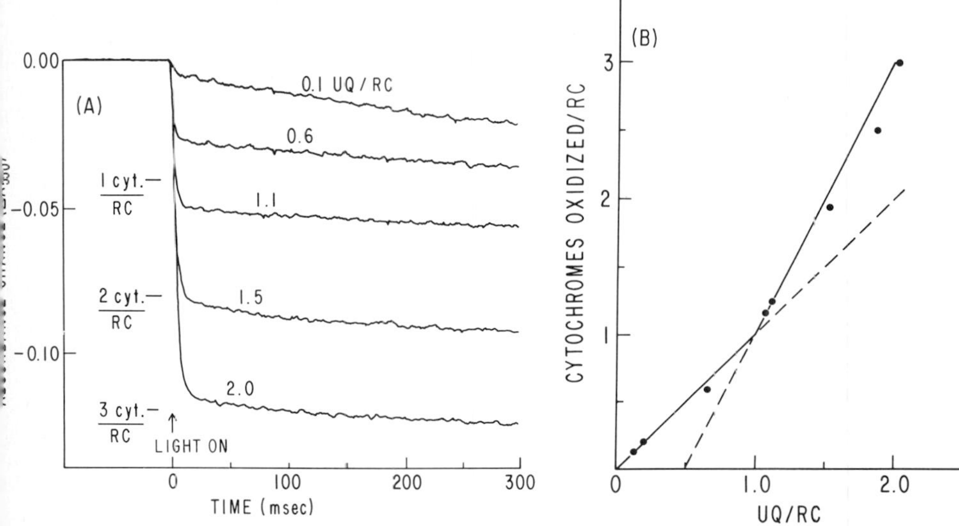

Fig. 3. Cytochrome photooxidation assay. (A) Optical absorption change at 550 nm versus time caused by cytochrome *c* oxidation by reaction centers containing different mole fractions of UQ-50. Reaction centers (2.0 μM in 0.025% LDAO, 10 mM Tris-Cl, pH 8, at T = 23 °C) with 20 μM reduced cytochrome *c* (horse heart) were illuminated with near IR light (Tungsten lamp filtered by 2 cm of H_2O and a Corning CS 2-64 filter giving $I \simeq 800$ mW/cm²) in an optical kinetics spectrometer (*51*). The number of cytochrome *c* oxidized per reaction center was determined from the relation

$$\frac{\text{cyt. } c \text{ oxidized}}{\text{RC}} = 14.4\,(\Delta A_{550\text{ nm}}/A_{802\text{ nm}})$$

The UQ-50 content was determined chemically [Okamura, *et al.* (12)]. It was assumed that at the low detergent concentration used, all quinones are bound to the reaction center. (B) The plot of cytochromes oxidized per RC versus UQ per RC. The solid line is the curve expected if the primary quinone accepts one electron and the secondary quinone accepts two electrons, and the binding of Q_I is much stronger than Q_{II}.

tochrome oxidation is limited by the light intensity. The slower phase (see Fig. 3A) is caused by electrons escaping from reduced acceptors. Since Q_I normally accepts only one electron while Q_{II} accepts two electrons, we expect that reaction centers with one and two bound quinones can oxidize one and three cytochromes, respectively. This was observed experimentally for reaction centers containing a variable number of quinones (see Fig. 3B).

Quinone Structure–Function Relationships

Using the assay procedures described, we investigated the effect of the quinone structure on photochemical activity. Quinones solubilized in ethanol or the detergent lauryldimethylamine oxide (LDAO) were added to reaction centers depleted of quinone. The binding of quinone to the Q_I site was determined from room temperature photoactivity measurements. Estimates of the dissociation constants, $C_{1/2}$, were obtained from the amount of quinone required to produce half the maximum photoactivity. In view of the low solubility of many quinones, it is not clear whether an equilibrium binding constant was measured. Consequently, the data for $C_{1/2}$ should be taken as only qualitatively correct. The charge recombination times, τ_D, were determined from the decay of the optical changes at 865 nm. The results of a survey of various quinones are shown in Table I.

Binding to the *primary* site occurs with high affinity, especially for quinones with side chains (Table I). Generally, addition of a stoichiometric amount of these quinones ($\sim 1\ \mu M$) will result in a significant degree of reconstitution. Reaction center charge recombination times from Q_I vary greatly with the structure of the quinone (Table I). All quinones studied, except for anthraquinone, have a recombination time that increases with increasing temperature. In anthraquinone, the opposite temperature dependence is found, as expected for a thermally activated process. A detailed investigation of the temperature dependence of this reaction is currently underway and may give some insight into the mechanism of electron transfer.

The binding of quinone to the secondary site is weaker than to the primary site. In addition, the requirements for activity appear to be more specific. The observation of biphasic charge recombination kinetics after a saturating light pulse was observed only when ubiquinones with isoprenoid side chains were used to reconstitute reaction centers. An interesting observation is that the ubiquinone analog UQ-0-decyl does not induce slow phase recombination kinetics. The only difference

TABLE I

Binding Constants and Charge Recombination Times for Reaction Centers Reconstituted with Different Quinones[A]

Quinone	Primary site		Secondary site	
	$C_{1/2}{}^{B}$ (μM)	$\tau_D{}^{C}$ (sec)	$C_{1/2}{}^{D}$ (μM)	$\tau_D{}^{E}$ (sec)
UQ-50[F]	<1	0.12	~5	1.2
UQ-10	<1	0.17	~5	1.8
UQ-0-decyl	<1	0.24	>50	—
UQ-0	~10	0.18	>50	—
Vitamin K_1	<1	0.078	>50	—
Menadione	<1	0.16	>50	—
Duroquinone	<1	0.55	>50	—
Anthraquinone[F]	~10	0.01	>50	—

[A] Samples in 10 mM Tris-Cl, 0.025% to 0.2% LDAO, 1 mM EDTA, pH 8.0 $T \simeq 23$ °C.
[B] Concentration of quinone required to produce half-maximum amount of fast phase.
[C] Time required for the amplitude of the fast phase of the 865 nm absorbance change to decay to 1/e of its maximum value.
[D] Concentration of quinone required to produce half-maximum amount of slow phase.
[E] Time required for the amplitude of the slow phase of the 865 nm absorbance change to decay to 1/e of its maximum value.
[F] Quinone added in LDAO solution.

between this quinone and UQ-10 is that it contains a ten-carbon straight chain instead of a ten-carbon isoprenoid chain. Either UQ-0 decyl does not bind to the Q_{II} site or it binds and is not active. To distinguish between these two possibilities, UQ-0-decyl was tested for its ability to compete with UQ-50 for the Q_{II} site. The results (Fig. 4) show that UQ-0-decyl, when added to reaction centers containing UQ-50 in the Q_{II} site, causes a decrease in the amount of slow phase kinetics. This suggests that UQ-0-decyl binds at the Q_{II} site. The lack of photochemical activity with UQ-0-decyl could arise from either kinetic constraints (i.e., a significant increase in the electron-transfer time from Q_I to Q_{II}) or thermodynamic constraints (i.e., stabilization of $Q_I^-Q_{II}$ relative to $Q_IQ_{II}^-$). These results indicate the highly specific requirements imposed on the structure of Q_{II}.

Location of the Quinone Binding Sites

The location of the Q_I binding site has been determined by Marinetti *et al.* (*17*) using the technique of photoaffinity labeling. The photoaffinity label [H^3]2-azido-anthraquinone was added to quinone-depleted

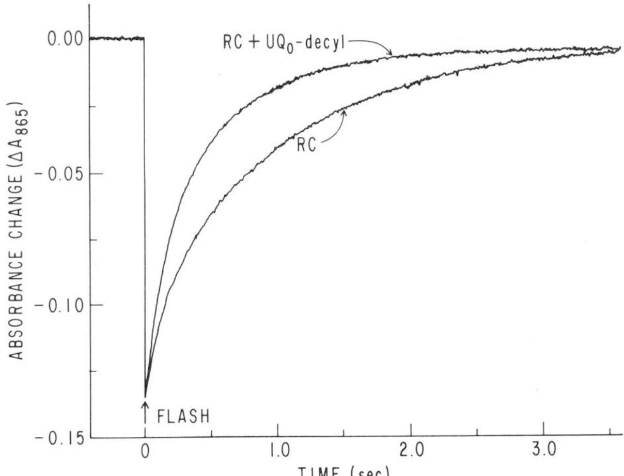

Fig. 4. The effect of UQ-0-decyl on the slow phase of the recombination kinetics. The lower curve (RC) was obtained as in Fig. 2A for reaction centers (1.1 μM) containing ~1.8 UQ-50. The upper curve shows the effect of adding UQ-0-decyl (17 μM) to these reaction centers. The increase in the decay rate indicates an inhibition of electron transfer from Q_I to Q_{II} believed to be caused by the successful competition of the photochemically inactive UQ-0-decyl for the Q_{II} binding site.

reaction centers and found to restore photochemical activity. Excess azido-anthraquinone was added to saturate the primary binding site and excess quinone was washed off by DEAE cellulose chromatography. Since at low detergent concentrations Q_I exchanges only slowly with the external medium, reaction centers containing anthraquinone in the Q_I binding site were obtained. Photolysis of this sample, followed by SDS polyacrylamide gel electrophoresis, showed a specific labeling of the M subunit (see Fig. 5), indicating that Q_I binds on or near M.

Photoaffinity labeling of the Q_{II} binding site has not been accomplished due to the more stringent structural requirements for the secondary quinone. Instead, an immunological approach has been used to locate this site. Antibodies against specific RC subunits were isolated using affinity chromatography (*18*). All of these were shown by radioimmunoassay to bind to reaction centers. These antibodies were tested to see if they could block electron transfer from Q_I to Q_{II} using the kinetic assay as described in the previous section (*19*). In order to prevent precipitation of the RC-antibody complex, Fab fragments made by papain digestion were used. The conditions of the assay (0.1% LDAO, 1 μM UQ-50, 0.2 μM RC) were chosen so that in the absence of antibodies 50% of the reaction centers showed the slow charge recombination kinetics due to

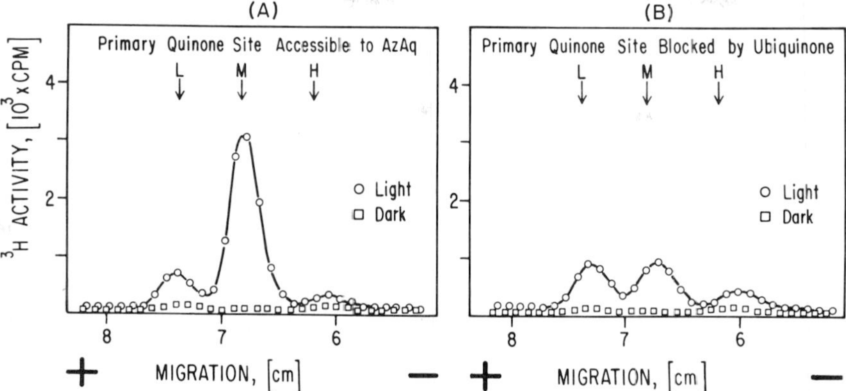

Fig. 5. SDS polyacrylamide gel electrophoretograms of reaction centers reconstituted and labeled with 2-azido[^{3}H]anthraquinone (AzAq). (A) Reaction centers depleted of UQ-50 and reconstituted with azido anthraquinone. (B) Reaction centers with UQ-50 in the Q_I site treated as in A. [From Marinetti *et al.* (*17*), used with permission.]

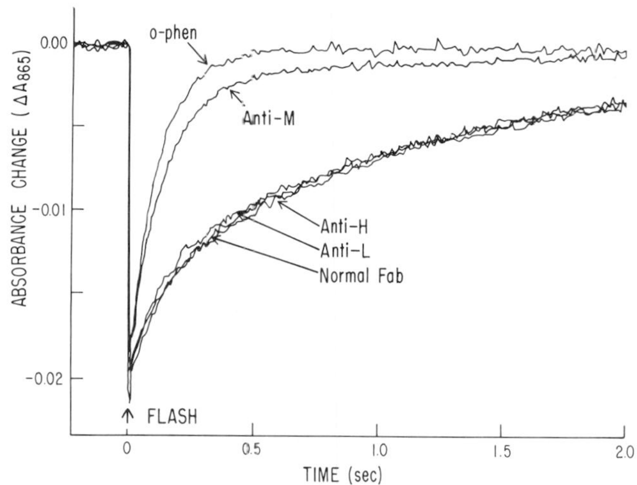

Fig. 6. Antibody inhibition of Q_{II} activity as measured by the kinetics of charge recombination (see Figs. 2A and 4). Fab fragments from affinity purified antibodies (3 μM) were added to reaction centers (0.2 μM) in the presence of 1 μM UQ-50 (0.1% LDAO, 10 mM Tris, 14 μM EDTA, pH 8.0). Fab made from normal antibodies was used as a control. Addition of anti-M Fab resulted in an increase in the charge recombination rate similar to that produced by 5 mM *o*-phenanthroline, a known inhibitor of electron transfer to Q_{II}. The simplest explanation of these results is that anti-M Fab inhibits the binding of Q_{II}.

electron transfer from Q_{II}^- to $(BChl)_2^+$. Addition of antibodies against the M subunit greatly decreased the amount of the slow component (see Fig. 6) while antibodies against the L or H subunits did not. This was true for all Fab preparations from different antisera (five M, three L, and two H antisera were tested). The inhibition of electron transfer occurred whether UQ-10 or UQ-50 was used as the secondary quinone. The simplest interpretation of these results is that binding of anti-M Fab at the Q_{II} site inhibits either binding or activity of the secondary quinone. Thus, both Q_I and Q_{II} appear to be bound on the M subunit. Further experiments are in progress to verify this. Although antibodies to the H subunit had little or no effect on the amplitude of the slow component, one of the two anti-H antisera produced a 20% increase in the rate of the slow phase. This suggests that the environment of Q_{II} is affected by the H subunit.

Further evidence of the possible role of the H subunit is provided by recent experiments involving dissociation and reconstitution of the H subunit. When the H subunit was dissociated from the reaction center, the remaining LM-pigment complex retained the primary photochemical reactants (including Fe^{2+}) and exhibited full photochemical activity (*3*, *20*). Electron transfer from Q_I in LM differs from that in reaction centers as shown by the photooxidation of cytochrome *c* (Fig. 7). Only one cytochrome *c* was rapidly oxidized by LM, even in the presence of excess UQ-50. When the H subunit was reconstituted with LM, the amount of rapidly oxidized cytochrome *c* was increased (*21*). This indicates that either the binding or the activity of Q_{II} is altered when the H subunit is removed.

Interaction of Quinone with Fe^{2+}

EPR The observation of a broad EPR signal (*22–24*) from the reduced primary acceptor (Fig. 8a) and the presence of Fe^{2+} in the reaction center led to the concept of an iron-quinone complex in which the electron is primarily localized on the quinone. The large width of the EPR signal is due to the magnetic interaction between Q_I^- and Fe^{2+}. This assignment is supported by several different observations (*3*) including the important observation of photochemical activity in iron-free reaction centers. In these preparations a new EPR signal (Fig. 8b) was observed (*25*) and identified as arising from a UQ anion radical (*26*). Subsequent work showed a broad EPR signal is also observed when Q_{II} is singly reduced (*27–29*), indicating that a similar magnetic interaction

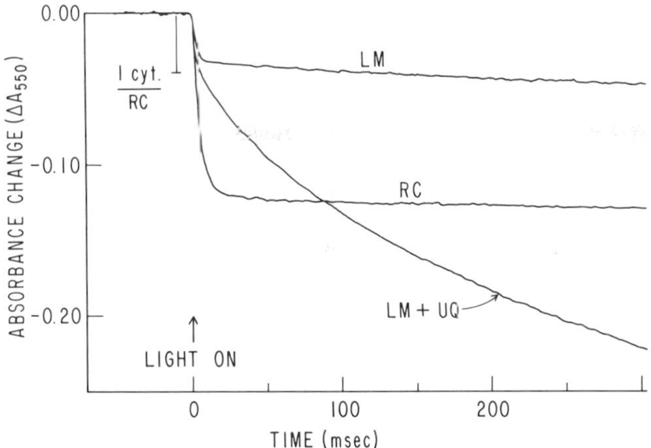

Fig. 7. Cytochrome photooxidation assay of the LM subunit. The LM subunits were prepared as described by Feher and Okamura (*3*). The cytochrome photooxidation assay was performed as in Fig. 2 and shows only one cytochrome *c* per reaction center oxidized during the fast phase (2.0 μM LM, 20 μM cytochrome *c*, 0.18% cholate, 0.2% deoxycholate, 1 mM EDTA, 10 mM Tris-Cl, pH = 8). When excess UQ-50 (30 μM) was added in deoxycholate, still only ~1 cytochrome *c* per reaction center was oxidized in the fast phase, although an increase in the slower rate of cytochrome oxidation was observed. The increase suggests that removal of the H subunit exposes Q_I^- to the exogenous UQ-50 which acts as an electron acceptor. The cytochrome photooxidation kinetics for reaction centers in LDAO is shown for comparison.

exists between Q_{II}^- and Fe^{2+}. A possible role of iron is to facilitate electron transfer from Q_I to Q_{II}, the "Iron-wire Hypothesis" (*12*). Support for this hypothesis has come from the observation by Blankenship and Parson (*14*) that reaction centers depleted of iron cannot transfer electrons from Q_I to Q_{II}. It should be noted that their procedure used to remove Fe^{2+} and our procedure used to dissociate H are very similar. Consequently, the loss of electron transfer from Q_I to Q_{II} which they observed may have been caused by loss of H rather than of Fe^{2+} (*21*).

MAGNETIC SUSCEPTIBILITY From the temperature dependence of the magnetization of reaction centers, the parameters (e.g., *g*-value, crystalline field splittings) characterizing the electronic structure of the paramagnetic Fe^{2+} were deduced (*30*). Since these parameters depend on the nature of the Fe^{2+} ligands, determination in reaction centers containing varying amounts of quinones can be used to determine whether the quinones are coordinated to the Fe^{2+}.

No changes in magnetization were observed between reaction cen-

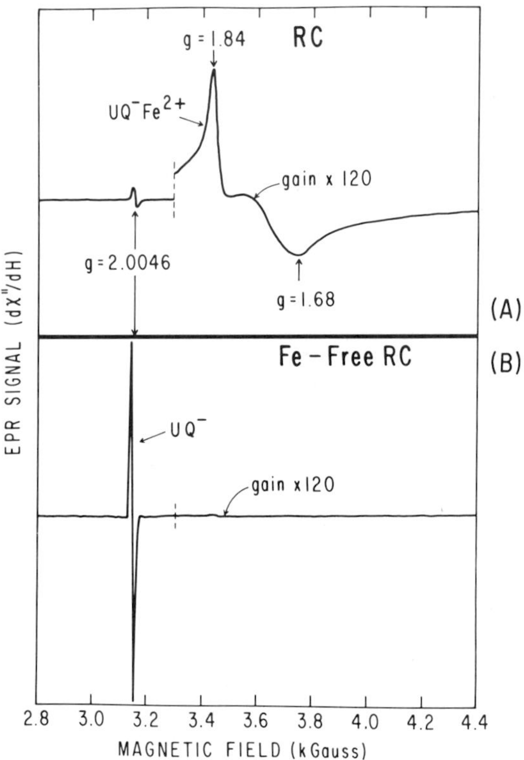

Fig. 8. EPR spectra of the reduced primary quinone in RCs (A) with reaction centers Fe and (B) without Fe. Both samples were reduced with 50 m*M* dithionite, 100 m*M* Tris-Cl, pH = 8, and quickly frozen to cryogenic temperatures. The Fe-free sample was prepared as described by Feher and Okamura (*3*). The spectrometer gain and microwave power were increased above 3.3 kG to observe the broad signal. The *g*-value (2.0046) of the narrow signal was obtained from a slower scan and is identical to that of $UQ\text{-}50^-$ (*24*). The broad EPR signal is thought to be due to Q_I^- magnetically coupled to Fe^{2+}; $\nu = 8.85$ GHz, $T = 2.1$ °K.

ters containing one and two quinones, indicating that Q_{II} is not coordinated to Fe^{2+}. However, changes occurred when Q_I was removed that can be ascribed to either a loss of ligation of Q_I to Fe^{2+} or to a conformational change of the protein. One argument in favor of a conformational change is that EPR data indicate that the interaction of Fe^{2+} with Q_I is approximately the same as with Q_{II}. Since we argued previously that Q_{II} is not a ligand of Fe^{2+}, we infer that neither is Q_I.

From the magnetization measurements of reduced reaction centers (i.e., $Q_I^-Fe^{2+}$), the magnetic exchange interaction, J, between the spin on Q_I^- and the spin on the Fe^{2+} was obtained (*30*). The small value of the

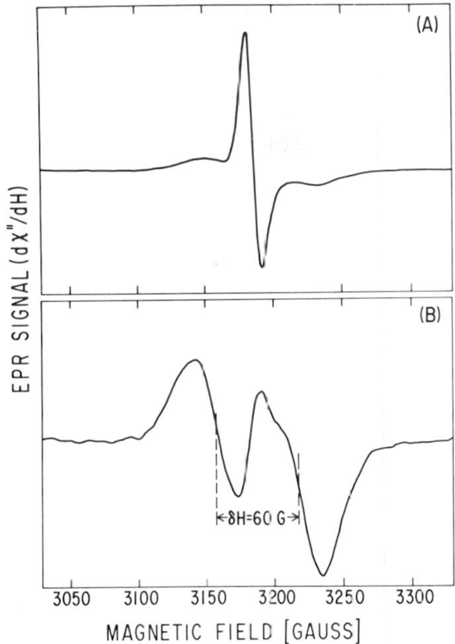

Fig. 9. EPR spectrum of reaction centers reconstituted with menaquinone (vitamin K_1) after rapid photoreduction at low redox potential. (A) At low power, (10^{-6} W, $T = 2.1$ °K, $\nu = 8.90$ GHz) the narrow singlet signal of reduced (BPh) appears. Additional lines appear on the wings. At high (10^{-2} W, $T = 2.1$ °K, $\nu = 8.90$ GHz) microwave power (B) the singlet signal is greatly reduced due to microwave saturation and the doublet signal predominates. The doublet signal is believed to be caused by an interaction between the (BPh) radical and the reduced menaquinone-Fe^{2+} complex. [From Okamura *et al.* (*43*), used with permission.]

interaction energy ($|J| < 1$ cm^{-1}) compared with those of other model compounds (*31*, *32*) provides an independent indication that Q_I is not coordinated to Fe^{2+}.

MOSSBAUER SPECTROSCOPY Reaction centers with different numbers of quinones were prepared from bacteria grown on a synthetic medium enriched in Fe^{57}. The two main quantities that are measured by Mossbauer spectroscopy are the isomer shift, δ, and the quadrupole splitting parameter, ΔE_Q. These quantities provide a sensitive measure of the iron environment. The values of δ and ΔE_Q were found to remain essentially unchanged when either one or both quinones were removed (*33*). These results indicate that either the quinones are not ligands of the iron, or that their removal is accompanied by a replacement with a similar ligand.

EXAFS The nature of the ligands around the Fe^{2+} were examined by extended x-ray fine-structure spectroscopy (EXAFS) (*34–36*). In this technique, the x-ray absorption spectrum close to the Fe^{2+} absorption edge (~7.2 keV) is measured. The spectrum shows oscillations caused by the scattering of photoelectrons from neighboring ligands. These oscillations provide a sensitive probe for investigating the number and nature of the ligands around Fe^{2+}. EXAFS spectra, obtained from reaction centers with one or two quinones showed no difference in number of ligands and average ligand distance from Fe^{2+} (<0.02 Å). In addition, no effect on the EXAFS spectrum was observed when orthophenanthroline was added. These results were interpreted to indicate that neither Q_{II} nor orthophenanthroline coordinates to Fe^{2+}. However, the EXAFS spectrum of reaction centers with no bound quinones showed a significant decrease ($\sim 12\%$) in amplitude. This reduction in amplitude may indicate Q_I coordination to Fe^{2+} or may be explained by a conformational change of the reaction center protein. In view of other experimental evidence, we favor the latter explanation.

Interaction of Quinone and Bacteriopheophytin

The primary quinone accepts an electron from reduced (BPh) in 10^{-10} sec (*37*, *38*). The rapid rate of this reaction implies a strong interaction between these species. The first evidence for an interaction between (BPh) and Q_I came from observation of an electrochromic shift in the optical spectrum of (BPh) when Q_I was reduced (*39–41*). Electrochromic shifts in the spectrum of (BPh) were also seen when Q_{II} was reduced (*41*). Evidence for the interaction between (BPh) and Q_I was first obtained in reaction centers from *C. vinosum,* where a doublet EPR signal was observed after photoreduction of (BPh) at low redox potential (*42*). Under the same conditions a doublet EPR signal was not observed in native reaction centers of *Rps. sphaeroides*. However, a doublet signal appeared (see Fig. 9) when the native Q_I (UQ-50) was replaced by Vitamin K_1, similar to the primary quinone in *C. vinosum* (*43*). The doublet splitting was assigned to an exchange interaction ($|J| = 60$ G) between $(BPh)^-$ and the MQ^-Fe^{2+} complex. This interaction arises from an overlap of electronic wavefunctions and is directly related to the rate of electron transfer between $(BPh)^-$ and Q_I^- according to a theory of thermally activated electron tunneling (*44*). The exchange interaction can be related to the distance between $(BPh)^-$ and Q_I^-. Using a simple model, this distance was estimated to be 7 to 10 Å (*43*). Other workers (*45*, *46*) have estimated similar distances.

Nature of the Quinone Binding Site

OPTICAL ABSORPTION SPECTROSCOPY One of the simplest probes of the quinone environment in reaction centers is its optical absorption spectrum. In reaction centers, optical difference spectra (reduced minus oxidized) of both Q_I and Q_{II} show an absorption peak near

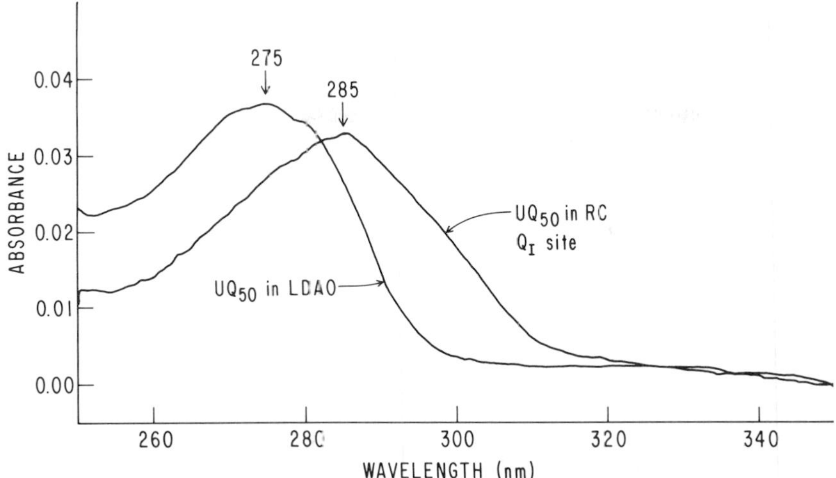

Fig. 10. Optical absorbance changes of UQ-50 in reaction centers and in detergent solution. The optical spectra were obtained with a CARY 14R spectrophotometer. The spectra of reaction centers (2.0 μM) in 1.0 ml buffer (0.025% LDAO, 10 mM Tris-Cl, pH 8) were taken before and after addition of a stoichiometric amount of quinone (6.5 μl of 320 μM UQ-50 in 10% LDAO). The reference compartment also contained a 2.0 μM reaction center solution in buffer. Also added was an equal amount of detergent solution with no quinone to eliminate differences due to concentration changes or detergent effects. The spectra before and after quinone addition were stored on a Nicolet 1180 computer and subtracted to give the spectra shown above. Each spectrum represents the average of four scans. The maximum absorbance in the sample was 0.6. The gains were adjusted so that the spectral resolution was always better than 1.2 nm. The spectrum of UQ-50 in LDAO was obtained by adding the same amount of UQ-50 to the buffer solution.

450 nm (*27*, *40*, *47*, *48*). This absorption peak is characteristic of the ubisemiquinone anion radical which, in methanol, has a peak near 445 nm (*49*, *50*). The similarity between the ubisemiquinone spectrum in reaction centers and *in vitro* provides another argument against quinone ligation to Fe^{2+}, since direct coordination would be expected to cause a larger change in the semiquinone spectrum. However, the UV difference spectra of the quinones in reaction centers are shifted to longer wavelengths (*47*, *48*). We reinvestigated these shifts by taking the difference spectrum of quinone-depleted reaction centers before and after addition of quinone. The spectrum of a stoichiometric amount of UQ-50 added to reaction centers is shown in Fig. 10, along with the spectrum of the same amount of UQ-50 added to buffer.

The spectrum of the quinone in the reaction center shows a long wavelength shift to 285 nm from ~275 nm in buffer. In addition, a sug-

TABLE II

Absorbance Maxima for UQ-50 in Various Solvents

Solvent	λ_{max} (nm)
Cyclohexane	272 ± 1
t-Butanol	274
Isopropanol	275
Ethanol	275
Acetic acid	275
1% LDAO in H_2O	276
Q_I in RC	285

gestion of a shoulder appears in the 300 nm region. Spectra of Q_{II} in the reaction center could not be easily obtained since an excess of quinone is required to saturate the site. However, preliminary results indicate that the Q_{II} spectrum shows a similar shift. When Vitamin K_1, a menaquinone, was added to reaction centers, small shifts to longer wavelengths were observed in all peaks. In addition, a marked shoulder appeared around 288 nm. To explain these spectral changes, measurements were made on UQ-50 dissolved in various solvents (Table II). The quinone absorption peak shifted to longer wavelengths as the polarity of the solvent was increased; this is indicative of a $\pi \rightarrow \pi^*$ transition (*51*). However, in none of the solvents was the shift nearly as large as that observed in reaction centers. This rules out a simple solvent effect. One possible explanation for the large shift is an interaction of the quinone with either (BPh), Fe^{2+}, or aromatic amino acid residues in the protein. Another possibility is the presence of large electric fields, perhaps resulting from a net charge or fixed dipoles, that stabilizes the polar excited state.

ENDOR Another spectroscopic probe of the quinone in the reaction center is electron nuclear double resonance (ENDOR) spectroscopy. ENDOR gives information about the spin density distribution of paramagnetic molecules. ENDOR spectra were obtained of ubisemiquinone in the Q_I site of reaction centers after removal of the Fe^{2+} (*52*). These spectra are similar but not identical to ENDOR spectra of ubisemiquinone obtained *in vitro* (*53*). It is not yet known whether these differences are due to specific features of the quinone binding site in the reaction center. Theoretical studies of the spin densities should help to provide an explanation (*54*).

SUMMARY AND DISCUSSION

We have briefly summarized some of the spectroscopic and structural properties of the two quinones bound to bacterial reaction centers. The primary quinone Q_I is tightly bound and is close to the M subunit as well as to the (BPh) that serves as its immediate electron donor. The distance between Q_I and (BPh), estimated from EPR measurements, is 7 to 10 Å. The secondary quinone Q_{II} is more weakly bound to the M subunit. Preliminary data suggests that the H subunit is also involved in providing an active binding site. Both Q_I and Q_{II}, as shown by EPR spectroscopy, are magnetically coupled to the Fe^{2+} and must, therefore, be in close proximity. However, magnetic susceptibility, EXAFS, Mossbauer, and optical absorption spectroscopy indicate that neither quinone is directly coordinated to the Fe^{2+}. One possible structure of the ferroquinone complex involves close association of both quinones with an aromatic ligand or ligands of Fe^{2+}, for example, tryosine or histidine residues. Such an association may explain the large wavelength shift observed in the UV absorption spectrum of both Q_I and Q_{II}. The unusual stability of the semiquinone state of Q_I, and to some extent Q_{II}, could arise from an aprotic environment which prevents protonation leading to two-electron reduction. Alternatively, the stabilization may be caused by the presence of fixed positive charges that may be provided by Fe^{2+}.

In conclusion, the bacterial reaction center provides a good example of a quinone-containing protein complex. Understanding the structure and function of quinone binding sites in reaction centers may lead to a better understanding of other quinone-protein complexes in biology.

ACKNOWLEDGMENTS

We would like to thank Chang-An Yu for providing samples of UQ-10 and UQ-0-decyl, Ed Abresch for preparing the reaction centers, Roger Isaacson for assistance in many of the measurements, and Gunars Valkirs for preparation and characterization of the antibodies. This work was supported by grants from the NSF (PCM 78-13699), and the NIH (GM 13191).

REFERENCES

1. Arnon, D. I., and Crane, F. L. (1965). *In* "Biochemistry of Quinones" (R. A. Morton, ed.), pp. 433–459. Academic Press, New York.
2. Loach, P. A. (1976). *Prog. Bioorg. Chem.* **4,** 89–192.

3. Feher, G., and Okamura, M. Y. (1978). *In* "The Photosynthetic Bacteria" (R. K. Clayton and W. R. Sistrom, eds.), pp. 349–386. Plenum, New York.
4. Dutton, P. L., Prince, R. C., and Tiede, P. M. (1978). *Photochem. Photobiol.* **28,** 939–949.
5. Okamura, M. Y., Feher, G., and Nelson, N. (1982). *In* Photosynthesis, Energy Conversion by Plants and Bacteria" (Govindjee, ed.). Academic Press, New York. To be published.
6. Parson, W. W. (1978). *In* "The Photosynthetic Bacteria" (R. K. Clayton and W. R. Sistrom, eds.), pp. 455–470. Plenum, New York.
7. Bolton, J. R. (1978). *In* "The Photosynthetic Bacteria" (R. K. Clayton and W. R. Sistrom, eds.), pp. 419–430. Plenum, New York.
8. Wraight, C. A. (1979). *Photochem. Photobiol.* **30,** 767–776.
9. Blankenship, R. E., and Parson, W. W. (1979). *Annu. Rev. Biochem.* **47,** 635–653.
10. Parson, W. W., and Ke, B. (1981). *In* "Integrated Approach to Plant and Bacterial Photosynthesis" (Govindjee, ed.). Academic Press, New York.
11. Cogdell, R. J., Brune, D. C., and Clayton, R. K. (1974). *FEBS Lett.* **45,** 344–347.
12. Okamura, M. Y., Isaacson, R. A., and Feher, G. (1975). *Proc. Natl. Acad. Sci. U.S.A.* **72,** 3491–3495.
13. Clayton, R. K., and Yau, H. F. (1972). *Biophys. J.* **12,** 867–881.
14. Blankenship, R. E., and Parson, W. W. (1979). *Biochim. Biophys. Acta* **545,** 429–444.
15. Parson, W. W. (1968). *Biochim. Biophys. Acta* **153,** 248–259.
16. Clayton, R. K., Fleming, H., and Szuts, E. Z. (1972). *Biophys. J.* **12,** 46–63.
17. Marinetti, T. D., Okamura, M. Y., and Feher, G. (1979). *Biochemistry* **18,** 3126–3133.
18. Valkirs, G. and Feher, G. (1981). *Biophys. J.* **33,** 18a.
19. Debus, R. J., Valkirs, G., Okamura, M. Y., and Feher, G. To be published.
20. Okamura, M. Y., Steiner, L. A., and Feher, G. (1974). *Biochemistry* **13,** 1394–1403.
21. Debus, R. J., Okamura, M. Y., and Feher, G. (1981). *Biophys. J.* **33,** 19a.
22. McElroy, J. D., Feher, G., and Mauzerall, D. C. (1970). *Biophys. Soc. Abstr.* **10,** 204.
23. Feher, G. (1971). *Photochem. Photobiol.* **14,** 373–388.
24. Dutton, P. L., Leigh, J. S., and Reed, D. W. (1973). *Biochim. Biophys. Acta* **292,** 654–664.
25. Loach, P. A., and Hall, R. L. (1972). *Proc. Natl. Acad. Sci. U.S.A.* **69,** 786–790.
26. Feher, G., Okamura, M. Y., and McElroy, J. D. (1972). *Biochim. Biophys. Acta* **267,** 222–226.
27. Wraight, C. A. (1977). *Biochim. Biophys. Acta* **459,** 525–531.
28. Wraight, C. A. (1978). *FEBS Lett.* **93,** 283–288.
29. Okamura, M. Y., Isaacson, R. A., and Feher, G. (1978). *Biophys. J.* **21,** 8a.
30. Butler, W. F., Johnston, D. C., Shore, H. B., Fredkin, D. R., Okamura, M. Y., and Feher, G. (1980). *Biophys. J.* **32,** 967–992.
31. Kessel, S. L., Emberson, R. M., Debrunner, P. G., and Hendrickson, D. N. (1980). *Inorg. Chem.* **19,** 1170–1178.
32. Coffman, R. E., and Buettner, G. R. (1979). *J. Phys. Chem.* **83,** 2387–2392.
33. Boso, B., Debrunner, P., Okamura, M. Y., and Feher, G., (1981). *Biochim. Biophys. Acta* **638,** 173–177.
34. Eisenberger, P. M., Okamura, M. Y., and Feher, G. (1980). *Fed. Proc., Fed. Am. Soc. Exp. Biol.* **39,** 1802.
35. Bunker, G., Stern, E. A., Blankenship, R. E., and Parson, W. W., (1982). *Biophys J.,* **37,** 539–551.
36. Eisenberger, P. M., Okamura, M. Y., and Feher, G. (1982). *Biophys J.,* **37,** 523–538.

37. Parson, W. W., Clayton, R. K., and Cogdell, R. J. (1975). *Biochim. Biophys. Acta* **387,** 265–278.
38. Fajer, J., Brune, D. C., Davis, M. S., Forman, A., and Spaulding, L. D. (1975). *Proc. Natl. Acad. Sci. U.S.A.* **72,** 4956–4960.
39. Loach, P. A. (1966). *Biochemistry* **5,** 592–600.
40. Clayton, R. K., and Straley, S. C. (1972). *Biophys. J.* **12,** 1221–1234.
41. Verméglio, A., and Clayton, R. K. (1977). *Biochim. Biophys. Acta* **461,** 159–165.
42. Tiede, D. M., Prince, R. C., and Dutton, P. L. (1976). *Biochim. Biophys. Acta* **449,** 447–467.
43. Okamura, M. Y., Isaacson, R. A., and Feher, G. (1979). *Biochim. Biophys. Acta* **546,** 394–417.
44. Hopfield, J. J. (1977). *In* "Electrical Phenomena at the Biological Membrane Level" (E. Roux, ed.), pp. 471–492. Elsevier, Amsterdam.
45. Peters, K., Avouris, P., and Rentzepis, P. M. (1978). *Biophys. J.* **23,** 207–217.
46. Gast, P., and Hoff, A. J. (1979). *Biochim. Biophys. Acta* **548,** 520–535.
47. Slooten, L. (1972). *Biochim. Biophys. Acta* **276,** 208–218.
48. Vermeglio, A. (1977). *Biochim. Biophys. Acta* **459,** 516–524.
49. Land, E. J., Simic, M., and Swallow, A. J. (1971). *Biochim. Biophys. Acta* **226,** 239–240.
50. Bensasson, R., and Land, E. J. (1973). *Biochim. Biophys. Acta* **325,** 175–181.
51. Morton, R. A. (1965). *In* "Biochemistry of Quinones" (R. A. Morton, ed.), pp. 23–66. Academic Press, New York.
52. Okamura, M. Y., Debus, R. J., Isaacson, R. A., Feher, G. (1980). *Fed. Proc., Fed. Am. Soc. Exp. Biol.* **39,** 1802.
53. Das, M. R., Conner, H. D., Leniart, D. S., and Freed, J. H. (1970). *J. Am. Chem. Soc.* **92,** 2258.
54. Coker, A., Mishra, K. C., and Das, T. P. (1980). *Fed. Proc., Fed. Am. Soc. Exp. Biol.* **39,** 1802.
55. McElroy, J. D., Mauzerall, D. C., and Feher, G. (1974). *Biochim. Biophys. Acta* **333,** 261–277.

Two Small Peptides from Complex II and Their Role in the Reconstitution of Q Reductase Activity and in the Binding of TTF

6

BRIAN A. C. ACKRELL
RONA R. RAMSAY
EDNA B. KEARNEY
THOMAS P. SINGER
GORDON A. WHITE
G. DENIS THORN

INTRODUCTION

During recent years unprecedented progress has been made in our understanding of the elusive problems of how succinate dehydrogenase is structurally bound and functionally linked to the rest of the respiratory chain. Until a few years ago the reconstitution of succinate oxidation via the respiratory chain from the soluble dehydrogenase had only been accomplished with crude preparations of succinate dehydrogenase and alkali-treated samples of the entire inner membrane. This precluded identification of the components needed for reconstitution and, therefore, involved in the functional binding of the enzyme to the membrane. A major advance toward the identification of these components came from a new procedure for isolation of succinate dehydrogenase (*1*). This procedure yielded a homogeneous enzyme capable of being fully reinserted into

Function of Quinones in Energy Conserving Systems

ISBN 0-12-701280-X

the inner membrane or into Complex II from which the succinate dehydrogenase had been extracted, with restoration of all the catalytic properties of the native system. Although still membranous, Complex II is far simpler in composition than the Keilin–Hartree preparations that had been used by others in reconstitution studies. Thus, the door was open to the positive identification of the factors required for recombination of dehydrogenase with membrane components to yield a catalytically competent, reconstituted system. Additional important steps were the identification of peptide components of Complex II and development of antibodies to succinate dehydrogenase and its large subunit (*2*, *3*). These developments led directly to the identification of two small peptides in Complex II required for (1) binding the dehydrogenase to the membrane, (2) electron transport to ubiquinone (Q), and (3) inhibition of succinate oxidation by thenoyltrifluoroacetone (TTF), as summarized below.

EXPERIMENTAL RESULTS

Isolation of Two Small Peptides from Complex II and Their Recombination with Succinate Dehydrogenase

Complex II contains four major polypeptides in constant and equimolar amounts (*2*), in addition to cytochrome *b* (*4*) and contaminating peptides from Complex III (*2*). Two of these, denoted as C_{II-1} and C_{II-2}, are the 70,000 and 30,000 molecular weight subunits of succinate dehydrogenase (*5*); the function of the other two, with molecular weights of 13,500 and 7,000, respectively (*2*), was only recently elucidated. We reported (*6*) the extraction of C_{II-3} and C_{II-4} from Complex II with 0.5% (w/v) Triton X-100 or deoxycholate, after previous treatment with 2 *M* sodium perchlorate to extract the succinate dehydrogenase. Figure 1A shows the SDS-polyacrylamide gel scan of a typical preparation of Complex II (*2*); the gel pattern of the detergent extract (Fig. 1B) shows that the peptides C_{II-3} and C_{II-4} were essentially free of contaminating peptides. Cytochrome *b* was present at low and variable concentrations and the preparation always contained substantial amounts of phospholipids, predominantly lecithin, but no ubiquinone (Table I).

It is known (*8*) that the incorporation of succinate dehydrogenase into alkali-extracted inner membrane (ETP) preparations results in increased activity in the succinate-phenazine methosulfate (PMS) reaction, the generation of succinate–Q reductase activity, and the concomitant loss of the "low K_m" ferricyanide reductase activity (*9*). This activity depends on the access of ferricyanide to the "HiPIP" iron–sul-

Fig. 1. Gel scans of (A) Complex II and (B) a preparation of peptides C_{II-3} and C_{II-4} following SDS-polyacrylamide gel electrophoresis according to Swank and Munkres (*7*).

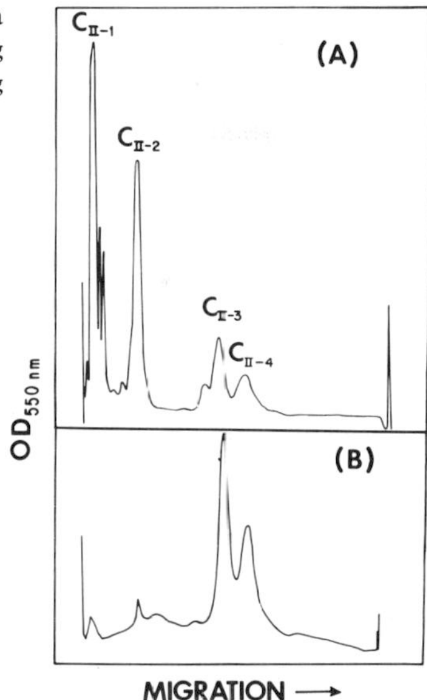

fur cluster (*10*), which is blocked when the dehydrogenase is incorporated into the membrane. Table II shows that the same activities appeared or disappeared on adding a limiting amount of soluble succinate dehydrogenase, prepared by our method (*1*), to the purified preparations of peptides C_{II-3} and C_{II-4} (*6*). The ratio of reactivities of the recombined sample with PMS and Q was higher than in untreated Complex II, but was nearly the same as in Complex II treated with the same concentration of Triton X-100 present in the peptide sample. The slightly higher value (2.2 versus 1.7) may reflect the presence, under these experimental conditions, of some reconstitutively inactive en-

TABLE I

Composition of Peptide Preparation

Component	Content	Molar ratio[A] of component to C_{II-3} and C_{II-4}
	nmole/mg	
SDH flavin	0	0
Total phosphorus	1250	25
Cytochrome *b*	2–6	0.04–0.12
Ubiquinone	0	0

[A] Based on a combined molecular weight for C_{II-3} and C_{II-4} of 20,000.

TABLE II

Reconstitution of Q Reductase and Cytochrome *c* Reductase

Sample	Succinate-acceptor reductase activities[A]				
	$Fe(CN)_6^{3-}$	PMS	DPB[B]	PMS:DPB	Cytochrome *c*
SDH	24	73	0		0
SDH + Triton X-100	20	68	0		0
SDH + peptides	0	90	40	2.2	0
Complex II	0	36	40	0.9	
Complex II + Triton X-100	0	31	18	1.7	
SDH + peptides + Complex III					9.4[C]

[A] μmole succinate oxidized per min per mg of succinate dehydrogenase or Complex II; acceptor concentrations: $Fe(CN)_6^{3-}$, 200 μ*M*; PMS, 1.08 m*M*; DPB, 35 μ*M*; cytochrome *c*, 80 μ*M*.

[B] DPB, 2,3-dimethoxy-5-methyl-6-pentyl-1,4-benzoquinone, a measure of Q reductase activity.

[C] Antimycin sensitive.

zyme that contributes to PMS reductase but not to Q reductase activity.

The increased reactivity with PMS (30%) on combining the dehydrogenase with the two peptides, and the simultaneous disappearance of the "low K_m" ferricyanide activity was exactly as expected from inserting the dehydrogenase into alkali-treated membranes (*9*). This emphasizes that assays with artificial electron acceptors are extremely useful for judging the nativity of the enzyme or its reconstitution activity. In the PMS assay, the turnover number of the enzyme in the recombined sample was 14,000 to 15,000 at 38 °C, close to the value observed in the best preparations of Complex II (*11*) but lower than in ETP or mitochondria (22,000). In addition to the evidence that all catalytic activities of unresolved Complex II were quantitatively restored, the mixture showed the same sensitivity to TTF and carboxanilides as Complex II, with inhibition kinetics characteristic of Complex II.

Table II shows that addition of highly purified Complex III to the peptide preparation-dehydrogenase mixture elicited high succinate-cytochrome *c* reductase activity that was antimycin sensitive and strictly dependent on the presence of peptides C_{II-3} and C_{II-4}.

Anaerobic titrations of reconstitutively active succinate dehydrogenase with the peptide preparation indicated (*6*) that 3 to 5 mole each of C_{II-3} and C_{II-4} were required per mole of flavoprotein to restore Q-reductase activity to the same level shown by unresolved Complex II in Triton X-100 (turnover number = 5,500 at 38 °C). Since Complex II contains equimolar amounts of the two small peptides and succinate dehydrogenase, it is possible that some of the extracted peptides C_{II-3} and C_{II-4} have aggregated and, therefore, are inaccessible to the dehydrogenase or in some other manner rendered inactive in the course of isolation.

Individual Roles of Peptides C_{II-3} and C_{II-4}

Since it has proved impossible by a variety of procedures to separate the two peptides in native form, we resorted to an indirect approach to ascertain their respective roles in the binding of succinate dehydrogenase and of TTF, and in the restoration of Q reductase activity. The peptide preparation was subjected to limited digestion with chymotrypsin, resulting in the disappearance of the band corresponding to C_{II-3} on polyacrylamide gel electrophoresis, without any evident change in the content or R_f value of C_{II-4} (Fig. 2). Chymotryptic digestion re-

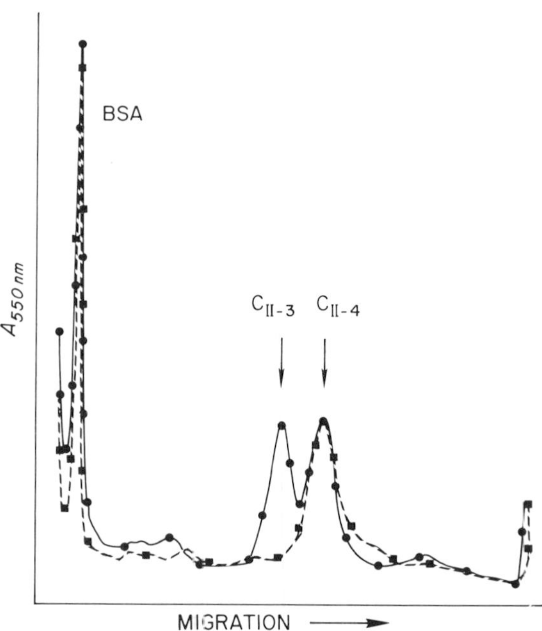

Fig. 2. Gel scans of the preparation of C_{II-3} and C_{II-4} before (●—●) and following (■—■) a 2 h treatment (22 °C) with chymotrypsin (1:50, pH 7.0). Bovine serum albumin was included in each sample prior to SDS-polyacrylamide gel electrophoresis (*7*) to serve as an internal standard for staining.

sulted in progressive loss of succinate–Q reductase activity (Fig. 3), and the inability of the digested peptide sample to mask the "low K_m" ferricyanide activity of soluble succinate dehydrogenase. Although, as documented below, the chymotrypsin-treated peptide preparation was still able to combine with succinate dehydrogenase.

The same limited digestion with chymotrypsin (0.02 mg/mg protein at 20 °C for 2 h) of Complex II had no effect on either subunit of succinate dehydrogenase, or on peptide C_{II-4}, while C_{II-3} was incompletely digested, yielding a shorter (~9,000 dalton) peptide. Under these conditions TTF-sensitive Q reductase activity was not lost and a peptide preparation isolated from this partially digested sample conferred normal Q reductase activity on soluble succinate dehydrogenase. If the peptide preparation containing C_{II-4} and partially digested C_{II-3} was treated with a second aliquot of chymotrypsin (0.02 mg/mg for 2 h), digestion of C_{II-3} proceeded to completion with total loss of the ability to elicit Q reductase activity.

These experiments show that C_{II-3} is essential for normal electron transfer from succinate dehydrogenase to ubiquinone. It must be located in the membrane very near the reaction site of ferricyanide in the "low K_m" ferricyanide assay (presumably the HiPIP Fe–S cluster), since the masking of this activity hinges on the presence of C_{II-3}.

To establish that the small peptides physically combine with the de-

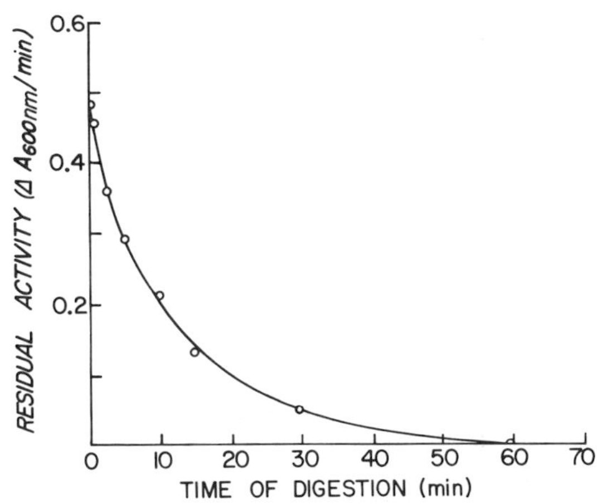

Fig. 3. Loss of activity of the peptide preparation during chymotrypsin treatment. Samples of the peptide preparation were removed from the digestion mixture at various time intervals, treated with phenylmethylsulfonylfluoride to inhibit proteolysis, mixed with an excess of succinate dehydrogenase, and assayed for succinate–DPB reductase activity as described in footnotes to Table II (from Ackrell *et al.* [6], used with permission).

hydrogenase in reconstitution experiments, advantage was taken of the fact that an antibody to the large subunit of the dehydrogenase (*3*) precipitates C_{II-3} and C_{II-4} only when they are combined with the dehydrogenase, as in Complex II or reconstituted samples. Figure 4A shows the SDS-polyacrylamide gel pattern of an immunoprecipitate from a sample of the peptide preparation that was recombined with a substoichiometric amount of succinate dehydrogenase; C_{II-3} and C_{II-4} were present in the immunoprecipitate. In contrast, when the dehydrogenase was omitted (Fig. 4B), no trace of these peptides was seen. This experiment clearly establishes the physical combination of C_{II-3} and C_{II-4} with succinate dehydrogenase under the conditions of reconstitution.

When the C_{II-3} component of the peptide preparation was destroyed with chymotrypsin, and the preparation was combined with succinate dehydrogenase and precipitated with antibody, the gel pattern of the precipitate was that seen in the solid line of Fig. 4C. The dashed line represents the pattern of a precipitate obtained using a control γ-globulin fraction; this caused sedimentation of a minor amount of the hydrophobic C_{II-4} sample. (Note that C_{II-4} is much less soluble in the absence of

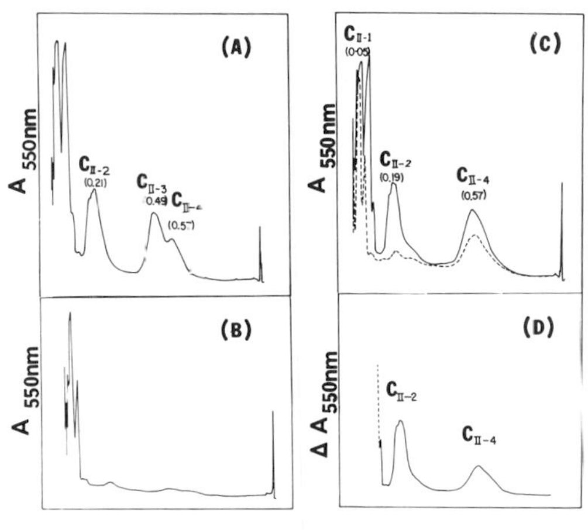

Fig. 4. Immunoprecipitation of enzyme–peptide mixtures before and after chymotryptic digestion of the peptides. The peptide preparation was treated with chymotrypsin (1:50) at pH 7.0 (22 °C) for 2 h. The peptide samples were mixed with a substoichiometric amount of succinate dehydrogenase, treated with antibody, and the resultant precipitates analyzed by SDS-polyacrylamide electrophoresis (*7*). (A) untreated peptide preparation plus succinate dehydrogenase plus antibody; (B), untreated peptide preparation plus antibody; (C), chymotrypsin-treated peptide preparation plus succinate dehydrogenase, plus antibody (—) or control γ-globulin fraction (---); (D), scan obtained by subtracting dashed from solid line in (C). Figures in parentheses are R_f value relative to dye front.

C_{II-3}). Figure 4D indicates the difference between the two patterns of Fig. 4C, showing that C_{II-4} is specifically precipitated by the antibody to the large subunit of succinate dehydrogenase in the recombined sample, even though C_{II-3} was absent and the recombined preparation devoid of ubiquinone reductase activity. Thus, peptide C_{II-4}, unable alone to confer Q reductase activity on succinate dehydrogenase, nevertheless combines with the enzyme.

Role of C_{II-3} and C_{II-4} in the Inhibition of Succinate Oxidation by TTF and Carboxanilides

Previous studies in this laboratory established that a series of oxathiin derivatives (carboxanilides or carboxins) are potent and highly specific inhibitors of succinate dehydrogenase in membrane preparations, but have no effect on the soluble, purified enzyme. These derivatives act at the same site and in the same manner as TTF, have an inhibition site on the oxygen side of the enzyme, and despite the low K_i values of carboxanilides, combine reversibly with the enzyme (*12*, *13*).

To localize their specific combining site, (i.e., the site responsible for the inhibition), and particularly to establish whether the small peptides are a part of this site, it was necessary to bind the inhibitors in a covalent linkage to Complex II, dissociate the complex into its constituent peptides, and identify which components formed an adduct with the inhibitor. To this end we synthesized a tritiated carboxanilide (3′-azido-5,6-dehydro-2-methyl-1,4-oxathiin-3-carboxanilide) for photoaffinity labeling experiments (*14*). This compound proved to be a potent, noncompetitive inhibitor of the succinate–Q reductase activity of Complex II (K_i = 2.5 μM at 38 °C) acting in the same manner as TTF. Figure 5 is a Dixon plot of the pattern of inhibition by mixtures of TTF and the azidocarboxin. The series of parallel lines indicate mutual exclusion from the inhibition site, in accord with the general observation (*13*) that TTF displaces carboxanilides from the specific binding site responsible for the inhibition (although not from nonspecific sites).

The ability of TTF to displace carboxanilides from the specific binding site provides a means of quantitating the amount of inhibitor bound at the specific site. Equilibrium dialysis was conducted at a series of [^{3}H]azidocarboxin concentrations, both in the presence and absence of a large excess of unlabeled TTF, as in Fig. 6. When TTF was present (solid symbols) only nonspecific binding was operative and, since this cannot be easily saturated, a linear relation was obtained between free and bound inhibitor. In the absence of TTF, both specific and non-

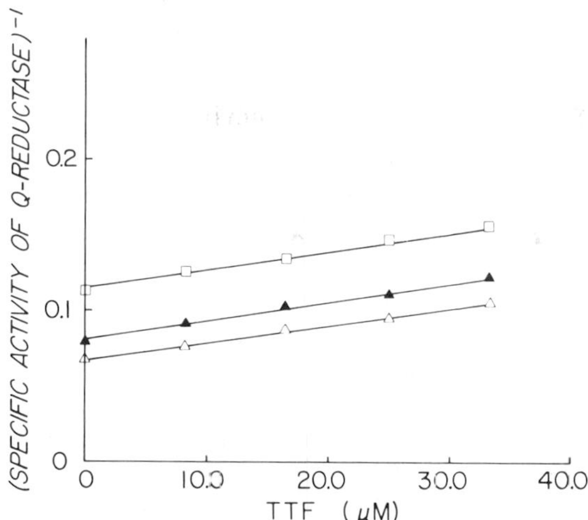

Fig. 5. Inhibition of the succinate–Q reductase activity of Complex II by mixtures of TTF and azidocarboxin. Activity (38 °C) was determined in the presence of DPB (30 μM), TTF as shown, and zero (△), 1.7 (▲), and 4.5 (□) μM concentrations of azidocarboxin.

specific binding occurred, resulting in a biphasic binding curve (Fig. 6, open symbols). The difference between the two curves represents specific binding.

Figure 6A represents the results of such equilibrium dialysis experiments on unresolved Complex II. The amount of specific binding corresponded to 0.7 mole of azidocarboxin per mole of succinate dehydrogenase. Neither purified succinate dehydrogenase nor the two small peptides alone bound the inhibitor specifically, but on recombination of the two preparations the specific binding site was restored (Fig. 6B) in accord with the reappearance of carboxanilide- and TTF-sensitive Q reductase activity. The K_D values in the unresolved (~0.5 μM at 4 °C) and reconstituted preparations (1.2 μM at 16 °C), calculated from Scatchard plots of the data in Fig. 6, were in satisfactory agreement with each other when corrected for the difference in temperature, as well as with the K_i values calculated from kinetic experiments (*14*). These results reinforce the conclusion that the enzyme and peptides C_{II-3} and C_{II-4} assume the correct orientation on recombination.

Implicit in these data is the conclusion that the two small peptides are essential for specific binding of carboxanilides and direct evidence for their involvement in this binding site will be presented. Since selective digestion of C_{II-3} with chymotrypsin abolished the specific binding site (Fig. 6C), it may be concluded that this peptide is essential for electron transport from succinate dehydrogenase to Q and specific

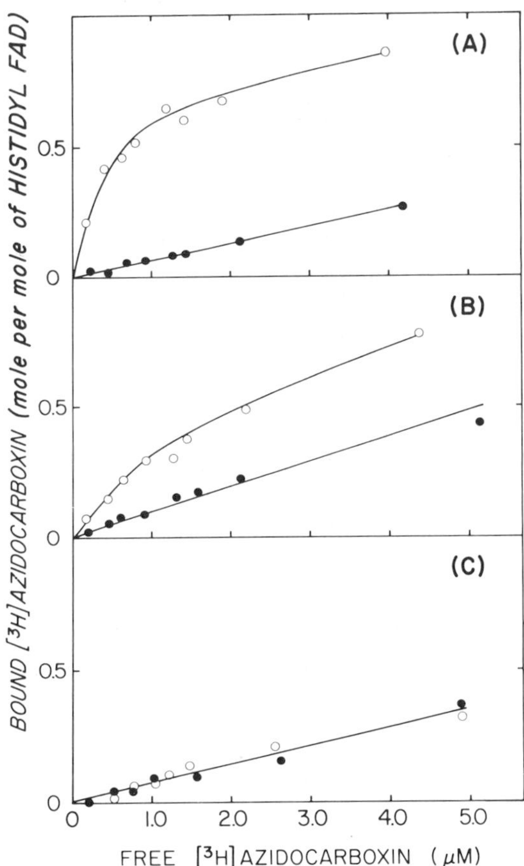

Fig. 6. Binding of azidocarboxin to Complex II (A), mixtures of succinate dehydrogenase plus peptide preparation (B), and succinate dehydrogenase plus chymotrypsin-treated peptide preparation (C). Equilibrium dialysis experiments were conducted in the dark according to Coles *et al.* (*13*), in the presence (●) and absence (○) of 500 μM TTF.

binding of TTF and carboxanilides. Whether C_{II-4} plays a direct role in either event cannot be decided from data presented, but the results of photoaffinity labeling experiments showed that C_{II-4} is located either at or in the immediate vicinity of the inhibitor binding site.

Photoaffinity labeling was conducted by irradiating membrane preparations at a series of azidocarboxin concentrations, with and without TTF present, followed by SDS-polyacrylamide electrophoresis to separate the individual components for radioactivity counting. A serious limitation of this technique was that the ultraviolet light required to convert the azidocarboxin to the nitrene caused gradual destruction of the succinate–Q reductase activity of Complex II. Thus, the time of irradiation and extent of labeling had to be curtailed. Moreover, the concentration of labeled inhibitor had to be kept very low to minimize covalent binding at spurious sites. Coupled with the relatively low specific radioactivity of the azidocarboxin, these significant sources of error make quantitation of the affinity label in individual peptide components difficult. Also, since TTF absorbs ultraviolet light, it might be expected to lower the yield of nitrene from azidocarboxin bound at unspecific sites. Despite these limitations, the technique provided an unambiguous qualitative identification of the components near the specific binding site, as well as an approximate measure of the amount of inhibitor covalently bound to the individual components of Complex II.

Table III presents the distribution of radioactivity in the various components of Complex II, calculated from a photoaffinity labeling experiment in which ~90% of the Q reductase activity was irreversibly inhibited following irradiation. It is clear that little, if any, specific binding to either dehydrogenase subunit occurred, but a larger amount of inhibitor became covalently bound to peptides C_{II-3} and C_{II-4}. Although incorporation into these two peptides is presented together because their separation on the gel was not quite complete, it was quite clear that both of these peptides were extensively labeled in this and all other similar experiments. In addition to the peptides, extensive labeling was found in the nonstaining portion of the gel where phospholipids migrate. On a molar basis, the extent of phospholipid labeling was quite low, but because of their high concentration in Complex II relative to the peptides, a major fraction of the carboxanilide was recovered in the phospholipid band.

In view of the technical limitations previously mentioned, one would

TABLE III

Photoaffinity Labeling of Complex II

Component	nmole Azidocarboxin bound[A]		
	50 μM Azidocarboxin	50 μM Azidocarboxin + 2 mM TTF	Difference
70*s*	0.26	0.16	0.10
30*s*	0.13	0.03	0.10
C_{II-3} + C_{II-4}	0.91	0.51	0.40
Phospholipids	2.97(0.06)[B]	1.21(0.02)[B]	1.76(0.04)[B]

[A] per 250 μg protein (≃1 nmole histidyl flavin).
[B] per mole of phospholipid, assuming 52 mole per mole flavin (*6*).

not expect quantitative agreement between the concentration of irreversibly inhibited enzyme (~90% of the total) and the mole fraction of components that become labeled. Nevertheless, 0.4 mole of inhibitor/mole of C_{II-3} + C_{II-4} seems low, unless one considers that ~0.5 mole/mole of Complex II was also the value obtained in equilibrium dialysis experiments with azidocarboxin and with another carboxanilide, 2,4,5-trimethyl-3-carboxanilinofuran (*13*). The reason for this seemingly substoichiometric binding is not yet clear.

Since identification of the components that form the specific binding site of carboxanilide and of TTF is of great interest, confirmation of these findings was sought by another technique that does not depend on displacement of the carboxanilide by TTF. Specific and nonspecific binding may be distinguished by the shape of the binding curves in photoaffinity labeling experiments conducted in the absence of TTF. If only nonspecific binding occurs, the dependence of binding on the added azidocarboxin concentration will be linear; if specific binding also occurs, the initial part of the binding curve will be hyperbolic. Figure 7 shows that binding to the small and large subunits of the enzyme was strictly linear, while binding to C_{II-3} plus C_{II-4} and phospholipids

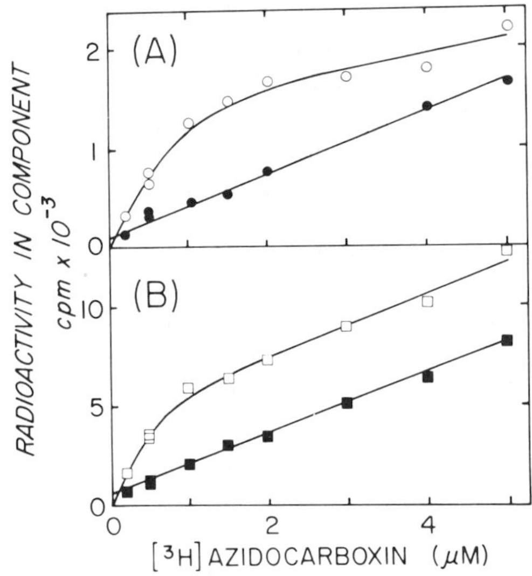

Fig. 7. Photoaffinity labeling of Complex II components. Samples of Complex II, which had been irradiated with UV light in the presence of [^{3}H]azidocarboxin, were subjected to SDS-polyacrylamide gel electrophoresis (*7*) to resolve the constituent peptides and phospholipids for counting purposes. Incorporation of label is shown for the small (●) and large (■) subunits of succinate dehydrogenase, C_{II-3} and C_{II-4} taken together (○), and phospholipids (□) [from Ramsay, *et al.* (*14*) used with permission].

was not. These data confirm the important conclusion that the two small peptides and phospholipids are components of the specific binding site, while the subunits of succinate dehydrogenase itself, although required for specific binding, are sufficiently far removed from the nitrene group of the specifically bound label that they do not become covalently labeled. Possibly the dehydrogenase effects a conformation change in the small peptides which is necessary to accommodate the inhibitor, just as the small peptides, on binding to the dehydrogenase, might alter its conformation to enable it to pass electrons to endogenous Q.

DISCUSSION

The purified preparation of the two molecular weight peptides from Complex II, C_{II-3} and C_{II-4}, and reconstitutively active, pure succinate dehydrogenase (*1*) when mixed anaerobically combine into a tight complex, as shown by the fact that an antibody to the large subunit of the enzyme precipitates both dehydrogenase subunits as well as the two small peptides. Because of this behavior, the term "binding peptides" seems appropriate for C_{II-3} and C_{II-4}. The combination with dehydrogenase reestablishes the full succinate-Q reductase activity of unresolved Complex II and also abolishes the "low K_m" ferricyanide activity of the soluble enzyme. This is as expected, since this reaction site is not available to ferricyanide when dehydrogenase is incorporated into the membrane (*9*). The restored Q reductase activity is fully sensitive to TTF and carboxanilides. These agents also inhibit competitively the reduction of PMS by succinate in the reconstituted sample, as in unresolved Complex II. Thus, it appears that the reconstituted complex behaves exactly like the original in all known respects.

Our studies on the individual functions of C_{II-3} and C_{II-4} demonstrated that both combine with the dehydrogenase but only peptide C_{II-3}, because it can be selectively destroyed by chymotryptic digestion, is known to be essential for Q reduction and for specific binding of carboxanilides and TTF. An interesting point that emerged from studies of the chymotrypsin digested preparation: combination with the dehydrogenase does not result in the disappearance of "low K_m" ferricyanide reductase activity. Since C_{II-3} thus appears to be implicated in masking the ferricyanide reaction site, this peptide must be juxtaposed to the HiPIP center of the enzyme in the membrane. Further, since labeling with [^{35}S]-diazobenzenesulfonate indicates that C_{II-3} is exposed on the cytoplasmic surface, while the dehydrogenase is exposed on the matrix side (*3*), it follows that Complex II spans the inner membrane.

ACKNOWLEDGMENTS

This investigation was supported by the National Institutes of Health Program Project No. HL-16251, the National Science Foundation (PCM 78-23716 and PCM 78-22935), and the Veterans Administration.

REFERENCES

1. Ackrell, B. A. C., Kearney, E. B., and Coles, C. J. (1977). *J. Biol. Chem.* **252,** 6963–6965.
2. Capaldi, R. A., Sweetland, J., and Merli, A. (1977). *Biochemistry* **16,** 5707–5710.
3. Merli, A., Capaldi, R. A., Ackrell, B. A. C., and Kearney, E. B. (1979). *Biochemistry* **18,** 1393–1400.
4. Ziegler, D. M., and Doeg, K. A. (1962). *Arch. Biochem. Biophys.* **97,** 41–50.
5. Davis, K. A., and Hatefi, Y. (1971). *Biochemistry* **10,** 2509–2516.
6. Ackrell, B. A. C., Ball, M. B., and Kearney, E. B. (1980). *J. Biol. Chem.* **255,** 2761–2769.
7. Swank, R. T., and Munkres, K. D. (1971). *Anal. Biochem.* **39,** 462–477.
8. Ackrell, B. A. C., Kearney, E. B., and Singer, T. P. (1977). *J. Biol. Chem.* **252,** 1582–1588.
9. Vinogradov, A. D., Gavrikova, E. V., and Goloveshkina, V. G. (1975). *Biochem. Biophys. Res. Commun.* **65,** 1264–1269.
10. Beinert, H., Ackrell, B. A. C., Vinogradov, A. D., Kearney, E. B., and Singer, T. P. (1977). *Arch. Biochem. Biophys.* **182,** 95–106.
11. Mowery, P. C., Steenkamp, D. J., Ackrell, B. A. C., Singer, T. P., and White, G. A. (1977). *Arch. Biochem. Biophys.* **178,** 495–506.
12. Ackrell, B. A. C., Kearney, E. B., Coles, C. J., Singer, T. P., Beinert, H., Wan, Y.-P., and Folkers, K. (1977). *Arch. Biochem. Biophys.* **182,** 107–117.
13. Coles, C. J., Singer, T. P., White, G. A., and Thorn, G. D. (1978). *J. Biol. Chem.* **253,** 5573–5578.
14. Ramsay, R. R., Ackrell, B. A. C., Coles, C. J., Singer, T. P., White, G. A., and Thorn, G. D. (1981). *Proc. Natl. Acad. Sci. U.S.A.* **78,** 825–828.

Ubiquinone Binding Protein in the Cytochrome b-c_1 Complex

7

C. -A. YU
L. YU

INTRODUCTION

Although the participation of ubiquinone (Q)[1] in the mitochondrial respiratory chain has long been established through reconstitution studies of functional complexes and chemical identification (*1–3*), the details of the reaction mechanism and the identity of the active molecular species are yet to be clarified. Until the recent discovery of a Q-binding protein (*4*), ubiquinone was generally thought to exist as a mobile electron carrier, shuttling the electron (proton) between electron-transfer complexes (*5*, *6*). The existence of a Q-binding protein was deduced from repeated observation of a stoichiometric ratio between Q and cytochrome c_1 in the isolated succinate-cytochrome *c* reductase (SCR). It was observed, in over a hundred batches of SCR, that Q stimulated enzyme activity only when the preparation contained less than a stoichiometric amount of Q to cytochrome c_1 (*7*). This observation led to the proposal of the existence of a specific Q-binding protein(s) in SCR. This Q-binding protein theory was further strengthened by the observation that removal of phospholipids

[1] Abbreviations: Q (ubiquinone), Q_0 (2,3-dimethoxy-5-methyl-1,4-benzoquinone), QP_s, QP_c, and QP_n (Q-binding proteins functioning in succinate-Q reductase, ubiquinone-cytochrome *c* reductase, and NADH-Q reductase, respectively), SCR (succinate-cytochrome *c* reductase), QCR (ubiquinol-cytochrome *c* reductase), SQR (succinate-ubiquinone reductase), SCR_d or *b*-c_1III_d (phospholipid and Q depleted preparation of SCR or *b*-c_1III complex), SDS (sodium dodecyl sulfate), NAPA (*N*-4-azido-2-nitrophenyl-β-alanine), TMPOC(2,2,5,5-tetramethyl-3-pyrrolin-1-oxyl-3-carboxylic acid).

Function of Quinones in Energy Conserving Systems

ISBN 0-12-701280-X

from SCR by repeated ammonium sulfate precipitation in the presence of 0.5% sodium cholate (*7*), was not concurrent with the removal of Q. The differences between the properties of our soluble cytochrome b-c_1 complex (*8*) and the well-known Complex III (*9*) in reconstitution with succinate dehydrogenase (SDH) to form SCR, and loss of the ability of soluble b-c_1 complex to reconstitute with SDH upon treatment with chymotrypsin also indicated the involvement of a specific protein in the succinate-Q reductase (SQR).

The recent successful isolation of a protein (*4*, *10*) that binds Q and can convert soluble, purified succinate dehydrogenase into a succinate-Q reductase confirmed the participation of a specific Q-binding protein in this region of the electron-transfer chain (*11*). The participation of a Q-binding protein in SQR has been independently confirmed by several laboratories (*12–14*).

During the course of our isolation of Q-binding proteins of succinate-Q reductase (QP_s) from the soluble b-c_1 complex (*8*, *14*), it was observed that the Q associated with cytochrome b-c_1 complex was evenly distributed between the crude QP_s fraction and purified b-c_1III complex[2] (ubiquinol-cytochrome *c* reductase, QCR), and that the latter was inactive in reconstitution with SDH to form SQR. This result indicates that the Q associated with the purified QCR is different from that in QP_s.

The protein responsible for Q binding in QCR is called QP_c (*15*, *16*) to distinguish it from QP_s, and from QP_n, the protein involved in NADH–Q reductase (*18*). The existence of QP_c was further substantiated by the fact that a high concentration of Q-free radical was detected by EPR measurements when the QCR complex was reduced by succinate in the presence of catalytic amounts of SQR. Formation of a ubisemiquinone radical was phospholipid dependent and sensitive to proteolytic digestion (*16*). These results confirmed that QP_c is indeed a protein. Evidence for QP_c is, however, only indirect, and does not identify which protein(s) in the QCR is responsible for QP_c activity in this electron-transport region. Recently, in continuation of our studies of Q proteins we employed a photoaffinity labeling technique using a functionally active ^{14}C- labeled Q derivative to identify the Q-binding protein (*19*). We have also performed preliminary studies of Q binding, using Q- and phospholipid-depleted QCR and radioisotope-labeled and spin-labeled Q analogs. These studies are reported in this paper.

[2] Cytochrome b-c_1III complex, also known as b-c_1III is a highly purified ubiquinol-cytochrome *c* reductase that is different from soluble cytochrome b-c_1 complex reported previously (*8*). Soluble cytochrome b-c_1 complex contains QP_s and is active in reconstitution with SDH to form SQR. Cytochrome b-c_1III has only QCR activity, and is not active in reconstitution with SDH.

EXPERIMENTAL RESULTS

Properties of QP_c

At pH 9 about 60% of the ubiquinone present (2–3 nmole/mg protein) in highly purified cytochrome b-c_1III Complex can be detected by EPR as a ubisemiquinone radical[3] ($g = 2.0045$), when the enzyme is reduced by succinate in the presence of a catalytic amount of SQR, at room and liquid nitrogen temperatures. Formation of the ubisemiquinone radical is concurrent with the reduction of cytochrome b, after the complete reduction of cytochrome c_1. The rate of ubisemiquinone radical formation is comparable to that of reduction of cytochrome c_1 or cytochrome b, and is dependent on the amount of SQR used. However, the maximal concentration of ubisemiquinone radical formed is independent of the amount of SQR used, and dependent on the amount of succinate present.

A direct correlation between enzymatic activity of b-c_1III and ability to generate ubisemiquinone radical was observed. Addition of thenoyltrifluoracetone (TTFA) to the system not only prevented the formation of ubisemiquinone radical by blocking electron flow from succinate to Q, but also abolished the ubisemiquinone radical that was already formed. Addition of TTFA to the system also prevented the reduction of cytochrome b and caused the oxidation of prereduced cytochrome b. The reoxidation of cytochrome b induced by TTFA was oxygen dependent. Antimycin A also diminished the ubisemiquinone radical, but in this case cytochrome b remained reduced, indicating that antimycin A either directly competes with Q for a binding site, or indirectly causes a conformational change of the enzyme and decreases or inhibits Q binding.

Formation of ubisemiquinone radical in b-c_1III was also achieved by addition of sufficient ubiquinol. The redox titration of b-c_1III in the presence of redox dyes revealed the ubisemiquinone radical has a maximal concentration at around 76 mV. This was estimated to be between 30 to 40 percent of the total Q found in the complex at pH 8.0, indicating that the two partial Q reactions, $Q/Q^{\overline{\cdot}}$ $(Q/QH\cdot)$ and $Q^{\overline{\cdot}}/Q^{-2}(QH\cdot/QH_2)$ have rather close midpoint potentials. When a fumarate and succinate mixture was used as substrate, maximal ubisemiquinone radical formation was observed as the fumarate-to-succinate ratio approached 100. The ubisemiquinone radical concentration is pH de-

[3] We thank Drs. G. Feher, M. Y. Okamura, and Mr. R. A. Isaacson for their help in determining the precise g value of QP_c.

pendent with no ubisemiquinone radical detected below pH 7.0. The concentration of ubisemiquinone radical increased with the pH of the solution up to pH 9, and then rapidly decreased. Since the maximum ubisemiquinone radical concentration was found at alkaline pH, the radical is assumed to be in the ubisemiquinone anion form. Electron spin relaxation behavior of the ubisemiquinone radical, as examined by microwave power saturation studies, indicates that the ubisemiquinone radical is somewhat isolated from other paramagnetic centers.

Formation of the ubisemiquinone radical is phospholipid dependent. Phospholipid- and Q-depleted b-c_1III could use the Q associated with QP_s to form ubisemiquinone radical upon addition of phospholipids and a reducing system. This result suggests that the binding between Q and QP_s differs from that between Q and QP_c, possibly involving different parts of the Q molecules. A possibility also exists that under the experimental conditions, QP_c binds Q more tightly than does QP_s. Stronger binding of Q to QP_c is also suggested by the fact that the Q–QP_c complex forms ubisemiquinone radical easily, while ubisemiquinone radical in the isolated QP_s has yet to be detected.

Synthesis of Photoaffinity and Spin-Labeled Q Derivatives for Studies of Q Proteins

From the properties of QP_c previously mentioned, it is clear that identification of QP_c through conventional isolation and reconstitution methods will not be an easy task due to the close association of QP_c with phospholipids. Therefore, a different approach to identification of QP_c was begun without awaiting the development of an isolation procedure. We adapted the photoaffinity labeling technique, and synthesized a functionally active ^{14}C-labeled Q derivative to identify possible Q-binding protein(s) in the cytochrome b-c_1III complex.

Scheme I shows the key steps in the synthesis of the radioisotope labeled 2,3-dimethoxy-5-methyl-1,4-benzoquinone (Q_0). Low molecular weight analogs and their 6-(ω-bromoalkyl) derivatives of Q were synthesized essentially according to Wan *et al.* (*20*). The 6-(ω-bromoalkyl) derivative was then converted to the 6-(ω-hydroxyalkyl) derivative (Scheme II) and esterified with photoaffinity (Scheme III) and spin (Scheme IV) labels. The synthesis of the photoaffinity label, arylazido-β-alanine, was carried out according to Jeng and Guillory (*21*). One of the photoaffinity-labeled Q derivatives synthesized was 2,3-dimethoxy-5-methyl-6-[10-(4-azido-2-nitroanilinopropionoxy)-decyl]-1,4-benzoquinone (Q_0C_{10}NAPA). The spin label used was 2,2,5,5-tetramethyl-3-pyrrolin-1-oxy-3-carboxylic acid. The spin labeled Q derivative synthesized was 2,3-dimethoxy-5-methyl-6-[10-[3-(2,2,5,5,-tetramethyl-3-pyrrolin-1-oxy-3-carboxyl)]-decyl]-1,4 benzoquinone. Both esterifications proceeded well with dicyclohexylcarbodiimide and pyridine at room temperature.

The spectral properties of $Q_0C_{10}NAPA$ are given in Fig. 1. $Q_0C_{10}NAPA$ has millimolar extinction coefficients of 31 and 5 at 275 nm and 450 nm, respectively. Upon reduction by $NaBH_4$, the absorption at 275 nm decreased by about one third, with a millimolar extinction coef-

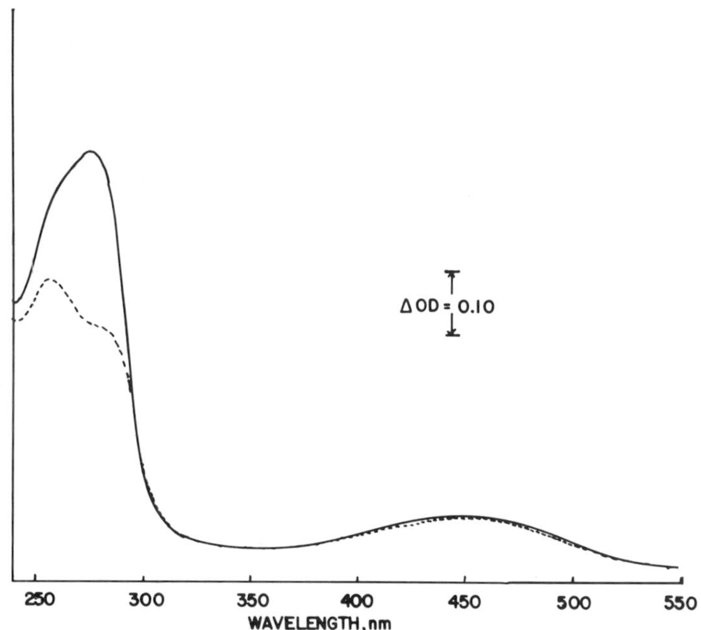

Fig. 1. Absorption spectra of Q_0C_{10}NAPA. Q_0C_{10}NAPA was diluted to 21 μM in 95% EtOH, and the spectra were measured in a Cary spectrophotometer Model 14, at room temperature. 1.0 cm light path cuvettes were used. The dotted spectrum was made after the sample was reduced by $NaBH_4$.

ficient of 19. The absorption at 450 nm decreased only slightly. This effect was expected as the main contribution to absorption at 450 nm is from the NAPA moiety, which is not affected by reduction. Using the absorption data given by Jeng and Guillory (*21*) and that of Q_0C_{10} or Q_2, Q_0C_{10}NAPA was found to be, as expected, composed of 1 molecule of Q_0C_{10} and 1 molecule of NAPA.

The spectral properties of Q_0C_{10}TMPOC (Fig. 2) are rather simple and are similar to Q_2 or Q_0C_{10} as the contribution to absorption by TMPOC is negligible. The absorption maximum at 278 nm decreases dramatically and shifts to 290 nm upon reduction. The difference between the millimolar extinction coefficients of the oxidized and reduced forms of Q_0C_{10}TMPOC equals that reported for Q_2.

Effectiveness of Synthesized Q Derivatives in Electron-Transfer Reactions

For comparison, both the overall activity of Q in the electron-transfer reactions in SCR and the activity of Q as electron acceptor in SQR

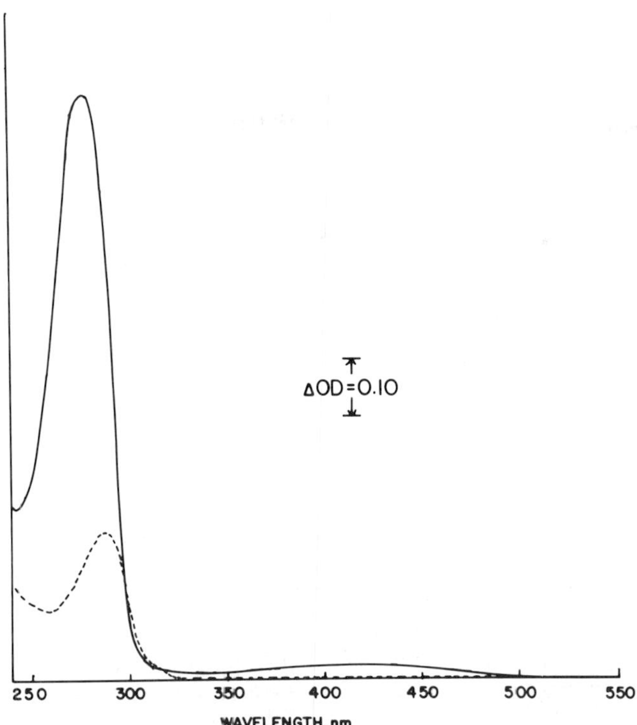

Fig. 2. Absorption spectra of Q_0C_{10}TMPOC. The same conditions as Fig. 1 were used except the concentration of the sample was 65 μM.

were determined; the results are given in Table I. Experimentally, for determination of overall activity in SCR, Q analogs at various concentrations were mixed with phospholipid- and Q-depleted SCR (SCR_d), and the mixture was then replenished with phospholipids. The SCR_d was prepared by repeated ammonium sulfate precipitation in the presence of 0.5% sodium cholate. The reconstituted SCR activity was then assayed after 0.5 to 1 hour incubation at 0 °C. The maximal activity was expressed as a percentage of the activity obtained when Q_2 was used. The electron acceptor activity of Q was followed by the Q mediated succinate-DCIP reductase activity of Complex II. The activity of Q analogs in SCR is dependent on the length of the 6-alkyl carbon chain up to 10 carbons. Modification of the ω-position has very little effect on activity, as ester linked or ω-bromo derivatives are as active as unmodified alkyl side chain analogs, with the exception of the ω-hydroxylalkyl derivative of Q, which shows very little activity in SCR but is fully active as an electron acceptor of Complex II. Introduction of a bulky group, such as an arylazido or spin label also produced no signif-

TABLE I

Relative Effectiveness of Q Analogs in the Electron-Transfer Reaction of Succinate-Q and Succinate-cytochrome *c* Reductases

	SQR[A]		SCR[B]	
Analogs	Activity, %	Concentration, μM	Activity, %	Concentration, Q/*b*
Q_2	100	17	100	5
Q_0	0	30	0	30
Q_1	83	16	10	15
$Q_0(CH_2)_4Cl$	100	17	17	12
$Q_0(CH_2)_5H$	98	14	24	12
$Q_0(CH_2)_5Br$	99	13	26	12
$Q_0(CH_2)_{10}H$	111	17	125	5
$Q_0(CH_2)_{10}Br$	115	17	130	5
$Q_0(CH_2)_{10}OH$	80	20	11	12
$Q_0(CH_2)_{10}NAPA$	94	17	96	7
$Q_0(CH_2)_{10}TMPOC$	91	24	108	7

[A] The activity was followed by the Q mediated succinate-DCIP reductase activity of Complex II. The concentration refers to the final concentrations of Q analogs in the assay mixture.

[B] The activity refers to that restored from the Q- and phospholipid-depleted succinate-cytochrome *c* reductase upon addition of Q analogs and phospholipid. The concentration represents mole of Q analogs used per mole of cytochrome *b* in the reconstituted system.

icant change in electron-transfer activity. The full electron-transfer activity of $Q_0C_{10}NAPA$ permitted the use of this photoaffinity-labeled Q derivative in identification of QP_c in the *b*-c_1III complex.

Identification of Ubiquinone-Binding Protein (QP_c) in the Cytochrome *b*-c_1III Complex

To identify the Q-binding protein, $Q_0C_{10}NAPA$ was prepared with the ^{14}C label at the carboxyl group of the β-alanine moiety. When the ^{14}C-labeled $Q_0C_{10}NAPA$ was mixed with freshly prepared phospholipid- and Q-depleted cytochrome *b*-c_1III complex and illuminated for 20 minutes with a 300 watt spot light at approximately 8 °C, $Q_0C_{10}NAPA$ became covalently linked to the cytochrome *b*-c_1III complex. The distribution of radioactivity among the proteins of the complex was revealed by SDS polyacrylamide gel electrophoresis. Figures 3 and 4 show the protein and ^{14}C distributions of the cytochrome *b*-c_1III complex on an electrophoretic gel column. Two peaks of radioactivity were observed,

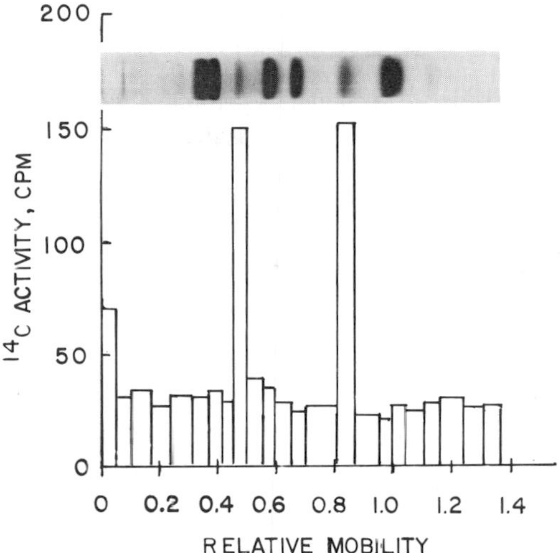

Fig. 3. Distribution of ^{14}C-radioactivity among the proteins of the cytochrome b-c_1III complex, in the SDS polyacrylamide gel electrophoresis system of Weber and Osborn (*22*). Twenty three stained gel columns (0.5 × 8 cm) were sliced into fractions according to the protein band location. Each fraction was pooled and weighed before being mixed with two weight volumes of 7.5 *N* HCL. The samples were sealed and hydrolyzed at 120 °C for 8 hours. The hydrolysates were cooled, the gels removed, and dried at 100 °C. The residues were redissolved in 0.4 ml of water, mixed with 10 ml of complete counting cocktail, Budget Solve, and the ^{14}C activity of each fraction was determined. The relative mobility is expressed using that of cytochrome *c* as one.

in both the Weber and Osborn (*22*) and the Swank and Munkres (*23*) systems.

To avoid confusion resulting from variations in experimental conditions and investigator preferences on the assignment of molecular weights to protein bands in the SDS polyacrylamide gel column, we have reported the mobility relative to a known protein such as cytochrome *c*, in addition to the apparent molecular weight, which we previously assigned (*24*). For example, R_c^{wo} and R_c^{sm} stand for the mobility relative to cytochrome *c* in the Weber–Osborn and the Swank–Munkres gel systems, respectively. One of the radioactivity peaks was located in the protein with R_c^{wo} of 0.841 and R_c^{sm} of 1.220, which has an apparent molecular weight of 17,000 (*24*) or 15,000 (*25*), depending on the investigator. This protein has been identified as one of the heme *b* associated proteins (*26*). The second radioactivity peak was located at R_c^{wo} of 0.475 and R_c^{sm} of 0.32. This protein has been identified as a cytochrome *b* with a reported molecular weight of 37,000 (*26*). Since the distribution of the radioactivity is very specific, it is safe to assume

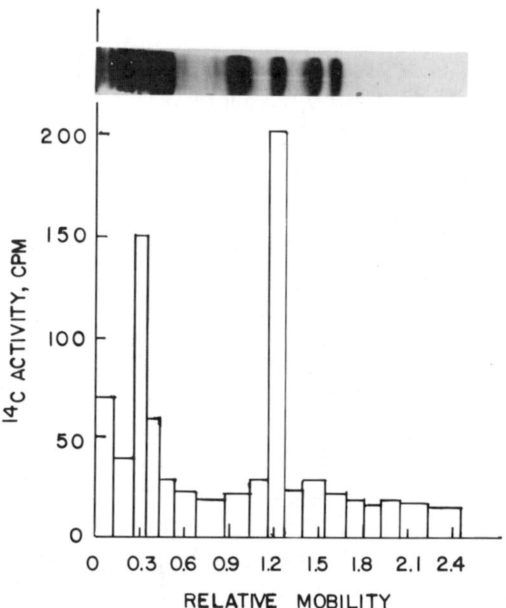

Fig. 4. Distribution of ^{14}C radioactivity among the proteins of the cytochrome b-c_1III complex in the SDS polyacrylamide gel electrophoresis system of Swank and Munkres. The same conditions were used as in Fig. 3 except that the dissociation of protein and SDS gel electrophoresis were carried out according to Swank and Munkres (*23*).

that the proteins with relative mobilities, R_c^{wo} of 0.841 and 0.475 or R_c^{sm} of 1.22 and 0.32 are responsible for Q binding (QP_c). Whether the Q is bound to both proteins, sandwiched between them, or bound to one of the two proteins and the binding site is physically close to the other one, can not be asserted definitely. Furthermore, from the fact that a close association and redox equilibrium exist between the ubisemiquinone radical and cytochrome b, it is also possible that Q serves as a heme ligand of one or both of the cytochromes b or is even sandwiched between them.

Effect of Antimycin A on Q_0C_{10}NAPA Binding in the Cytochrome b-c_1III Complex

When the phospholipid- and Q-depleted cytochrome b-c_1III complex was treated with antimycin A prior to addition of Q_0C_{10}NAPA and phospholipid, the radioactivity covalently bound to the protein after photolysis was not significantly different from that observed without antimycin A treatment. No significant change in the radioactivity distribution pattern was found in the protein bands with R_c^{wo} of 0.475 and 0.841. These results indicate that antimycin A does not bind to the same site as Q. This is, in fact, consistent with the report (*17*) that the antimycin A binding protein is a small-molecular-weight protein present in Complex III. According to Das Gupta and Rieske (*27*), the molecular weight of the antimycin A binding protein is 11,000, which is much smaller than that of a protein with a R_c^{wo} of 0.841, or a R_c^{sm} of 1.22.

Effect of Phospholipids on Q_0C_{10}NAPA Binding in the Cytochrome *b*-c_1III Complex

Phospholipid- and Q-depleted preparations are used in studies of Q binding to avoid the complications imposed by phospholipid, as Q is a hydrophobic molecule and associates with phospholipid. Since phospholipids are functionally and structurally (*28*) essential in the native complex, studies of specific Q binding in the absence of phospholipid may undermine the significance of the results. Therefore, we also carried out the illumination of the cytochrome *b*-c_1III complex in the presence of phospholipid and determined the distribution of radioactivity among the proteins of this enzyme complex. A slightly lower (10%) recovery of radioactivity in the proteins was obtained. The distribution pattern, however, was the same as that obtained in the absence of phospholipid. These results indicate that the Q_0C_{10}NAPA binding to a specific protein is not altered by phospholipid. This is consistent with the fact that addition of Q to the depleted cytochrome *b*-c_1III complex must be made prior to the addition of phospholipid to restore full activity (*7*). The slight decrease in radioactivity recovery could be attributed either to a lower photoreaction efficiency in the presence of phospholipid or, less likely to competition for the photoactivated nitrene between phospholipid and protein. Since the decrease in the amount of radioactivity recovered in the SDS polyacrylamide gel column is so small, its significance is questionable.

Interaction of Delipidated *b*-c_1III Complex with Spin-Labeled Q Derivatives

Although positive results were obtained from studies of Q binding to delipidated SCR, *b*-c_1III complex, and Complex II using radioisotope-labeled Q analogs, determination of Q-binding constants is complicated by the solubility of the Q analogs in the system. The solubility of the Q analogs is, as expected, detergent dependent.

An alternate approach to the study of Q binding to delipidated b-c_1III complex is by the use of spin-labeled Q derivatives. It is expected that, if binding occurs, some immobilization of the spin label will be observed. Spectrum A in Fig. 5 shows the EPR scan of a spin-labeled Q derivative, Q_0C_{10}TMPOC[4], bound to delipidated b-c_1III complex in

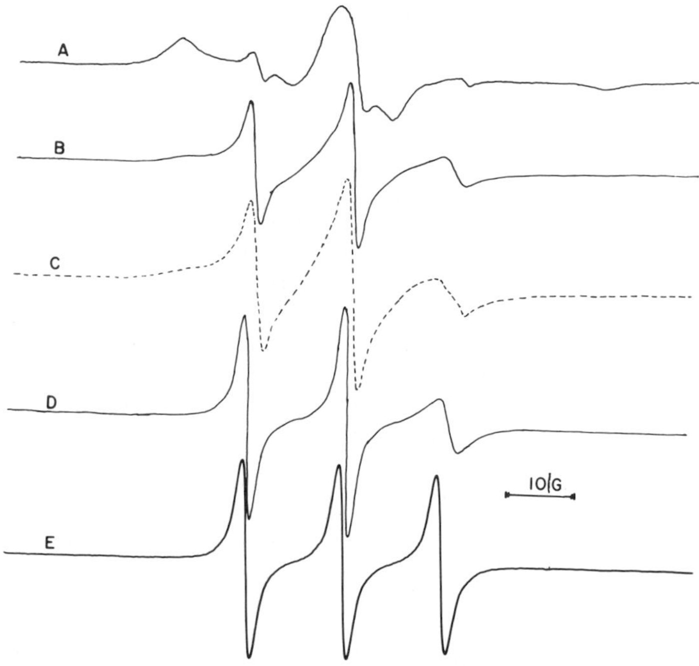

Fig. 5. EPR spectra of spin-labeled Q derivative, Q_0C_{10}TMPOC, in different environments. The spectra were taken on a Varian E-4 EPR spectrophotometer at room temperature, with the instrument settings as follows: field modulation frequency, 100 kHz, microwave power, 10 mW, microwave frequency 9.5 GHz, modulation amplitude 1 G, time constant 0.3 sec, scan rate 12.5 G/min. Spectrum A represents Q_0C_{10}TMPOC in delipidated b-c_1III complex, 13 mg/ml; spectrum B, same as spectrum A plus 1% SDS; spectrum C, same as spectrum A plus phospholipid, azolectin 0.2 mg/ml protein; spectrum D, in 2% sodium cholate. Spectra A to D are in 50 mM Na-K-phosphate buffer, pH 7.4, containing 20% glycerol. Spectrum E shows Q_0C_{10}TMPOC dissolved in 95% EtOH. The concentrations of Q_0C_{10}TMPOC used and the receiver gain settings were: 140 μM, RG 2×10^3; 120 μM, RG 8×10^2; 140 μM, RG 2×10^3; 93 μM, RG 8×10^2; 87 μM, RG 4×10^2; for spectra A to E, respectively.

[4] Q_0C_{10}TMPOC was slowly reduced by succinate in the presence of SCR. The rate of reduction is easily followed by noting the decrease in the intensity of the EPR signal. Q_0C_{10}TMPOC undergoes hydrolysis slowly in alcoholic solvents.

50 mM phosphate buffer, pH 7.4, containing 20% glycerol. Spin-labeled Q was immobilized upon interaction with delipidated protein, as indicated by the broadening of the EPR spectrum, compared to spectrum B obtained under the same conditions but in the presence of 1% SDS, or to spectrum D obtained in the buffer, with 2% sodium cholate in the absence of protein.

However, when phospholipids are added to the delipidated b-c_1III complex after the addition of Q_0C_{10}TMPOC, the environment of the spin label becomes more mobile, (indicated by the sharpening of the EPR spectrum C). It is not as mobile as spin label in buffer containing detergent (spectrum D) or ethanol (spectrum E). The EPR spectrum of Q_0C_{10}TMPOC in b-c_1III containing phospholipids is very similar to that observed when Q_0C_{10}TMPOC is added to phospholipid vesicles. These results suggest that in the presence of phospholipids, the spin label, or the alkyl side chain of Q, is preferentially bound to, or is extended into, the lipid environment. Such an environment is either provided by the added phospholipids alone, or by a combination of phospholipid and protein. The protein may undergo some conformational change upon the addition of phospholipids (*27*).

It should be mentioned that addition of phospholipids to the delipidated b-c_1III complex after addition of Q restores full enzymatic activity to the complex as well as the ability to generate the ubisemiquinone free radical. Therefore, phospholipids do not dislodge Q from its binding site. The identical photoaffinity labeling patterns of cytochrome b-c_1III in the presence and absence of phospholipids (as described in the previous section) also indicate that Q remains on its binding site. One possible explanation for the observation that spin label becomes mobile upon addition of phospholipids is that, in the presence of phospholipids, the alkyl side chain becomes mobile on the surface of QP_c, thus showing a more mobile EPR spectrum. The strong binding site of the Q molecule may depend mainly on the benzoquinone ring. The lack of enzymtic sensitivity to modifications of the alkyl or prenyl side chains also supports the above deduction.

Since Q_0C_{10}TMPOC is not soluble in the buffer system used, which lacks detergent, it shows very little EPR absorbance in the absence of the b-c_1III complex or detergent. The intensity of the EPR signal in the low field is proportional to the amount of Q_0C_{10}TMPOC bound to the protein. A titration of Q_0C_{10}TMPOC against delipidated b-c_1III complex is shown in Fig. 6. The maximum intensity was obtained when Q_0C_{10}TMPOC reached the same concentration as cytochrome b in the system. However, the significance of this stoichiometric relationship remains to be verified as the delipidated b-c_1III complex was prepared through ammonium sulfate precipitation in the presence of 0.5% sodium cholate, and the residual

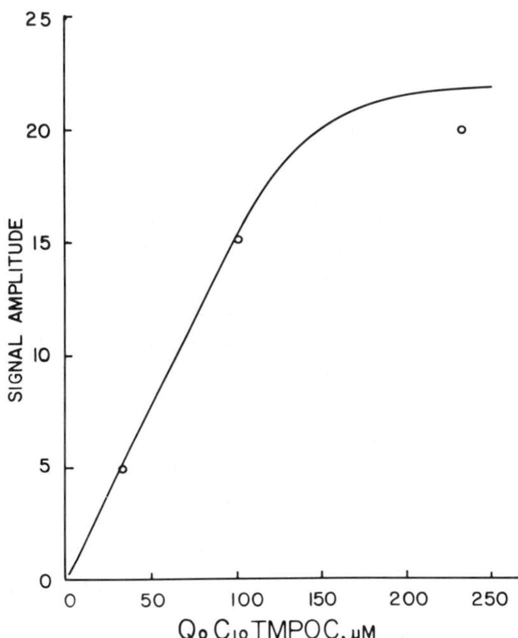

Fig. 6. Titration of phospholipid- and Q-depleted b-c_1III complex with Q_0C_{10}TMPOC. Aliquots of 0.3 ml of delipidated b-c_1III, 13 mg/ml in 50 mM phosphate buffer pH 7.4, containing 20% were mixed with various amounts of Q_0C_{10}TMPOC. The EPR signal was measured under the same conditions as those used for spectrum A in Fig. 5.

amount of cholate present in the system, with its accompanying effect on binding has not yet been studied.

DISCUSSION

Of the three suggested Q proteins (QP_s, QP_c, and QP_n) in the mitochondrial electron transport system, QP_s has received the most attention. Its existence has been confirmed by many laboratories using various isolation methods with different purities, and denoted by different terminologies. Isolation procedures for QP_c and QP_n have not yet been developed. The intact cytochrome b-c_1III complex is still the most purified, active preparation. Fractionation beyond that has resulted in inactive proteins. The use of photoaffinity labeling techniques to identify the protein responsible for QP_c is only the first step toward actual isolation. The properties of QP_c and the conditions required for the formation of the ubisemiquinone radical indicate that QP_c is closely associated with phospholipids. Thus far successful isolation and reconstitution studies of electron-transport components are of components not closely involved with phospholipids; none of the components strongly associated with phospholipids in the mitochondrial respiratory chain have yet been isolated in active form. This is probably due to the lack of information about the interaction between protein and phospholipids. The isolated protein components might actually be in an active form, and show a lack of enzymatic activity due to a lack of proper lipid involvement. With this in mind, and with knowledge of the identity of QP_c as *b*-cytochrome proteins, future isolation attempts should stress protein–lipid interactions.

Although the photoaffinity labeling study demonstrated clearly that QP_c may be identical to one or both of the cytochrome *b* proteins, the possibility exists that neither of these two *b* proteins is responsible for QP_c, although this is unlikely. This remote possibility stems from the fact that the photoaffinity label was attached to the alkyl side chain of the Q molecule, at which location modification produces the least change in enzymatic activity. The spin-label studies showed that the end of the 6-alkyl chain is in a fairly mobile state when phospholipids are present, while the photoaffinity label experiments conducted in the presence and absence of phospholipids gave identical results. Placement of the photoaffinity label closer to the benzoquinone ring of the Q molecule may give a more definite conclusion. At present, we can state only that QP_c could be one or both of the cytochrome *b* proteins, or a protein closely associated with cytochrome *b* or could even be sandwiched between the two cytochrome *b* proteins. The final proof of QP_c must await successful isolation and reconstitution.

Since two cytochrome *b* proteins have been indicated as QP_c in this study, it is worthwhile to reiterate the abnormal behavior of these two proteins on SDS gel electrophoresis. The mobility of the cytochrome *b* protein with R_c^{wo} of 0.475 is very dependent on the detergent present in the sample. Under some conditions it moves faster and merges into the cytochrome c_1 band, and in other conditions it moves slowly enough to overlap with the second band of the *b*-*c*III complex. Several investigators were not able to observe the existence of this protein band in SDS polyacrylamide gel electrophoresis, probably due to this behavior. The smaller cytochrome *b* protein, R_c^{wc} of 0.841, also showed abnormal mobility, although it generally moved slower if the bile salt was removed completely before the protein was dissociated with SDS. The electrophoretic patterns of *b*-c_1III complex, as shown in Figs. 3 and 4, are readily reproducible in our laboratory if the sample is dialyzed extensively before treatment with

2 mg SDS/mg protein in Na-phosphate buffer containing 1% β-ME. Both cytochrome *b* proteins stain poorly in the Weber–Osborn system.

The use of spin-labeled Q derivatives in the study of Q binding and interaction between bound Q and free Q has the advantage of high sensitivity but may suffer from a lack of specificity, especially when the spin label is placed on the alkyl side chain of Q, where interaction with phospholipids is inevitable. Titration of the delipidated b-c_1III complex with Q_0C_{10}TMPOC shows a clear saturation point. In the presence of phospholipids, however, the results are less pronounced. Even in the absence of phospholipids, interpretation of saturation concentration should be approached with caution, as residual detergents are present in the system. Perhaps placement of the spin label at the benzoquinone ring will avoid the complications with phospholipids, if such a spin-labeled Q derivative is functionally active, or is a potent inhibitor of Q. A more detailed investigation on Q binding in the interaction between the spin label and ubisemiquinone radical formed during reduction is currently in progress in our laboratory.

ACKNOWLEDGMENTS

We express our gratitude to Joy Steidl and Paul Haley for their technical assistance. This work was supported by the grants from the National Institute of Health (GM 26292), and from the National Science Foundation (PCM 78-01394).

REFERENCES

1. Green, D. E. (1962). *Comp. Biochem. Physiol.* **4,** 81–122.
2. Depierre, I. W., and Ernster, L. (1977). *Annu. Rev. Biochem.* **40,** 201–262.
3. Crane, F. L. (1977). *Annu. Rev. Biochem.* **46,** 439–469.
4. Yu, C. A., Yu, L., and King, T. E. (1977). *Biochem. Biophys. Res. Commun.* **78,** 259–265.
5. Green, D. E., and Baum, H. (1969). "Energy and the Mitochondrion," p. 205. Academic Press, New York.
6. Klingenberg, M. (1968). *In* "On Biological Oxidations" (T. P. Singer, ed.), pp. 3–54. Wiley (Interscience), New York.
7. Yu, L., Yu, C. A., and King, T. E. (1978). *J. Biol. Chem.* **253,** 2657–2663.
8. Yu, C. A., Yu, L., and King, T. E. (1974). *J. Biol. Chem.* **249,** 4905–4910.
9. Rieske, J. S. (1967). *In* "Oxidation and Phosphorylation" (R. W. Estabrook and M. E. Pullman, eds.), Methods in Enzymology, Vol. 10, pp. 239–244. Academic Press, New York.
10. Yu, C. A., and Yu, L. (1980). *Biochemistry* **19,** 3579–3585.

11. Yu, C. A., Yu, L., and King, T. E. (1977). *Biochem. Biophys. Res. Commun.* **79,** 939–946.
12. Ackrell, B. A. C., Ball, M. B., and Kearney, E. B. (1980). *J. Biol. Chem.* **255,** 2761–2769.
13. Hatefi, Y., and Galante, Y. M. (1980). *J. Biol. Chem.* **255,** 5530–5537.
14. Vinogradov, A. D., Gavrikov, V. G., and Gavrikova, E. V. (1980). *Biochim. Biophys. Acta* **592,** 13–27.
15. Yu, C. A., and Yu, L. (1980). *Biochim. Biophys. Acta* **591,** 409–420.
16. Yu, C. A., Nagaoka, S., Yu, L., and King, T. E. (1978). *Biochem. Biophys. Res. Commun.* **82,** 1070–1078.
17. Yu, C. A., Nagoaka, S., Yu, L., and King, T. E. (1980). *Arch. Biochem. Biophys.* **204,** 59–70.
18. King, T. E., Yu, L., Nagaoka, S., Widger, W. R., and Yu, C. A. (1978). *In* "Frontiers of Biological Energetics" (P. L. Dutton, J. S. Leigh, and A. Scarpar, eds.), Vol. 1, pp. 174–182. Academic Press, New York.
19. Yu, C. A., and Yu, L. (1980). *Biochem. Biophys. Res. Commun.* **96,** 286–292.
20. Wan, Y. P., Williams, R. H., Folkers, K., Leung, K. H., and Racker, E. (1975). *Biochem. Biophys. Res. Commun.* **63,** 11–15.
21. Jeng, S. J., and Guillory, R. J. (1975). *J. Supramol. Struct.* **3,** 448–468.
22. Weber, K., and Osborn, M. (1969). *J. Biol. Chem.* **244,** 4406–4412.
23. Swank, R. T., and Munkres, K. D. (1971). *Anal. Biochem.* **39,** 462–477.
24. Yu, L., Yu, C. A., and King, T. E. (1977). *Biochim. Biophys. Acta* **495,** 232–247.
25. Nelson, B. D., and Gellerfords, P. (1978). *In* "Biomembranes: Part D: Biological Oxidations, Mitochondrial and Microbial Systems" (S. Fleischer and L. Packer, eds.), Methods in Enzymology, Vol. 53, pp. 80–91. Academic Press, New York.
26. Yu, C. A., Yu, L., and King, T. E. (1975). *Biochem. Biophys. Res. Commun.* **66,** 1194–1200.
27. DasGupta, U., and Rieske, J. S. (1973). *Biochem. Biophys. Res. Commun.* **54,** 1247–1253.
28. Yu, C. A., and Yu, L. (1980). *Biochemistry* **19,** 5715–5720.

Complex III Isolated in Triton X-100 as a Valuable System in Studying the Mechanism of Energy Conversion

8

G. VON JAGOW
W. D. ENGEL
H. SCHÄGGER
W. F. BECKER

INTRODUCTION

In this contribution we describe the composition and function of a Complex III prepared by a new method. Instead of the anionic detergent cholate, we used the nonionic detergent Triton X-100, and instead of purification by salt fractionation, we employed hydroxyapatite chromatography followed by gel chromatography. The complex prepared by this method shows a protein composition and functional properties quite similar to the cholate complex. It possesses a high ubiquinol:cytochrome *c* reductase activity and translocates protons across a phospholipid membrane.

In addition to the salt fractionation method (*1*) and the hydroxyapatite method described here (*2*), an affinity chromatography technique (*3*) has recently been used to isolate Complex III. This method has been described in detail only for *Neurospora crassa*. Isolation from beef heart mitochondria should be possible however, provided yeast cytochrome *c* is linked through its cysteine residue to the Sepharose 6B (*4*).

What are the advantages of the hydroxyapatite method? The prepara-

Function of Quinones in Energy Conserving Systems

ISBN 0-12-701280-X

tive expense is lower than for the salt fractionation procedure and the yield is higher than for the affinity method. All kinds of scales can be performed on hydroxyapatite, from a small-scale procedure starting with only a few mg of mitochondrial protein up to a large-scale procedure starting with 10 g of mitochondrial protein (*5*).

ISOLATION PROCEDURE AND COMPOSITION OF THE TRITON COMPLEX

Initially, when using the hydroxyapatite method (*6*), we stabilized the complex by adding antimycin. At the end of the procedure, the inhibitor was still bound to the protein. The redox potentials of the heme centers indicated a functional state of the preparation. Apart from the fact that electron flow was blocked by the bound antimycin, the complex was devoid of the iron–sulfur protein, a component essential for electron transfer from ubiquinol to cytochrome *c* (*7*). The iron–sulfur protein was cleaved from the complex when the complex was bound to hydroxyapatite. The dissociation seemed to be induced by the large amounts of Triton X-100 in contact with the complex during the chromatographic process. The detergent not only removed iron–sulfur protein, but also stripped the complex of lipids and ubiquinol. Thus, to isolate a complete and enzymatically active complex, we had to avoid stabilization of the conformation by antimycin and modify the hydroxyapatite chromatography. Our procedure was modified by the application of a hydroxyapatite batch step (*2*). The fractions obtained in this step were only 60% pure and had to be purified subsequently by gel chromatography. The final yield amounted to 30%.

Electrophoresis of the preparation revealed 8 bands indicating molecular weights of 49, 47, 30, 30, 25, 12, 10, and 6 kdalton (Fig. 1). The 30 kdalton band consists of cytochromes *b* and c_1. The 25 kdalton band represents the iron–sulfur protein. The complex has a molecular weight of 250,000, but is isolated in a dimeric state with a molecular weight of 500,000, as shown by ultracentrifugal studies. The specific detergent binding amounts to 0.2 g of Triton X-100/g of protein. The investigations described in the following were performed on this dimeric system.

ELECTRON TRANSFER ACTIVITY OF THE TRITON COMPLEX

When the complex is not an integral part of the membrane, *i.e.*, when no vectorial processes are possible, it catalyzes merely an electron trans-

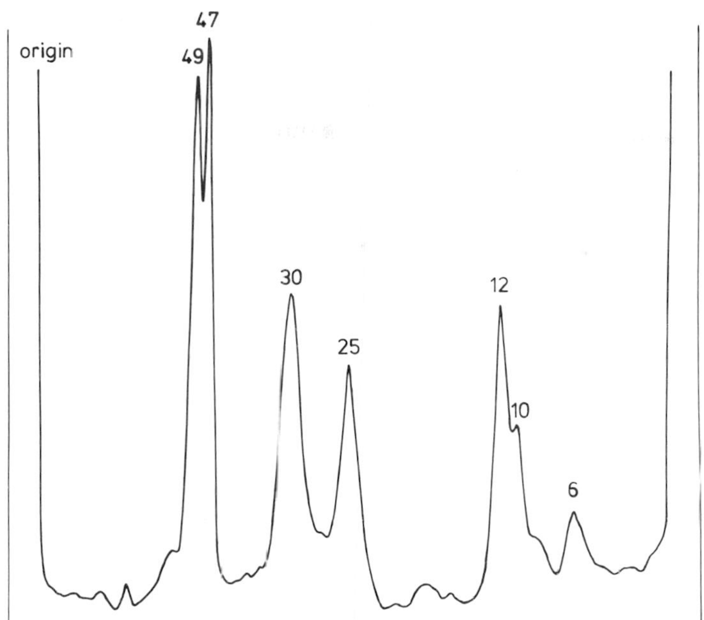

Fig. 1. Polypeptide pattern of the reductase isolated in Triton X-100. Electrophoresis was performed according to Neville and Glossman (*25*). The numbers indicate the molecular weights in kdalton.

fer from ubiquinol via *b*-type cytochromes, the iron–sulfur protein and cytochrome c_1, and finally to cytochrome *c*. The sum of the various consecutive reactions can be expressed as follows

$$QH_2 + 2 \text{ cytochrome } c^{3+} \longleftrightarrow Q + 2 \text{ cytochrome } c^{2+} + 2H^+ \quad (1)$$

In mammalian mitochondria, the physiological quinone is ubiquinone-10, but isolated beef heart complex also accepts electrons from quinones with short isoprenoid side chains. In the latter case, catalytic activity can be measured without detergent in the test buffer.

The reductase-dependent reduction of cytochrome *c* is inhibited by antimycin; less than 3% of the original activity remains. The reductase shows a maximal molecular activity with ubiquinol-2 as electron donor, whereas the K_M value is apparently independent of side chain length (Fig. 2 and Table I). Duroquinol is a less suitable substrate for the enzyme, but its oxidation can also be completely inhibited by antimycin.

Ubiquinols with long side chains can only be solubilized in buffer containing Triton X-100. The ubiquinol moves into the Triton micelles. When measuring electron-transfer activity under these conditions, Triton X-100 seems to act as an inhibitor competing with ubiquinol for the binding site.

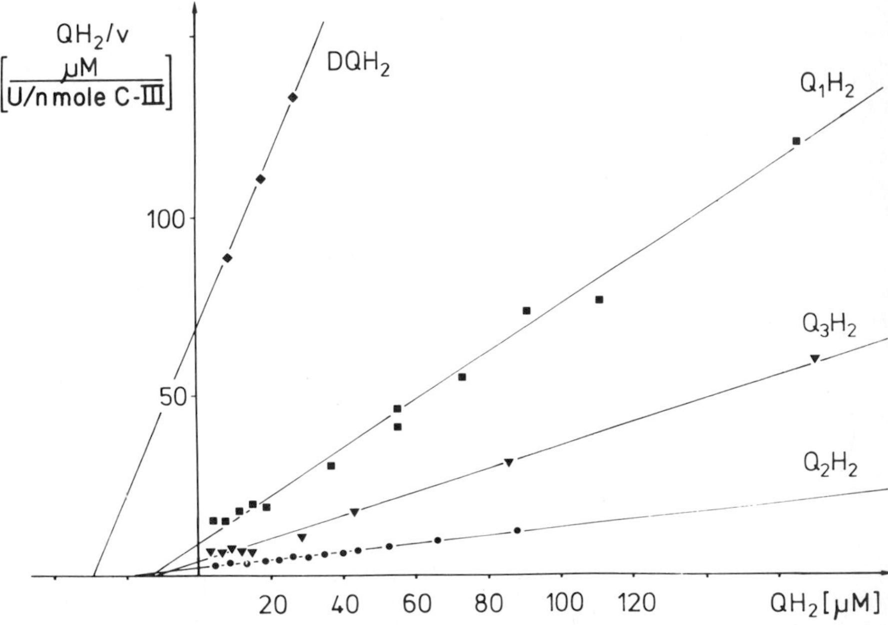

Fig. 2. Hanes plot for different quinols. The buffer (*26*) contained no detergent.

A more plausible explanation than competitive inhibition of the enzyme is that there is competition between the Triton micelle and the enzyme for the quinol. However, of probable occurrance is an additional reaction step preceding the normal Michaelis–Menten kinetics, in which the Triton micelle, loaded with ubiquinone, collides with the enzyme (Fig. 3). With regard to this collision step note that the enzyme is held in solution through binding of one Triton micelle.

The whole kinetic event can be described by the "surface dilution

TABLE I

Kinetic Constants of the Ubiquinol : Cytochrome *c* Reductase Isolated in Triton X-100

Quinol	K_M^*	V	Molecular activity
	(μM)	(U/nmole complex III)	(1/sec)
Q_1H_2	14 ± 6	1.5 ± 0.2	25
Q_2H_2	17 ± 7	8.9 ± 1.1	149
Q_3H_2	12 ± 6	3.1 ± 0.2	52
DQH_2	30 ± 8	0.4 ± 0.1	7

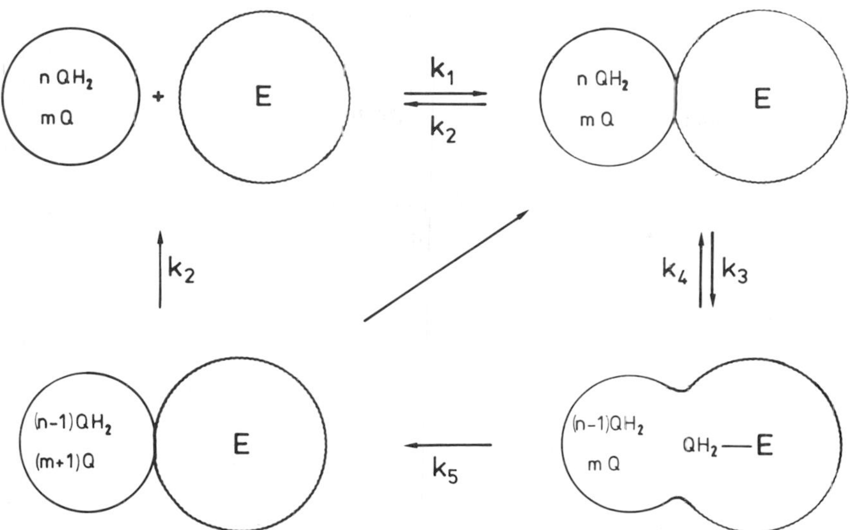

Fig. 3. Scheme of the reaction sequence of long-chain ubiquinol oxidation in a micellar test system. In a first step the ubiquinol-containing Triton micelle (○) collides with the enzyme micelle (◯). The following steps are analogous to conventional enzyme kinetics. The ubiquinol-containing micelle and the enzyme micelle, still connected after the decomposition of the Michaelis complex, may react once again (diagonal arrow) or dissociate into free micelles.

scheme'' developed by Dennis to analyze phospholipase activities (*8*). As in Complex III, these enzymes work with a water-insoluble substrate, namely phospholipids. The velocity of the first reaction step of the ubiquinol: cytochrome *c* reductase activity depends on the sum of the surface areas of the mixed ubiquinol–Triton micelles. Since we cannot yet calculate this area exactly, the amount of disposable binding area was replaced in a first approximation by the total concentration of Triton X-100 plus ubiquinol $A = [TX] + [Q]$. In the second step the enzyme–substrate complex is formed. The velocity of this formation depends on the availability of ubiquinol in the mixed micelles. This availability has been expressed as the mole fraction of ubiquinol in the micelle, $B = [Q]/([Q] + [TX])$. Subsequently, in the third reaction step, the enzyme–substrate complex dissociates into product and free enzyme. The whole reaction is described by the following equation:

$$v = \frac{V(A)(B)}{K_s'K_M' + K_M'(A) + (A)(B)} \tag{2}$$

where $K_s' = k_2/k_1$ and $K_M' = (k_4 + k_5)/k_3$.

With ubiquinol-9 as substrate, the maximal activity of the complex amounts to 4.5 U/nmole cytochrome c_1 at 25 °C and pH 7. The K'_M value, expressed as mole fraction of ubiquinol in the ubiquinol-Triton micelle, is 0.025 (*2*).

Kinetic analysis of the reductase by the surface dilution scheme indicates that the detergent Triton is not an inhibitor of the enzyme, as mimicked when the reaction is described in terms of conventional enzyme kinetics (*9*, *10*), but a diluent of the substrate ubiquinol.

INHIBITION OF THE CATALYTIC ACTIVITY OF THE ISOLATED COMPLEX BY ANTIMYCIN AND 2-NONYL-4-HYDROXY-QUINOLINE-*N*-OXIDE

As mentioned previously, antimycin completely blocks electron-transport activity of the isolated complex. Figure 4 gives a dose–response curve for antimycin inhibition of the ubiquinol-9 : cytochrome *c* reductase activity. Besides inhibiting electron flow, antimycin induces a shift of

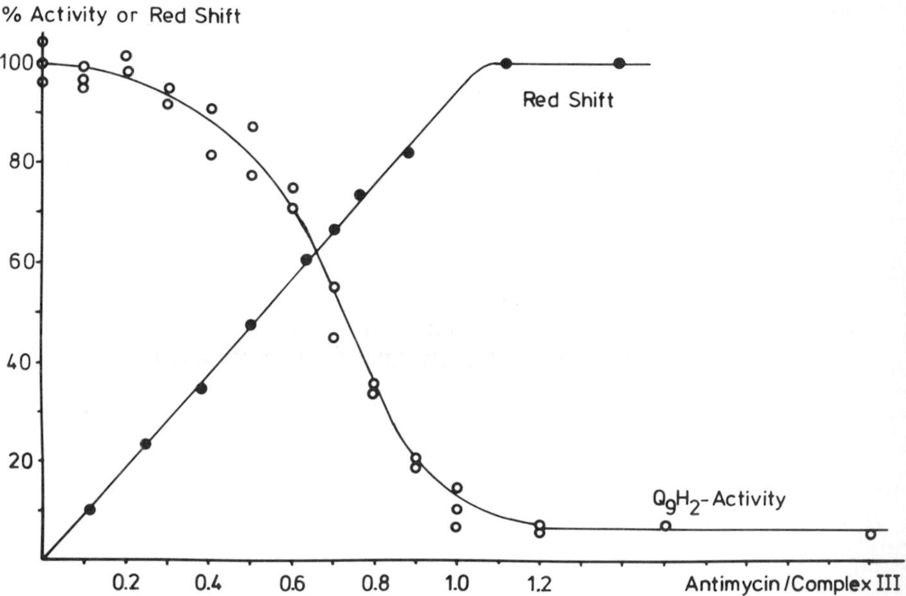

Fig. 4. Antimycin binding and inhibition curve of the isolated Complex III. Antimycin binding was measured by the red shift of the ferrocytochrome *b* α-band (cf. Fig. 5). The inhibition was measured at a constant ubiquinol-9 concentration of 120 μM and a Triton concentration of 2.5 mM. The complex was preincubated with the respective antimycin concentration for 2 minutes.

about 0.5–1.0 nm to the red of the α- and γ-absorbance bands of reduced cytochrome *b*, shown in Fig. 5. The maximum of the α red shift lies at 564 nm, that of the γ red shift at 433 nm. The γ signal is about five times larger than the α signal. Both signals can be used for a quantitative analysis of the binding of inhibitor to its point of action, as demonstrated in Fig. 4. Obviously, antimycin binds in a linear concentration dependence

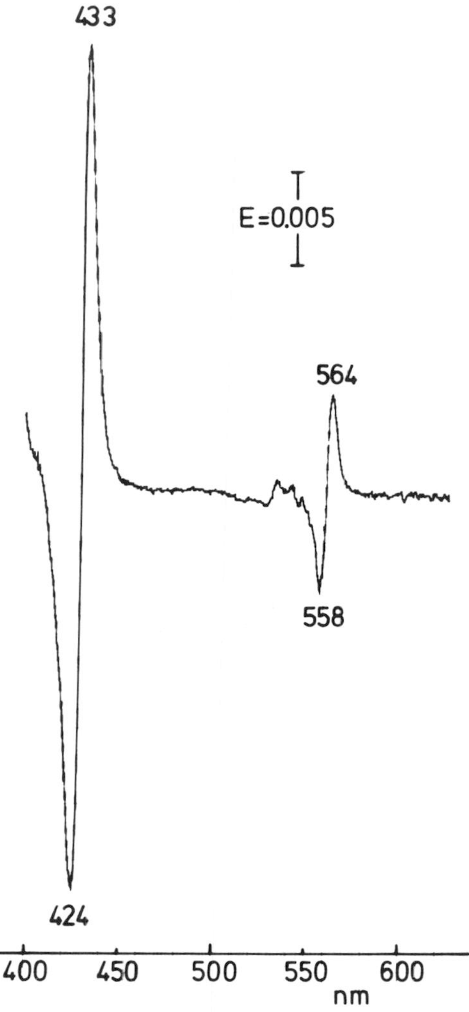

Fig. 5. Antimycin-induced α and γ red shift of ferrocytochrome *b* of Complex III. Complex III 2.3 μM was reduced by dithionite and antimycin was added in tenfold excess over the binding sites into the sample cuvette.

to its specific binding sites. The titer amounts to 1 mole of antimycin/mole of complex; this is the same titer as that required for complete inhibition of electron flow.

The sigmoidal antimycin titration curves of substrate oxidation in mitochondria and mitochondrial particles, (for example succinate or NADH oxidation), were explained by a kinetic model that assumes the turnover rate of the dehydrogenases is usually lower than that of the cytochromes (*11*). In contrast to this interpretation, it was proposed that under certain conditions cooperative effects may also play a role (*12*). Since the isolated complex is a dimer with two antimycin binding sites and the antimycin titration curve of the reductase activity is sigmoidal (Fig. 4), it was tempting to speculate that the two monomers work cooperatively in the process of electron transfer. However, with long side chain ubiquinols as substrates, the micelle–micelle collision step (Fig. 3) may act as a rate-limiting step of the whole reaction. If this occurred, a sigmoidal titration curve would also result, since the collision step would only be ineffective when both halves of the dimer were inhibited. Therefore we performed experiments in which we varied the ubiquinol concentration at each point of antimycin inhibition. Thus an extrapolation to ubiquinone saturation became possible. In these experiments a linear antimycin titration curve for ubiquinol-9 oxidation was obtained (Fig. 6). The experiments were performed at constant Triton concentrations with a large excess of Triton over ubiquinol. Therefore equation (2) is reduced to

$$v = \frac{V[\mathrm{QH_2}]}{K_s'K_M' + K_M'[\mathrm{TX}] + [\mathrm{Q_9H_2}]} \tag{3}$$

since $A \approx [\mathrm{TX}]$, and $B \approx [\mathrm{Q_9H_2}]/[\mathrm{TX}]$. At constant Triton concentration an apparent $K_M{}^{\mathrm{x}} = K_M' \cdot [\mathrm{TX}] + K_M'K_{\mathrm{s}}$ is obtained. As shown in the lower part of Fig. 6, $K_M{}^{\mathrm{x}}$ is constant, which seems to indicate noncompetitive inhibition. This conclusion is not justifiable, however, since tightly binding inhibitors would in any case mimic this type of inhibition (*13*). We conclude that there is no cooperation between the two halves of the Complex III dimer during electron flow. Moreover, under the conditions of Fig. 4 the first reaction step is rate-limiting, and not the formation of the enzyme–substrate complex.

Electron flow of the isolated complex from ubiquinol-9 to cytochrome *c* can also be inhibited by alkyl-hydroxy-*N*-oxides, such as heptylhydroxy-*N*-oxide (HQNO). These inhibitors bind much less firmly to their specific binding sites than antimycin.

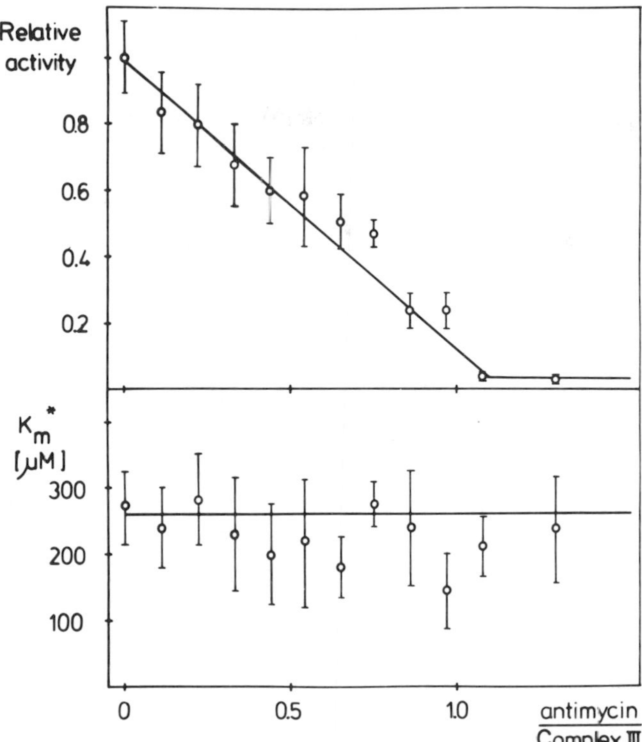

Fig. 6. Antimycin inhibition curve of the ubiquinol: cytochrome *c* reductase activity of isolated Complex III. Each point represents a *V* or $K_M{}^x$ value with standard error, obtained by a test series at different ubiquinol concentrations. The enzyme was preincubated with antimycin for 2 min. The Triton concentration in the test buffer was 2.5 m*M*.

The type of inhibition, whether competitive or not, was tested by kinetic experiments using increasing ubiquinol-9 concentrations at a constant inhibitor concentration. The typical picture of noncompetitive inhibition is obtained, as shown in Fig. 7. HQNO seems to bind to free enzyme as well as to the enzyme–substrate complex but does not seem to attack the binding site for the substrate ubiquinol. The K_i amounts to 13 μM.

It was shown in inhibitor competition experiments using a fluorescence quench technique, that antimycin displaces NQNO from its binding site (*14*). It has been concluded that antimycin and HQNO bind to the same site (*14*). If this is true, it seems reasonable to assume that antimycin, like HQNO, inhibits in a fashion noncompetitive with ubiquinol.

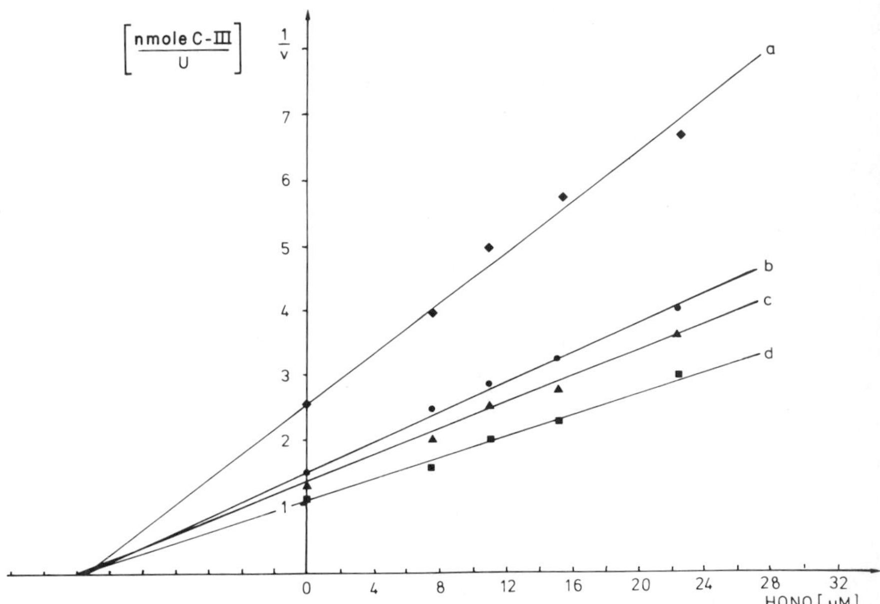

Fig. 7. Dixon plot of the HQNO inhibition. The ubiquinol concentrations were (a) 43 μM, (b) 86 μM, (c) 129 μM and (d) 172 μM. Test conditions as described in Fig. 4.

PROTON-TRANSFERRING ACTIVITY OF THE TRITON COMPLEX

The ubiquinol:cytochrome *c* reductase not only transfers electrons from ubiquinol to cytochrome *c*; it also translocates protons. When incorporated into a phospholipid membrane, protons are translocated from one side of the membrane to the other in a process coupled to electron flow. The mechanism of this translocation is still unknown, but a protonmotive ubiquinone cycle or a redox-driven proton pump has been suggested (see next section).

Previously proton translocation was measured both in mitochondria and reconstituted Complex III formed from phospholipid vesicles and a complex isolated using cholate as detergent (*15*, *16*). For measuring proton-translocating activity we incorporated our Triton X-100 Complex III into phospholipid vesicles by the cholate dialysis procedure. Similar to the complex isolated by salt fractionation with cholate as detergent, the Triton complex releases $2H^+$/cytochrome *c* reduced, as shown in Fig. 8. Only one of these protons is translocated electrogenically in an active process across the membrane and, therefore, flows back if uncoupler is added. The other proton is released to the outside in the course of the ubi-

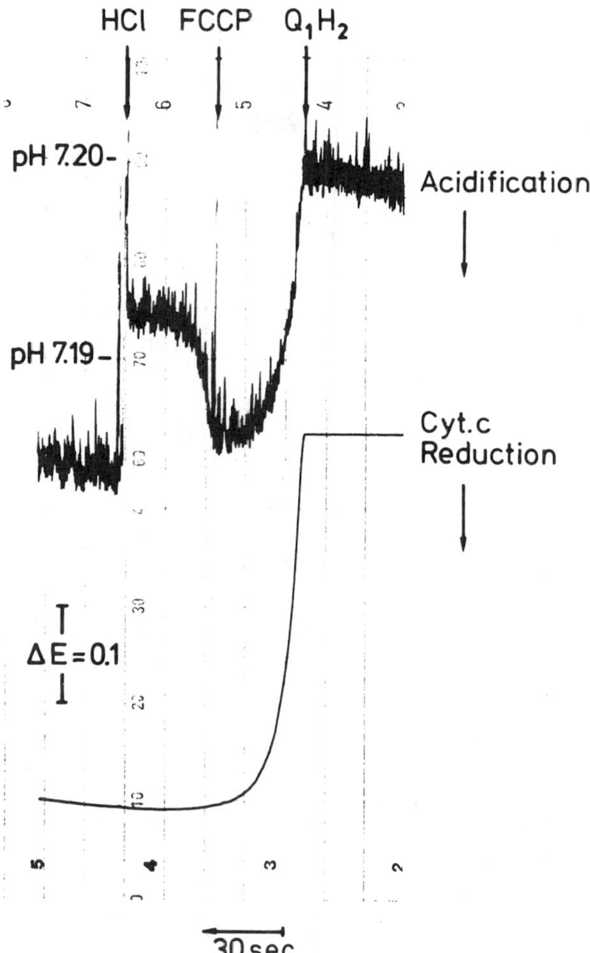

Fig. 8. Proton translocation by Complex III isolated in Triton X-100, when incorporated into soybean phospholipid vesicles by the cholate dialysis procedure. The final volume in the cuvette amounts to 2 ml, containing 1.1 μM Complex III incorporated in phospholipid vesicles. The test buffer consists of 100 mM KCl, 2 mM MOPS, 5 μM valinomycin and 30 μM cytochrome *c*. The reaction is started by the addition of 20 nmole ubiquinol-1. About 70 nmole of protons are released when 40 nmole of cytochrome *c* are reduced. About 30 nmole of protons flow back upon uncoupling by addtion of 5 μM FCCP. The calibration of the system is performed by addtion of 40 nmole HCl. A proton/electron stoichiometry of 1.75 is calculated for this experiment, indicative of an integer stoichiometry of $2H^+/e^-$.

quinol oxidation process. This reaction does not involve charge transfer across the membrane and, therefore, this proton does not flow back when uncoupler is added, according to the following equation.

$$QH_2 + 2 \text{ cytochrome } c^{3+} + 2H^+_{in} \longrightarrow Q + 2 \text{ cytochrome } c^{2+} + 4H^+_{out} \quad (4)$$

The described measurements are too preliminary to permit a distinction between the reaction mechanisms proposed so far. Further experiments with modified protein and/or substrate may elucidate details of the translocation mechanism. The present experiments show, however, that our complex isolated in Triton X-100 possesses all components necessary for proper function.

THE POSSIBLE MECHANISM OF PROTON TRANSPORT DRIVEN BY REDOX REACTIONS AT ENERGY CONVERSION SITE 2

The proton translocating mechanism proposed originally by Mitchell for energy conversion site 2 is a redox loop mechanism (*17*). This mechanism requires an alternating sequence of an electron carrier, a hydrogen carrier, and again an electron carrier. The hydrogen carrier has to accept the electron on the inner side of the mitochondrial membrane, accompanied by the uptake of a proton from the matrix face. Subsequently the hydrogen carrier has to transfer the electron onto the following electron carrier, which is located on the outer side of the membrane. This process is accompanied by release of the proton into the cytosol. Thus proton translocation is made possible by vectorial electron flow from the inner to the outer side of the mitochondrial membrane.

Later Mitchell (*18*, *19*) proposed a more complex kind of loop mechanism which he called the protonmotive Q cycle. Its existence would establish a transport of two protons across the membrane per one electron transported from ubiquinol to cytochrome *c*. Some experiments support the existence of this mechanism, but do not supply proof for the arrangement of ubiquinone on the substrate as well as on the oxygen side of cytochrome *b* of Complex III, i.e., an electron backflow from Complex III to Complex I by means of ubisemiquinone (*20*).

In our view there seem to be more indications for the existence of a proton transport by cytochrome *b* (*21*). This alternative was defined by us as the cytochrome *b* redox pump mechanism (*22*).

What are the indications gathered to date for the existence of such a cytochrome *b* proton pump? The oldest and nearly classical indication is the fact that there is a crossover point between cytochromes *b* and cytochrome c_1; i.e., when the mitochondrial membrane is brought from a deenergized to an energized state, cytochromes *b* are reduced, whereas cytochrome c_1 is oxidized. There are a number of further reactions indicating that the cytochromes *b* of Complex III are intimately involved in the process of energy conversion (*20*). Moreover, the two cytochromes *b* of Complex III possess what is called the Bohr effect, i.e., an amino acid residue, so far undefined, is present in a protonated state when cytochrome *b* is reduced and present in an unprotonated state when cytochrome *b* is oxidized (*23*, *24*). The sequence of the reversible reactions can be formulated as follows

$$\text{cyt. } b^{2+}\text{RH}^+ \xleftrightarrow{\ e^-\uparrow\ } \text{cyt. } b^{3+}\text{RH}^+ \xleftrightarrow{\ H^+\uparrow\ } \text{cyt. } b^{3+}\text{R} \qquad (5)$$

However, the concept of a cytochrome *b* proton pump lacks conclusive experimental support. Further experiments are needed to determine whether the protons are translocated by a Q-loop or a proton pump mechanism.

ACKNOWLEDGMENTS

This work was supported by a grant of the Deutsche Forschungsgemeinschaft to G.v.J. and by a fellowship of the Studienstiftung des Deutschen Volkes to W.D.E. We are indebted to Eisai Co., Ltd., Tokyo, Japan for generous gifts of different ubiquinones and to Hoffmann–La Roche, Basel, Switzerland, for the generous gift of ubiquinone-9. The skillful assistance of Miss C. Michalski is gratefully acknowledged.

REFERENCES

1. Hatefi, Y. (1978). *In* "Biomembranes: Part D: Biological Oxidations, Mitochondrial and Microbial Systems" (S. Fleischer and L. Packer, eds.), Methods in Enzymology, Vol. 53, p. 35–40. Academic Press, New York.

2. Engel, W. D., Schägger, H., and von Jagow, G. (1980). *Biochim. Biophys. Acta* **592,** 211–222.

3. Weiss, H., Juchs, B., and Ziganke, B. (1978). *In* "Biomembranes: Part D: Biological Oxidations, Mitochondrial and Microbial Systems" (S. Fleischer and L. Packer, eds.), Methods in Enzymology, Vol. 53, pp. 99–112. Academic Press, New York.
4. Bill, K., Casey, R. P., Broger, L., and Azzi, A. (1980). *FEBS Lett.* **120,** 248–250.
5. von Jagow, G., Schägger, H., Riccio, P., Klingenberg, M., and Kolb, H. J. (1977). *Biochim. Biophys. Acta* **462,** 549–558.
6. von Jagow, G., Schägger, H., Engel, W. D., Riccio, P., Kolb, H. J., and Klingenberg, M. (1978). *In* "Biomembranes: Part D: Biological Oxidations, Mitochondrial and Microbial Systems" (S. Fleischer and L. Packer, eds.), Methods in Enzymology, Vol. 53, pp. 92–98. Academic Press, New York.
7. Trumpower, B. L., and Edwards, C. A. (1979). *J. Biol. Chem.* **254,** 8697–8706.
8. Deems, R. A., Eaton, B. R., and Dennis, E. A. (1975). *J. Biol. Chem.* **250,** 9013–9020.
9. Engel, W. D., Schägger, H., and von Jagow, G. (1979). *In* "Metalloproteins" (U. Weser, ed.), pp. 185–193. Thieme, Stuttgart.
10. Weiss, H., and Wingfield, P. (1979). *Eur. J. Biochem.* **99,** 151–160.
11. Kröger, A., and Klingenberg, M. (1973). *Eur. J. Biochem.* **34,** 358–368.
12. Berden, J. A., and Slater, E. C. (1972). *Biochim. Biophys. Acta* **256,** 199–215.
13. Williams, J. W., and Morrison, J. F. (1979). *In* "Enzyme Kinetics and Mechanism," Part A (D. L. Purich, ed.), Methods in Enzymology, Vol. 63, pp. 437–467. Academic Press, New York.
14. von Ark, G. (1980). Ph.D. Thesis, Univ. of Amsterdam.
15. Guerrieri, F., and Nelson, B. D. (1975). *FEBS Lett.* **54,** 339–342.
16. Leung, K. H., and Hinkle, P. C. (1975). *J. Biol. Chem.* **250,** 8467–8471.
17. Mitchell, P. (1961). *Nature* (*London*) **191,** 144–148.
18. Mitchell, P. (1975). *FEBS Lett.* **56,** 1–6.
19. Mitchell, P. (1975). *FEBS Lett.* **59,** 137–139.
20. von Jagow, G., and Sebald, W. (1980). *Annu. Rev. Biochem.* **49,** 284–314.
21. von Jagow, G., and Engel, W. D. (1980). *Angew. Chem.* **92,** 684–700; *Angew. Chem., Int. Ed. Engl.* **19,** 659–675.
22. von Jagow, G., and Engel, W. D. (1980). *FEBS Lett.* **111,** 1–5.
23. Urban, P. F., and Klingenberg, M. (1969). *Eur. J. Biochem.* **9,** 519–525.
24. Wilson, D. F., Erecinska, M., Leigh, J. S., and Koppelmann, M. (1972). *Arch. Biochem. Biophys.* **151,** 112–121.
25. Neville, D. M., and Glossman, H. (1974). *In* "Biomembranes," Part B (S. Fleischer and L. Packer, eds.), Methods in Enzymology, Vol. 32, pp. 92–102. Academic Press, New York.
26. Rieske, J. S. (1967). *In* "Oxidation and Phosphorylation" (R. W. Estabrook and M. E. Pullman, eds.), Methods in Enzymology, Vol. 10, pp. 239–245. Academic Press, New York.

*Effects of 5-*n*-Undecyl-6-hydroxy-4,7-dioxobenzothiazole on the Reduction–Oxidation Reactions of the Cytochrome* b-c_1 *Segment of Mammalian Mitochondria*

9

JOHN R. BOWYER

INTRODUCTION

Ubiquinone is essential for electron transport from succinate to cytochrome *c* in mammalian mitochondria (for review, see *1*) and a considerable amount of thermodynamic and structural information is now available on ubisemiquinone species bound in the succinate–cytochrome *c* oxidoreductase segment (*2*, *3*). However, because the reaction kinetics of ubiquinone cannot be conveniently studied spectrophotometrically, its mechanistic role is poorly characterized and has been largely deduced from studies on behavior of the cytochromes.

One approach to studying the reactions of ubiquinone is to disrupt them using inhibitory analogs. 5-*n*-Undecyl-6-hydroxy-4,7-dioxobenzothiazole (UHDBT), first synthesized by Friedman *et al.* (*4*), inhibits oxidation of succinate by yeast mitochondria, and a preliminary analysis of its effect on the redox poise of the cytochromes suggested that the inhibitor acted in the *b*-c_1 segment of this system (*5*). Subsequently, Bowyer *et al.* (*6*) showed that UHDBT inhibits electron transfer from the Rieske iron–sulfur cluster (characterized by a g_y resonance at 1.90 in its EPR spectrum (*7*)) to photooxidized cytochrome c_2 in the photosynthetic bacterium *Rho-*

Function of Quinones in Energy Conserving Systems

ISBN 0-12-701280-X

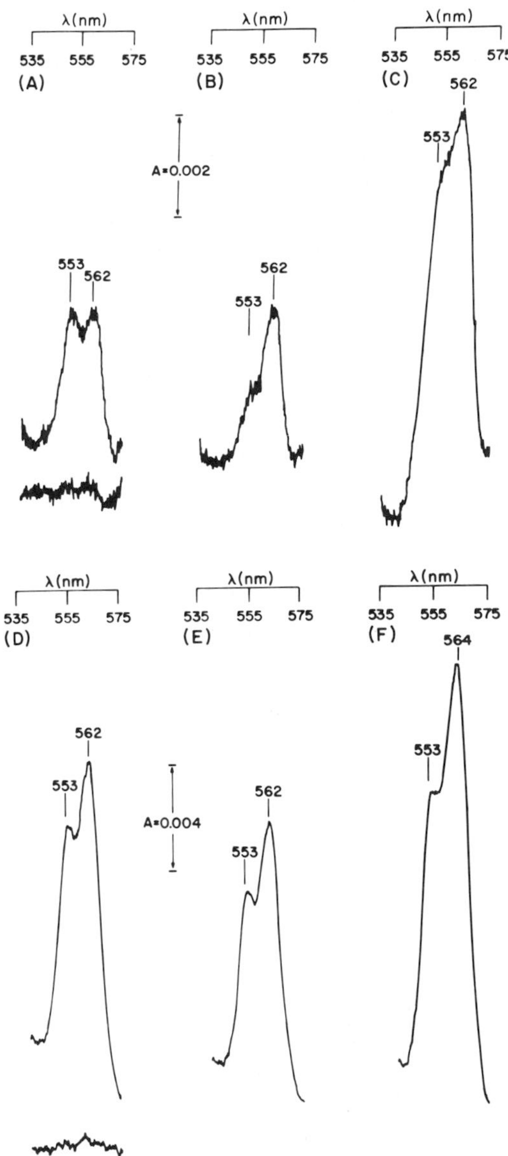

Fig. 1. Effect of UHDBT on the oxidation–reduction poise of cytochromes *b* and c_1 during turnover in cytochrome *c* depleted mitochondria (A–C) and in isolated succinate-cytochrome *c* reductase complex (D–F). Rat liver mitochondria depleted of cytochrome *c* (prepared according to Jacobs and Sanadi (*12*)) were suspended to approximately 0.2 μM cytochrome c_1 ($\sim$2 mg/ml protein) in 120 m*M* KCl, 10 m*M* potassium phosphate, 20 m*M* glycylglycine, 225 m*M* mannitol, 75 m*M* sucrose, 5 m*M* $MgCl_2$, 1 μM carbonylcyanide-*m*-chlorophenylhydrazone at pH 7.0 in an open stirred cuvette (mixing time $\sim$ 1 sec) at ambient temperature. Spectra were recorded using an Aminco DW2a dual wavelength spectrophotometer with a 2 nm band pass. The bottom tracing in (A) is the baseline of the fully oxidized cytochromes. Spectra were then recorded after adding 15 m*M* succinate (A), followed by

dopseudomonas sphaeroides, which has a cytochrome b-c_1 type complex analogous to that found in mammalian mitochondria. In this organism, UHDBT also inhibits photoreduction of cytochrome b, both in the presence and absence of antimycin, at a site distinct from the photochemical reaction center quinone-acceptor complex (*8*, *9*). More recently, Trumpower and Haggerty (*10*) showed that in mammalian mitochondria, UHDBT is a highly potent inhibitor of electron transfer from ubiquinol to cytochrome c through the b-c_1 segment; however, it does not inhibit electron transfer from succinate to ubiquinone.

The purpose of this paper is to report that several of the effects of UHDBT on reduction–oxidation (redox) reactions of the mitochondrial succinate-cytochrome c reductase complex may be linked to an interaction of UHDBT with the iron–sulfur protein (“Rieske cluster”) of this segment (*11*).

EXPERIMENTAL RESULTS

Effect of UHDBT on the Steady State Redox Poises of the Cytochromes During Ongoing Electron Flow from Succinate to Oxygen

When UHDBT is added to respiring rat liver mitochondria partially depleted of cytochrome c in order to resolve cytochrome c_1, there is a partial oxidation of ferrocytochromes c and c_1, but no change in the redox poise of cytochrome b (Figs. 1A–C).

To demonstrate the effect of UHDBT on the redox poises of cytochromes in isolated succinate-cytochrome c reductase complex during ongoing electron flow, a succinate oxidase system was reconstituted by mixing cytochrome c and cytochrome c oxidase with isolated reductase complex. The spectra of Figs. 1D–F show that addition of UHDBT resulted in an oxidation of cytochrome c and cytochrome c_1

---

5 μM UHDBT (B). The spectrum in (C) is of the fully reduced cytochromes after addition of 1 μM antimycin, 0.25 mM KCN and dithionite. UHDBT was synthesized as in Friedman *et al.* (*4*) with minor modifications (*10*).

Isolated succinate-cytochrome c reductase complex (prepared according to Trumpower and Simmons (*13*)) was suspended to 0.35 μM cytochrome c_1 in 100 mM sodium phosphate, 0.5 mM EDTA pH 7.2 and 0.1 μM cytochrome c plus 0.12 μM cytochrome oxidase (prepared according to Yonetani (*14*)) were added. The bottom tracing in (D) is the baseline of the fully oxidized cytochromes. The spectra were obtained after adding 5 mM succinate (D), followed by UHDBT to 1 μM (E) and 0.25 mM KCN, dithionite and antimycin to 1 μM (F).

(although to a smaller extent than that seen in mitochondria), and a partial oxidation of cytochrome *b* with no shift in the α band peak position. The oxidation of the cytochrome *c* and cytochrome c_1 (but not cytochrome *b*) was reversed by the addition of KCN. These effects of UHDBT differ from those of antimycin (*15*) in that antimycin leads to an increased reduction of cytochrome *b*, and shifts its α band peak position to the red. If UHDBT was added after antimycin, it diminished the extent of cytochrome *b* reduction in both mitochondria and reductase complex. The results suggest that UHDBT inhibits electron transfer to cytochrome c_1 but has more complex effects on cytochrome *b*.

Effects of UHDBT on Reduction of Cytochrome c_1 by Succinate in Isolated Succinate-Cytochrome *c* Reductase Complex

The upper set of traces in Figs. 2A–B shows the absorption change monitored at 553–539 nm on addition of succinate to oxidized reductase complex in the absence and presence of UHDBT. The change at

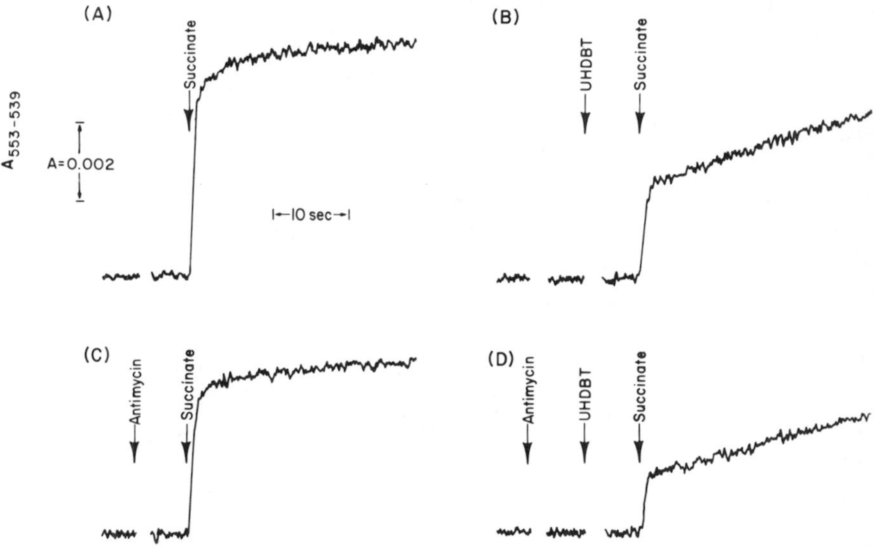

Fig. 2. Effect of UHDBT on the reduction of cytochrome c_1 by succinate in isolated succinate-cytochrome *c* reductase complex in the absence and presence of antimycin. Reductase complex was suspended to 0.35 μM cytochrome c_1 in 100 mM sodium phosphate, 0.5 mM EDTA pH 7.2. The oxidation–reduction poise of the cytochromes was checked by scanning the spectrum, and if necessary, a minimal amount ($\leq 2\ \mu M$) of ferricyanide was added to oxidize any reduced cytochromes. The tracing in (A) shows the absorption change at 533–539 nm, resulting primarily from reduction of cytochrome c_1, after addition of 5 mM succinate. The other tracings show the same reaction as in (A) except that 1 μM UHDBT was added in (B), 1 μM antimycin was added in (C), and 1 μM UHDBT with 1 μM antimycin were added in (D).

553–539 nm monitors primarily cytochrome c_1, but there is some overlap from cytochrome *b*. In the presence of UHDBT (~3 mole per mole of cytochrome c_1) there is a fast phase of cytochrome c_1 reduction followed by a markedly slower phase. In the presence of antimycin, (Figs. 2C–D) the total absorption change is diminished both in the presence and absence of UHDBT. This was attributed to the decreased spectral overlap from cytochrome *b* resulting from a spectral red shift induced by binding of antimycin (see, e.g., *15*). In the absence of UHDBT, the rate of cytochrome c_1 reduction is not significantly affected by antimycin (*15*) and the effect of UHDBT is very similar to that in the presence of antimycin. The extent of the fast phase in the presence of UHDBT with or without antimycin was variable, making it difficult to obtain a reliable concentration dependency for the UHDBT effect. However, the extent was maximal when a small excess of ferricyanide, beyond that required to oxidize the cytochromes, was added. This suggests that UHDBT is not inhibitory when a component with a midpoint redox potential higher than that of cytochrome c_1 is oxidized before the addition of succinate.

A further indication that the efficacy of UHDBT is dependent on the redox poise of the complex is shown by studying the effect of UHDBT on the reductase-catalyzed reduction of a small molar excess of ferricytochrome *c*. Traces (A) and (B) in Fig. 3 show the effect of UHDBT when added to oxidized reductase complex and a small molar excess of ferricytochrome *c* before addition of succinate to initiate reduction. There is a marked reduction phase before any inhibition by UHDBT is observed. However, when reduction of cytochrome *c* is initiated by the addition of ferricytochrome to reduced complex in the presence of succinate and UHDBT, there is an immediate inhibition (Figs. 3C–D).

Effect of UHDBT on Reduction of Cytochrome *b* by Succinate in Isolated Succinate Cytochrome *c* Reductase Complex

There appear to be at least two pathways by which succinate reduces cytochrome *b*. One pathway is not inhibited by antimycin but is eliminated by removal of the iron–sulfur protein (*16*; see also Edwards and Bowyer, Paper 10, Chapter V, this volume) or by previous reduction of the iron–sulfur cluster and cytochrome c_1 (*17*). The other pathway is inhibited by antimycin but is not affected by removal of the iron–sulfur protein or reduction of the iron–sulfur cluster and cytochrome c_1. By appropriate choice of conditions, it is possible to test the effect of UHDBT on the two different pathways for cytochrome *b* reduction.

Traces A and B in Fig. 4 show the effect of UHDBT (~3 mole per

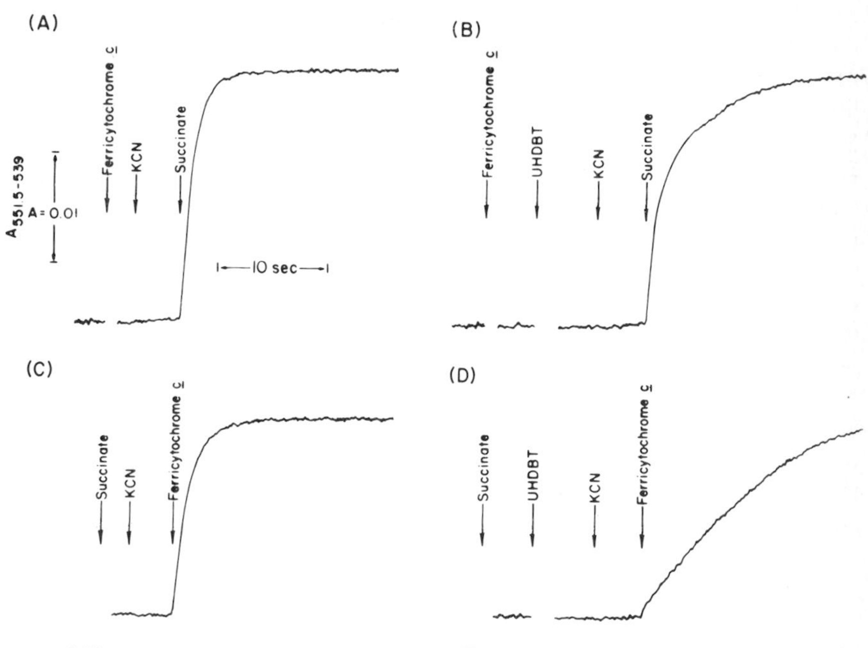

Fig. 3. Effect of UHDBT on the reduction of cytochromes c_1 plus c by succinate. Reductase complex was suspended to 0.35 μM cytochrome c_1 as in Fig. 2. For the experiments in (A) and (B), ferricytochrome c was added to 1.5 μM. UHDBT was then added to 1 μM where indicated (B) followed by 0.25 mM KCN and 5 mM succinate to start the reduction reactions. For the experiments in (C) and (D), 5 mM succinate was added, and where indicated (D) UHDBT was added to 1 μM. After adding 0.25 mM KCN, the reduction of cytochrome c was started by addition of 1.5 μM ferricytochrome c.

mole of cytochrome c_1) on cytochrome b reduction in the presence of antimycin when cytochrome c_1 is oxidized before addition of succinate. In the absence of UHDBT, approximately 70% of the dithionite-reducible cytochrome b is reduced. The effect of UHDBT is very similar to its effect on cytochrome c_1 reduction but there is also a diminution in extent by 20–30%. The extent of the fast phase could again be increased by addition of a slight excess of ferricyanide or ferricytochrome c before addition of UHDBT and succinate.

Traces C and D in Fig. 4 show the effect of UHDBT on cytochrome b reduction in the absence of antimycin, when cytochrome c_1 is reduced before addition of succinate. At the same concentration of UHDBT used in the previous experiments, there is no obvious effect on the rate of cytochrome b reduction, but there is a diminution in extent by 25–30%. At concentrations of UHDBT $\geq 10\ \mu M$ (≥ 30 mole per

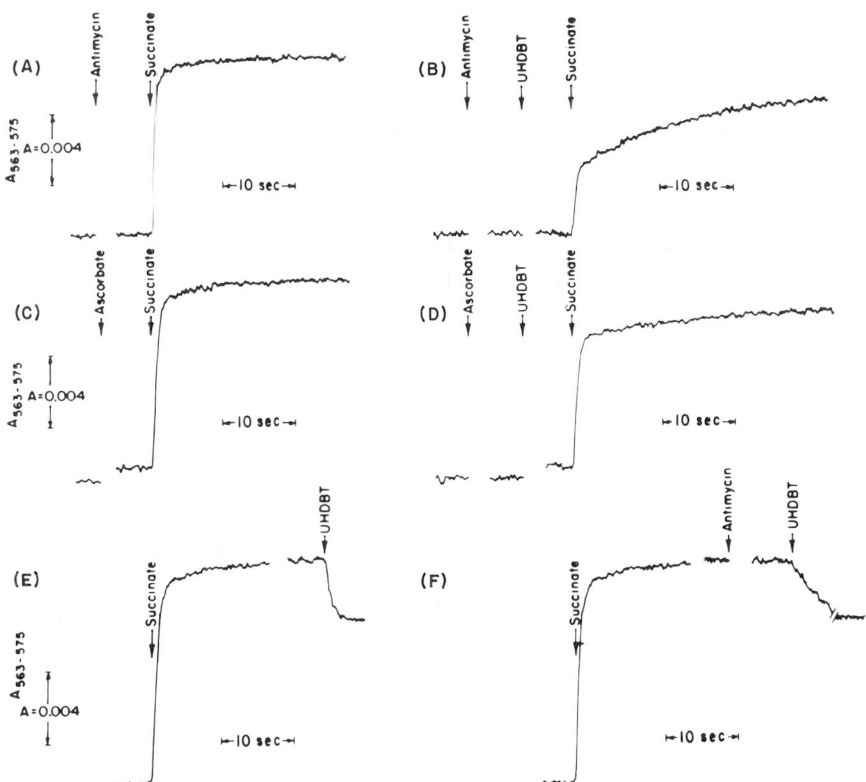

Fig. 4. Reduction of cytochrome *b* by succinate in isolated succinate–cytochrome *c* reductase complex. Reductase complex was suspended as in Fig. 2. Reduction was monitored at 563–575 nm. For the experiments in (A) and (B), a minimum amount of ferricyanide was added to oxidize any reduced cytochromes, followed by 1 μM antimycin, and where indicated, 1 μM UHDBT. Reduction was initiated by addition of 5 mM succinate. For the experiments in (C) and (D), a minimal amount of ascorbate was added to reduce cytochrome c_1, but not cytochrome *b*, as determined spectrophotometrically. 1 μM UHDBT was then added where indicated (D), followed by 5 mM succinate. For the experiments in (E) and (F), a minimal amount of ferricyanide was added to oxidize any reduced cytochromes, followed by 5 mM succinate. Where indicated, UHDBT was added to 1 μM and antimycin to 1 μM.

mole of cytochrome c_1) there is some inhibition of the rate of cytochrome *b* reduction as well as a further diminution in extent. However, reduction by DBH, an analog of ubiquinol-2, is not inhibited even at 40 μM UHDBT.

If UHDBT is added after succinate, there is a very rapid oxidation of about 25% of the ferrocytochrome *b*-562 (Fig. 4E), so that the level of reduction is close to that reached when succinate is added after UHDBT. The rate of the UHDBT-induced cytochrome *b* oxidation is diminished markedly by antimycin, although the extent is increased (Fig. 4F).

Effect of UHDBT on the Oxidant-Induced Reduction of Cytochrome *b*

As noted earlier, when succinate is added to reductase complex in the presence of antimycin, with cytochrome c_1 already reduced, no cytochrome *b* reduction occurs. Reoxidation of cytochrome c_1 leads to cytochrome *b* reduction (oxidant-induced cytochrome *b* reduction). UHDBT inhibits the oxidant-induced reduction of cytochrome *b* (Figs. 5A–B), but not the oxidation of cytochrome c_1 (using ferricyanide as oxidant) as in Fig. 5C (*18*). Spectra recorded after the addition of ferricyanide indicate that cytochrome *b*-562 is the major component reduced. However, if cytochrome *b*-562 is reduced by succinate before addition of antimycin, ferricyanide addition leads to reduction of cytochrome *b*-566. UHDBT also inhibits this oxidant-induced reduction.

Effects of UHDBT on the Redox Reactions of Succinate-Cytochrome *c* Reductase Complex Depleted of Iron–Sulfur Protein

By using iron–sulfur protein-depleted complex, it is possible to determine whether some of the effects of UHDBT are independent of any interaction with this protein. When UHDBT was added to depleted reductase in which cytochrome *b* had been reduced by succinate in the absence of antimycin (cytochrome c_1 remains oxidized) a rapid oxidation of ~25% of the cytochrome *b* was induced, and if antimycin was added before the UHDBT the rate of the oxidation was dramatically slowed. This indicates that the same changes observed in intact reductase are independent of the redox poise of cytochrome c_1 and do not require an interaction of UHDBT with the iron–sulfur protein. These effects of UHDBT are responsible for the oxidation of cytochrome *b* in the succinate-oxidase system.

DISCUSSION

Several of the effects of UHDBT on the redox reactions of succinate-cytochrome *c* reductase complex were similar to, but less pronounced

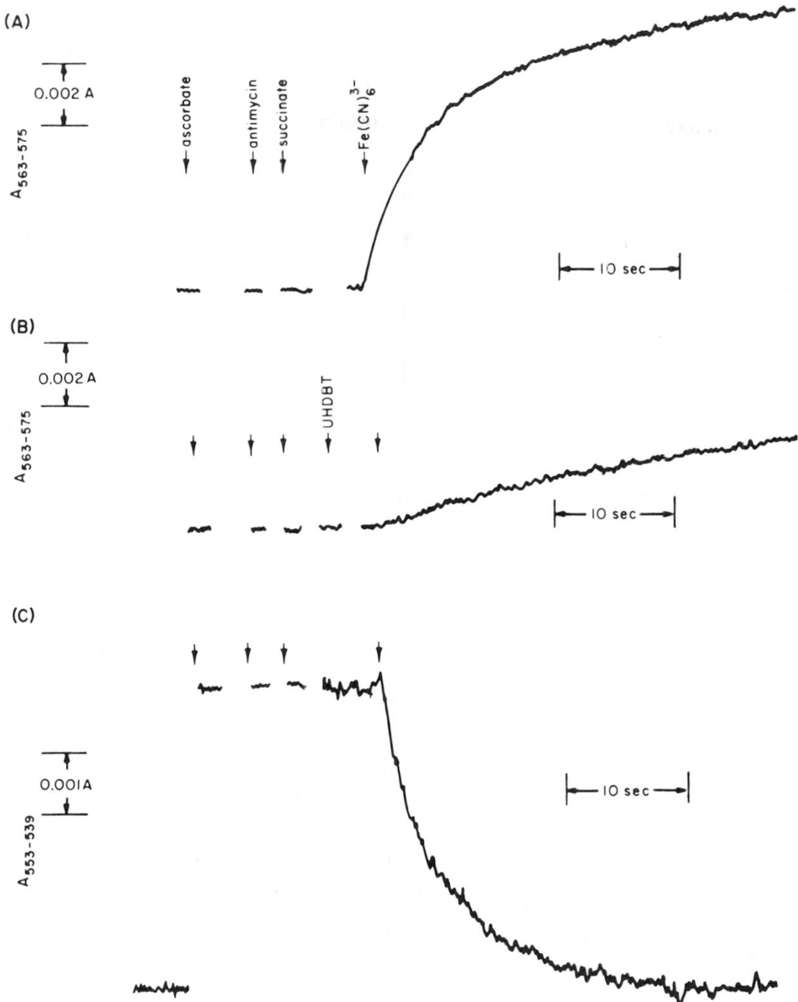

Fig. 5. Inhibition by UHDBT of the oxidant-induced reduction of cytochrome *b* in isolated succinate-cytochrome *c* reductase complex. Reductase complex was suspended to 0.36 μM cytochrome c_1 as in Fig. 2. For the reaction shown by tracing (A), ascorbate was added to reduce cytochrome c_1, followed by 3 μM antimycin and 5 mM succinate. Reduction of cytochrome *b* was then induced by addition of 4 μM ferricyanide 2 minutes after addition of the succinate. The reaction shown in tracing (B) was performed under the same conditions, except that 1 μM UHDBT was added prior to addition of ferricyanide. Tracing (C) shows the oxidation of cytochrome c_1 during the oxidant-induced reduction of cytochrome *b* in the absence of UHDBT. The reaction was carried out under conditions identical to those of Fig. 5A. [From (*18*); reproduced with permission.]

than, removal of the iron–sulfur protein. Removal of the protein eliminated reduction of cytochrome c_1 by succinate, and, in the presence of antimycin, reduction of cytochromes *b* (both *b*-562 and *b*-566) and c_1(*16*; see also Edwards and Bowyer, Paper 10, Chapter V, this volume). It also eliminated the oxidant-induced reduction pathway (*19*). These effects were interpreted in terms of a Q-cycle scheme in which the Rieske iron–sulfur cluster catalyzed electron transfer from ubiquinol to cytochrome c_1, generating ubisemiquinone, which acts as a specific reductant for cytochrome *b* in the presence of antimycin (*16*, *20*).

In photosynthetic bacteria, UHDBT inhibits electron transfer from the reduced iron–sulfur cluster to ferricytochrome c_2 (possibly via a bound cytochrome analogous to cytochrome c_1 (*21*)) regardless of the redox state of the ubiquinone species thought to reduce the iron–sulfur cluster (*8*). It seems likely, however, that UHDBT interacts with the iron–sulfur protein at the ubiquinol reduction site. If UHDBT has the same site of action in the mitochondrial reductase complex, it would inhibit reduction of cytochrome c_1 by succinate. Rapid cytochrome c_1 reduction was however, not fully inhibited even at $\geq 10\ \mu M$ UHDBT.

UHDBT raises the midpoint redox potential (at pH 7) of the iron–sulfur cluster from 280 mV to 350 mV in *Rps. sphaeroides* (*6*) and in reductase complex (Bowyer, Edwards, Ohnishi, and Trumpower, *J. Biol. Chem.*, in press). Therefore, it is possible that in the presence of UHDBT, the iron–sulfur cluster may be partially reduced when cytochrome c_1 is fully oxidized. If UHDBT forms an inhibitory complex only with reduced iron–sulfur protein, a fast phase of cytochrome c_1 reduction (and cytochrome *b* reduction) would occur in those centers where the iron–sulfur cluster was oxidized before addition of succinate, provided that the rate of electron transfer to cytochrome c_1 from iron–sulfur clusters reduced after the addition of succinate was faster than formation of the inhibitory complex. In centers in which the iron–sulfur cluster was reduced before addition of succinate (UHDBT bound), the extent of cytochrome *b* reduction would be diminished (due to the limitation of the ubiquinol oxidizing pool to cytochrome c_1), and the rate of reduction would be no faster than the inhibited rate of electron transfer from the iron-sulfur protein to ferricytochrome c_1. The model predicts that UHDBT does not inhibit electron transfer from ubiquinol to the oxidized iron–sulfur protein; this reaction is under study.

The inhibition by UHDBT of the oxidant-induced reduction of cytochrome *b* with ferricyanide as oxidant indicates that UHDBT inhibits oxidation of the iron–sulfur cluster by ferricytochrome c_1 or ferricyanide. The effects of UHDBT on cytochrome *b* reduction are complicated by effects that do not involve an interaction with the iron–sulfur protein.

These effects are not understood, but may indicate a change in the thermodynamic properties of the ubiquinone species, which oxidizes/reduces cytochrome *b*-562 in an antimycin-sensitive reaction.

ACKNOWLEDGMENT

This work was supported by NIH Grant GM 20379 to B. L. Trumpower, in whose laboratory it was carried out.

REFERENCES

1. Trumpower, B. L. (1981). *J. Bioenerg. Biomembr.* **13,** 1–24.
2. Ohnishi, T., and Trumpower, B. L. (1980). *J. Biol. Chem.* **255,** 3278–3284.
3. Salerno, J. C., and Ohnishi, T. (1980). *Biochem. J.* **192,** 769–781.
4. Friedman, M. D., Stotter, P. L., Porter, T. H., and Folkers, K. (1973). *J. Med. Chem.* **16,** 1314–1316.
5. Roberts, H., Choo, W. H., Smith, S. C., Marzuki, S., Linnane, A. W., Porter, T. H., and Folkers, K. (1978). *Arch. Biochem. Biophys.* **191,** 306–315.
6. Bowyer, J. R., Dutton, P. L., Prince, R. C., and Crofts, A. R. (1980). *Biochim. Biophys. Acta* **592,** 445–460.
7. Rieske, J. S., MacLennan, D. H., and Coleman, R. (1964). *Biochem. Biophys. Res. Commun.* **15,** 338–344.
8. Bowyer, J. R., and Crofts, A. R. (1978). *In* "Frontiers of Biological Energetics" (P. L. Dutton, J. S. Leigh, and A. Scarpa, eds.), Vol. 1, pp. 326–333. Academic Press, New York.
9. Bowyer, J. R., and Crofts, A. R. (1981). *Biochim. Biophys. Acta* **636,** 218–233.
10. Trumpower, B. L., and Haggerty, J. G. (1980). *J. Bioenerg. Biomembr.* **12,** 151–164.
11. Trumpower, B. L., and Edwards, C. A. (1979). *J. Biol. Chem.* **254,** 8697–8706.
12. Jacobs, E. E., and Sanadi, D. R. (1960). *J. Biol. Chem.* **235,** 531–534.
13. Trumpower, B. L., and Simmons, Z. (1979). *J. Biol. Chem.* **254,** 4608–4616.
14. Yonetani, T. (1967). *In* "Oxidation and Phosphorylation" (R. W. Estabrook and M. E. Pullman, eds.), Methods in Enzymology, Vol. 10, pp. 332–335. Academic Press, New York.
15. Bowyer, J. R., and Trumpower, B. L. (1981). *J. Biol. Chem.* **256,** 2245–2251.
16. Trumpower, B. L. (1976). *Biochem. Biophys. Res. Commun.* **70,** 73–80.
17. Trumpower, B. L., and Katki, A. (1975). *Biochem. Biophys. Res. Commun.* **65,** 16–23.
18. Bowyer, J. R., and Trumpower, B. L. (1980). *FEBS Lett.* **115,** 171–174.
19. Bowyer, J. R., Edwards, C. A., and Trumpower, B. L. (1981). *FEBS Lett.* **126,** 93–97.
20. Mitchell, P. (1976). *J. Theor. Biol.* **62,** 327–367.
21. Wood, P. M. (1980). *Biochem. J.* **189,** 385–391.

10

Reconstitution of Electron-Transfer Reactions of the Mitochondrial Respiratory Chain with Iron–Sulfur Protein of the Cytochrome b-c_1 *Segment*

CAROL A. EDWARDS
JOHN R. BOWYER

INTRODUCTION

The iron–sulfur protein of the cytochrome *b*-c_1 segment of the mitochondrial respiratory chain has been purified in a reconstitutively active form from succinate-cytochrome *c* reductase complex (*1*). Purified iron–sulfur protein restores succinate-cytochrome *c* reductase and ubiquinol-cytochrome *c* reductase activities to reductase complex depleted of the iron–sulfur protein with guanidine and cholate (*1*, *2*). The iron–sulfur protein is not required for succinate dehydrogenase nor for succinate-ubiquinone reductase activities of the reconstituted reductase complex. Thus, resolution and reconstitution of the iron–sulfur protein has shown that it is required for electron transfer from ubiquinol to cytochrome *c* in the mitochondrial respiratory chain (*1*, *2*). This paper describes oxidation–reduction reactions of cytochromes *b* and c_1 in the depleted and reconstituted reductase complex and the function of the iron–sulfur protein in these reactions.

Function of Quinones in Energy Conserving Systems

ISBN 0-12-701280-X

EXPERIMENTAL RESULTS

Purified iron–sulfur protein restores electron transfer activities to reductase complex depleted of these activities with guanidine and cholate (*1*, *2*). The results in Fig. 1 show the reconstitution of succinate-cytochrome *c* reductase (solid lines) and ubiquinol-cytochrome *c* reductase (dashed lines) activities to depleted reductase complex with purified iron–sulfur protein. If the iron–sulfur protein is added back to the depleted complex in the presence of succinate dehydrogenase, it restores both succinate and ubiquinol-cytochrome *c* reductase activities (Fig. 1A). Both of these reconstituted activities are inhibited by antimycin (results not

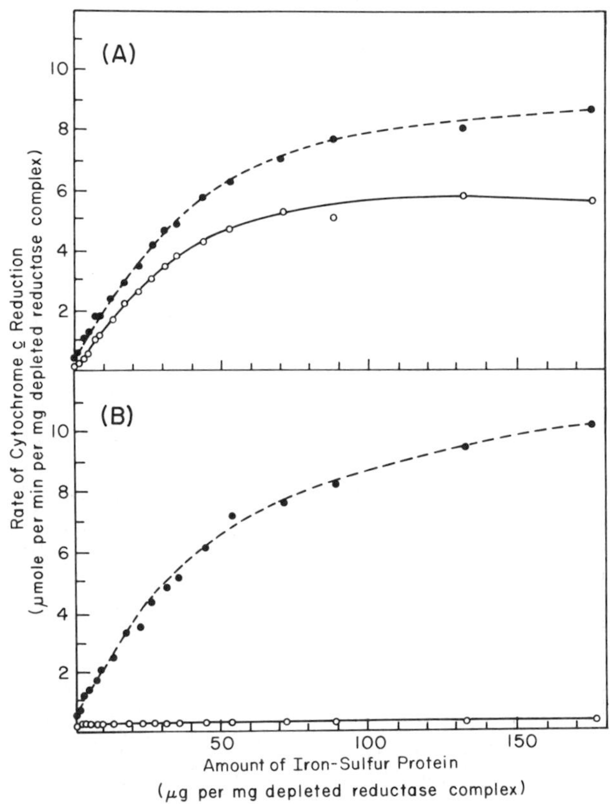

Fig. 1. Reconstitution of succinate-cytochrome *c* reductase (—○—○—) and ubiquinol-cytochrome *c* reductase (– –●– –●– –) activities to depleted reductase complex by purified iron–sulfur protein in the presence (A) and absence (B) of succinate dehydrogenase. [Reproduced from Trumpower and Edwards (*1*), with permission.]

shown). If succinate dehydrogenase is omitted, the iron–sulfur protein reconstitutes ubiquinol-cytochrome *c* reductase activity (Fig. 1B).

The iron–sulfur protein is required for electron transfer within the b-c_1 segment to cytochrome c_1, but not to cytochrome *b*, as shown in Fig. 2. When succinate is added to depleted reductase complex there is rapid reduction of approximately 75% of the cytochrome *b*, as shown in the kinetics trace, Fig. 2A. The amount of cytochrome *b* reduced by succinate is identical to that in the parent reductase complex, and is not changed on addition of iron–sulfur protein. From the spectrum to the right, taken immediately after the kinetics trace, it can be seen that no cytochrome c_1 was reduced upon addition of succinate to the depleted complex.

Reconstitution of the depleted complex with iron–sulfur protein allows rapid reduction of cytochrome c_1 upon addition of succinate (Fig. 2B). Succinate present in the incubation mixture for the reconstitution was removed with centrifugal gel columns (*3*) prior to dilution of the sample for spectroscopy. The experiments in Figs. 2A and 2B were performed with different concentrations of reductase complex in the reaction cuvette. In both cases 70 to 75 percent of the dithionite reducible cytochrome *b* was reduced by succinate.

The iron–sulfur protein is not required for reduction of cytochrome *b* in depleted reductase complex, but is required for reoxidation of cytochrome *b* by cytochrome *c* plus cytochrome *c* oxidase. The spectra in Fig. 3 show the effect of increasing amounts of cytochrome *c* plus cytochrome *c* oxidase on depleted reductase complex which has been reconstituted with iron–sulfur protein. The amount of cytochrome *b* reduced in the reconstituted complex decreases in the presence of increasing amounts of cytochrome *c* plus cytochrome *c* oxidase (Figs. 3A–D). The iron–sulfur protein is not required for reoxidation of cytochrome c_1 by cytochrome *c* plus cytochrome *c* oxidase, and the apparent decrease in absorption of cytochrome c_1 in Figs. 3A–D, is due to decreased spectral overlap from cytochrome *b* as the amount of reduced cytochrome *b* decreases. The apparent increase in the amount of cytochrome *b* reduced upon reconstitution with iron–sulfur protein (Figs. 3A–D) over control spectra (Figs. 3E–H) is due to spectral overlap from cytochrome c_1 which is also reduced upon reconstitution with iron–sulfur protein. The slight decrease in amount of reduced cytochrome *b* in the control spectra (Figs. 3E–H) upon addition of cytochrome *c* plus cytochrome *c* oxidase correlates with the small amount of iron–sulfur protein present in the depleted complex after extraction with guanidine and cholate, as demonstrated by catalytic activity of the depleted complex. From these results the iron–sulfur protein appears to function in a linear sequence of electron transfer between the *b* cytochromes and cytochrome c_1.

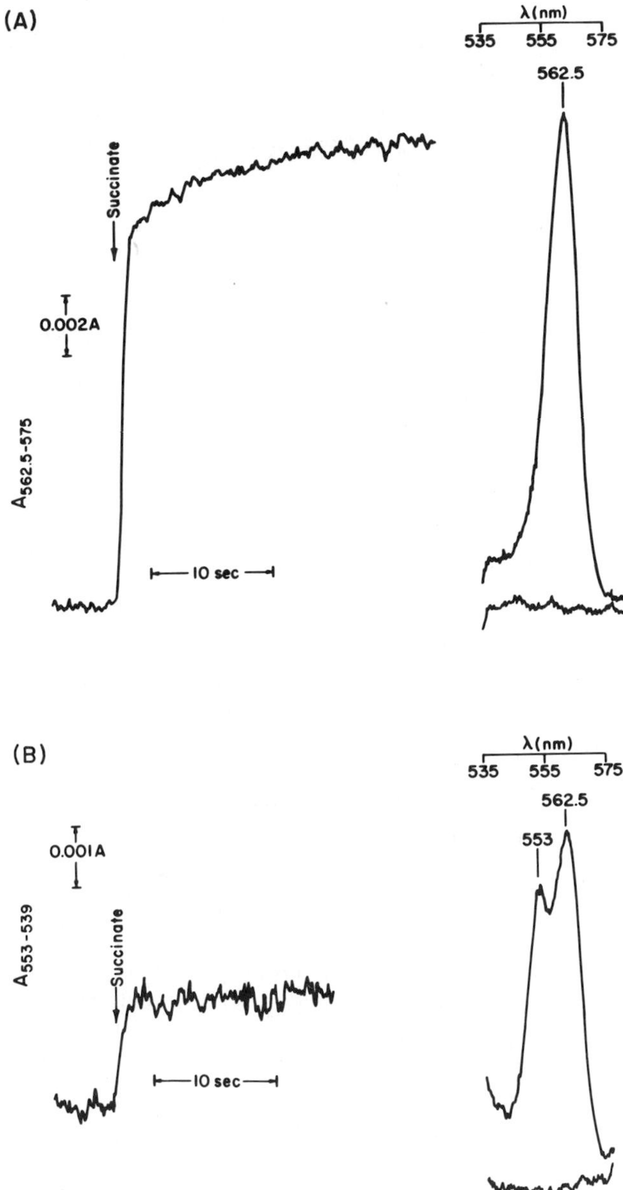

Fig. 2. Reduction of cytochromes *b* and c_1 by succinate in depleted reductase complex before (A) and after (B) reconstitution with iron–sulfur protein. The spectra to the right of the kinetics traces show the oxidation–reduction poise of the cytochromes after reduction by succinate; baselines recorded before addition of succinate are shown underneath each spectrum.

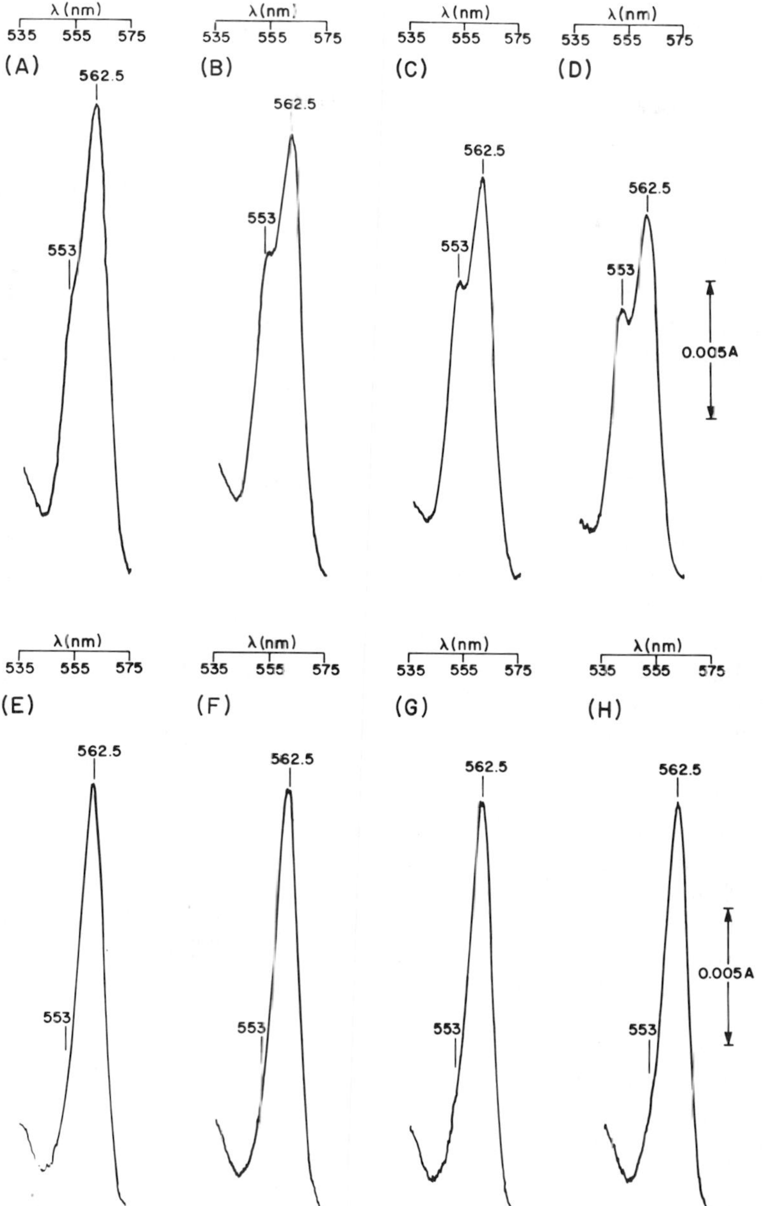

Fig. 3. Absorption spectra showing reoxidation of cytochrome *b* in depleted reductase complex reconstituted with iron–sulfur protein by increasing concentrations of cytochrome *c* plus cytochrome *c* oxidase (A–D). The spectra in (E–H) are controls for the experiments in (A–D), respectively, but without iron–sulfur protein.

Antimycin inhibits the rapid reduction of cytochrome *b* in the depleted complex. The amount of cytochrome *b* reduced by succinate in depleted reductase complex decreases in the presence of increasing amounts of antimycin, as shown in Figs. 4 and 5. The tracing in Fig. 4A shows the absorbance change upon reduction of 0.73 nmole of cytochrome *b* in the absence of antimycin. The tracings in Figs. 4B–G show the same reaction as in 4A, except that increasing amounts of antimycin were present. In Fig. 4G no rapid reduction of cytochrome *b* occurred in the presence of an amount of antimycin stoichiometric to the cytochrome c_1 content.

The stoichiometry of the antimycin inhibition is shown in Fig. 6; 1.1 mole of antimycin per mole of cytochrome c_1 gives complete inhibition of cytochrome *b* reduction in the depleted reductase complex. This agreed with previous work on Complex III (*4*, *5*) and the parent reductase complex (*6*), in which amounts of antimycin stoichiometric to the cytochrome c_1 content inhibited catalytic activity of the complex.

In the presence of an amount of antimycin at least stoichiometric with the cytochrome c_1 content, succinate was unable to reduce either cytochrome *b* or cytochrome c_1 in reductase complex depleted of iron–sulfur protein. When the iron–sulfur protein was reconstituted to the depleted reductase complex, addition of succinate in the presence of antimycin caused reduction of both cytochromes *b* and c_1.

The kinetics tracings in Fig. 7 show rapid reduction of cytochrome *b* by succinate in the presence of antimycin in depleted and reconstituted reductase complex. Addition of succinate to depleted reductase complex in the presence of antimycin caused no rapid reduction of cytochrome *b* (Fig. 7A).

When the depleted reductase complex was reconstituted with iron–sulfur protein, addition of succinate in the presence of antimycin caused rapid reduction of cytochrome *b*; the extent of cytochrome *b* reduction increased with increasing amounts of iron–sulfur protein reconstituted to the depleted complex (Figs. 7B–E). Corresponding absorption spectra recorded immediately after each kinetics tracing in Fig. 7 are shown in Fig. 8.

The extent of cytochrome *b* reduction as iron–sulfur protein was reconstituted to depleted reductase complex closely paralleled the restoration of ubiquinol-cytochrome *c* reductase activity upon reconstitution of iron–sulfur protein to the depleted complex, as shown in Fig. 9. Samples were reconstituted with iron–sulfur protein and succinate was removed with centrifugal gel columns prior to spectroscopy. The catalytic activity of the samples used for kinetics traces (Fig. 7) and absorption spectra (Fig. 8) was measured after oxidation of the cytochromes with potassium ferricyanide ($\leq 3\ \mu M$) for obtaining baseline spectra, but before addition of

Fig. 4. Reduction of cytochrome *b* by succinate in depleted reductase complex in the absence (A) and in the presence of increasing amounts of antimycin (B–G). Absorption spectra showing reduced cytochrome *b* for each reaction are shown in Fig. 5.

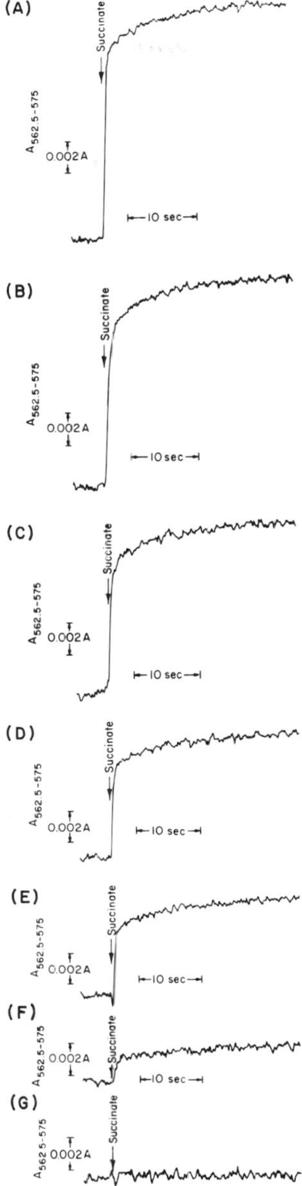

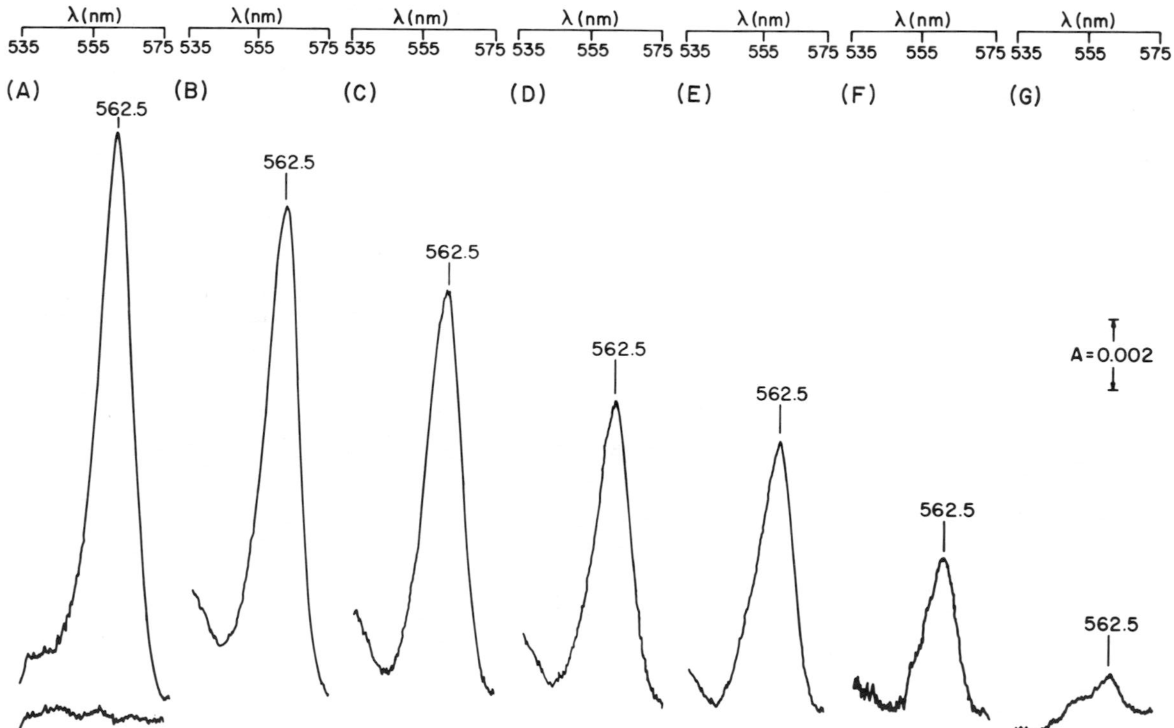

Fig. 5. Absorption spectra showing reduction of cytochrome *b* by succinate in depleted reductase complex in the presence of increasing amounts of antimycin. The spectra (A–G) were obtained immediately at the end of the reactions shown in Fig. 4(A–G), respectively. The bottom tracing in (A) is the baseline of the fully oxidized cytochromes.

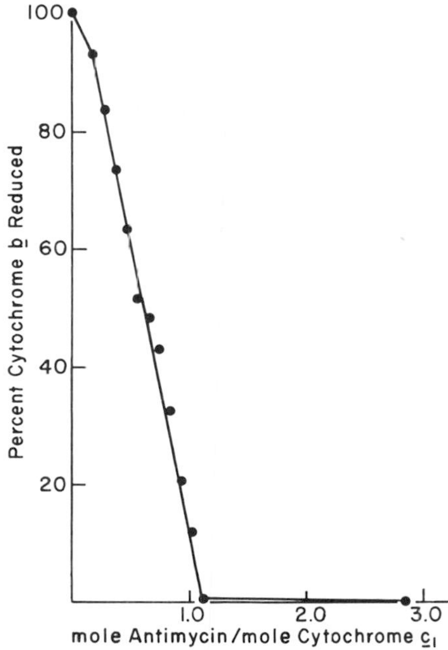

Fig. 6. Effect of antimycin on reduction of cytochrome *b* by succinate in depleted reductase complex. The data are from the tracings of rapid cytochrome *b* reduction shown in Fig. 4. In the absence of antimycin, 0.73 nmole of cytochrome *b* was reduced by succinate.

antimycin and succinate. The abscissa in Fig. 9 refers to moles of iron–sulfur protein per mole of cytochrome c_1 in the depleted reductase complex in the incubation mixture for the reconstitution. The ordinates refer to catalytic activity and extent of cytochrome *b* reduction of the sample recovered from the centrifugal gel column and diluted for spectroscopic measurements. The amount of cytochrome *b* reduced was the percentage of the dithionite-reducible cytochrome *b* recovered from the gel column. The catalytic activity is expressed as units per nmole of depleted reductase complex (based on cytochrome c_1 content) recovered from the gel column. The small amount of cytochrome *b* reduced by succinate in the presence of antimycin (Fig. 8A) was consistent with the slight amount of ubiquinol–cytochrome *c* reductase activity exhibited by the depleted complex (Fig. 9). Where 25 percent of the cytochrome *b* was reduced, catalytic activity equal to 22 percent (4.5 units/nmole) of native reductase complex activity was restored.

In native reductase complex, antimycin dramatically slowed the rate of cytochrome *b* reduction by succinate, but only when cytochrome c_1 was already reduced. Oxidation of cytochrome c_1 resulted in a fast reduction

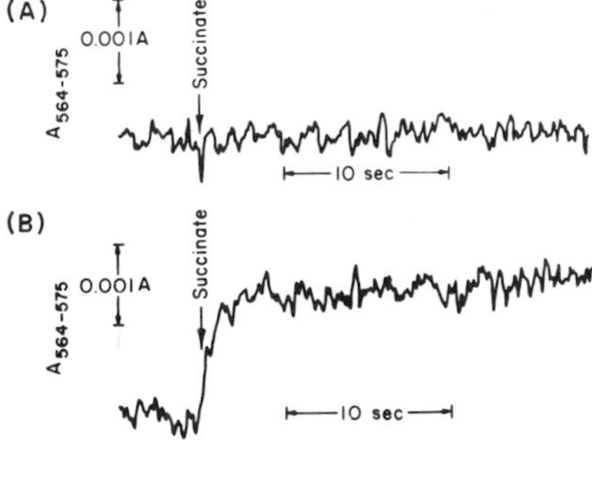

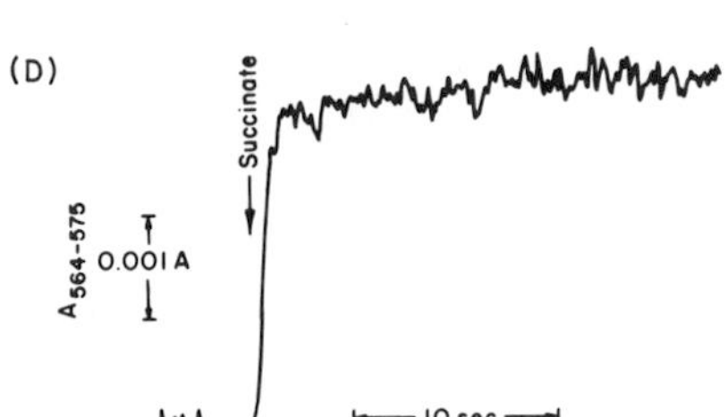

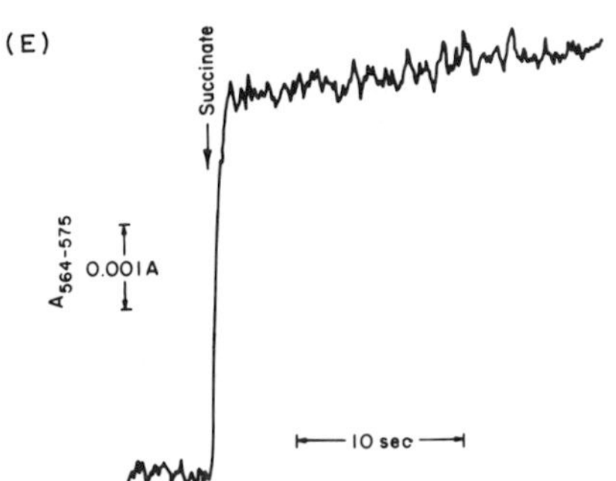

(E)
Succinate
A564-575
0.001A
10 sec

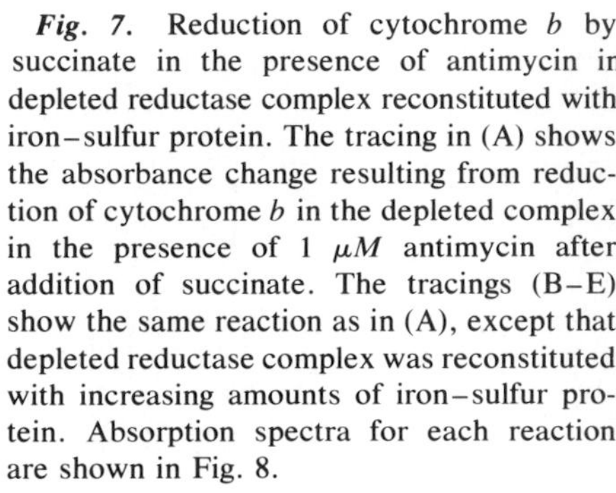

Fig. 7. Reduction of cytochrome *b* by succinate in the presence of antimycin in depleted reductase complex reconstituted with iron–sulfur protein. The tracing in (A) shows the absorbance change resulting from reduction of cytochrome *b* in the depleted complex in the presence of 1 μM antimycin after addition of succinate. The tracings (B–E) show the same reaction as in (A), except that depleted reductase complex was reconstituted with increasing amounts of iron–sulfur protein. Absorption spectra for each reaction are shown in Fig. 8.

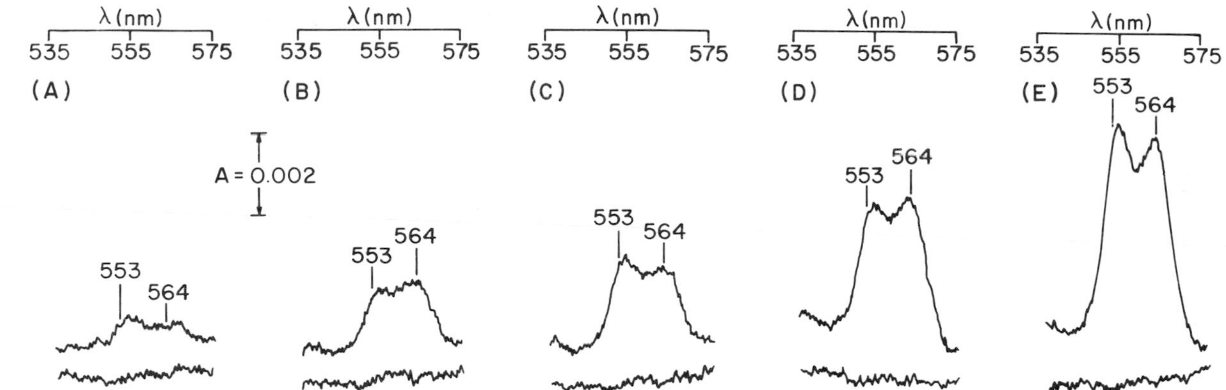

Fig. 8. Absorption spectra showing reduction of cytochromes b and c_1 by succinate in depleted reductase complex reconstituted with increasing amounts of iron–sulfur protein in the presence of antimycin. The spectra (A–E) were obtained immediately at the end of the reactions shown in Figs. 7(A–E), respectively; baselines recorded before addition of succinate are shown underneath each spectrum.

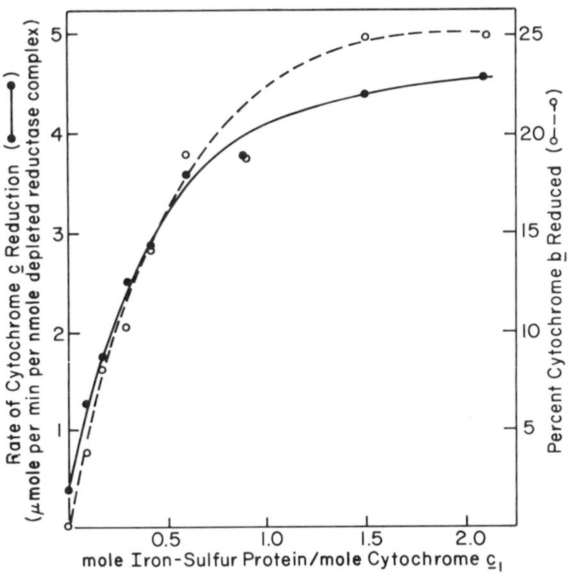

Fig. 9. Relationship between amount of ubiquinol-cytochrome *c* reductase activity reconstituted and amount of cytochrome *b* reduced by succinate in the presence of antimycin in depleted reductase complex reconstituted with iron–sulfur protein. The ubiquinol-cytochrome *c* reductase activity of the reconstituted complex used in the reactions shown in Fig. 7 was measured immediately after the cytochromes were oxidized, prior to addition of antimycin and succinate. Antimycin was added, then the sample was reduced by succinate, and the spectrum recorded as shown in Fig. 8.

of cytochrome *b* (*6*, *13*). The kinetics traces in Fig. 10 and the spectra in Fig. 11 show that the iron–sulfur protein is required for this oxidant-induced reduction of cytochrome *b*. After addition of ascorbate (to reduce cytochrome c_1), antimycin and succinate to the depleted complex, addition of cytochrome *c* plus cytochrome *c* oxidase resulted in oxidation of 60 percent of the cytochrome c_1, but no reduction of cytochrome *b* (Fig. 11A) (*7*). The small absorbance change measured at 564–575 nm was due to a turbidity change from addition of the cytochrome *c* plus oxidase (Figs. 10B, 11B).

After reconstitution of the depleted complex with iron–sulfur protein, cytochrome *b* reduction occurred after addition of cytochrome *c* plus oxidase; the extent of cytochrome *b* reduction increased as increasing amounts of iron–sulfur protein were reconstituted to the depleted complex (Figs. 10C and 10D; 11C and 11D). The extent of cytochrome *b* reduction was consistent with the data in Figs. 7 and 8, after considering that in native reductase only about 85 percent of the succinate reducible cytochrome *b* was reduced by the oxidant-induced pathway (*13*). The

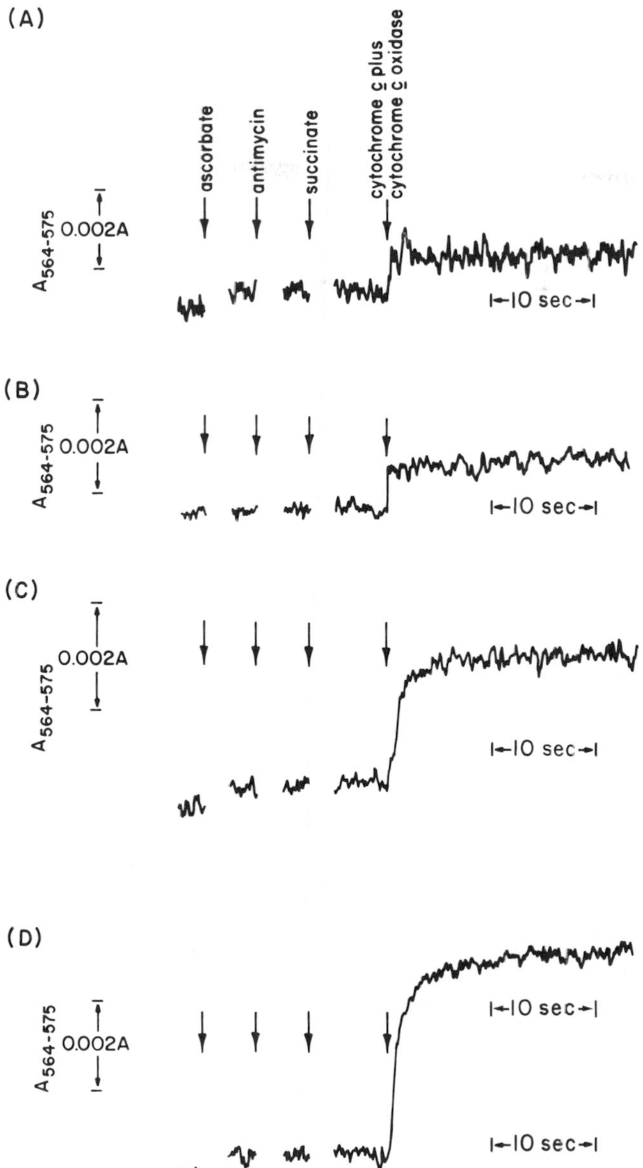

Fig. 10. Oxidant-induced reduction of cytochrome *b* in depleted reductase complex reconstituted with iron–sulfur protein. Ascorbate was added to reduce cytochrome c_1, followed by antimycin and succinate. The tracing in (A) shows the absorption change resulting from the addition of cytochrome *c* plus cytochrome *c* oxidase to depleted complex. In (B), the depleted complex was omitted to demonstrate the absorption change due to addition of the cytochrome *c* plus cytochrome *c* oxidase. The experiments in (C) and (D) were performed with depleted reductase complex previously incubated with 0.43 and 1.51 mole of iron–sulfur protein per mole of cytochrome c_1, respectively. [From Bowyer *et al.* (*7*), with permission.]

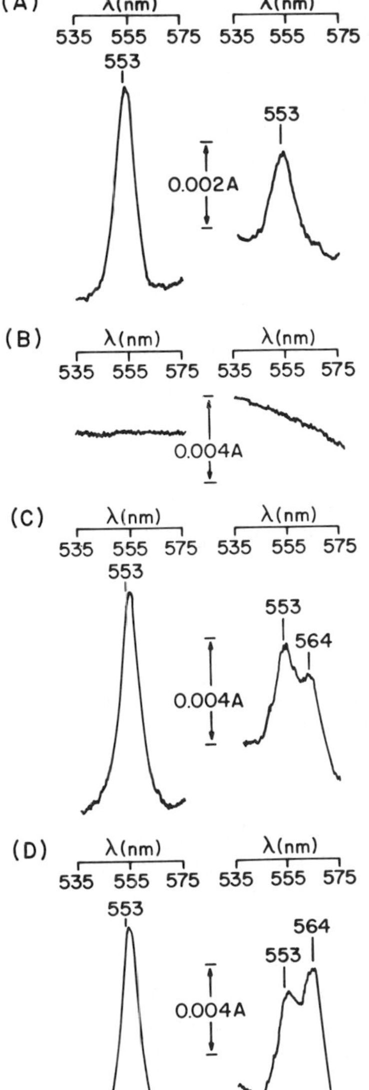

Fig. 11. Absorption spectra showing the oxidant-induced reduction of cytochrome *b* in depleted reductase complex reconstituted with iron–sulfur protein. The spectra (A–D) were obtained immediately at the end of the reactions shown in Fig. 10(A–D), respectively. [From Bowyer *et al.* (7), with permission.]

spectra in Figs. 11C and 11D indicate a greater amount of cytochrome *b* reduction than is reflected in the kinetics traces owing to slow cytochrome *b* reduction while switching from kinetics to scanning mode.

The iron–sulfur protein is required for reduction of both cytochromes *b* and c_1 in the presence of antimycin. The kinetics tracings in Fig. 12 show rapid reduction of cytochrome c_1 by succinate in the presence of antimycin upon reconstitution of iron–sulfur protein to depleted reductase complex. The tracing in Fig. 12A is of depleted reductase complex before reconstitution with iron–sulfur protein and shows no reduction of cytochrome c_1. Tracings 12B–E show increasing amounts of cytochrome c_1 were reduced upon reconstitution with increasing amounts of iron–sulfur protein. Absorption spectra showing reduction of both cytochromes *b* and c_1 by succinate in the presence of antimycin are shown in Fig. 8. The extent of c_1 reduction and the restoration of catalytic activity as iron–sulfur protein is reconstituted to depleted reductase complex reappear in parallel (Fig. 13). The equivalent of approximately 16 percent (3.3 units/nmole) of native reductase-complex activity is restored where 15 to 16 percent of the cytochrome c_1 in the sample was reduced.

The iron–sulfur protein was also required for reduction of cytochrome c_1 and, in the presence of antimycin, for reduction of cytochrome *b* by DBH,[1] as shown in Table I. Cytochrome *b* was rapidly reduced by DBH in the depleted reductase complex in the absence of iron–sulfur protein. However, there was no rapid reduction of cytochrome c_1 by DBH in the absence of iron–sulfur protein. The reduction of cytochrome *b* by DBH in the depleted complex was inhibited by antimycin. Upon reconstitution of iron–sulfur protein to the depleted complex, both cytochromes *b* and c_1 were reduced by DBH in the absence and presence of antimycin. In the presence of antimycin, reduction of the cytochromes in the reconstituted complex was not as extensive as in its absence (Table I). Antimycin has been shown to facilitate removal of the iron–sulfur protein from the complex (*8*) and may displace some of the reconstituted iron–sulfur protein. The amounts of cytochromes *b* and c_1 reduced by DBH in the reconstituted reductase complex in the presence of antimycin represent 27 and 16 percent, respectively, of the cytochromes present during the reaction. This is in close agreement with the results in Figs. 9 and 13 where reduction of the cytochromes by succinate in reconstituted reductase complex in the presence of antimycin was measured.

In the absence of antimycin, the iron–sulfur protein appears to be required for reduction of cytochrome c_1, but not for reduction of cytochrome *b*, and the electron-transfer sequence, succinate → Q → cy-

[1] DBH, a synthetic analog of ubiquinone-2 having a decyl side chain.

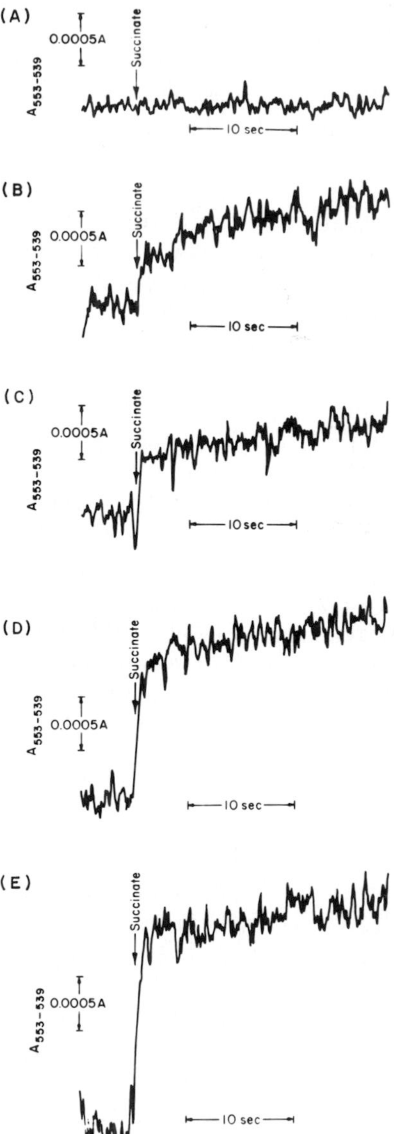

Fig. 12. Reduction of cytochrome c_1 by succinate in the presence of antimycin in depleted reductase complex reconstituted with iron–sulfur protein. The tracing in (A) shows the absorbance change resulting from reduction of cytochrome c_1 in the depleted complex in the presence of 1 μM antimycin after addition of succinate. The tracings (B–E) show the same reaction as in (A) except that the depleted reductase complex was reconstituted with increasing amounts of iron–sulfur protein.

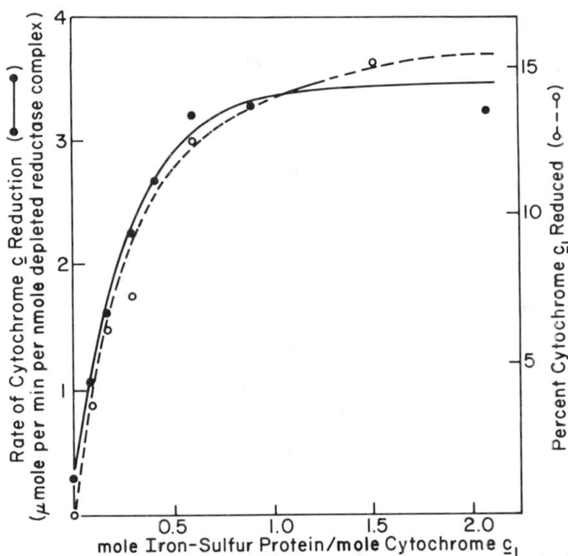

Fig. 13. Relationship between amount of ubiquinol-cytochrome *c* reductase activity reconstituted and amount of cytochrome c_1 reduced by succinate in the presence of antimycin in depleted reductase complex reconstituted with iron–sulfur protein. The ubiquinol-cytochrome *c* reductase activity of the reconstituted complex used in the reactions shown in Fig. 12 was measured immediately after the cytochromes were oxidized, prior to addition of antimycin and succinate. Antimycin was added, then the sample was reduced by succinate and the spectrum recorded as shown in Fig. 8.

TABLE I

Reduction of Cytochromes *b* and c_1 by DBH in Depleted Reductase Complex Reconstituted with Purified Iron–Sulfur Protein in the Presence and Absence of Antimycin[A]

Additions to depleted reductase comples	Cytochrome *b* (nmole)	Cytochrome c_1 (nmole)
None	0.752	0
Antimycin	0.075	0
Iron–sulfur protein	0.872	0.338
Iron–sulfur protein + antimycin	0.489	0.182

[A] Reductase complex was suspended to 0.78 μM cytochrome c_1. When indicated, 1 μM antimycin and 1.77 μM iron–sulfur protein were added. Reduction of cytochromes *b* and c_1 was monitored at 562.5–575 nm and 553–539 nm, respectively, as shown in Fig. 2.

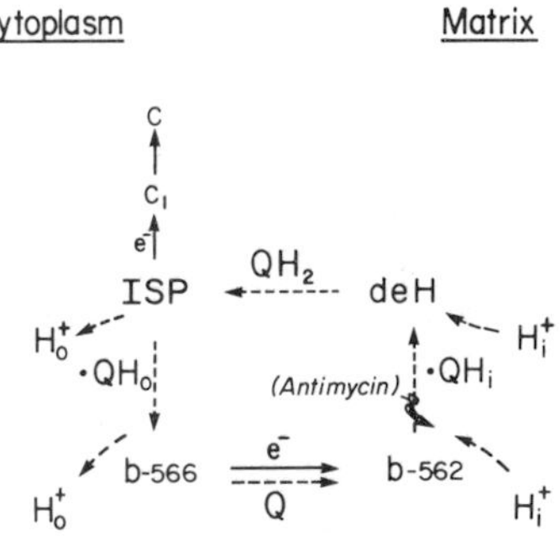

Fig. 14. Protonmotive Q-cycle mechanism of electron transfer showing the proposed function of the iron–sulfur protein (ISP) of the b-c_1 segment. The diagram represents the Q cycle operating under steady-state conditions of electron transfer from a dehydrogenase (deH) to the c cytochromes. The dashed arrows designate the mobilities of ubiquinol, ubisemiquinone, and ubiquinone between the centers at which they react, and the release and uptake of protons associated with these reactions. The solid arrows indicate electron-transfer reactions not involving ubiquinone. The subscripts i and o refer to the inner (matrix) and outer (cytoplasmic) sides of the mitochondrial membrane. [From Bowyer and Trumpower (*10*), with permission.]

tochrome $b \rightarrow$ iron–sulfur protein $\rightarrow$ cytochrome c_1 seems adequate. In the presence of antimycin, the iron–sulfur protein appears to be required for reduction of both cytochromes b and c_1. These results indicate that the function of the iron–sulfur protein cannot be explained by a simple linear arrangement of electron carriers in the cytochrome b-c_1 segment. An explanation for the function of the iron–sulfur protein is that it is a ubiquinol/cytochrome c_1-ubisemiquinone/cytochrome b oxidoreductase in a protonmotive Q-cycle mechanism (*9*) as shown in Fig. 14 and discussed below.

DISCUSSION

The iron–sulfur protein restores electron transfer between ubiquinol and cytochrome c in isolated succinate-cytochrome c reductase complex that has lost cytochrome c reductase activities from extraction with guanidine and cholate. Electron transfer from succinate to ubiquinone and from ubiquinol to cytochrome c can be restored specifically and independently by succinate dehydrogenase and the iron–sulfur protein of the b-c_1 segment, respectively. If both of these proteins are included, succinate-cytochrome c reductase activity is reconstituted. This reconstitution of the iron–sulfur protein shows for the first time that this iron–sulfur protein is required for electron transfer between ubiquinol and cytochrome *c* in the succinate-cytochrome *c* reductase segment of the respiratory chain.

The iron–sulfur protein is required for electron transfer within the *b*-c_1 segment to cytochrome c_1 and, under certain conditions, is also required for reduction of cytochrome *b*. From experiments with oxidation factor before it was identified as the iron–sulfur protein (*11*), the protein was shown to be necessary for reduction of cytochrome c_1 by durohydroquinone or succinate (*9*, *12*). Using purified iron–sulfur protein and depleted reductase complex, we have followed these reactions kinetically to gain insight into the sequence of electron-transfer reactions. Using either succinate or the quinone analog, DBH, as substrate, the iron–sulfur protein was found to be essential for reduction of cytochrome c_1 and reoxidation of cytochrome *b*; this could be explained by the linear electron-transfer sequence, succinate → Q → cytochrome *b* → iron–sulfur protein → cytochrome c_1. However, in the presence of antimycin the iron–sulfur protein is essential for reduction of both cytochromes *b* and c_1. In the reconstituted complex, both cytochromes *b* and c_1 are reduced, and the extent of reduction increases with increasing amounts of iron–sulfur protein. These results cannot readily be incorporated into a linear sequence of electron-transfer reactions.

The effect of antimycin on the depleted reductase appears to be the same as on native succinate-cytochrome *c* reductase complex (*6*, *10*, *13*), Complex III (*4*, *5*), and mitochondria (*5*, *10*). Binding of antimycin shifts the absorption maximum of the α band of ferrocytochrome *b* slightly toward the red in the depleted complex, and the binding is stoichiometric to the cytochrome c_1 content in the depleted complex. Antimycin interacts with the complex such that it facilitates dissociation of the iron–sulfur protein from the complex (*1*, *4*, *8*). In the presence of antimycin, rebinding of iron–sulfur protein to the depleted complex may be decreased by the same mechanism, since reconstitution of the iron–sulfur protein does not seem to be as effective in its presence as in its absence, as shown by the extent of reduction of the cytochromes. The catalytic activity restored to depleted reductase complex with reconstitution of the iron–sulfur protein closely parallels the extent of reduction of the cytochromes in the presence of antimycin. This finding indicates that, in the presence of antimycin, the reaction is due to that portion of the reconstituted complex that has catalytic activity.

The function of the iron–sulfur protein and a mechanism of electron transfer through the cytochrome *b*-c_1 complex can be described by a protonmotive Q-cycle mechanism as shown in Fig. 14, in which the iron–sulfur protein functions as a ubiquinol/cytochrome c_1-ubisemiquinone/cytochrome *b* oxidoreductase. The concept of the Q cycle was first introduced by Mitchell (*14*) and has since been more fully developed (*15*, *16*). In the specific formulation shown in Fig. 14 (*6*, *9*, *10*), it was postulated that the protein, now identified as the iron–sulfur protein of the *b*-c_1 segment (*1*, *11*), fulfills an important function in oxidation–reduction reactions of ubiquinone.

According to this mechanism the iron–sulfur protein would transfer one electron from ubiquinol to cytochrome c_1, thus liberating a proton and generating the semiquinone reductant for cytochrome *b*-566. The ubisemiquinone formed remains bound to the iron–sulfur protein or a closely associated polypeptide, thus stabilizing the semiquinone and preventing dismutation. An electron is transferred from *b*-566 to *b*-562 and the quinone moves to the matrix side of the membrane where it is reduced with uptake of two electrons and two protons that it will carry across the membrane. The Q cycle can run in reverse from the dehydrogenase to cytochromes *b*-562 and *b*-566, but not through to cytochrome c_1. The ubisemiquinone formed on reduction of ubiquinone by *b*-566 or on oxidation of ubiquinol by the iron–sulfur protein will not serve as reductant for the iron–sulfur protein and cytochrome c_1. Thus the iron–sulfur protein participates in the energy conserving reactions at the second coupling site of mitochondrial oxidative phosphorylation, whereby two protons are released at the outer surface of the inner mitochondrial membrane for each electron passing through the *b*-c_1 segment. The iron–sulfur protein has a unique function in that it diverts two electrons from ubiquinol to two different acceptors and at the same time participates in the electron-transfer reactions, translocating protons outward across the mitochondrial membrane, thus fulfilling the requirements of the chemiosmotic hypothesis (*17*).

The effects of antimycin are consistent with the protonmotive Q-cycle mechanism of electron transport. In this mechanism antimycin inhibits oxidation of ferrocytochrome *b-562* by ubiquinone (*14*), causing a classical "crossover" effect on the steady-state level of reduction of the cytochromes. Antimycin does not inhibit reduction of cytochrome c_1 by succinate in isolated succinate-cytochrome *c* reductase complex under conditions where the respiratory chain complex undergoes one oxidation–reduction turnover (*10*).

The oxidant-induced reduction of cytochrome *b* and the requirement for iron–sulfur protein are readily explained in terms of the Q cycle. Reduction of cytochrome c_1, and presumably of the iron–sulfur cluster, eliminates the electron sink necessary to generate the ubisemiquinone reductant for cytochrome *b*-566 when the "reversed" pathway for cytochrome *b*-562 reduction is blocked by antimycin. Oxidation of cytochrome c_1, and presumably of the iron–sulfur cluster, generates the required oxidant for ubiquinol. The demonstration that this process requires the iron–sulfur protein eliminates a specific model in which oxidation of cytochrome c_1 leads to a change in the redox properties of cytochromes *b* by permitting their reduction by ubiquinol (*18*).

To function in the proposed manner, the iron–sulfur protein must be located on or near the cytoplasmic surface of the inner mitochondrial membrane and must be closely structurally associated with cytochrome c_1 and a *b* cytochrome. The iron–sulfur protein has been shown to be accessible to impermeant probes at the cytoplasmic surface of the inner membrane (*9*, *19*). Using the methodology of resolution and reconstitution in combination with EPR spectroscopy, it should be possible to test whether the iron–sulfur protein stabilizes ubisemiquinone.

ACKNOWLEDGMENT

This work was supported by NIH Research Grant GM 20379 to B. Trumpower.

REFERENCES

1. Trumpower, B. L., and Edwards, C. A. (1979). *J. Biol. Chem.* **254,** 8697–8706.
2. Trumpower, B. L., Edwards, C. A., and Ohnishi, T. (1980). *J. Biol. Chem.* **255,** 7487–7493.
3. Penefsky, H. (1977). *J. Biol. Chem.* **252,** 2891–2899.
4. Rieske, J. S., Baum, H., Stoner, C. D., and Lipton, S. H. (1967). *J. Biol. Chem.* **242,** 4854–4866.
5. Slater, E. C. (1973). *Biochim. Biophys. Acta* **301,** 129–154.
6. Trumpower, B. L., and Katki, A. (1975). *Biochem. Biophys. Res. Commun.* **65,** 16–23.
7. Bowyer, J. R., Edwards, C. A., and Trumpower, B. L. (1981). *FEBS Lett.* **126,** 93–97.
8. Rieske, J. S. (1976). *Biochim. Biophys. Acta* **456,** 195–247.
9. Trumpower, B. L. (1976). *Biochem. Biophys. Res. Commun.* **7,** 73–80.
10. Bowyer, J. R., and Trumpower, B. L. (1981). *J. Biol. Chem.* **256,** 2245–2251.
11. Trumpower, B. L., and Edwards, C. A. (1979). *FEBS Lett.* **100,** 13–16.
12. Nishibayashi-Yamashita, H., Cunningham, C., and Racker, E. (1972). *J. Biol. Chem.* **247,** 698–704.
13. Bowyer, J. R., and Trumpower, B. L. (1980). *FEBS Lett.* **115,** 171–174.
14. Mitchell, P. (1975). *FEBS Lett.* **56,** 1–6.
15. Mitchell, P. (1975). *FEBS Lett.* **59,** 137–139.
16. Mitchell, P. (1976). *J. Theor. Biol.* **62,** 327–367.

17. Mitchell, P. (1968). "Chemiosmotic Coupling and Energy Transduction." Glynn Res., Bodmin, England.
18. Chance, B. (1974). *In* "Dynamics of Energy-Transducing Membranes" (L. Ernster, R. W. Estabrook, and E. C. Slater, eds.), pp. 553–578. Elsevier, Amsterdam.
19. Gellefors, P., and Nelson, B. D. (1977). *Eur. J. Biochem.* **80,** 275–282.

VI PLASTOQUINONE IN CHLOROPLAST ELECTRON-TRANSFER REACTIONS

The Function of Plastoquinone in Electron Transfer

1

B. R. VELTHUYS

INTRODUCTION

The function of plastoquinone (PQ) in photosynthesis has been well understood for many years (Fig. 1A). It transfers H (H-pairs) between electron-transfer sites located at opposite sides of the thylakoid membrane. Thus, plastoquinone allows the electron-transfer chain to act as an energy-conserving (*1*) proton pump (*2*). However, since the manner in which plastoquinone accomplishes this H transfer is still uncertain, several mechanisms are possible.

One possibility is that all plastoquinone is immobilized at binding sites (Fig. 1B). According to this scheme, H transfer is accomplished by the transmission of H-pairs along plastoquinone chains (also called "strands" because of evidence that the individual chains are interconnected (*2*)). These chains would start with the secondary photosystem II acceptor, called "B" or "R" (*5–8*). The primary photosystem II acceptor (called "Q" (*3*, *4*) or "X-320" (*2*)) is also a plastoquinone molecule but it is not reduced beyond the PQ^- state. The final plastoquinone of the chains is called "U" (*9*, *10*).

The extreme alternative view is that plastoquinone molecules are not required to transfer H-pairs to one another (Fig. 1C). Instead, the typical 2H transfer by plastoquinone would be: (1) the release of $PQ(H_2)$ from photosystem II or any other plastoquinone reducing enzyme, (2) its diffusion through the membrane, and (3) its rebinding at an oxidizing enzyme. According to this scheme, neither R nor U, and possibly not even Q, are permanently fixed molecules. The main advantage of this type of hypoth-

Function of Quinones in Energy Conserving Systems

ISBN 0-12-701280-X

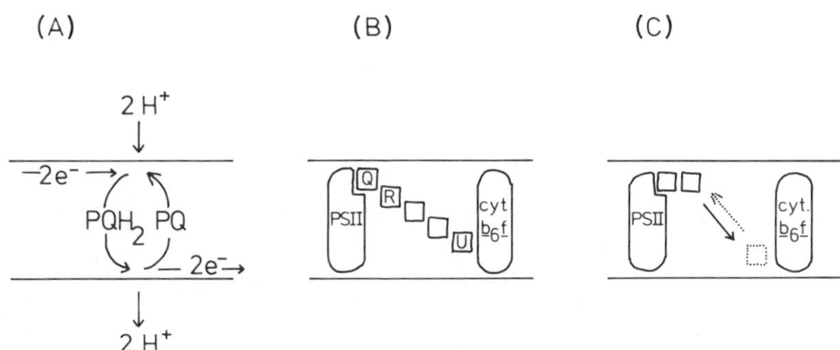

Fig. 1. Transmembrane H transfer by PQ (A) may involve the transmission of H-pairs through a PQ chain (B), or the diffusion of PQ and PQH_2 between the active sites of electron-transfer enzymes (C).

esis is that it assumes (not more than) a very loose spatial relationship between the different reaction sites acting on plastoquinone and thus agrees with the evidence for a fractional (non-integral) and variable stoichiometry of the various enzyme complexes in chloroplasts (*11*). A second advantage of the hypothesis is that it helps to interpret the susceptibility of plastoquinone-involving reactions to inhibitors. This report primarily will illustrate this second argument.

EXPERIMENTAL RESULTS AND DISCUSSION

The Reduction of Plastoquinone by Photosystem II

The assumption that plastoquinone is frequently released and rebound from the sites at which it is chemically modified allows us to propose a simple explanation for the action of inhibitors: they inhibit because they bind to the reaction center while the plastoquinone binding site is empty, and by occupying (part of) this binding site, inhibitors prevent the rebinding of plastoquinone.

Since the reduction by photosystem II of the primary plastoquinone, Q, is highly resistant to inhibition (see however, *12*), there still is little reason to hypothesize that Q exchanges frequently with free plastoquinone. The reoxidation of Q^- by secondary plastoquinone, however, is inhibited quite readily by a variety of substances. Therefore, the suggestion that secondary plastoquinone is frequently released, and exchanged for free plastoquinone (or inhibitor), seems quite reasonable.

One must admit, however, that secondary plastoquinone can be released only when it is fully reduced or fully oxidized. Plastoquinone in its intermediate reduction state, the semiquinone form, is retained by the reaction center. Presumably this is accomplished partly by a stabilization of the PQ^-, e.g., by electrical interaction with positively charged groups. In addition, since the lifetime of R^- is very long (*11*), we must assume that kinetic factors, such as those caused by a negatively charged barrier that must be crossed by plastoquinone, also play a role in the retention of PQ^-.

The retention of PQ^- by the reaction center, and the competition between inhibitor and PQ, are strikingly revealed by two closely interrelated phenomena: 1) inhibitor-induced reduction of Q, first observed by Clayton (*13*), and 2) Q^--induced release of inhibitor, first observed by Delrieu (*14*; see also *11*).

The inhibitor-induced reduction of Q is dependent on the presence of secondary semiquinone (R^-) (*6*). It is represented by step b′ → c seen in Fig. 2. The original interpretation of the phenomenon was that the inhibitor, bound to the reaction center, ''decreases the midpoint potential of R to a point below that of Q'', i.e., makes R harder to reduce than Q (*6*). In a way this interpretation still is valid, except we now assume that when inhibitor is bound, R is nonexistent and thus impossible to reduce.

The Q^--induced release of inhibitor is explained as follows (refer to Fig. 2): The equilibrium constant of the conversion a → c should not depend on the intermediate state (b or b′). Thus, if step b → c requires an f-times higher concentration of electrons than step a → b′, then step b′ → c must require an f-times higher concentration of inhibitor than step a → b. In other words, the deposition of an electron into the photosystem II acceptor complex will change the half-saturating concentration for binding of inhibitor by (at room temperature) a factor of 10 per

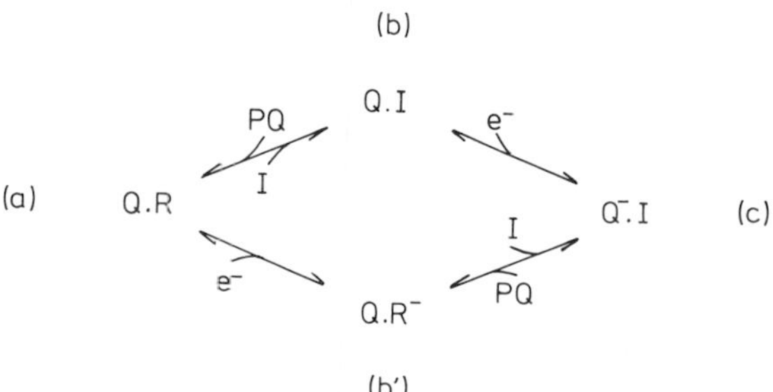

Fig. 2. Conversions of the acceptor complex of photosystem II induced by electrons and inhibitors (I) of PQ reduction.

60 mV difference in the midpoint potential of the redox couples Q/Q^- and R/R^-.

If we accept that inhibitors replace secondary plastoquinone, it is of interest to know the (maximum) binding rate of inhibitor so that we can deduce the plastoquinone release rate. Some useful information on this matter is reproduced in Fig. 3; this illustrates the phenomenon of inhibitor-induced reduction of Q with the inhibitors DCMU (A) and *o*-phenanthroline (B). The Q reduction was monitored as a rise of the fluorescence yield of chlorophyll. Whereas Q is a quencher of this fluorescence, Q^- is not (*3*). The kinetics of the phenomenon are clearly dependent on the concentration and identity of the inhibitor. The fastest kinetics were observed in the experiment with *o*-phenanthroline. At 1 m*M*, the highest concentration used, the half-time of Q reduction was ~0.3 sec. Even without extrapolating to an infinite inhibitor concentration, we can see that the release of plastoquinone must be quite fast. The release of plastoquinone from centers that contain one electron can occur only from the state $Q^-{\cdot}R$; this is applicable only about 1/15th of the time (*16*). Consequently, a 0.3 sec half-time for the inhibitor-induced fluorescence rise suggests that $PQ(H_2)$ is released from centers in state $Q{\cdot}R$ or state $Q{\cdot}RH_2$ with a half-time of less than 20 msec. This half-time compares well with the measured ~10 msec half-time for

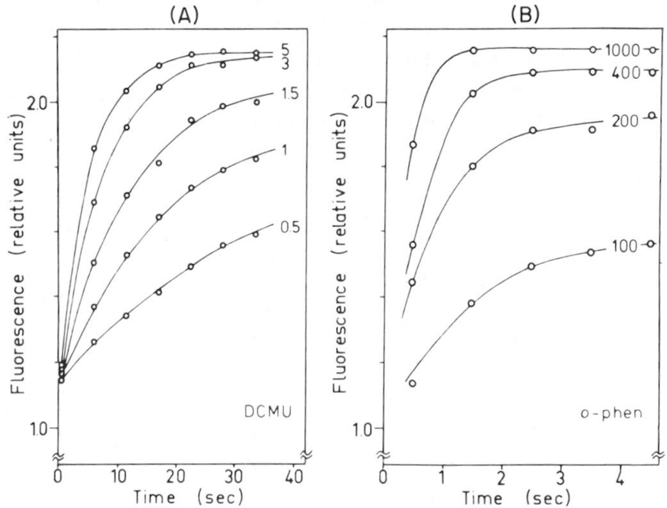

Fig. 3. Kinetics of the fluorescence rise (= Q reduction) caused by binding of DCMU or *o*-phenanthroline to R^--containing photosystem II centers. Final concentration of the inhibitors indicated in μM; chlorophyll concentration, 12.5 μM. [From Velthuys (*15*) used with permission.]

the "oxidation of RH_2 by PQ" (*17*, *18*). This oxidation, then, very likely is the release of PQH_2 followed by the binding of PQ.

The Oxidation of PQH_2 by the Cytochrome b_6f Complex

Three lines of evidence suggest that the ready inhibition of PQH_2 oxidation results from competition between $PQ(H_2)$ and inhibitor: (1) Many of the compounds that inhibit plastoquinone reduction (e.g., *o*-phenanthroline, diphenylamine, and 2-heptyl-4-hydroxyquinoline-*N*-oxide) also inhibit PQH_2 oxidation; (2) with dibromothymoquinone (DBMIB), one of the most extensively studied inhibitors of PQH_2 oxidation, the inhibition is relieved by adding $PQ(H_2)$ (*19*); and (3) the assumption of inhibition by competition is adequate to explain the effect of the inhibitors on the behavior of the PQH_2 oxidizing enzyme (the cytochrome b_6f complex.) Further discussion of this point follows.

The inhibition of PQH_2 oxidation by DBMIB is illustrated in Fig. 4. In the experiment described, PQH_2 oxidation was (conveniently) monitored by measuring the internal proton release; only a single flash was applied so that virtually all proton release of the control (Fig. 4A) was due to PQH_2 oxidizing electron flow (*20*). When the measurement was repeated in the presence of 1 μM DBMIB, the PQH_2 oxidation was slowed, but not completely inhibited (Fig. 4B). At increasing concentrations of inhibitor, the rate of H^+-releasing electron flow was decreased further, as in Fig. 4C (*21*).

Additional experiments show that the slowed electron transfer (as observed in Fig. 4B) is caused by a slowed catalysis by all enzymes

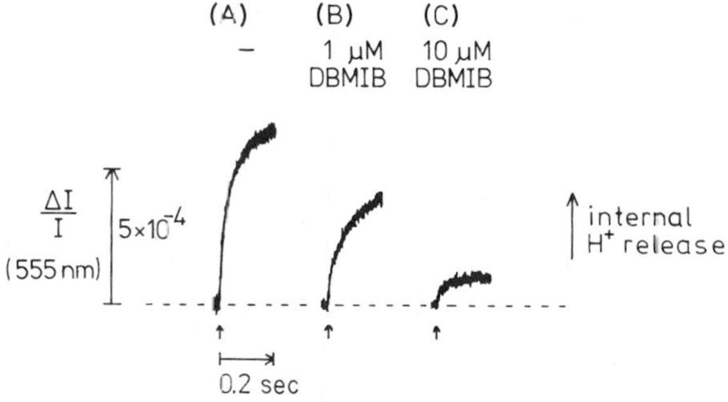

Fig. 4. Proton release inside chloroplast thylakoids, measured with neutral red [as in Velthuys (*20*)]. The dark-adapted chloroplasts were illuminated by a single flash, 20 sec after addition of DBMIB. Chlorophyll concentration, 100 μg/ml; optical path length, 2 mm.

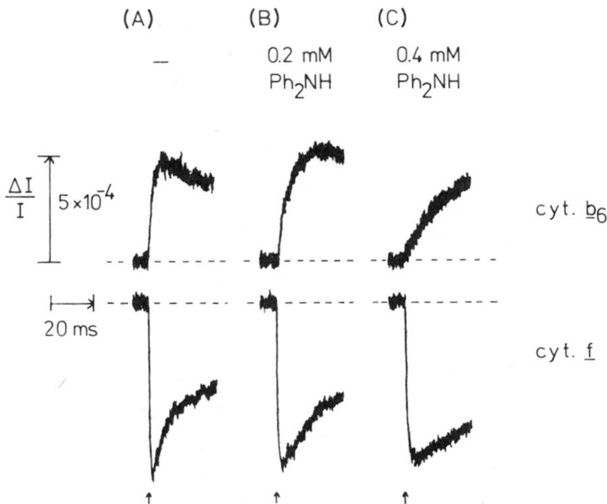

Fig. 5. Cytochrome b_6 and cytochrome *f* redox changes (reduction upward) [measured as in Velthuys (*24*)]. The pre-illuminated chloroplasts (as in *22*) were illuminated by a single flash, 20 sec after the addition of DCMU (10 μM), methylviologen (100 μM), gramicidin D (2 μM), and diphenylamine (as indicated).

rather than by a fraction of uninhibited enzymes making multiple turnovers. The observations on which this conclusion is based are essentially the same for DBMIB (not shown) and for some other less well-known inhibitors (Figs. 5 and 6). Unlike DBMIB, these inhibitors are not redox carriers; therefore preventing complications. The experiment of Fig. 5 consisted of prereducing PQ by pre-illumination (as in *22*), after which a photosystem I acceptor, a specific photosystem II inhibitor, and various concentrations of a PQH_2 oxidation inhibitor, diphenylamine (*23*), were added. Subsequently, photosystem I was activated by a short (few μsec) flash, to oxidize some plastocyanin and thus induced electron flow through the cytochrome b_6f complex. Oxidized plastocyanin oxidized cytochrome *f*, which led to oxidation of PQH_2 and turnover (i.e., reduction followed by oxidation) of cytochrome b_6 (*11*, *24*). Both cytochrome *f* and cytochrome b_6 were monitored.

The cytochrome *f* recordings of Fig. 5 show that the rereduction of cytochrome *f* is slowed by the inhibitor. This information is equivalent to that provided by the H^+ measurements and, since the cytochrome *f* molecules of different complexes can probably (via plastocyanin) rapidly equilibrate with one another (*11*), does not disprove a dichotomous behavior of the cytochrome complexes. More significantly, however, cytochrome b_6 behaved qualitatively similar to cytochrome *f*: the amplitude of the changes was not affected, but the kinetics was slowed. It is

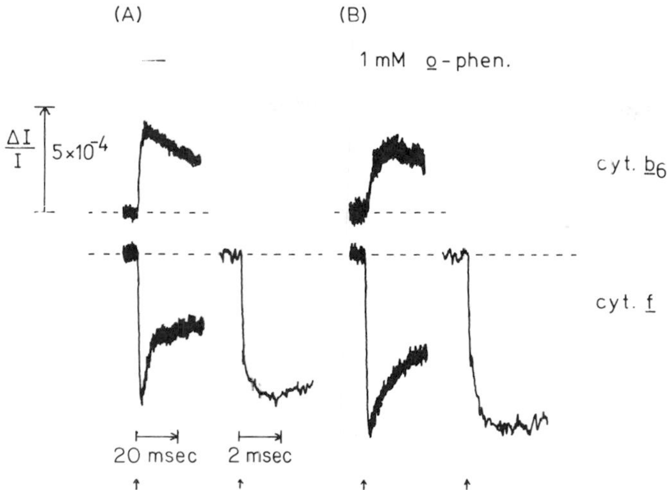

Fig. 6. Same measurements as for Fig. 5, with *o*-phenanthroline.

evident from observations such as these (see also *25*) that the inhibitors examined cause cytochrome b_6f complexes to catalyze more slowly.

The effect of inhibitor resembles the effect of a high redox potential of the plastoquinone pool (*24*), i.e., inhibitor and PQ inhibit in the same manner. Their effects may be interpreted thus: at any one moment, a fraction of the enzymes binds PQ or inhibitor rather than PQH_2 and therefore cannot perform a PQH_2 oxidation. This effect slows the apparent turnover rate of the whole ensemble of enzymes, indicating that the exchange of PQ or inhibitor for PQH_2 and vice versa is very fast, faster than the intrinsic turnover rate of the enzyme.

Some nuances in the effect of different inhibitors may be caused by differences in the inhibitor release rate from the active site of the enzyme. The inhibitor *o*-phenanthroline may not be released as quickly as diphenylamine. It, too, decreases the apparent turnover rate of the cytochrome b_6f complex; additionally, it apparently increases the extent of cytochrome *f* oxidation (Fig. 6). A similar observation was made with DBMIB (*26*). This effect was interpreted as evidence for an inhibitable rapid oxidation, via cytochrome *f*, of the Rieske iron–sulfur center (*26*). Strong support for this suggestion was recently obtained with the parallel system of photosynthetic bacteria (*27*). If we take into account the observations of Fig. 5 (and similar earlier observations (*26*, *27*)) and our earlier conclusions, then (1) it is the Rieske center that binds and oxidizes PQH_2, and (2) the Rieske center is prevented from reacting with cytochrome *f* while its plastoquinone binding site is occupied. Indeed, this latter property of the center seems essential for it to act as a one-equivalent acceptor in generating PQ^- from PQH_2.

ACKNOWLEDGMENT

This research was supported in part by U.S.D.A. (Competitive Research Grant Office) Grant 5901-0410-8-0179-0.

REFERENCES

1. Mitchell, P. (1961). *Nature (London)* **191,** 144–148.
2. Witt, H. T. (1975). *In* "Bioenergetics of Photosynthesis" (Govindjee, ed.), pp. 495–554. Academic Press, New York.
3. Duysens, L. N. M., and Sweers, H. E. (1963). *In* "Studies on Microalgae and Photosynthetic Bacteria" (Japanese Society of Plant Physiologists, eds.), pp. 353–372. Univ. of Tokyo Press, Tokyo.
4. Van Gorkom, H. J. (1974). *Biochim. Biophys. Acta* **347,** 439–442.
5. Bouges-Bocquet, B. (1973). *Biochim. Biophys. Acta* **314,** 250–256.
6. Velthuys, B. R., and Amesz, J. (1974). *Biochim. Biophys. Acta* **333,** 85–94.
7. Pulles, M. P. J., Van Gorkom, H. J., and Willemsen, J. G. (1976). *Biochim. Biophys. Acta* **449,** 536–540.
8. Mathis, P., and Haveman, J. (1977). *Biochim. Biophys. Acta* **451,** 167–181.
9. Bouges-Bocquet, B. (1980). *Int. Congr. Photosynth. 5th, Halkidiki, Greece* Abstr., p. 82.
10. Hind, G., Shahak, Y., and Crowther, D. (1980). *Int. Congr. Photosynth. 5th, Halkidiki, Greece* Abstr., p. 255.
11. Velthuys, B. R. (1980). *Annu. Rev. Plant Physiol.* **31,** 545–567.
12. Golbeck, J. H. (1980). *Arch. Biochem. Biophys.* **202,** 458–466.
13. Clayton, R. K. (1969). *Biophys. J.* **9,** 60–76.
14. Delrieu, M. J. (1978). *Plant Cell Physiol.* **19,** 1447–1456.
15. Velthuys, B. R. (1976). Ph.D. Thesis, Univ. of Leiden.
16. Diner, B. A. (1977). *Biochim. Biophys. Acta* **460,** 247–258.
17. Diner, B. A. (1975). *Proc. Int. Congr. Photosynth., 3rd, Rehovot, 1974* pp. 589–601.
18. Van Best, J. A., and Duysens, L. N. M. (1975). *Biochim. Biophys. Acta* **408,** 154–163.
19. Trebst, A., and Reimer, S. (1973). *Biochim. Biophys. Acta* **305,** 129–139.
20. Velthuys, B. R. (1980). *FEBS Lett.* **115,** 167–170.
21. Ausländer, W., and Junge, W. (1975). *FEBS Lett.* **59,** 310–315.
22. Velthuys, B. R. (1978). *Proc. Natl. Acad. Sci. U.S.A.* **76,** 6031–6034.
23. Bashford, C. L., Prince, R. C., Takamiya, K., and Dutton, P. L. (1979). *Biochim. Biophys. Acta* **545,** 223–235.
24. Velthuys, B. R. (1979). *Proc. Natl. Acad. Sci. U.S.A.* **76,** 2765–2769.
25. Olsen, L. F., Telfer, A., and Barber, J. (1980). *FEBS Lett.* **118,** 11–17.
26. Koike, H., Satoh, K., and Katoh, S. (1978). *Plant Cell Physiol.* **19,** 1371–1380.
27. Bowyer, J. R., Dutton, P. L., Prince, R. C., and Crofts, A. R. (1980). *Biochim. Biophys, Acta* **592,** 445–460.

The Electrogenic Loop in Green Algae and Higher Plants: Carriers Involved, Relation to the Plastoquinone Pool, Coupling to the Transfer Chain

2

BERNADETTE BOUGES-BOCQUET

INTRODUCTION

According to Mitchell's theory (*1*), components of photosynthetic electron-transfer chains are located in the membrane such that photoreactions generate a transmembrane electric field. The fast increase of the absorption at 515 nm is a linear measure of this transmembrane electric field (*2*). The increase in absorption at 515 nm due to charge separation takes place less than 20 nsec after the photochemical reaction (*3*). An additional increase after flashes in the millisecond range has been observed in *Chlorella* cells (*4*, *5*) and in chloroplasts (*6*–*8*). According to the terminology in Joliot and Delosme (*5*), the fast absorption increase at 515 nm due to charge separation will be referred to as phase *a*, the slow increase in the millisecond time range as phase *b*.

The 515 nm absorption increase due to phase *b* cannot be distinguished from the 515 nm absorption increase due to phase *a*; both originate from an electrochromic effect that reflects a transmembrane electric field (*5*, *9*, *10*). For example, both can be used for phosphorylation (*5*). One must conclude that phase *b* as well as phase *a* originate from charges crossing the membranes.

Function of Quinones in Energy Conserving Systems

ISBN 0-12-701280-X

What is the reaction that generates phase *b*? It could be either (A) the end of delocalization of charges separated by the photoreactions or (B) additional charges crossing the membrane. Three arguments favor the second hypothesis:

1. The amplitude of phase *b* at its maximum is identical to the amplitude of the phase *a* generated by photosystem I (*5*, *11*, *12*).
2. Electron transfer from photosystem II to photosystem I having similar charge separation and similar phase *a*, can be obtained with or without phase *b* (*12*).
3. The compounds involved in the reaction giving rise to phase *b* (*11*–*13*) (see next sections), are very similar to the compounds involved in the reaction giving rise to the slow electrochromic rise (phase III) in bacteria (*14*). Phase III was shown to be due to an additional charge crossing the membrane (for review, see *15*).

In this paper, I shall consider phase *b* to be an additional charge crossing the membrane in the millisecond time range after photoreactions. I shall try to answer the following questions: (a) What carriers are involved in this reaction? (b) What is the driving force for the charge that crosses the membrane? (c) How is this reaction related to the electron-transfer chain? and (d) Is the electrogenic path obligatory for transfer from photosystem II to photosystem I and for cyclic transfer around photosystem I?

RESULTS AND DISCUSSION

Organization of the Plastoquinone Pool

Unless specified, all results presented in this section were obtained with dark-adapted cells illuminated by one short xenon flash. The optical absorption changes induced by the flash were then measured with a time response of 1 μsec (*16*). Under these conditions, the absorption increase at 515 nm exibited two phases, phase *a* due to the photoreactions and phase *b* in the millisecond time range.

Phase *b* was decreased by oxidants and restored by reductants (*11*, *12*). It was thus dependent upon the reduced state of a compound that I will refer to as U (*12*, *13*). Table I indicates that the half-time of phase *b* was independent of the redox state of U, and that phase *b* could also be decreased at very low potential with no modification of its half-time. Phase *b* behaved like a first-order process in *Chlorella* cells. This was also observed by Joliot and Delosme (*5*) using non-saturating flashes.

TABLE I

Amplitude and Half-Time of Phase *b* in *Chlorella* Cells in Different Redox Conditions[A]

Additions	Amplitude of the electrogenic reaction	Half-time of the electrogenic reaction
DAD[B] 5×10^{-5} *M* $K_3Fe(CN)_6$ 2×10^{-4} *M*	25%	3 msec
None	80%	3 msec
$Na_2S_2O_4$ 2×10^{-4} *M*	100%	3 msec
$Na_2S_2O_4$ 2×10^{-2} *M*	30%	3 msec

[A] Average of 64 flashes fired on the same sample every 3.2 sec
[B] DAD: 2,3,5,6-tetramethyl-*p*-phenylenediamine.

Thus, phase *b* took place within a solid structure, and U must have been present at a concentration similar to that of other components of the structure. The amplitude of phase *b* after dark adaptation in the presence of DCMU could be taken as an indicator of the redox state of U. After dark adaptation, I estimated the concentration of the reduced plastoquinone pool, using the fluorescence gush as an index (*17*), to be around 12% when phase *b* reached 80% of its maximum amplitude. At a given pH, it was possible to determine an equilibrium constant between U and the plastoquinone pool (PQ). If U is assumed to be a two-electron carrier,

$$K = \frac{[U_{red}][PQ_{ox}]}{[U_{ox}][PQ_{red}]} = \frac{0.8 \times 0.88}{0.2 \times 0.12} \sim 30. \tag{1}$$

The difference between the midpoint potential of (U_{ox}, U_{red}) and (PQ_{ox}, PQ_{red}) would be

$$\frac{RT}{2F} \ln K \sim 40 \text{ mV} \qquad \text{at } 25\ ^\circ\text{C}. \tag{2}$$

U could thus receive electrons from the plastoquinone pool (*7*) because of its higher potential.

When photosystem II was active, the amplitude of phase *b* during a series of closely spaced flashes oscillated with period 2 (Fig. 1, and Velthuys (*19*) with chloroplasts). In Fig. 1, the amplitude of the oscillations was small because the system was close to the equilibrium between B and B^-. The oscillating pattern was suppressed by DCMU (*18*). This indicated that, in the absence of DCMU, the electrons arriving on U (the reductant required for the electrogenic phase) originated

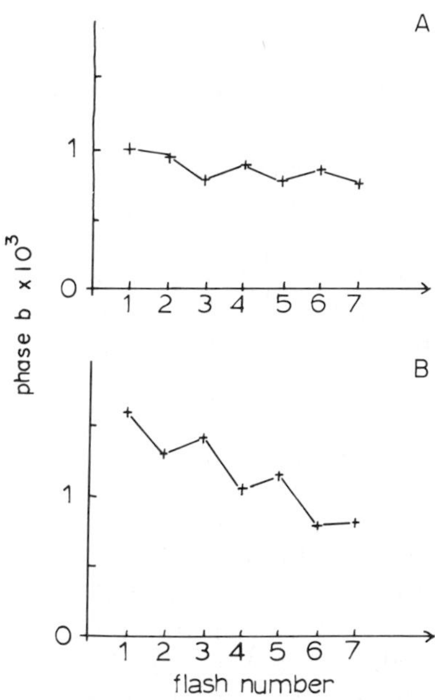

Fig. 1. Amplitude of phase *b* (corrected for the following electric field decay) during a series of flashes spaced by 125 msec on *Chlorella* cells. Average of 256 series fired on the same sample once every 6.4 sec. (A) Benzoquinone, 2×10^{-5} *M*; (B) Dithionite, 10^{-4} *M*.

from the secondary acceptor of photosystem II, the special quinone B (*20, 21*).

Transfer from B to U occurred in 10 msec or less (*13*); thus, U was part of the acceptor pool located between photosystem II and photosystem I. This acceptor pool exhibited heterogeneous kinetic behavior (*22–24*). A fraction of the pool reacted within less than 30 msec with the primary acceptor of photosystem II, while the remaining fraction was reduced by photosystem II within several hundred milliseconds. The fast-reacting fraction corresponded to a capacity of 5 electrons, including the primary stable acceptor Q, when measured according to oxygen or fluorescence gush (~5 in Joliot (*22*); ≤5 in Radmer and Kok (*23*); 5 or 6 in Forbush and Kok (*24*)).

The plastoquinone pool detected by spectroscopic measurements at 260 nm behaved like the acceptor pool detected using fluorescence or oxygen measurements as an index (*25*). The fraction reacting in less than 30 msec induced an absorption change $\Delta I/I$ of 6×10^{-4} for a chlorophyll concentration of 5×10^{-5} *M* and an optical path of 0.12 cm, in

Chlorella cells (*25*). This change included the change due to the primary acceptor, the semiquinone Q. Using an extinction coefficient of 15 mM^{-1} cm^{-1} for Q–Q^-, and 14 mM^{-1} cm^{-1} for PQ–PQH_2 at 260 nm (*26*), a flattening factor of 2.6 for *Chlorella* cells at 260 nm (*27*), and a ratio of chlorophyll/photosystem II center of 600, one can calculate that the absorption change induced by 30 msec of strong illumination (*25*) (once the contribution of the reduced primary acceptor has been substracted) corresponds to the full reduction of two other quinones per photosystem II center. This suggests that the pool reduced by photosystem II in less than 30 msec would be exclusively composed of quinones.

This pool contained Q, the primary stable acceptor of photosystem II and B, the secondary acceptor (*20*, *21*). The rate of reduction of U by photosystem II in 10 msec or less (*13*) indicated that U was also part of this pool and that U would be a quinone. This identification of U as a quinone accounts for the inhibitory effect of 2,5-dibromothymo-3-methyl-6-isopropyl-1,4-benzoquinone (DBMIB), a quinone antagonist, on phase *b* (*11*). The main path of electron and proton transfer between photosystem II and photosystem I would involve only these three specialized quinones, Q, B, and U. The reduction of the bulk of the plastoquinone pool by photosystem II would occur only after strong illumination, being slower than the electron transfer from photosystem II to U.

This situation seems similar to that in bacteria where only three quinones are required for the transfers (*28*, *29*): Q_I, the primary stable acceptor; Q_{II}, the secondary acceptor equivalent to B; and Z, a special quinone equivalent to U and linked to an electrogenic phase in the millisecond time range. Z has a potential 40 mV more positive than the ubiquinone pool (*30*, *31*).

Carriers Related to the Slow Electrogenic Phase

In the preceding section, three ways of varying principally the amplitude of phase *b* have been mentioned: (1) addition of an oxidant, (2) addition of a reductant, and (3) addition of DBMIB. A fourth way will be reported in the next section: pre-illumination with a few flashes when phosphorylation is inhibited. These four treatments for varying the amplitude of phase *b* can be used to correlate photoinduced redox changes in the path, including phase *b*.

Bouges-Bocquet (*11*) correlated the existence of phase *b* with a reduction of the photooxidized donors of photosystem I (cytochrome *f* and plastocyanin) in the millisecond time range after a photochemical reaction. In addition, the similarity of the rates of the electrogenic reaction and the reduction of the photooxidized donors showed that the two reactions are closely related. Since the donors of photosystem I are in equilibrium and have low equilibrium constants, it is not yet clear which donor is closely related to the electrogenic reaction. Both cytochrome *f* and the Rieske protein are present in the cytochrome *b-f* particle (*32*, *33*) (the structure in which the electrogenic reaction very likely occurs), so this donor could be either cytochrome *f* or the Rieske protein if it mediates the electrons to cytochrome *f* (*34–37*). It has not yet definitely been proved that the Rieske protein mediates the electrons to cytochrome *f* in chloroplasts and green algae. However, since this proof exists for bacteria with very similar structure (*38*), I will assume in the following discussion that the Rieske protein is the photosystem I donor closely related to the electrogenic reaction.

The correlation between the reduction of photooxidized donors of photosystem I in the millisecond range and phase *b* has been confirmed for chloroplasts by several laboratories (*7*, *10*). I observed a new optical signal in *Chlorella* cells (Fig. 2) (*18*). Let us call C the compound whose redox change is revealed by this signal. Table II indicates that redox changes of C correlate with the existence of phase *b* and suggests that C is involved in the section of the electron-transfer chain that includes the electrogenic process. Thus, it was of interest to compare the kinetics of phase *b* of the electrochromic effect (already correlated

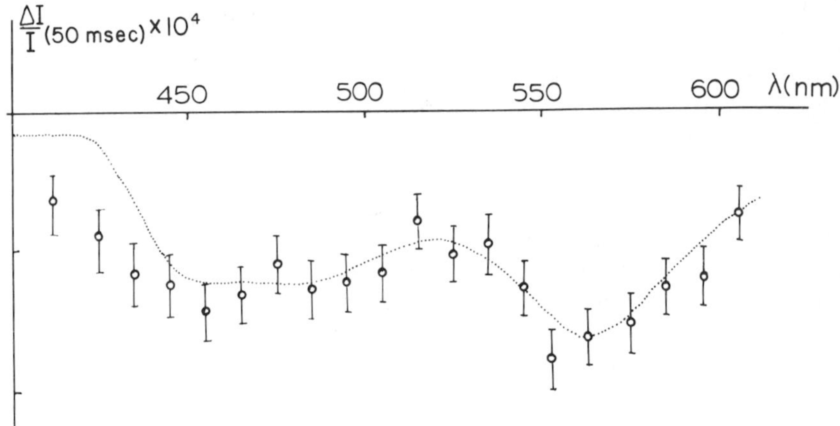

Fig. 2. (○), absorption changes 50 msec after a flash on dark adapted *Chlorella cells*, as a function of the wavelength. Average of 100 flashes. NH_2OH^{-4} *M* + $DCMU^{-5}$ *M*. The electrochromic effect was corrected. (····), reduced minus oxidized spectrum of the Rieske protein, drawn from Rieske *et al.* (*39*).

TABLE II

Phase *b* and Signal Due to the Compound C in Dark-Adapted *Chlorella* Cells with Tri-*N*-butyltin Chloride 10^{-5} *M*.

	Additions	Phase *b*	C reduction
1st Flash	None	+	+
	DAD[A] 5×10^{-5} *M*; $K_3Fe(CN)_6$ 2×10^{-4} *M*	−	−
	DAD 5×10^{-5} *M*; $K_3Fe(CN)_6$ 2×10^{-4} *M*; $Na_2S_2O_4$ 2×10^{-2} *M*	+	+
	DBMIB[B] 10^{-5} *M*	−	−
	DBMIB[B] 10^{-5} *M*; $Na_2S_2O_4$ 2×10^{-2} *M*	−	−
9th Flash	None	−	−

[A] DAD: 2,3,5,6-tetramethyl-*p*-phenylenediamine.
[B] DBMIB: 2,5-dibromo-3-6-isoproyl-1,4-benzoquinone.

with the reduction of cytochrome *f*) with the kinetics of C. Figure 3 indicates that C is bleached in the millisecond time range simultaneously with reduction of cytochrome f^+ and phase *b*. The difficulty of these measurements however does not allow more precise kinetics. The absorption changes due to C disappear with a half-time of ~40 msec.

A comparison of the difference spectrum of C (Fig. 2, open circles) with the difference spectrum of Rieske iron–sulfur protein (*39*) (corrected for the "particle flattening effect" due to the cells (*40–42*) (Fig. 2, dotted line) reveals a close resemblance of the two spectra, indicating that C could be a new iron–sulfur protein. The potential of C, which is oxidized in the dark even when the plastoquinone pool of midpoint potential around 100 mV is mainly reduced, excludes the possibility that C is the Rieske protein, the midpoint potential of which is between +290 mV (*43*) and +330 mV (*33*).

The slow electrogenic reaction occurs in the same time range and correlates with the reduction of both C and a photooxidized donor of photosystem I. Given the fact that U, the reductant required for this slow electrogenic reaction, is a quinone, i.e., a two-electron carrier, all these results can be interpreted by the following chain of reactions:

1. Photooxidation of the Rieske protein, or eventually cytochrome *f*, initiates the reaction by reducing the quinol UH_2 to its low potential semiquinone form UH or U^- (midpoint potential in the order of −150 mV for plastosemiquinone (*44*)).

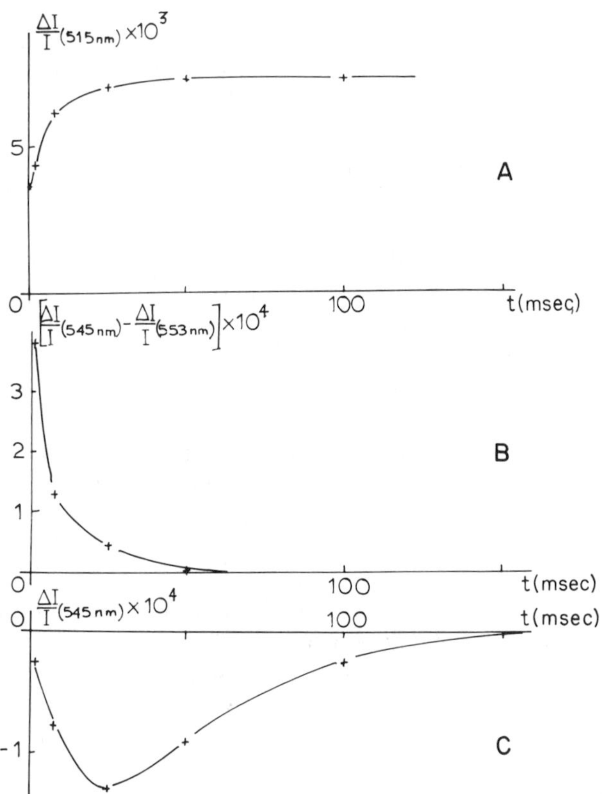

Fig. 3. Absorption changes of difference of absorption changes as a function of the time after a flash on dark-adapted *Chlorella* cells (after correction for the electrochromic effect). Average of 100 flashes. NH_2OH 10^{-4} M + DCMU 10^{-5} M. (A) Electrochromic effect, (B) Cytochrome f^+, (C) C^-.

2. This low potential semiquinone then reduces C in a reaction associated with a charge crossing the membrane against the transmembrane electric field already built by the photoreactions, thus increasing the transmembrane electric field and generating phase *b*.

Two types of reactions might account for charges crossing the membrane. Either C is at the outside part of the membrane and UH becomes deprotonated inside the thylakoid before transferring an electron to C, or a proton channel allows C^-, buried in the membrane, to be protonated from the outside medium.

The mechanism of the electrogenic reaction could be summarized by the following reactions

$$UH_2 + \text{Fe-S}^+ \rightleftharpoons UH + H^+_{int} + \text{Fe-S}, \tag{3}$$

$$UH + C(+H_{ext}\ ?) \rightleftharpoons U + H^+_{int} + C^-_{ext}\ (\text{or CH}) \tag{4}$$

where Eq. (4) is the electrogenic reaction.

What is the role of cytochrome b_6 in these reactions? In bacteria it is generally thought that the electrogenic phase involves some cytochrome *b* molecules (*45*). In chloroplasts, two cytochrome b_6 molecules are closely associated with the structure responsible for phase *b* (*32*) and several authors have proposed that cytochrome b_6 is involved in the electrogenic reaction (*46*, *47*). The experimental finding that cytochrome b_6 reduction oscillates with period 2 (*46*) as in bacteria (*48*), clearly shows that cytochrome b_6 can be reduced by the semiquinone U^-. However, in the most precise experiments with intact chloroplasts, where phase *b* was close to its maximum value (*10*), and in my experiments with *Chlorella*, it was not possible to correlate a reduction of cytochrome b_6 with phase *b*. In addition, after one flash on dark-adapted chloroplasts phase *b* was insensitive to antimycin (*18*).

Additional experiments are required before drawing definite conclusions about the role of cytochrome b_6 in the electrogenic reaction. Two types of hypothesis can be proposed to account for the apparently contradictory results published in the literature:

1. Cytochrome b_6 may be reduced by U^- when C is not available; in this case there is no electrogenic reaction.
2. A cytochrome b_6 molecule is involved in the electrogenic reaction, but is protected against antimycin. Its redox changes in intact chloroplasts or whole cells would not be detectable because its reoxidation is faster than its reduction.

Another question is how C^- (or CH) is reoxidized. All the experiments where I have tried to observe a charge-accumulating process, storing two charges before rereducing a plastoquinone, have failed. It seems more likely that a dismutation process occurs. Such a dismutation was observed in the acceptor of photosystem I, the flavoprotein, ferredoxin-NADP-reductase (*49*). The rapidity of the dismutation of this flavoprotein, $t_{1/2} = 30\ \mu sec$ (*49*) suggests that photosystem I centers are aggregated at least two by two and that a dismutation process in C, or further on in this part of the electron-transfer chain, is likely.

The mechanism described in this paper starts like Mitchell's Q cycle (*50*) and would behave like a Q cycle if the electron from C came back on U with an associated hydrogen crossing the membrane from the outside towards the inside. In this case, one would expect the ratio H^+/e under stationary light conditions to be 3. This ratio has indeed been found equal to 3 under low light excitation (*51*), suggesting that a Q-cycle takes place in intact systems.

The transfer chain between photosystem II and photosystem I could be tentatively represented by the following simplified scheme.

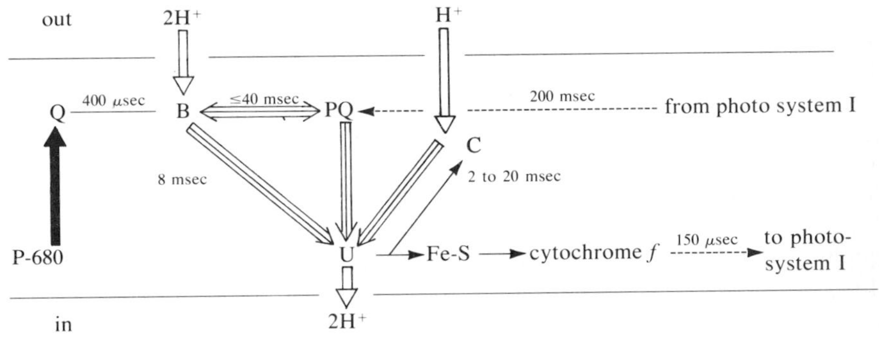

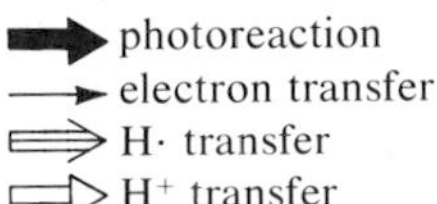

Q, B, and U, three specialized quinones.
PQ, the bulk of the plastoquinone pool.
Fe-S, Rieske iron–sulfur protein C (likely iron–sulfur protein).

Coupling of the Electrogenic Phase to the Electron-Transfer Chain

The oscillations with period 2 of phase *b* (Fig. 1) (*9*) indicate that the electrogenic loop is located in the main linear pathway between photosystem II and photosystem I. The still efficient generation of phase *b* after 2 or 3 flashes in the presence of DCMU indicates that U may also receive electrons from the acceptors of photosystem I when photosystem I is working under cyclic conditions.

However, the amplitude of phase *b* decreased during a series of flashes and was low under stationary conditions with high flash frequency (*5*). Figure 4 (see also *12*) presents the amplitude of phase *b* under stationary conditions as a function of flash frequency. The same batch of cells was used (either immediately after harvesting (○), where phosphorylation was rapid (*52*), or after exposure to dim light (●), where phosphorylation was slow (*52*) giving results of a larger transmembrane electric field and pH gradient, corresponding to a smaller phase *b*. This decrease in amplitude of phase *b* at high transmembrane

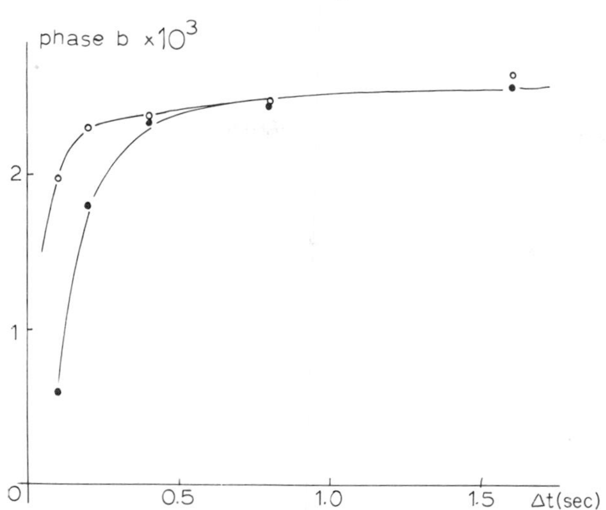

Fig. 4. Amplitude of phase *b* for an average of 64 flashes fired on the same sample as a function of the time Δt between two consecutive flashes, in *Chlorella* cells with dithionite 2×10^{-4} *M*. (○), cells with fast electric field decay; (●), cells with slow electric field decay.

electric field or pH gradient could be interpreted as an inhibition for energetic reasons: the free energy available would no longer be sufficient to draw electrons across the membrane.

Figure 5B gives further evidence for limitation of the amplitude of phase *b* due to energetic reasons. At low pH, the midpoint potential of (U, UH_2) becomes too high and the reaction

$$UH_2 + \text{Fe-S}^+ + C + (H^+_{ext}\ ?) \rightleftharpoons U + 2H^+_{int} + \text{Fe-S} + C^-_{ext}\ (\text{or CH}) \quad (5)$$

is no longer exergonic.

At pH 7, 50% of the reaction is achieved after a single flash; the free energy of this reaction would be around zero (see more detailed analysis in *12*)

$$E_1 + E_2 - 2E_U \sim 0 \qquad \text{at pH 7} \quad (6)$$

where E_1, E_2 and E_U are, respectively, the midpoint potentials of the couples (Fe-S^+, Fe-S), (C_{ox}, C_{red}), and (U, UH_2) at pH 7. E_U is in the order of 140 mV (see above; see also *12*) and E_1 is in the order of −310 V (*33*, *43*). Thus, E_2 would be in order of −30 mV, which is consistent with the reduction of C by the semiquinone U^- of potential around −150 mV (*44*).

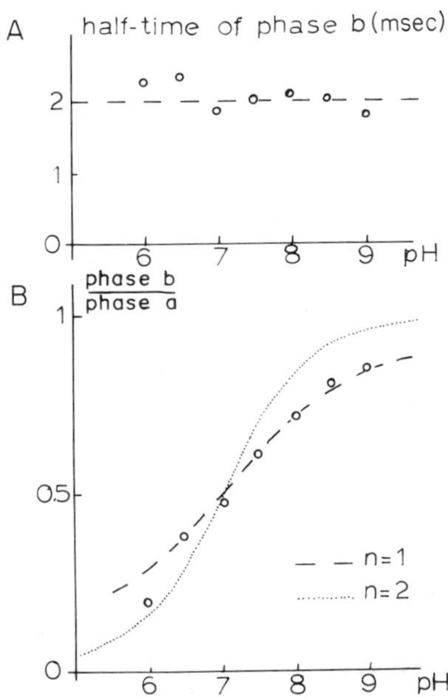

Fig. 5. (A) Half-time, and (B) relative amplitude of phase *b* in spinach chloroplasts as a function of pH. Dithionite concentration was adjusted at each pH value to get the maximum amplitude of phase *b*. Average of 64 flashes fired on the same sample every 6.4 sec.

When phase *b* is impeded for energetic reasons, reduction of the photooxidized donors of photosystem I slows (*11*). However, this reduction may be accelerated upon addition of NH_4Cl, and Fig. 6 indicates that reduction of cytochrome *f* is possible when the electrogenic reaction is impeded for energetic reasons. Thus, electron transfers from photosystem II to photosystem I do not necessarily involve the electrogenic reaction and at high pH gradient or a high transmembrane electric potential, the H^+/e ratio will drop to a value close to 2. This was indeed observed by Rathenow and Rumberg (*51*), who detected an H^+/e ratio of 3 under low light excitation, but of 2 at high intensity. In *Chlorella* cells, the critical value of the light intensity above which the H^+/e ratio drops from 3 to 2 is around one photon per reaction center every 200 msec (Fig. 5).

The optimization of energy conversion by plants would thus be performed by reactions, coupled at low proton gradient, allowing a high stoichiometry of translocated protons, and uncoupled at high proton gradient, allowing efficacious electron transfer from water to NADP.

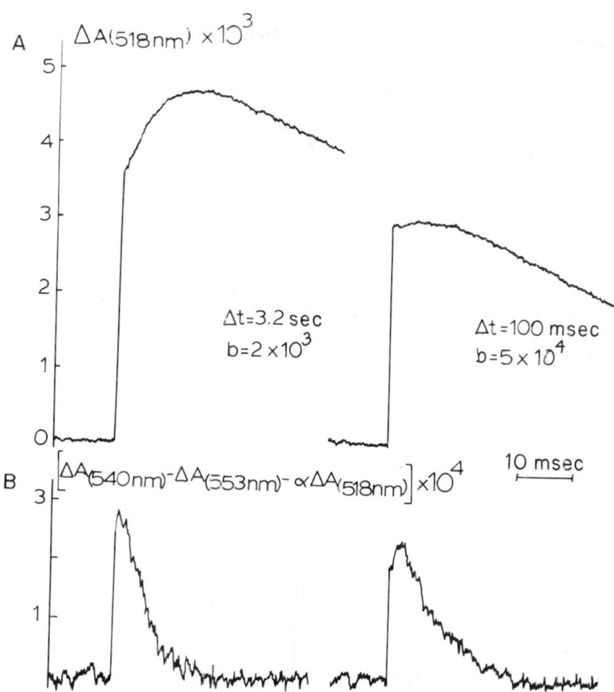

Fig. 6. Absorption changes at (A) 518 nm, and (B) cytochrome *f* kinetics after one flash in *Chlorella* cells with NH_4Cl 10^{-1} *M*. Average of 64 flashes (A) and 256 flashes (B) fired on the same sample every 3.2 sec (left curves) or every 100 msec (right curves).

ACKNOWLEDGMENTS

This work was supported by the Centre National de la Recherche Scientifique and an NSF grant from U.S.–France exchange program.

REFERENCES

1. Mitchell, P. (1966). "Chemiosmotic Coupling in Oxidative and Photosynthetic Phosphorylation." Glynn Res., Bodmin, England.
2. Junge, W., and Witt, H. T. (1968). *Z. Naturforsch., Teil B* **23,** 244–254.
3. Wolff, C., Buchwald, H. E., Rüppel, H., and Witt, H. T. (1969). *Z. Naturforsch., Teil B* **24,** 1038–1041.
4. Witt, H. T., and Moraw, R. (1959). *Z. Phys. Chem.* (*Weisbaden*) **20,** 254–282.
5. Joliot, P., and Delosme, R. (1974). *Biochim. Biophys. Acta* **357,** 267–284.

6. Horvath, G., Droppa, M., Mustardy, L., and Faludi-Daniel, A. (1978). *Planta* **141,** 239–244.
7. Velthuys, B. R. (1978). *Proc. Natl. Acad. Sci. U.S.A.* **75,** 6031–6034.
8. Slovacek, R. E., and Hind, G. (1978). *Biochem. Biophys. Res. Commun.* **84,** 901–906.
9. Diner, B., and Joliot, P. (1976). *Biochim. Biophys. Acta* **423,** 479–498.
10. Crowther, D., and Hind, G. (1980). *Arch. Biochem. Biophys.* **204,** 568–577.
11. Bouges-Bocquet, B. (1977). *Biochim. Biophys. Acta* **462,** 371–379.
12. Bouges-Bocquet, B. (1981). *Biochim. Biophys. Acta* **635,** 327–340.
13. Bouges-Bocquet, B. (1981) *In* "Photosynthesis II–Electron Transport and Photophosphorylation" (G. Akoyunoglou, ed.) Vol. II, pp. 19–27.
14. Jackson, J. B., and Dutton, P. L. (1973). *Biochim. Biophys. Acta* **325,** 102–113.
15. Wraight, C. A., Cogdell, R. J., and Chance, B. (1978). In "The Photosynthetic Bacteria" (R. K. Clayton and W. R. Sistrom, eds.), pp. 471–512. Plenum, New York.
16. Joliot, P., Béal, D., and Frilley, B. (1980). *J. Chim. Phys.* **77,** 209–216.
17. Goldbeck, J. H., and Kok, B. (1979). *Biochim. Biophys. Acta* **547,** 347–360.
18. Bouges-Bocquet, B. (1980). *FEBS Lett.* **117,** 54–58.
19. Velthuys, B. R. (1980). *FEBS Lett.* **115,** 167–170.
20. Bouges-Bocquet, B. (1973). *Biochim. Biophys. Acta* **314,** 250–256.
21. Velthuys, B. R., and Amesz, J. (1974). *Biochim. Biophys. Acta* **333,** 85–94.
22. Joliot, P. (1965). *Biochim. Biophys. Acta* **102,** 116–134.
23. Radmer, R., and Kok, B. (1973). *Biochim. Biophys. Acta* **314,** 28–41.
24. Forbush, B., and Kok, B. (1968). *Biochim. Biophys. Acta* **162,** 243–253.
25. Stiehl, H. H., and Witt, H. T. (1969). *Z. Naturforsch., Teil B* **24,** 1588–1598.
26. Amesz, J. (1977). *Encycl. Plant Physiol., New Ser.* **5,** 238–246.
27. Amesz, J., Van den Engh, G. J., and Visser, J. W. M. (1972). *Photosynth., Two Centuries Its Discovery Joseph Priestley, Int. Congr. Photosynth. Res., 2nd Stresa, Italy, 1971,* Vol. 1 pp. 419–430.
28. Baccarini-Melandri, A., and Melandri, B. A. (1977). *FEBS Lett.* **80,** 459–464.
29. Bowyer, J. R., Baccarini-Melandri, A., Melandri, B. A., and Crofts, A. R. (1978). *Z. Naturforsch., Teil C* **33,** 704–711.
30. Cogdell, R. J., Jackson, J. B., and Crofts, A. R. (1972). *Bioenergetics* **4,** 413–429.
31. Prince, R. C., and Dutton, P. L. (1977). *Biochim. Biophys. Acta* **462,** 731–747.
32. Nelson, N., and Neumann, J. (1972). *J. Biol. Chem.* **247,** 1817–1824.
33. Rich, P. R., Heathcote, P., Evans, M. C. W., and Bendall, D. S. (1980). *FEBS Lett.* **116,** 51–56.
34. Malkin, R., and Posner, H. B. (1978). *Biochim. Biophys. Acta* **501,** 552–559.
35. White, C. C., Chain, R. K., and Malkin, R. (1978). *Biochim. Biophys. Acta* **502,** 127–137.
36. Whitmarsh, J., and Cramer, W. A. (1979). *Proc. Natl. Acad. Sci. U.S.A.* **76,** 4417–4420.
37. Koik, H., Satoh, K., and Katoh, S. (1978). *Plant Cell Physiol.* **19,** 1371–1380.
38. Bowyer, J. R., Dutton, P. L., Prince, R. C., and Crofts, A. R. (1980). *Biochim. Biophys. Acta* **592,** 445–460.
39. Rieske, J. S., Maclennan, D. H., and Colemann, R. (1964). *Biochem. Biophys. Res. Commun.* **15,** 338–344.
40. Duysens, L. N. M. (1956). *Biochim. Biophys. Acta* **19,** 1–12.
41. Pulles, M. P. J. (1978). Ph.D. Thesis, Univ. of Leiden.
42. Bouges-Bocquet, B. (1978). *FEBS Lett.* **94,** 95–99.
43. Malkin, R., and Aparicio, R. J. (1975). *Biochem. Biophys. Res. Commun.* **63,** 1157–1160.

44. Rich, P. R., and Bendall, D. S. (1980). *Biochim. Biophys. Acta* **592,** 506–518.
45. Jackson, J. B., and Crofts, A. R. (1971). *Eur. J. Biochem.* **18,** 120–130.
46. Velthuys, B. R. (1979). *Proc. Natl. Acad. Sci. U.S.A.* **76,** 2765–2769.
47. Olsen, L. F., Telfer, A., and Barber, J. (1980). *FEBS Lett.* **118,** 11–17.
48. De Grooth, B. G., Van Grondelle, R., Romijn, J. C., and Pulles, M. P. J. (1978). *Biochim. Biophys. Acta* **503,** 480–490.
49. Bouges-Bocquet, B. (1978). *FEBS Lett.* **85,** 340–344.
50. Mitchell, P. (1975). *FEBS Lett.* **85,** 340–344.
51. Rathenow, M., and Rumberg, B. (1980). *Ber. Bunsenges. Phys. Chem.* **84,** 1059–1062.
52. Diner, B. A., and Joliot, P. (1976). *Biochim. Biophys. Acta* **423,** 479–498.

Inhibitors of Plastoquinone Function as Tools for Identification of Its Binding Proteins in Chloroplasts 3

W. OETTMEIER
U. JOHANNINGMEIER
A. TREBST

INTRODUCTION

The role of plastoquinone in photosynthetic electron flow has been established for some time. Plastoquinone is considered to connect the acceptor side of photosystem II located on the outer membrane surface with the donor side of photosystem I located on the inner membrane side. In this vectorial electron flow across the membrane, plastoquinone carries hydrogens and is thus responsible for one of the "coupling sites" of photophosphorylation. Plastoquinone is also the hydrogen carrier accounting for the coupling in physiological (ferredoxin-catalyzed) cyclic electron flow. For recent reviews on photosynthetic electron flow and plastoquinone function, see Velthuys (*1*) and Trebst (*2*).

In connecting several photosystems II with several photosystems I, plastoquinone is also considered to be an "electron buffer" and a branching point between the photosystems (*3*). This agrees with the evidence that there is about 7–8 times more plastoquinone present in the membrane than electron-transport chains. Nevertheless, it might be discussed whether there exists protein-bound plastoquinone rather than, or in addi-

Function of Quinones in Energy Conserving Systems

ISBN 0-12-701280-X

tion to, free plastoquinone "solubilized" in the membrane (see Hauska and Hurt, Paper No. 1, Chapter III, this volume; see also *4*). The primary acceptor of photosystem II, named Q by Duysens, is a specially bound plastoquinone that undergoes only one electron transfer (*5*). By studying inhibitors of photosystem II, another redox component carrying two electrons was proposed as a secondary quencher of photosystem II, called B (*6*) or R (*7*). B (or R) may be another special plastoquinone, separate from the main pool (*8*). But this special plastoquinone could also be visualized as plastoquinone from the main pool, which attaches to the B-protein when it is reduced by photosystem II via Q. In addition to this binding site for plastoquinone at its reducing site or B-protein, there should be another binding site, for plastoquinol at its oxidizing site, possibly at the cytochrome b_6/cytochrome f/Fe-S complex. New results on inhibitors of photosynthetic electron flow have led to the identification of polypeptides and protein complexes involved in plastoquinone function.

EXPERIMENTAL RESULTS

Inhibitors on the Reducing Site of Plastoquinone

Inhibitors at the acceptor side of photosystem II include a number of commercial herbicides represented by DCMU. These herbicides inhibit electron flow between Q and the main plastoquinone pool and affect the midpoint potential of the new carrier, designated B (*6*) or R (*7*). This new carrier is possibly also a plastoquinone (*8*). DCMU has many analogs (*2*) that affect photosynthetic electron flow in the same way and are thought to interact with the membrane at an identical site. Tischer and Strotmann (*9*) established that analogs of DCMU compete effectively with each other for binding to the membrane. Figure 1 demonstrates the binding of radioactive metribuzin, a triazinone herbicide, to the thylakoid membrane. Metribuzin has a binding constant of 26 n*M* and binds about 2 nmole per mg chlorophyll (i.e., about one inhibitor per 500 chlorophyll or one inhibitor per electron-transport chain. If analogs like metamitrone or DCMU are added, the radioactive metribuzin is displaced from the membrane (*9*, *10*), as shown in Fig. 2.

Recently we described another group of effective electron flow inhibitors, functioning in an identical manner as DCMU (*11*, *12*). However, this group of substituted phenols has quite a different chemistry from the DCMU inhibitor family. Nevertheless, the displacement technique indicates that the two groups of inhibitors compete for the same binding site,

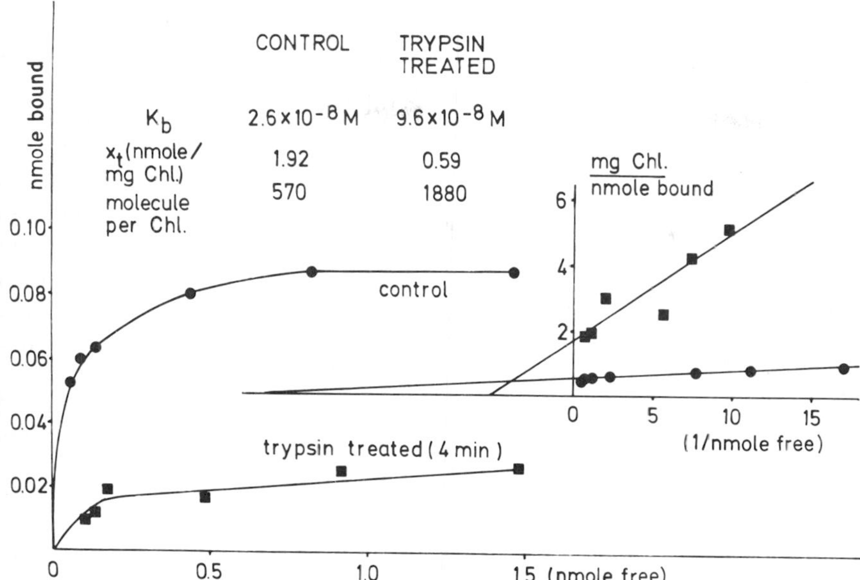

Fig. 1. Binding of radioactive metribuzin (4-amino-6-*t*-butyl-3-methylthio-1,2,4-triazin-5-one; formula, see Fig. 8) to washed thylakoid membranes before and after trypsin treatment (4 min). Inset: Double reciprocal plot of binding data for calculation of binding constant (abscissa intercept) and number of binding sites (ordinate intercept).

because phenols like bromonitrothymol or ioxynil also replace metribuzin (*10*) as in Fig. 2. Because of the different chemistry, but seemingly identical interaction with the membrane, a common binding area on an unidentified protein with overlapping yet special binding sites for the two types of inhibitors was postulated (*12*, *13*).

There are also distinct differences between the two groups; for example, analogs of the same chemical group replace each other competitively, whereas they displace others noncompetitively (*14*). Another difference is that inhibition of electron flow by the phenol inhibitors has a time lag (*10*) as well as a difference in inhibition efficiency after trypsin treatment (*15*), as will be discussed.

The protein nature of the binding area on the membrane of inhibitors of plastoquinone function (i.e., of DCMU) was shown by its sensitivity to trypsin. As Regitz and Ohad (*16*) and then Renger *et al.* (*17*) demonstrated, electron flow is severly inhibited by trypsin treatment, except for the ferricyanide Hill reaction. Even more importantly, this ferricyanide Hill reaction becomes DCMU insensitive (*16*, *17*). Renger proposed a model of a trypsin-sensitive protein "shield" above the accep-

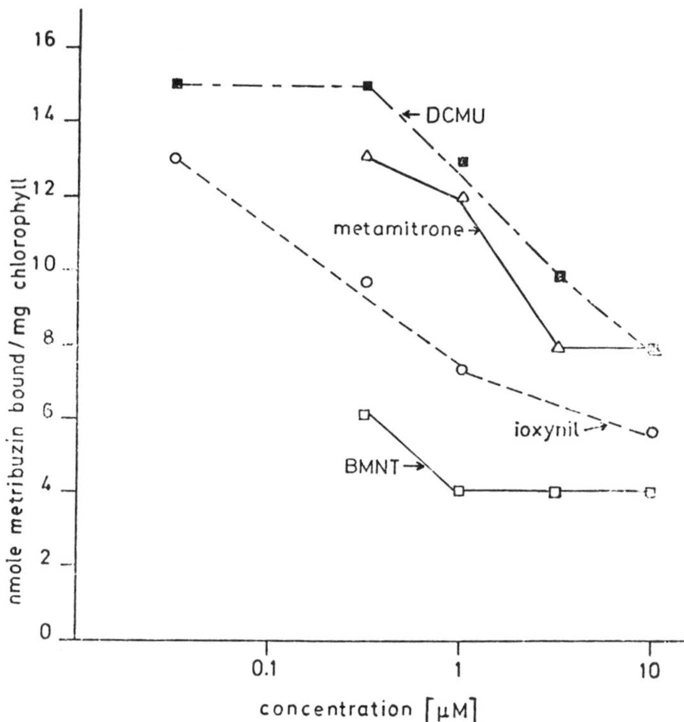

Fig. 2. A radioactive inhibitor of the DCMU type binds to the membrane and is replaced by the addition of analogs of similar chemistry, as DCMU, metamitrone (6-phenyl-4-amino-3-methyl-1,2,4-triazin-5-one) and also by analogs of different chemistry, as phenol inhibitors (bromonitrothymol or ioxynil).

tor side of photosystem II and plastoquinone carrying the DCMU binding site (*18*). This model was refined to indicate a B-protein with a property of a Q → plastoquinone oxidoreductase (*12*, *13*).

Mild trypsin treatment affects membrane bound electron flow at one site (component) only; mild trypsin also affects the coupling factor, the light harvesting chlorophyll protein, and the peripheral ferredoxin–NADP reductase (*2*). This is expected from the inhibition pattern: electron flow from water to methylviologen, and also to an acceptor of photosystem II is quickly inactivated (*19*). As summarized in Fig. 3 and Table I, trypsin treatment does not inhibit electron flow from electron-donor systems to photosystem I, neither from DAD to methylviologen via plastocyanin or from durohydroquinone to methylviologen via the cytochrome b_6/cytochrome f complex after a short time exposure. This cytochrome complex appears to participate, because of the sensitivity

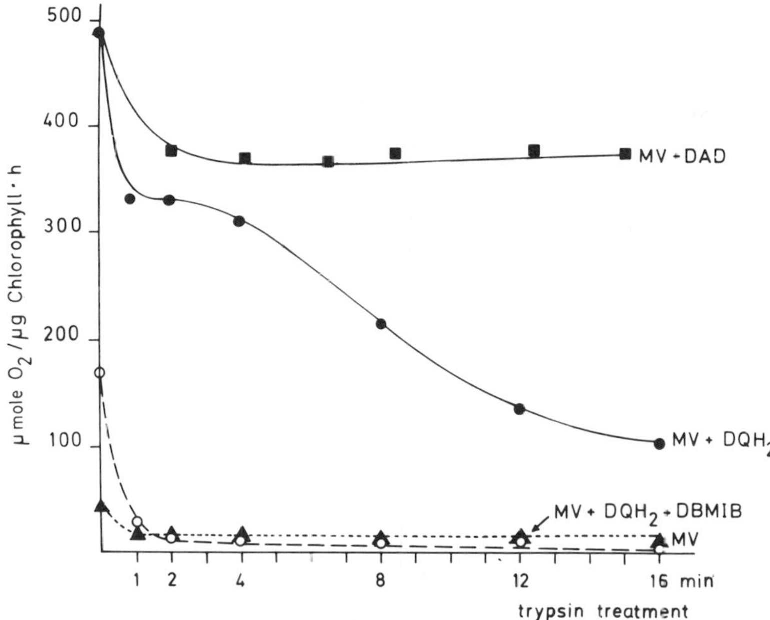

Fig. 3. Effect of trypsin treatment on electron flow from water to methylviologen (inhibition), from DAD to methylviologen (no inhibition) and from durohydroquinone to methylviologen (inhibition only after longer trypsin treatment). The durohydroquinone donor system is DBMIB sensitive.

of this donor system to DBMIB and its analog DNP–INT; this will be discussed subsequently in greater detail. Table I also demonstrates that the durohydroquinone donor system can be inhibited by the plastoquinone antagonist DNP–INT, and is not sensitive to 4 min trypsin treatment, in contrast to water as electron donor. Also the TMPD bypass (*20*) from durohydroquinone to methylviologen is not inhibited by trypsin treatment. Thus, it follows that electron flow from water via photosystem II to Q is not impaired by (short time) trypsin treatment, because oxygen evolution in a ferricyanide Hill reaction remained active: electron flow onto photosystem I through plastoquinone and the DBMIB inhibition site is also not affected. This is summarized in the scheme of Fig. 5. Only after longer incubation is durohydroquinone oxidation also affected, probably due to a trypsin effect on the cytochrome *b*/cytochrome *f* complex (Fig. 3).

Binding studies indicate that the trypsin effect on DCMU sensitivity of a Hill reaction is due to an effect on the DCMU binding site (*14*, *19*). Figures 1 and 4 show trypsin treatment lowers the number of binding

TABLE I

Effect of Trypsin Treatment on a DNP–INT Sensitive Durohydroquinone Donor System for Photosystem I[A]

e Donor	Methylviologen as acceptor, uncoupled chloroplasts μmole oxygen taken up/mg chlorophyll and hour	
	Control	4 min Trypsin
Water	145	33
Durohydroquinone	335	367
Durohydroquinone + 5 μM DNP–INT	59	16
TMPD bypass durohydroquinone + 5 μM DNP–INT + 0.1 μM TMPD	686	564

[A] The TMPD bypass (from DQH_2 to MV)[B] is not very trypsin sensitive. The electron-flow system from water to MV is inactivated by trypsin.

[B] Abbreviations: DQH_2, durohydroquinone; MV, methylviologen.

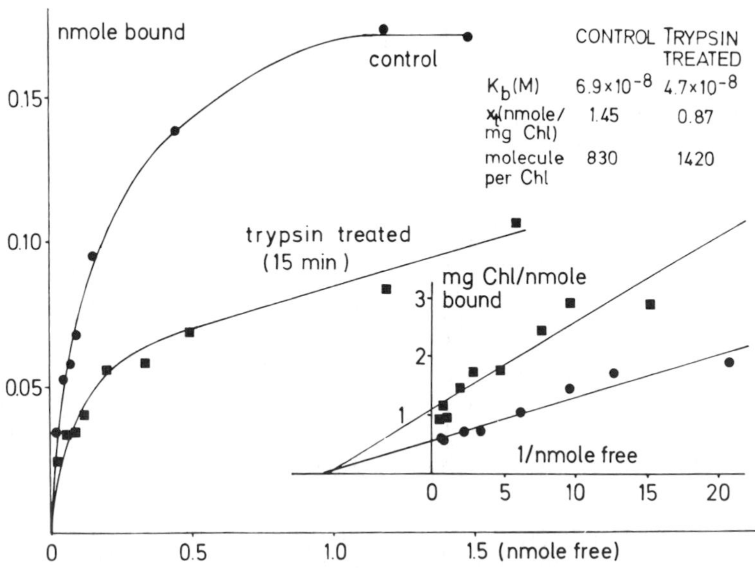

Fig. 4. Binding of radioactive i-dinoseb (2,4-dinitro-6-isobutyl-phenol) to washed thylakoid membranes before and after trypsin treatment (15 min). Inset: Same as in Fig. 1.

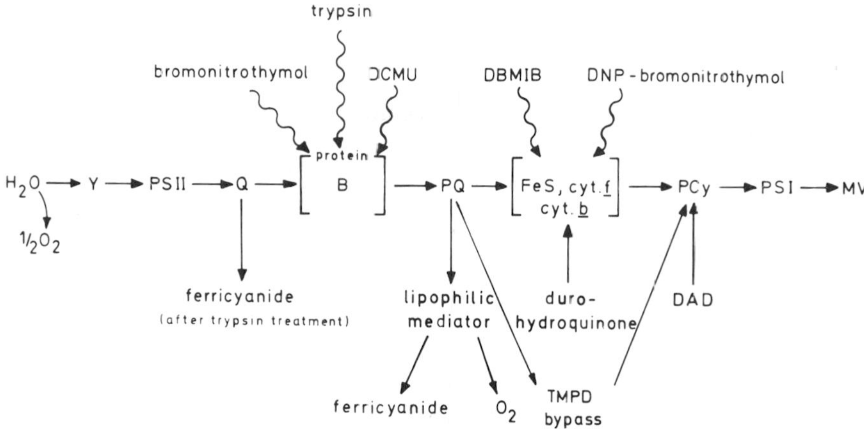

Fig. 5. Scheme of photosynthetic electron flow indicating the point of action of inhibitors on the reducing site (DCMU and nitrophenols), and the oxidizing site (DBMIB and DNP–INT) of plastoquinone. Artificial donor and acceptor systems and the pathway of the TMPD bypass are also indicated.

sites for both metribuzin and dinoseb, from an analysis of binding data in control and trypsin-treated chloroplast membranes. However, the differences between phenolic and DCMU-type herbicides, as already stressed, also becomes evident. After trypsin treatment, the binding constant for metribuzin is drastically increased whereas that of dinoseb is virtually unchanged.

Identification of the herbicide-binding protein among the 50 membrane peptides came from recent studies with inhibitory radioactive photoaffinity labels. Oettmeier *et al.* synthesized 2-azido-4-nitro-6-isobutylphenol labeled with tritium in the butyl group (*21*, *22*). After strong illumination this inhibitor becomes covalently attached to the membrane. Polyacrylamide gel electrophoresis showed several peptide bands are labeled, in addition to that of lipids and free pigments (*21*, *22*, *22a*). Two labeled bands disappeared when the membrane was preincubated with either dinoseb or DCMU. As Fig. 6 indicates, these two bands, which represent the herbicide binding peptides, have molecular weights between 41 kdalton and 17 kdalton. These results, using a labeled phenol inhibitor should be compared with results using an azido-triazine reported at the 5th International Congress on Photosynthesis this year (*23*). The polypeptide of Pfister, *et al.* (*23*, *23a*), predominantly labeled, and not seen in membranes from triazine-resistant plants, has a molecular weight of 32 kdalton (*23*, *23a*). This peptide is

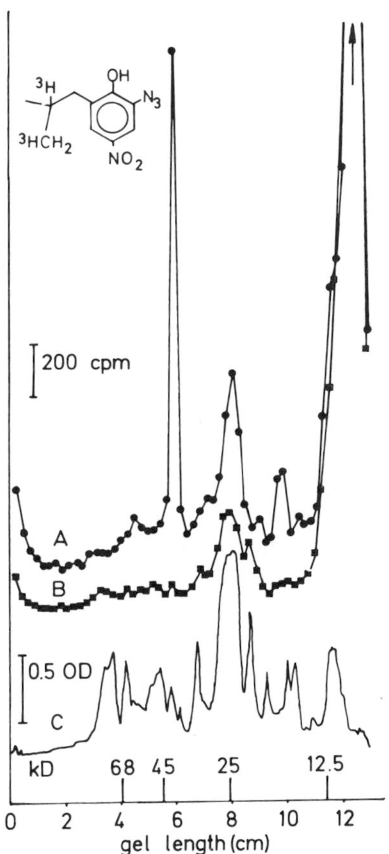

Fig. 6. Polypeptide pattern, carrying radioactivity after the membrane was illuminated with the radioactive photoaffinity label. The peaks at 34 to 40 and 20 kdalton disappeared when the membrane was preincubated with i-dinoseb.

the "rapidly turning over" photogene polypeptide of 32 kdalton, shown to be trypsin labile in parallel to the loss of DCMU sensitivity (*24*).

Inhibitors of the Oxidizing Site of Plastoquinone

Ten years ago we described dibromothymoquinone (DBMIB) as an effective inhibitor of plastoquinone function. At low concentrations ($< 1\ \mu M$) it inhibits only the oxidation of plastoquinone; above 1 μM it also inhibits its reduction. DBMIB does not inhibit photosystem II-driven Hill reactions; actually these reactions were established and are now defined this way. This also gave the first evidence for the site of inhibition as indicated in Fig. 5. By using the new inhibitor it could be shown that the photoreduction of photosystem II acceptors is coupled to ATP formation; and the second coupling site in photophosphorylation at the water splitting reaction was defined (for review, see *2*). DBMIB does inhibit the durohydroquinone donor systems (*25*) (see Fig. 5).

Furthermore, a TMPD bypass around the DBMIB inhibition site is possible in which a catalytic amount of TMPD connects the reducing side of photosystem II with the oxidizing side of photosystem I (*20*). This bypass is coupled with almost the same P/e_2 ratio as the control system (*20*). This indicates that the cytochrome b_6/cytochrome *f* complex is not essential for coupled vectorial non-cyclic electron flow (but is for cyclic electron flow). From the inhibition pattern it was concluded that DBMIB affects the binding of plastoquinol at its oxidation site. This site is presumably on the cytochrome b_6/cytochrome *f* complex, substantiated by finding that plastoquinol-1-plastocyanin oxidoreductase activity of the purified complex is DBMIB sensitive (*26*). It was then shown that a nonheme iron protein, the Rieske center, is specifically affected by DBMIB (*27*). The Rieske center is part of the cytochrome complex. However, the conclusion, that plastoquinol is oxidized by the Rieske center does not necessarily follow.

Recently we described (*12*, *28*) another group of inhibitors affecting plastoquinol oxidation at low concentrations similar to DBMIB. However, this group of dinitrophenyl ethers of halogen substituted nitrophenols has quite a different chemistry; in particular they have no redox properties, in contrast to DBMIB (Fig. 7). Their inhibitor pattern is identical to that of DBMIB (*2*, *28*). Hauska *et al.* recently showed that the DBMIB-sensitive plastoquinol-1-plastocyanin oxidoreductase activity of the purified cytochrome b_6/cytochrome *f* complex is affected also by DNP–INT (*29*). We studied the binding of a radioactive DNP–INT, but did not obtain a satisfactory binding curve, due to low solubility in water and high unspecific binding of the very lipophilic inhibitor (*22*).

DBMIB and DNP–INT inhibit cyclic electron flow with ferredoxin as the catalyst but not with artificial catalysts such as PMS or DAD (*30*). This indicates that plastoquinone may or may not participate in cyclic electron flow. With Hauska, we introduced the concept of artificial energy coupling, because ATP formed in the DBMIB-insensitive cyclic system is due to proton translocation across the membrane via added catalysts rather than via plastoquinol in native coupling (*31*). The plastoquinone inhibitors were useful in recent studies of high cyclic photophosphorylation rates in intact chloroplasts (*32*).

The DBMIB-sensitive lipophilic durohydroquinone donor system (*25*)

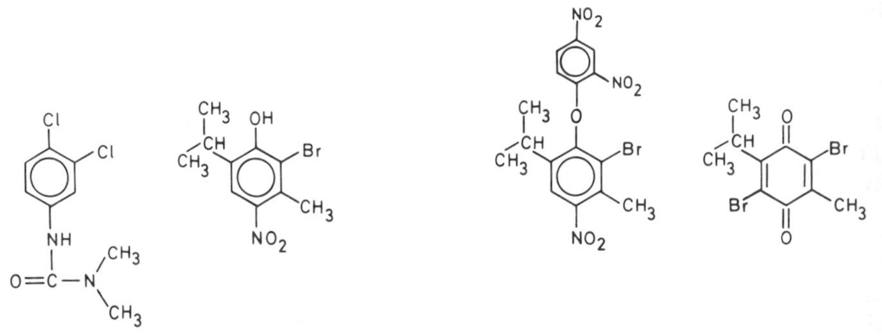

DCMU | BNT Bromonitrothymol | DNP-BNT Dinitrophenylether of bromonitrothymol | DBMIB

DCMU site | DBMIB site(s)

Fig. 7. Chemical formulas of inhibitors at the reducing site of plastoquinone (B protein, DCMU and BNT), and of inhibitors of the oxidizing site of plastoquinone (cytochrome *f* complex, DBMIB and DNP-BNT). DNP–INT has an iodo substitution instead of the bromo in DNP–BNT.

was mentioned previously. In this system it is assumed that durohydroquinone and plastoquinol feed electrons into the same site and the site may also be blocked by DBMIB. Binder and Selman (*33*) recently described a DNP–INT-sensitive donor system for photosystem I with hydrophilic donors. Assuming these donors do not gain access to the lipophilic binding site of plastoquinol, DQH_2 and DBMIB inside the membrane, this may indicate that they react via plastoquinone or B, which are accessible from the outside as indicated by the trypsin experiments.

Inhibitors that Affect Both Reduction and Oxidation of Plastoquinone

As already stated, the inhibitors of plastoquinol oxidation, DBMIB and DNP–INT, also inhibit the reduction of plastoquinone at higher concentrations (*2*, *20*, *28*). This may easily be explained by the assumption that there is some similarity in the chemistry of the binding sites for plastoquinone when it is reduced at the B protein and when it is oxidized by the cytochrome *f* complex. The binding of DBMIB to the B-protein has recently been shown in a direct way by Bowes and Crofts (*34*). In spite of DBMIB and DNT–INT binding to the B-protein at higher concentration, they do not compete for the binding site of

DCMU as shown in the binding studies with radioactive metribuzin or dinoseb (*10*, *14*, *19*, *22*).

Conversely, some of the phenol inhibitors at the DCMU binding site also affect plastoquinol oxidation. The inhibitors in Table II all inhibit oxygen evolution at 0.1 μM and (except for DBMIB) are DCMU analogs. Phenols with an alkyl side chain also effectively inhibit the durohydroquinone donor system. Neither phenyl-substituted nitrophenol nor ioxynil, dinoseb, nor DCMU, show a marked additional inhibitory effect at the DBMIB site.

UHDBT (5-*n*-undecyl-6-hydroxyl-4,7-dioxobenzothiazole, see Fig. 8) was recently introduced as a powerful inhibitor of ubiquinone function in respiration (*35*) and of photosynthetic electron flow in bacterial chromatophores (*36*). UHDBT also effectively inhibits photosynthetic electron flow in chloroplasts (*22*). Its site of action is on the B protein, (i.e., on the reducing site of plastoquinone) as expected from the inhibition pattern. Inhibition of photosynthetic electron transport from water

TABLE II

Effect of Inhibitors on the Photoreduction of Methylviologen in the Durohydroquinone Donor System[A]

Inhibitor	Concentration (μM)	Inhibition in percent of control
Dibromothymoquinone (DBMIB)	1	90
(2,5-dibromo-3-methyl-6-isopropyl-benzoquinone)		
Bromonitrothymol	1	40
(2-bromo-4-nitro-3-methyl-6-isopropyl-phenol)	10	62
Iodonitrothymol	1	56
	10	70
2-Bromo-4-nitro-3-methyl-6-*t*-butyl-phenol	1	83
DCMU	1	17
	10	8
Metribuzin	1	38
	10	28
2-Bromo-4-nitro-6-phenyl-phenol	10	4
Dinoseb	10	30
(2,4-dinitro-6-*sec*-butyl-phenol)		
Ioxynil	1	20
(2,6-diiodo-4-cyano-phenol)	10	10

[A] In addition to a DCMU-like inhibitory effect at low concentrations some also inhibit at high concentrations like DBMIB.

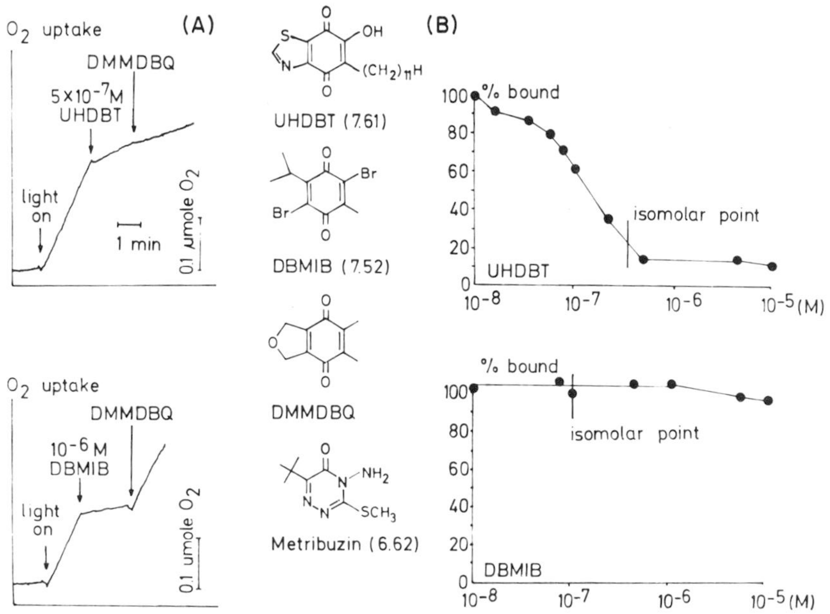

Fig. 8. Inhibition of photosynthetic electron flow from H_2O to methylviologen by UHDBT, not reversed by the photosystem II acceptor DMMDBQ in contrast to DBMIB (A). Displacement of radioactive metribuzin by UHDBT, but not by DBMIB from the thylakoid membrane (B).

to methylviologen by UHDBT cannot be restored by addition of the photosystem II acceptor DMMDBQ (2,3-dimethyl-5,6-methylene-dioxy-1,4-benzoquinone), as it can with DBMIB (Fig. 8). Furthermore, the durohydroquinone donor system is insensitive to UHDBT in contrast to DBMIB, and there is no TMPD bypass for UHDBT (Fig. 9). The same conclusion can be drawn from displacement experiments. UHDBT efficiently displaces metribuzin from the thylakoid membrane, again in contrast to DBMIB (Fig. 8). Only at higher concentrations does UHDBT also affect the oxidizing side of plastoquinone, as also indicated by Hauska *et al.* (*29*) from measurements of isolated cytochrome b_6/f complex activity.

DISCUSSION

Inhibitors of plastoquinone function are helpful in clarifying the present concept of the interaction of plastoquinone with membrane polypeptides.

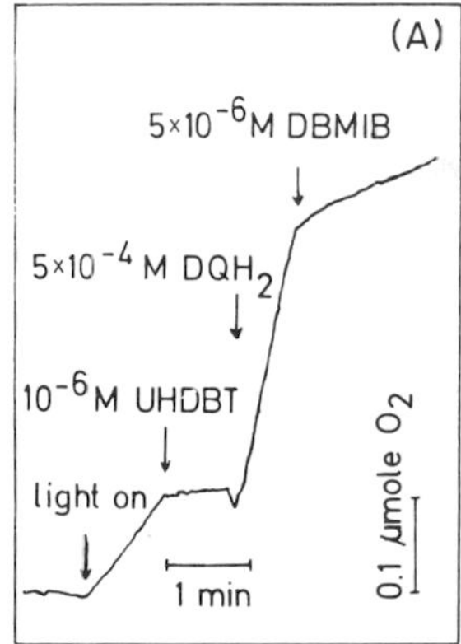

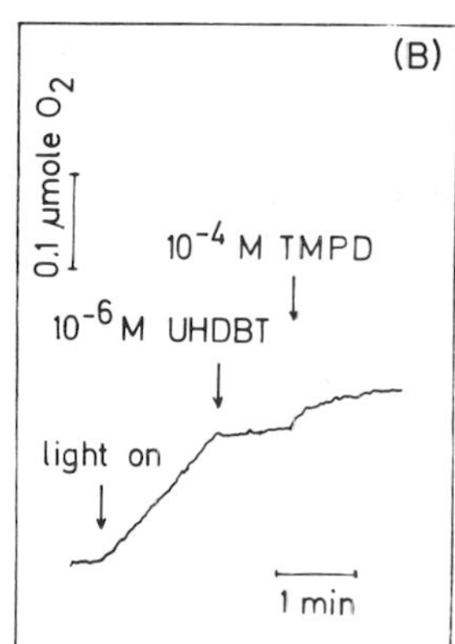

Fig. 9. Insensitivity of the durohydroquinone (DQH_2) photosystem I donor system to UHDBT as compared to DBMIB (A). Inhibition of photosynthetic electron transport by UHDBT is not bypassed by TMPD (B). Methylviologen was used as electron acceptor.

The assumption is that inhibitors of plastoquinone function may interact with the same protein that carries plastoquinone binding sites. The inhibitors may compete with plastoquinone (or quinol) for the same binding site or the effect may be allosteric. It is technically easier to identify inhibitor binding peptides than quinone binding sites and this strategy was used in identifying the peptides involved in plastoquinone reduction. Characterization of the peptide of the DCMU (triazine) and dinoseb binding protein showed it to be a part of the B protein (*21*–23). Triazine and phenol binding of the preferentially labeled peptides are not identical, because the molecular weight in two experiments by us (*21*–*22a*) was 41 kdalton, and by Arntzen *et al.* (*23*, *23a*), as well as in the experiments of Edelman (*24*) and Horton (*37*) with trypsin was 32 kdalton. These two polypeptides are interacting closely with one another on the matrix side of the thylakoid membrane because of their trypsin sensitivity and mutual replacement from the membrane (*22a*).

These peptides cover the special plastoquinone Q of the primary acceptor of photosystem II and convey onto another plastoquinone, called B, the property of a two-electron carrier. This shielding protein (*18*) has the property of a Q → plastoquinone oxidoreductase. It is assumed to carry binding sites for the inhibitors as well as for plastoquinone. This binding site for plastoquinone is probably not identical with that of the inhibitors, though this might follow from the UHDBT experiment. Certainly, they can interact allosterically with each other. Among the polypeptides of the B protein is the "rapidly turning over" photogene 32 kdalton peptide (*23*, *23a*, *24*, *38*, *39*) identical with the peak D peptide (*40*) or D-1 protein

(*41*). It is formed only in the light and degraded in the light (*24*, *38*). It is chloroplast coded (*38*, *39*, *41*–*43*) and probably carries the mutations in the herbicide-resistant plants studied by Pfister *et al.* (*13*, *23*, *23a*); it might also carry a CO_2 binding site (*44*). Its properties indicate that it may be a regulatory protein of the electron flow system, though neither the regulator nor any physiological evidence of this is known. Information on this aspect should appear in the near future, following recent progress in identification of the inhibitor-binding polypeptides (reviewed in *45*) with emphasis also on the importance in herbicide research.

Similarly, though much less defined, is the binding site for plastoquinol and inhibitors of the DBMIB type at the protein that carries the plastoquinol oxidation site. It is assumed to be a subunit of cytochrome b_6/cytochrome *f* complex with the Rieske nonheme iron center. The isolated complex still has the property of a plastoquinol–plastocyanin oxidoreductase (*26*) and retains its inhibitory sensitivity (*26*, *29*). The exact location of the plastoquinol binding site and the precise peptides involved are not clear. This plastoquinol binding site is also assumed to be occupied by inhibitors like DBMIB or DNP–INT, which interfere competitively with plastoquinone. This site is not necessarily the Rieske center as suggested (*27*). The increase in signal height but lack of shift of the g-value of the ESR signal of the Rieske center upon addition of DBMIB to the thylakoid membrane (*27*) is seen in the cytochrome b_6/*f* complex (*46*). Radioactive DNP–INT does react with the membrane and the complex (*22*), but this has not yet yielded information on the exact binding site (peptide).

The scheme in Fig. 10 summarizes the view that plastoquinone interconnects two integral proteins, that each have binding sites for oxidized and reduced plastoquinone, respectively. These proteins also have inhibitor binding sites for DCMU and analogs at the B protein as part of photosystem II, and at the cytochrome b_6/*f* complex for DBMIB-type inhibitors. It is interesting that the high activity in a TMPD bypass (including proton translocation) around the DBMIB inhibition site and, therefore, around the cytochrome b_6/*f* complex (*20*) (see Fig. 5) indicates that a large integral complex is mechanistically not that essential for cyclic electron flow from photosystem I to photosystem II in principal. Therefore, the invention by nature of the complex signals additional roles for it. One of them is its function in cyclic electron flow where it channels the electrons from ferredoxin back to the donor side of photosystem I. The cytochrome b_6/*f* complex is essential for the concept of a Q cycle and for a third electrogenic step, indicated for example, by an additional slow rise of the electric field (*47*) or the higher proton/electron stoichiometry (*1*).

As mentioned previously, cyclic electron flow is sensitive to the inhibitors of plastoquinol oxidation (DBMIB) but not to those of plastoquinone

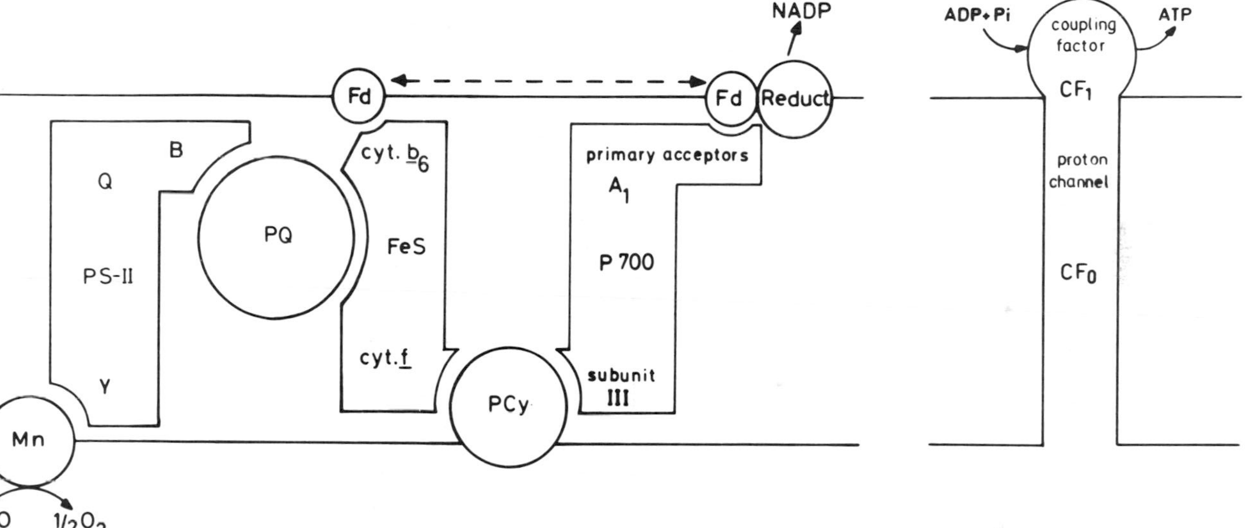

Fig. 10. Schematic architecture of the photosynthetic electron flow system consisting of integral protein complexes (photosystem II, cytochrome b_6/cytochrome *f* complex and photosystem I) interconnected by small carriers (plastoquinone, plastocyanin) and peripheral complexes (water splitting system, NADP reductase) in the thylakoid membrane.

reduction (DCMU). Therefore, the mechanism of plastoquinone reduction in cyclic electron flow is quite different from that of noncyclic electron flow. An understanding of the DBMIB interaction with the membrane and the cytochrome b_6/f complex will probably be very essential for understanding the functional details.

ACKNOWLEDGMENTS

We acknowledge the technical assistance of M. Meyer and K. Masson and the support by Deutsche Forschungsgemeinschaft.

REFERENCES

1. Velthuys, B. R. (1980). *Annu. Rev. Plant Physiol.* **31,** 545–567.
2. Trebst, A. (1980). *In* "Photosynthesis and Nitrogen Fixation," Part C (A. San Pietro, ed.), Methods in Enzymology, Vol. 69, pp. 675–715. Academic Press, New York.
3. Stiehl, H. H., and Witt, H. T. (1969). *Z. Naturforsch., Teil B* **24,** 1588–1598.
4. Futami, A., Hurt, E., and Hauska, G. (1979). *Biochim. Biophys. Acta* **547,** 583–596.
5. Knaff, D. B., Malkin, R., Myron, J. C., and Stoller, M. (1977). *Biochim. Biophys. Acta* **459,** 402–411.
6. Bouges-Bocquet, B. (1973). *Biochim. Biophys. Acta* **314,** 250–256.
7. Velthuys, B. R., and Amesz, J. (1974). *Biochim. Biophys. Acta* **333,** 85–94.
8. Pulles, M. P. J., van Gorkom, H. J., and Willemsen, J. G. (1976). *Biochim. Biophys. Acta* **449,** 536–540.
9. Tischer, W., and Strotmann, H. (1977). *Biochim. Biophys. Acta* **460,** 113–125.
10. Reimer, S., Link, K., and Trebst, A. (1979). *Z. Naturforsch., Teil C* **34,** 419–426.
11. Trebst, A., Reimer, S., Draber, W., and Knops, H. J. (1979). *Z. Naturforsch., Teil C* **34,** 831–840.
12. Trebst, A., and Draber, W. (1979). *In* "Advances in Pesticide Science" (H. Geissbühler, ed.), Part 2, pp. 223–234. Pergamon, Oxford.
13. Pfister, K., Radosevich, S. R., and Arntzen, C. J. (1979). *Plant Physiol.* **64,** 995–999.
14. Oettmeier, W., and Masson, K. (1980). *Pestic. Biochem. Physiol.* **14,** 86–97.
15. Böger, P. (1979). *Z. Naturforsch., Teil C* **34,** 1015–1020.
16. Regitz, G., and Ohad, I. (1976). *J. Biol. Chem.* **251,** 247–252.
17. Renger, G., Erixon, K., Döring, G., and Wolff, C. (1976). *Biochim. Biophys. Acta* **440,** 278–286.
18. Renger, G. (1976). *Biochim. Biophys. Acta* **440,** 287–300.
19. Trebst, A. (1979). *Z. Naturforsch., Teil C* **34,** 986–991.
20. Trebst, A., and Reimer, S. (1973). *Z. Naturforsch., Teil C* **28,** 710–716.
21. Oettmeier, W., Masson, K., and Johanningmeier, U. (1980). *FEBS Lett.* **118,** 267–270.
22. Oettmeier, W., Masson, K., and Johanningmeier, U. (1981). "Photosynthesis" (G. Akoyunoglou, ed.) Vol. 6, 585–594.

22a. Oettmeier, W., Masson, K., and Johanningmeier, U. (1982). *Biochim. Biophys. Acta,* **679,** 376–383.
23. Pfister, K., Steinback, K. E., and Arntzen, C. J. (1981) "Photosynthesis" (G. Akoyunoglou, ed.) Vol. 6, 595–606.
23a. Pfister, K., Steinback, K. E., Gardner, G., and Arntzen, C. J. (1981). *Proc. Natl. Acad. Sci. USA* **78,** 981–985.
24. Mattoo, A. K., Pick, U., Hoffman-Falk, H., and Edelman, M. (1981). *Proc. Natl. Acad. Sci. USA* **78,** 1572–1576.
25. Izawa, S., and Pan, R. L. (1978). *Biochem. Biophys. Res. Commun.* **83,** 1171–1177.
26. Wood, P. M., and Bendall, D. S. (1976). *Eur. J. Biochem.* **61,** 337–344.
27. Chain, R. K., and Malkin, R. (1979). *Arch. Biochem. Biophys.* **197,** 52–56.
28. Trebst, A., Wietoska, H., Draber, W., and Knops, H. J. (1978). *Z. Naturforsch., Teil C* **33,** 919–927.
29. Hurt, E., and Hauska, G. (1981). *Eur. J. Biochem.* **117,** 591–599.
30. Böhme, H., Reimer, S., and Trebst, A. (1971). *Z. Naturforsch., Teil B* **26,** 341–352.
31. Hauska, G., and Trebst, A. (1977). *Curr. Top. Bioenerg.* **6,** 151–220.
32. Slovacek, R. E., Mills, J. D., and Hind, G. (1978). *FEBS Lett.* **87,** 73–76.
33. Binder, R. G., and Selman, B. R. (1980). *Biochim. Biophys. Acta* **592,** 314–322.
34. Bowes, J. M., and Crofts, A. R. (1981). *Arch. Biochem. Biophys.* **209,** 682–686.
35. Bowyer, J. R., and Trumpower, B. L. (1980). *FEBS Lett.* **115,** 171–174.
36. Bowyer, J. R., Tierney, G. V., and Crofts, A. R. (1979). *FEBS Lett.* **101,** 207–212.
37. Croze, E., Kelly, M., and Horton, P. (1979). *FEBS Lett.* **103,** 22–26.
38. Reisfeld, A., Jakob, K. M., and Edelman, M. (1978). *In* "Chloroplast Development" (G. Akoyunoglou *et al.,* eds.), pp. 669–674. Elsevier, Amsterdam.
39. Grebanier, A. E., Coen, D. M., Rich, A., and Bogorad, L. (1978). *J. Cell Biol.* **78,** 734–746.
40. Eaglesham, A. R. J., and Ellis, R. J. (1974). *Biochim. Biophys. Acta* **335,** 396–407.
41. Chua, N. H., and Gillham, N. W. (1977). *J. Cell Biol.* **74,** 441–452.
42. Ellis, R. J., and Barraclough, R. (1978). *In* "Chloroplast Development" (G. Akoyunoglou *et al.,* eds.), pp. 185–194. Elsevier, Amsterdam.
43. Herrmann, R. G., and Possingham, J. V. (1980). *In* "Chloroplasts" (J. Reinert, ed.), pp. 45–96. Springer-Verlag, Berlin and New York.
44. Govindjee, and van Rensen, J. J. S. (1978). *Biochim. Biophys. Acta* **505,** 183–213.
45. Trebst, A. (1981) "Photosynthesis" (G. Akoyunoglou, ed.) Vol. 6, 507–520.
46. Rich, P. R., Heathcote, P., Evans, M. C. W., and Bendall, D. S. (1980). *FEBS Lett.* **116,** 51–56.
47. Crowther, D., Mills, J. D., and Hind, G. (1979). *FEBS Lett.* **98,** 386–390.

Polypeptide Determinants of Plastoquinone Function in Photosystem II of Chloroplasts

4

C. J. ARNTZEN
S. C. DARR
J. E. MULLET
K. E. STEINBACK
K. PFISTER

INTRODUCTION

The primary and secondary stable electron acceptors of the photosystem II (PS II) reaction center are protein bound plastoquinones (*1–3*). These enzymes are designated as Q and B, respectively. Each electron carrier displays unique characteristics (such as stable semiquinone anion states) ascribed to special microenvironments within the chloroplast membrane. This microenvironment is most probably determined by the proteins that create the quinone binding sites within the PS II complex (*4*).

Over the last two decades, several research groups have attempted to characterize the polypeptide components of PS II. Through analysis of PS II mutants, detergent-derived PS II submembrane preparations, and membranes of partially developed chloroplasts, it is now recognized that eight to ten polypeptides form the "native" PS II complex (recently reviewed in *5*, *6*). Some of these polypeptides bind chlorophyll while others presumably act as the apoproteins of electron-transport components. It has been our goal to identify those polypeptides of the PS II complex that

Function of Quinones in Energy Conserving Systems

ISBN 0-12-701280-X

determine the properties of the plastoquinone electron carriers Q and B. A major tool in our studies has been the use of specific inhibitors of photosynthetic electron transport between Q and B. We will discuss several approaches to utilization of these inhibitors that aid in identification of the plastoquinone-binding polypeptides.

HERBICIDES AS QUINONE-FUNCTION ANTAGONISTS

Several classes of commercial herbicides act as inhibitors of photosynthetic electron transport on the reducing side of photosystem II. The triazines (i.e., atrazine, 2-chloro-4-ethylamino-6-isopropylamino-*s*-triazine), substituted ureas (i.e., DCMU, 3-(3,4-dichlorophenyl)-1,1-dimethylurea), and certain nitrophenols (i.e., dinoseb; 2-*sec*-butyl-4,6-dinitrophenol), as well as several other chemical families of herbicides (amides, pyridazinones, uracils, etc.) all competitively bind to the chloroplast membrane (*7–9*). Non-covalent binding of any one of these inhibitors causes a block in electron transfer from Q to B (*7*). The onset of herbicide-induced inhibition of electron transport is accompanied by reversed electron flow from B^- to Q indicating a shift in the redox potential of the quinone cofactor of B with respect to that of *Q* (*4, 7, 10, 11*). On the basis of these and other data we hypothesized that herbicides such as diuron (DCMU) or atrazine bind to the polypeptide(s) controlling the environment of the quinone cofactor of B (*7*). Therefore, identification of the herbicide-binding protein should lead to identification of the apoprotein of B.

Herbicide-Resistant Chloroplasts

Reports have been published of localized populations of weedy species that are resistant to the triazine class of herbicides. Chloroplasts isolated from these resistant weeds have electron-transport activity that exhibits normal inhibition by diuron and dinoseb but is insensitive to atrazine (*7, 12*). Measurement of binding of radioactive herbicides in the resistant chloroplasts demonstrated selective loss of binding affinity for the triazine family of herbicides. Normal affinity was maintained for other types of herbicides. The genetic alteration causing resistance is, therefore, a subtle change in the herbicide-binding protein which affects the binding of one class of herbicides but not another. Triazine-resistant chloroplasts are valuable tools for analysis of herbicide receptor proteins.

Photoaffinity Labeling of the Herbicide Receptor Protein

The association of diuron or atrazine with chloroplast membranes is via high affinity, but noncovalent binding. When chloroplast membranes are solubilized by anionic detergents suitable for use in polypeptide separation such as sodium dodecyl sulfate (SDS), any herbicide–protein associations are lost. To overcome this dissociation, we adopted the use of azido-labeled atrazine analogs (2-azido-4-ethylamino-6-isopropylamino-*s*-triazine). Incubation of this azido-atrazine with chloroplast membranes under UV illumination resulted in formation of a highly reactive nitrene (*13, 14*) that binds covalently to the membranes. In preliminary experiments we found that prior to UV treatment, azido-atrazine exhibited normal noncovalent binding and inhibition of electron-transport functions.

Using SDS polyacrylamide gel electrophoresis we identified the polypeptide that binds the azido-atrazine. Azido-[^{14}C] atrazine was incubated with chloroplast thylakoids and treated with UV illumination; the sample was then subjected to electrophoresis and the [^{14}C]-labeled polypeptides identified using autoradiography. The results, shown in Fig. 1, indicate that polypeptides of 32–34 kdaltons are labeled in susceptible chloroplasts. In resistant chloroplasts, however, no polypeptides acquire a radioactive label indicating a complete loss of the triazine binding site.

Inheritance of Kinetic Properties of B

It has been demonstrated that herbicide-resistant chloroplasts exhibit altered kinetics of electron flow between Q^- and B (*4*, *7*). Using the rate of decay of room temperature chlorophyll fluorescence after a single, saturating flash in resistant versus susceptible chloroplasts, it was shown that electron flow between Q^- and B is approximately ten-fold slower in resistant chloroplasts. This is probably due to an alteration in the microenvironment of B in resistant chloroplasts (*4*). We used genetic analysis to test whether the two features of resistant membranes (modified kinetic properties of Q^- to B electron flow, and loss of triazine binding) correspond to a single inheritable trait.

In 1979 Souza Machado *et al.* (*15*) demonstrated that "whole plant" atrazine resistance was inherited through the female parent. We analyzed the progeny of reciprocal crosses between atrazine-resistant and susceptible plants and found that atrazine resistance at the thylakoid level is also inherited through the female parent (*16*). In addition, we examined the rate of Q^- to B electron transport (assayed by chlorophyll fluorescence decay) in resistant *x* susceptible progeny. The results are shown in Fig. 2.

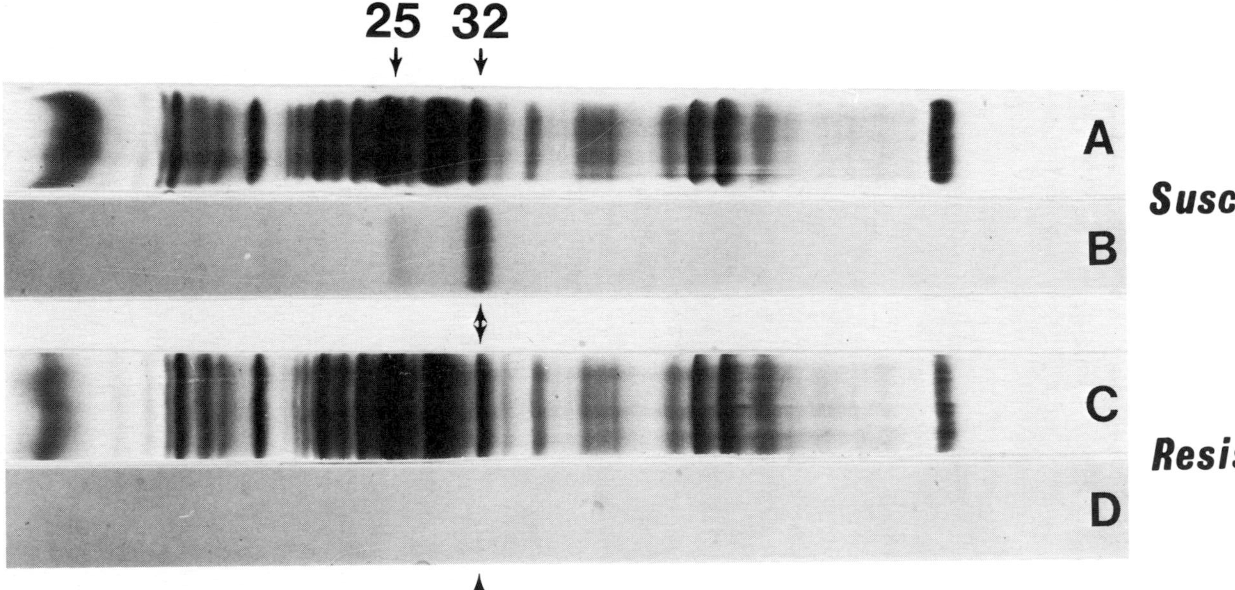

Fig. 1. Photoaffinity labeling of the triazine binding polypeptide in herbicide susceptible and resistant chloroplast membranes. Lanes A: membrane thylakoids separated using SDS-PAGE and stained with coomassie blue. Lanes B: autoradiogram of Lanes A showing location of [^{14}C]-labeled polypeptides. Numbers on left express molecular weight in kdalton. A slight amount of label was detected with a polypeptide of 25 kdalton; this is a light-harvesting pigment protein of PS II. Other experiments (*14*) demonstrate that it is not required for herbicide binding.

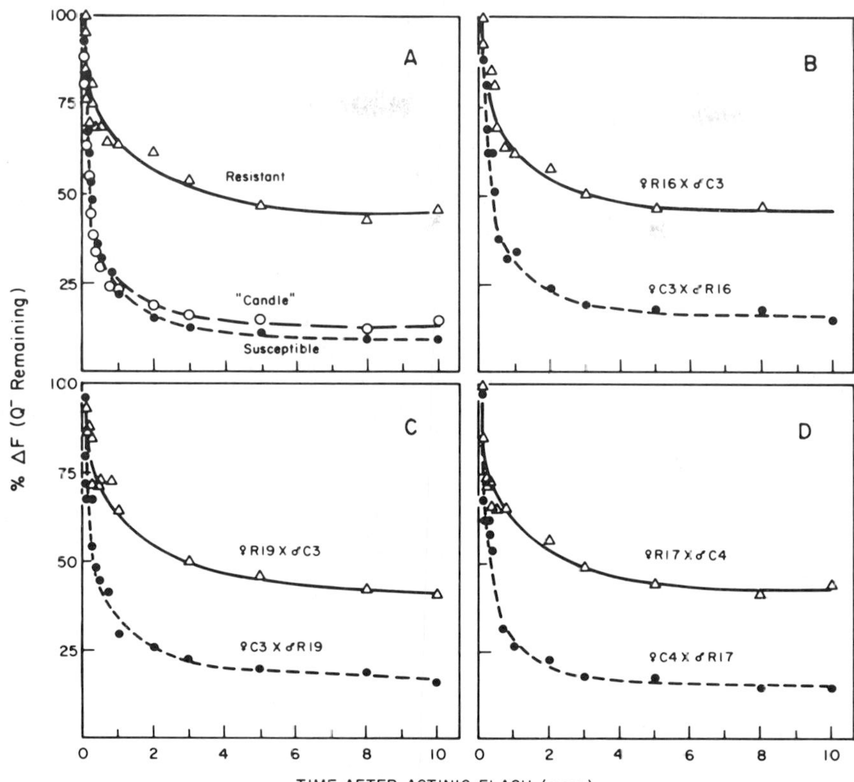

Fig. 2. Room temperature chlorophyll fluorescence decay. (A) The percent of variable fluorescence remaining after a saturating flash is shown for the parent biotypes of *Brassica campestris* used in reciprocal crosses. "Candle" is an herbicide-susceptible horticultural variety. (B–D) Similar data for progeny of complementary reciprocal crosses. The labels for each individual designate the parent plants used in the cross (i.e., R16 = resistant individual number 16; C = "candle".) Slow decay of fluoresence indicates a reduced rate of electron transfer between Q and B. In every case examined, this trait was inherited through the female parent as was atrazine-resistance. We suggest that a modification of the apoprotein of B is responsible for both phenomena.

In every case, the plants exhibited a fluorescence decay rate similar to that of the female parent. Therefore, atrazine resistance and altered kinetic activity of the Q–B electron acceptor pair are inherited in parallel. We believe that both are caused by an alteration in a PS II protein component which serves as the apoprotein of B and contains the triazine binding site.

Trypsin Alteration of PS II Proteins

To test the concept that the herbicide receptor is a protein of the PS II complex which functions as an electron carrier, we subjected isolated triazine-susceptible chloroplast thylakoids to trypsin for various periods of time. The samples obtained were analyzed by measuring binding of radioactive herbicides and by assays of PS II-dependent electron transport. The data, shown in Fig. 3, indicate the number of binding sites (on a chlorophyll basis) and the rate of DCPIP (2,6-dichlorophenol-indophenol) photoreduction were decreased in parallel by protease action. In other samples (data presented in *17*), electron flow from water to silicomolybdate (an electron acceptor at the level of Q) was only slightly affected by trypsin action. Based upon the use of several electron acceptors, we concluded that trypsin inhibition of PS II function is primarily due to loss of the activity of B. In total, the studies utilizing trypsin modification of thylakoids suggest that the apoprotein of B is the herbicide receptor.

Action of Herbicides upon Quinone Function in PS II Submembrane Particles

The results of many previous studies specify methods for isolating PS II-enriched submembrane preparations. Herbicide activity in PS II parti-

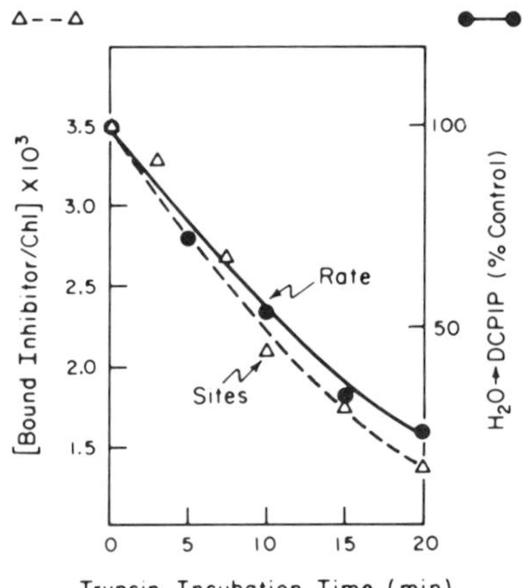

Fig. 3. Trypsin effects on inhibitor binding and electron transport assays. Parallel decreases in inhibitor binding (left axis) and electron transport from H_2O to the acceptor DCPIP (right axis) are graphed *versus* the time of trypsin incubation prior to measurement. Herbicide binding measurements utilized [^{14}C]-atrazine. Electron transport was measured under high light intensities in uncoupled chloroplasts. Other details of experimentation are described in Steinback *et al.* (*17*).

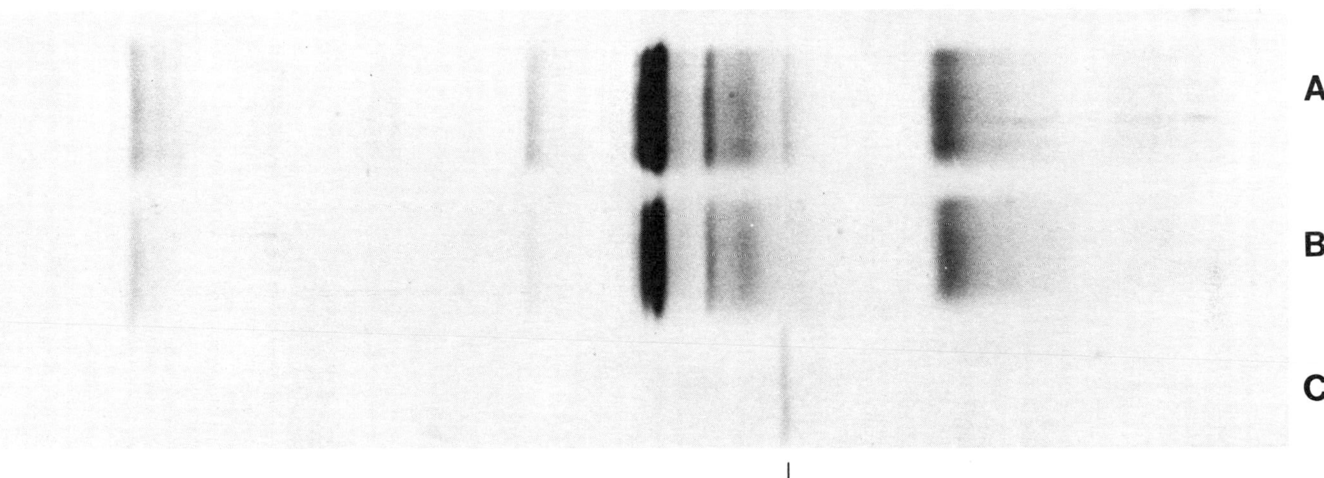

Fig. 4. Isolation of the 32 kdalton protein. Lane A: SDS-polyacrylamide gel electrophoresis of the photosystem II submembrane particle. Lane B: The polypeptides recovered in the PS II particle after cholate/urea washing. Lane C: The 32 kdalton protein retrieved from the top of the sucrose gradient.

cles prepared by these techniques was found to be decreased or absent (*18*, *19*). To further the evaluation of herbicide-receptor proteins, we developed a modified procedure to prepare a more "native" PS II submembrane particle that retains herbicide sensitivity (for details, see *6*). Isolation of this submembrane fragment allowed us to characterize the polypeptides in the complex with respect to their interaction with the various classes of photosynthesis inhibitors.

The effects of a range of herbicide concentrations on PS II-mediated photoreduction of DCPIP were tested in our PS II particles and in isolated thylakoids (data in *6*). Diuron and atrazine showed reduced binding affinity in the particles (i.e., higher concentrations of inhibitor were required to observe inhibition). Binding sites for these herbicides were still present, however, since 90% inhibition was observed at high inhibitor levels.

To correlate herbicide binding with particular polypeptides within the PS II complex, selective protein extraction was accomplished via further detergent treatment of the particles. An example of the data obtained is shown in Fig. 4. The PS II protein complex (lane A) was depleted of a 32 kilodalton polypeptide (lane B) via treatment with cholate/urea followed by sucrose gradient fractionation. The purified 32 kdalton polypeptide was recovered from the upper portion of the sucrose gradient (lane C).

Photochemical properties of PS II particles depleted of the 32 kdalton polypeptide are shown in Table I. Only a small reduction in the control rate of electron transport was observed as the polypeptide was lost. In contrast, the inhibitory activity of diuron and atrazine added to the depleted particles was dramatically reduced (i.e., removal of the 32 kdalton polypeptide corresponded to loss of inhibitor binding sites).

TABLE I

Herbicide Inhibition of Photosystem II Activity (DPC → DCPIP) After Selective Extraction of PS II Particles[A]

Sample	Control rate	+ 10^{-4} *M* Atrazine rate (% inhibition)	+ 10^{-4} *M* Diuron rate (% inhibition)
Control (untreated PS II particles)	310	93 (70)	40 (87)
Cholate/urea extracted PS II	274	260 (5)	181 (33)

[A] Rate = μmole DCPIP reduced mg Chl^{-1} h^{-1}

CONCLUSIONS

In a series of publications (*4*, *7*, *9*, *12*, *14*, *16*, *17*), we have characterized the mode of action of various herbicides that act as quinone-function antagonists in the PS II complex. This brief review has presented some of the data that lead us to suggest that the binding site for triazine herbicides is the apoprotein of B.

This paper has emphasized two points. First, a polypeptide of 32,000 dalton contains the binding site for PS II-directed herbicides such as the triazines. This was deliniated by use of the photoaffinity label, azido-atrazine (*6*, *14*) (Fig. 1) and selective polypeptide deletions from isolated PS II particles, resulting in the appearance of herbicide-resistant electron transport (*6*) (Table I and Fig. 4). Second, the polypeptide to which the triazine herbicides bind is the same membrane component that regulates functional properties of the protein-quinone B. This conclusion was supported by a parallel loss of binding sites and functional activity of B in trypsin-treated membranes (*17*) (Fig. 3) and by parallel inheritance patterns of triazine binding affinity and kinetic properties of Q^- to B electron transfer (*16*) (Fig. 2).

The 32,000 dalton protein has now been isolated in purified form (*6*) (Fig. 4). This preparation should allow exploration of the amino acid determinates that regulate high-affinity herbicide binding and, hopefully, also allow direct measurement of quinone-protein association and factors influencing this interaction.

ACKNOWLEDGMENTS

This research was supported, in part by USDA/BARD grant 80-79 and by funds from the MSU Agricultural Experiment Station. We thank Ms. Jan Watson for excellent technical assistance.

REFERENCES

1. Amesz, J., and Duysens, L. N. M. (1977). *In* "Primary Processes of Photosynthesis" (J. Barber, ed.), pp. 149–185. Elsevier, Amsterdam.

2. Pulles, M. P. J., Van Gorkom, H. J., and Gerben Willemsen, J. (1976). *Biochim. Biophys. Acta* **449,** 536–540.

3. Mathis, P., and Haveman, J. (1977). *Biochim. Biophys. Acta* **461,** 167–181.

4. Bowes, J., Crofts, A. R., and Arntzen, C. J. (1980). *Arch. Biochem. Biophys.* **200,** 303–308.
5. Diner, B. A., and Wollman, F. A. (1980). *Eur. J. Biochem.* **110,** 521–526.
6. Mullet, J. E., and Arntzen, C. J. (1981). *Biochim. Biophys. Acta* **635,** 236–248.
7. Pfister, K., and Arntzen, C. J. (1979). *Z. Naturforsch* **34c,** 996–1009.
8. Tischer, W., and Strotmann, H. (1977). *Biochim. Biophys. Acta* **460,** 113–125.
9. Pfister, K., Radosevich, S., and Arntzen, C. J. (1979). *Plant Physiol.* **64,** 995–999.
10. Velthuys, B. R., and Amesz, J. (1974). *Biochim. Biophys. Acta* **333,** 85–94.
11. Velthuys, B. R. (1976). Ph.D. Thesis, Univ. of Leiden.
12. Arntzen, C. J., Ditto, C. L., and Brewer, P. E. (1979). *Proc. Natl. Acad. Sci. U.S.A.* **76,** 278–282.
13. Bayley, H., and Knowles, J. R. (1977). *In* "Affinity Labeling" (W. B. Jakoby and M. Wilchek, eds.), Methods in Enzymology, Vol. 46, pp. 69–114. Academic Press, New York.
14. Pfister, K., Steinback, K. E., Gardner, G., and Arntzen, C. J. (1981). *Proc. Natl. Acad. Sci. U.S.A.* **78,** 981–985.
15. Souza Machado, V., Bandeen, J. D., Stephenson, G. R., and Lavigne, P. (1978). *Can. J. Plant Sci.* **58,** 977–981.
16. Darr, S. C., Souza Machado, V., and Arntzen, C. J. (1981). *Biochim. Biophys. Acta* **634,** 219–228.
17. Steinback, K. E., Pfister, K., and Arntzen, C. J. (1981). *Z. Naturforsch.* **36c,** 98–108.
18. Croze, E., Kelly, M., and Horton, P. (1979). *FEBS Lett.* **103,** 22–26.
19. Wessels, J. S. C., Van Alphen-Van Waveren, O., and Voorn, G. (1973). *Biochim. Biophys. Acta* **293,** 741–752.

Quinone Analogs and Their Interaction with the Chloroplast Electron-Transport Chain

5

RICHARD MALKIN
ROBERT CROWLEY

INTRODUCTION

The function of quinones in the electron-transport chain of chloroplasts is critical for understanding the mechanism of energy transduction in the photosynthetic membrane. Because of the inherent difficulties in studying quinone redox reactions directly, inhibitors that are quinone analogs have been used for probing involvement of quinones in various electron-transfer processes. One such inhibitor, 2,5-dibromo-3-methyl-6-isopropyl-*p*-benzoquinone (DBMIB)[1], was introduced by Trebst and co-workers (*1*) and used in studies of chloroplast energy-transducing sites and other membrane systems (*2–4*). A second quinone analog, UHDBT, was applied in studies with mitochondrial (*5*, *6*) and chromatophore membranes (*7*), but its effects in chloroplasts have not been investigated.

In this communication the modes of action of the quinone analogs, DBMIB and UHDBT, are considered. Our results indicate that both these inhibitors interact in the region of the Rieske iron–sulfur center and suggest a function for the iron–sulfur center in quinone redox reactions of chloroplasts.

[1] Abbreviations: DBMIB, 2,5-dibromo-3-methyl-6-isopropyl-*p*-benzoquinone; UHDBT, 5-*n*-undecyl-6-hydroxy-4,7-dioxobenzothiazole; DAD, diaminodurene; DCMU, 3-(3′,4′-dichlorophenyl)-1,1-dimethylurea.

Function of Quinones in Energy Conserving Systems

ISBN 0-12-701280-X

EXPERIMENTAL RESULTS

Interaction of DBMIB with the Chloroplast Electron-Transport Chain

DBMIB is an effective inhibitor of noncyclic electron transport from water to NADP in chloroplasts (*1*); under our experimental conditions, we found 50% inhibition at a concentration of 2.5×10^{-7} *M*. Our earlier results indicated that DBMIB can also act as an electron donor and reduce bound, high-potential electron carriers, such as cytochrome *f* (*8*). During this work an unusual effect of DBMIB on the Rieske iron–sulfur center was noted. As shown in Fig. 1, dark-adapted chloroplasts, in the presence of ascorbate, showed the EPR signal of the reduced Rieske iron–sulfur center at a *g*-value of 1.89. When DBMIB was added to the chloroplasts, the $g = 1.89$ signal was no longer present but a new signal at $g = 1.94$ appeared. This suggested that in the presence of DBMIB the component with the $g = 1.89$ signal (reduced Rieske center) was converted to a component with an EPR signal at $g = 1.94$. To test whether appearance of the $g = 1.94$ signal correlated with disappearance of the $g = 1.89$ signal, suboptimal amounts of DBMIB were added to chloroplast samples such that both signals could be observed. The results of such a study are shown in Fig. 2. Based on

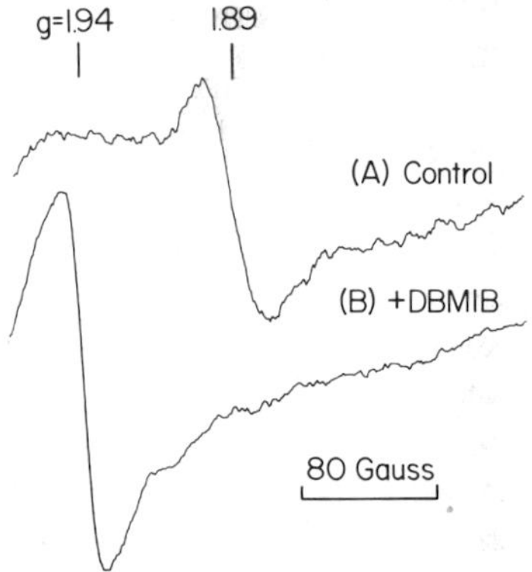

Fig. 1. Effect of DBMIB on the EPR spectrum of the chloroplast Rieske iron–sulfur center; (A) control; (B) +DBMIB. Chloroplast membranes at a chlorophyll concentration of 3 mg/ml were incubated in the dark with 5 m*M* sodium ascorbate and, where present, 30 μ*M* DBMIB. Samples were frozen to 77 °K and EPR spectra recorded at 15 °K with the following instrument settings: field, 3500 ± 250 gauss; microwave power, 5 mW; modulation amplitude, 10 gauss.

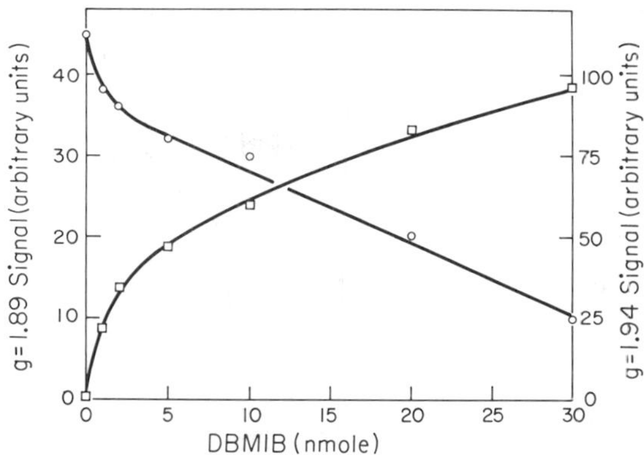

Fig. 2. Correlation of the appearance of the $g = 1.94$ EPR signal with the disappearance of the $g = 1.89$ EPR signal of the Rieske iron–sulfur center upon addition of DBMIB. Chloroplast membranes were incubated with ascorbate and the indicated amounts of DBMIB, as described in Fig. 1. EPR spectra were recorded as described in Fig. 1. The peak-to-trough amplitude of the $g = 1.89$ and 1.94 signals were measured for each sample. (—□—□—) $g = 1.94$, (—○—○—) $g = 1.89$.

the amplitude of the two EPR signals, it seemed that the $g = 1.94$ signal appeared as the $g = 1.89$ signal decreased in intensity. Therefore, it was concluded that the $g = 1.94$ EPR signal observed in the presence of the inhibitor represents a DBMIB-altered form of the Rieske iron–sulfur center.

In an attempt to characterize the altered Rieske center, measurements of temperature dependence and saturation properties were carried out. Both parameters showed marked changes relative to those for the unaltered signal at $g = 1.89$, indicating that a rather large change in the environment of the iron–sulfur center had occurred in the presence of DBMIB. We also found that treatments that have previously been shown to reverse the DBMIB inhibition of electron transfer (such as addition of dithiothreitol (*9*) or bovine serum albumin (*10*)) also cause a disappearance of the $g = 1.94$ signal and a reappearance of the signal at $g = 1.89$. These results indicate that the alteration of the iron–sulfur center is a manifestation of a specific reaction of DBMIB with the iron–sulfur center and is not related to a nonspecific effect of this compound.

Although the Rieske center had been altered by the addition of DBMIB, the ability of the modified center to undergo photooxidation in steady-state light was unaffected (Fig. 3). This finding suggests that the mode of action of DBMIB as an electron-transfer inhibitor may involve a change in the mechanism of the reduction of the iron–sulfur center, as opposed to any change in how the reduced center is oxidized.

We also tested other inhibitors reported to show a similar site of action as DBMIB in inhibiting electron transfer on the oxidizing side of plastoquinone. Compounds in this class include bathophenanthroline (*11*), trifluralin (*12*), DNP–INT (*13*) and several other ether derivatives (*14*). In none of our studies with these inhibitors was a *g*-value shift for the Rieske iron–sulfur center noted; this shift appears to be uniquely associated with DBMIB. In preliminary studies with other photosynthetic membrane systems that contain the Rieske center, such as *Rhodospirillum rubrum* and *Chromatium* chromatophores, have shown that DBMIB altered the EPR signal of the Rieske center in these as well as in chloroplasts.

Inhibition by UHDBT of Chloroplast Electron-Transport Reactions

UHDBT was found to be an inhibitor of electron transfer from water to noncyclic electron acceptors in chloroplasts. As shown in Fig. 4, 50% inhibition of NADP reduction was obtained at a UHDBT concentration of 1×10^{-7} *M*. UHDBT was a more effective inhibitor of this reaction than DBMIB when tested under the same conditions. The pat-

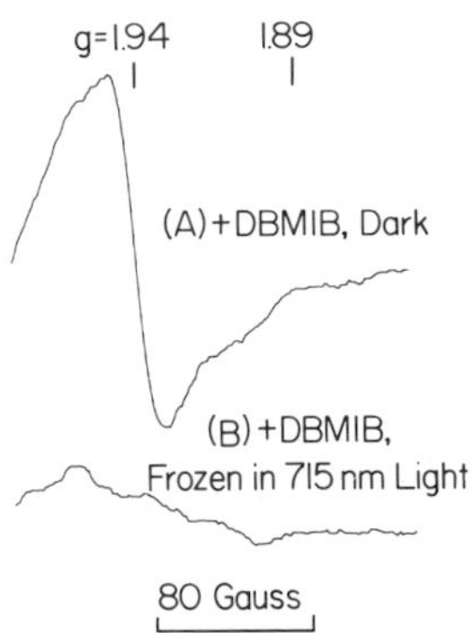

Fig. 3. Photooxidation of the DBMIB-altered Rieske iron–sulfur center in chloroplasts. The reaction mixture contained chloroplast membranes at a chlorophyll concentration of 3 mg/ml, 100 μM methyl viologen, 5 m*M* sodium ascorbate, 30 μM DBMIB and 25 μM DCMU. One sample (A) was dark-adapted and frozen to 77 °K while the second (B) was illuminated for 30 sec with 715 nm light at 20 °C and frozen in the light. EPR conditions were as in Fig. 1.

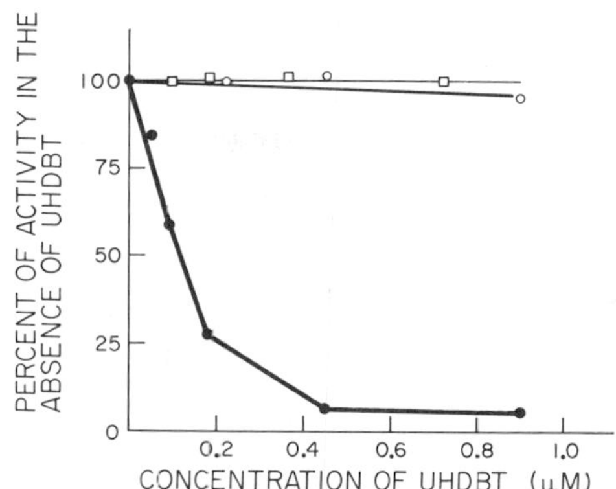

Fig. 4. Effect of UHDBT on electron transport with photosystem I and photosystem II electron donors. The reaction mixture contained 100 m*M* Tricine-KOH buffer (pH 8.3), 4 m*M* ADP, 5 m*M* $MgCl_2$, 2.5 m*M* potassium phosphate, chloroplast membranes (50 μg chlorophyll/ml), and, where present, 5 m*M* NADP, 10 μ*M* ferredoxin, 20 m*M* sodium ascorbate, 0.1 m*M* DAD and 0.5 m*M* durohydroquinone. With photosystem I electron donors, 1 μ*M* DCMU was present. The photoreduction of NADP was measured at 340 nm at room temperature with a Gilford spectrophotometer modified for side illumination. A 670 nm interference filter was used for actinic illumination. Durohydroquinone oxidation was measured with an oxygen electrode, as in Ref. *15*. In the absence of UHDBT, the rate of electron transfer (Q_{2e}) from water to NADP (●—●) was 250; from reduced DAD to NADP (○—○) was 85; and from durohydroquinone to O_2 (□—□) was 208.

tern of inhibition of UHDBT was insensitive to the state of coupling of the chloroplast membrane since the same concentrations of UHDBT gave half-maximal inhibition with either uncoupled or coupled chloroplasts.

Although electron transfer with water as electron donor was inhibited by UHDBT, a different behavior was observed with Photosystem I electron donors (Fig. 4). The photooxidations of reduced DAD and durohydroquinone, which utilize only Photosystem I, were not inhibited by UHDBT. The results with durohydroquinone are particularly interesting since oxidation of this donor previously had been shown to be sensitive to DBMIB (*15*, *16*). These results indicate that the two quinone analogs have different modes of inhibitory action.

Because of the previous demonstration that UHDBT acts in the cytochrome *b-c* region of the mitochondrial (*6*) and bacterial electron-transport chain (*7*), we attempted to localize its site of action in chloroplasts. The effect of UHDBT on the photooxidation of cytochrome *f* is shown in Figure 5. In the control sample with no added electron acceptor, cytochrome *f* undergoes a transient oxidation; the steady-state level then shifts to a more reduced level (Fig. 5A). In the presence of

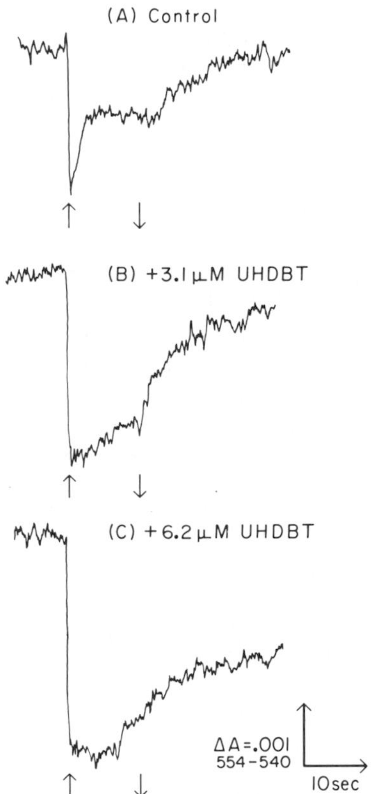

Fig. 5. Effect of UHDBT on chloroplast cytochrome *f* photooxidation. The reaction mixture (0.65 ml) contained chloroplast membranes (500 μg chlorophyll/ml), 100 m*M* Tricine-KOH buffer (pH 8.3), 5 m*M* $MgCl_2$, 2.5 m*M* ADP, and 2.5 m*M* potassium phosphate. To insure anaerobic conditions after equilibrating the reaction mixture with N_2, 10 m*M* glucose, 130 μg glucose oxidase, and 20 μg catalase were added. UHDBT was added at the indicated concentrations (A, B, and C). Cytochrome *f* oxidation was measured in 2 mm light path cuvettes in red light in an Aminco DW-2 spectrophotometer in the dual wavelength mode (554–540 nm). Upward arrows, light on; downward arrows, light off.

increasing amounts of UHDBT, two effects were noted: an increase in the extent of cytochrome oxidation and an inhibition in the reduction that occurs in the light. These observations indicate that the site of action of UHDBT is on the reducing side of cytochrome *f* as the behavior of the inhibitor is similar to that reported for DBMIB (*17*).

Because of the reported inhibition by UHDBT of the photooxidation of the Rieske iron–sulfur center in bacterial chromatophores (*18*), we considered the possibility of a similar effect in chloroplasts. As shown in Figs. 6A and 6B, the Rieske center underwent photooxidation in dark-adapted chloroplasts illuminated with 715 nm light. After addition of UHDBT, a slight narrowing of the EPR signal occured and a small shift in *g*-value from 1.890 to 1.895 was noted. The shift in *g*-value is similar to the observations made on the Rieske center in bacterial chromatophores. However, the results of Figs. 6C and 6D show that UHDBT does not inhibit the photooxidation of the iron–sulfur center.

UHDBT does inhibit the dark reduction of the Rieske center after illumination. In control chloroplasts, in the presence of an electron acceptor such as methyl viologen, the Rieske center was predominantly oxidized in steady-state red light and a reduction could be observed in the dark period following illumination. As shown in Fig. 7, the steady-state level of the Rieske center also was predominantly oxidized in the presence of UHDBT, but no subsequent reduction of the iron–sulfur center occurred in the dark period following illumination (compare

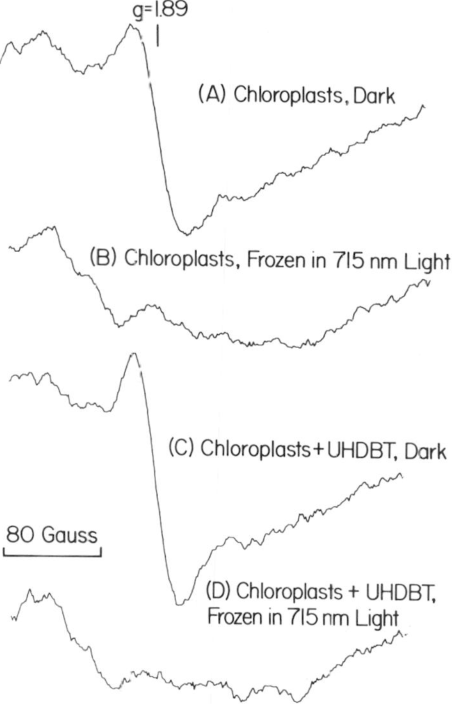

Fig. 6. Effect of UHDBT on the photooxidation of the chloroplast Rieske iron–sulfur center. Reaction conditions and illuminations were as in Fig. 3, except that UHDBT (50 μM) replaced DBMIB as the inhibitor. EPR conditions were as in Fig. 1.

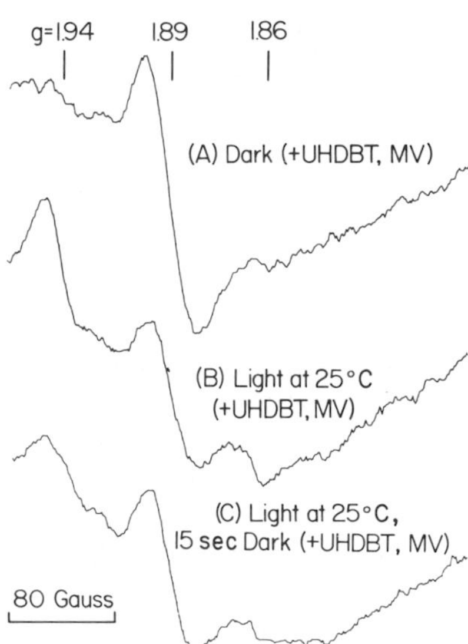

Fig. 7. Effect of UHDBT on the dark reduction of the chloroplast Rieske iron–sulfur center. The reaction mixture contained 100 m*M* Tricine-KOH buffer (pH 8.3), 100 μM methyl viologen, chloroplast membranes (3 mg chlorophyll/ml), and UHDBT (50 μM). The sample was (A) dark adapted, (B) illuminated with red light for 30 sec and frozen in the light, and (C) illuminated with red light for 30 sec, followed by a 15 sec dark period prior to freezing. EPR conditions were as in Fig. 1.

Figs. 7B and C). This experiment, taken in conjunction with the results of Fig. 6, indicates that UHDBT inhibits electron transfer from plastohydroquinone to the Rieske iron–sulfur center and has no effect on components on the oxidizing side of the iron–sulfur center.

Displacement of DBMIB by UHDBT

As shown by our results, both UHDBT and DBMIB interact with the chloroplast electron-transport chain in the region of the Rieske iron–sulfur center. We tested the idea that the sites of inhibition actually overlap by attempting to displace DBMIB by UHDBT. This experiment is feasible because of the EPR signal of the altered Rieske center that is observed in the presence of DBMIB. Addition of UHDBT to a sample containing DBMIB results in a shift of the EPR *g*-value from 1.94 to 1.89. We explained this observation was a displacement of DBMIB from its binding site at or near the Rieske center by UHDBT. Since UHDBT can displace DBMIB from this site, UHDBT must have a greater affinity for the inhibitor site than DBMIB; this agrees with our finding that UHDBT is two to three times more effective as an inhibitor of non-cyclic electron transfer.

DISCUSSION

The Rieske iron–sulfur center is a component common to the energy transducing membranes of chloroplasts, mitochondria, and the bacterial chromatophore (for a recent review, see *19*). Although detailed characterization of the site of function of this carrier is not complete in all these systems, the iron–sulfur center appears to function as the electron donor to a high potential *c*-type cytochrome. The strongest evidence for this comes from identification of the mitochondrial "oxidation factor" as the Rieske iron–sulfur protein (*20, 21*).

The mechanism of reduction of the Rieske iron–sulfur center is not yet understood, but the presumed electron donor in all the systems described here is a quinone; in chloroplasts this quinone would be plastoquinone. Our results indicating that quinone analogs such as DBMIB and UHDBT interact in the region of the Rieske iron–sulfur center are particularly noteworthy.

The effect of DBMIB on the iron–sulfur center is striking in that a large shift in the EPR spectrum occurs after the addition of the inhibitor. The EPR properties of the altered center, as measured by saturation and temperature dependencies, are also altered. Preliminary studies of the redox potential of the DBMIB-altered center indicate a large shift of E_m to a more negative value. All these findings suggest that DBMIB interacts strongly with the iron–sulfur center and, in fact, may directly affect the iron–sulfur prosthetic group.

The effects of UHDBT are not as striking as those noted in studies of UHDBT and the bacterial chromatophore Rieske center (*18*). A small *g*-value shift is observed, but we were unable to detect a change in E_m. UHDBT does interact with the center in a manner that displaces DBMIB, indicating some overlap of the binding sites of UHDBT and DBMIB.

In the presence of either inhibitor, the ability of the iron–sulfur center to undergo photooxidation is unaffected. We may presume that an alteration in the reductive pathway of the iron–sulfur center occurred, and we observed an inhibition of reduction by UHDBT. Since DBMIB serves as an electron donor for the Rieske center (*3*), it has not been possible to study the reductive pathway with this inhibitor.

We interpret our results with these inhibitors to be indicative of an interaction of quinone with the Rieske center under physiological conditions. Further studies of the interaction of such analogs with the chloroplast iron–sulfur center are anticipated to provide information on the nature of the electron-transfer sequence between quinone and the iron–sulfur center, relevant to all energy-transducing membranes.

ACKNOWLEDGMENTS

This work was supported in part by a grant from the National Institutes of Health (GM-20571). We would like to thank Dr. B. L. Trumpower and Dr. A. Trebst, respectively, for samples of UHDBT and DBMIB.

REFERENCES

1. Böhme, H., Reimer, S., and Trebst, A. (1971). *Z. Naturforsch., Teil B* **26,** 341–352.
2. Loschen, G., and Azzi, A. (1974). *FEBS Lett.* **41,** 115–117.
3. Melandri, B. A., Baccarini-Melandri, A., Lenaz, G., Bertoli, E., and Masotti, L. (1974). *J. Bioenerg.* **6,** 125–133.
4. Gromet-Elhanan, Z., and Gest, H. (1978). *Arch. Microbiol.* **116,** 29–34.
5. Roberts, H., Choo, W. M., Smith, S. C., Maryaki, S., Linnane, A. W., Porter, T. H., and Folkers, K. (1978). *Arch. Biochem. Biophys.* **191,** 306–315.
6. Trumpower, B. L., and Haggerty, J. G. (1980). *J. Bioenerg. Biomembr.* **12,** 151–164.
7. Bowyer, J. R., Tierney, G. V., and Crofts, A. R. (1979). *FEBS Lett.* **101,** 207–212.
8. Chain, R. K., and Malkin, R. (1979). *Arch. Biochem. Biophys.* **197,** 52–56.
9. Reimer, S., and Trebst, A. (1976). *Z. Naturforsch., Teil C* **31,** 103.
10. Robinson, H. H., Guikema, J. A., and Yocum, C. F. (1980). *Arch. Biochem. Biophys.* **203,** 681–690.
11. Bering, L., Dilley, R. A., and Crane, F. L. (1977). *Biochim. Biophys. Acta* **430,** 327–335.
12. Robinson, S. J., Yocum, C. F., Ikuma, H., and Hayashi, F. (1977). *Plant Physiol.* **60,** 840–844.
13. Trebst, A., Wietoska, H., Draber, W., and Knops, H. J. (1978). *Z. Naturforsch., Teil C* **33,** 919–927.
14. Bugg, M. W., Whitmarsh, J., Rieck, C. E., and Cohen, W. S. (1980). *Plant Physiol.* **65,** 47–50.
15. White, C. C., Chain, R. K., and Malkin, R. (1978). *Biochim. Biophys. Acta* **502,** 127–137.
16. Izawa, S., and Pan, L. (1978). *Biochem. Biophys. Res. Commun.* **83,** 1171–1177.
17. Böhme, H., and Cramer, W. A. (1971). *FEBS Lett.* **15,** 349–351.
18. Bowyer, J. R., Dutton, P. L., Prince, R. C., and Crofts, A. R. (1980). *Biochim. Biophys. Acta* **592,** 445–460.
19. Malkin, R., and Bearden, A. J. (1978). *Biochim. Biophys. Acta* **505,** 147–181.
20. Trumpower, B. L., and Edwards, C. A. (1979). *J. Biol. Chem.* **254,** 8697–8706.
21. Trumpower, B. L., Edwards, C. A., and Ohnishi, T. (1980). *J. Biol. Chem.* **255,** 7487–7493.

VII PROTONMOTIVE MECHANISMS OF QUINONE FUNCTION

The Function of Ubiquinone and Menaquinone in the Respiratory Chain of Escherichia coli

1

ROBERT W. JONES
PETER B. GARLAND

INTRODUCTION

At first sight the branched respiratory chain of *Escherichia coli* (*1*) appears to be a complicated system in which to study proton translocation. However, a closer examination shows this chain to be an attractive experimental system for such study. The polypeptide subunit composition of the respiratory complexes is much simpler than that of mitochondrial systems (*2*). The composition of the respiratory chain can be altered phenotypically by manipulation of growth conditions (*3*). There are also available many mutants that are deficient in specific components of the respiratory chain (*1*, *3*). Furthermore, recent developments in genetic manipulation (*4*) should allow for the amplification of respiratory proteins. Already, amplification of NADH dehydrogenase (*5*) and fumarate reductase (*6*) has been achieved, and lamda-transducing phages containing the fumarate reductase gene have been obtained (*7*). These considerations make the respiratory chain of *E. coli* a valuable tool also in the study of more general problems such as membrane biogenesis and regulation of enzyme synthesis.

E. coli contains both menaquinone-8 and ubiquinone-8. Mutants are available that lack either or both quinones (see *8*, and references therein). The use of such mutants has shown that ubiquinone is required for electron transfer to oxygen and nitrate, and that menaquinone is necessary for

Function of Quinones in Energy Conserving Systems

ISBN 0-12-701280-X

fumarate reduction. Ubiquinone appears to be the focal point of the respiratory chain: reducing equivalents from the substrate dehydrogenases are channelled into ubiquinone, and from there distributed into either nitrate reductase or the cytochrome systems. The dehydrogenases include NADH dehydrogenase, formate dehydrogenase, lactate dehydrogenase, α-glycerophosphate dehydrogenase, succinate dehydrogenase and hydrogenase. Two cytochrome systems are reduced by ubiquinol. One, terminating in cytochrome *o*, is synthesised during growth under conditions of high oxygen tension; the other is synthesised under conditions of low aeration and terminates in cytochrome *d* (*1*). The factors regulating synthesis of these alternative cytochrome systems are probably more complex than simply oxygen tension (*1*). In addition, synthesis of the dehydrogenases, and also of nitrate reductase, is regulated by growth conditions. Thus at any one time, not all pathways for reduction of ubiquinone and oxidation of ubiquinol are available. Menaquinone accepts electrons from NADH dehydrogenase, hydrogenase, anaerobically synthesised α-glycerophosphate dehydrogenase, or formate dehydrogenase, and funnels them into fumarate reductase.

In this laboratory we have chosen to study three segments of the anaerobic respiratory chains: hydrogenase, formate dehydrogenase, and the nitrate reductase complex. The subunit composition of these enzymes is known (*9–12*) and they can be explored using artificial electron donors and acceptors (*2*, *13–17*). In this paper we discuss quinone function in the respiratory chain of *E. coli* in view of our knowledge of the structural and functional orientation of these respiratory enzymes and the $\rightarrow H^+/2e$ stoichiometries measured in these systems. The data eliminate any requirement to postulate a protonmotive "Q cycle" of the type proposed to operate in mitochondria (*18*, *19*).

EXPERIMENTAL RESULTS

Formate Dehydrogenase

Of the enzymes studied, formate dehydrogenase is the least characterized. Formate-dependent reduction of exogenously added ubiquinone-1 drove proton translocation with an $\rightarrow H^+/2e^-$ stoichiometry of close to 2.0 (Table I). Although we know that the enzyme is transmembranous (*20*) and that formate is reduced at the cytoplasmic aspect of the cytoplasmic membrane (*15*, *21*, *22*), no firm conclusions could be drawn about the mechanism of this proton translocation (*15*).

TABLE I

→$H^+/2e^-$ Stoichiometries in Anaerobically Grown *E. coli*[A]

Reductant	Oxidant	→$H^+/2e^-$
Formate	Oxygen	3.62 ± 0.48(9)
Formate	Nitrate	1.69 – 3.68
Formate	Ubiquinone-1	1.58 ± 0.26(8)
Formate	Menadione	1.73 ± 0.29(6)
H_2	Fumarate	1.85 ± 0.15(7)
H_2	Nitrate	3.31 ± 0.29(4)
H_2	Ubiquinone-1	1.81 (2)
H_2	Menadione	1.67 ± 0.33(2)
Ubiquinol-1	Oxygen	2.28 ± 0.42(9)
Ubiquinol-1	Nitrate	1.49 ± 0.36(8)
Ubiquinol-1	Oxygen	[B]2.00 ± 0.21(9)
Ubiquinol-1	Nitrate	[B]1.86 ± 0.24(15)

[A] The data are taken from Jones *et al.* (*2*) and Jones (*14, 15*) with permission, where experimental details are also given. Values are expressed as mean ± S.D. for several separate cultures, the number of which is given in parentheses.

[B] Values were obtained with double-quinone mutant AN 384 (*8*).

Hydrogenase

Hydrogenase reduces benzyl viologen (BV^{++}) at the cytoplasmic aspect of the cytoplasmic membrane (*13*). The stoichiometric release of protons in the reaction

$$H_2 + 2BV^{++} \longrightarrow 2H^+ + 2BV^+ \qquad (1)$$

occurs at the periplasmic aspect of the membrane since the full rate and extent of the increase in pH (as measured with an extracellular pH electrode) is observed in the absence of a protonophore (*14*). To reduce benzyl viologen at the cytoplasmic aspect of the membrane and to liberate protons at the periplasmic aspect, the enzyme must be transmembranous. The transmembranous orientation of the enzyme was confirmed by covalent labeling with diazotized [^{125}I]diodosulfanilic acid and by progressive immunoabsorption studies (A. Graham and R. W. Jones, unpublished work). Figure 1A shows a scheme for H_2-dependent reduction of benzyl viologen. The scheme also predicts that the reaction is electron translocating, a point which needs further examination.

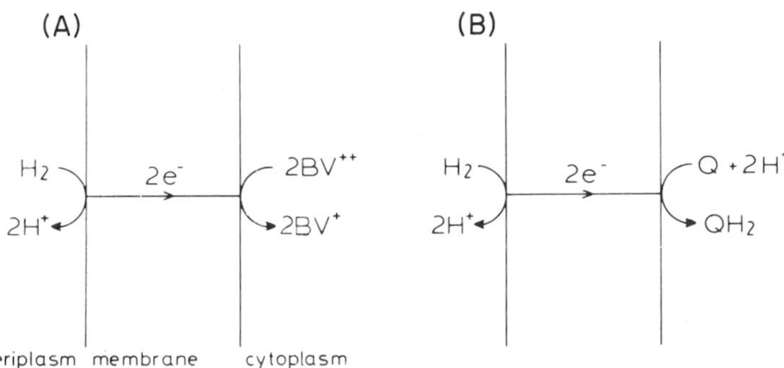

Fig. 1. Scheme for H_2-dependent reduction of (A) benzyl viologen and (B) menaquinone or ubiquinone. See text for details. Abbreviations: BV^{++} and BV^{+}, respectively, oxidized and radical species of benzyl viologen; Q, ubiquinone or menaquinone; QH_2, ubiquinol or menaquinol.

When H_2-saturated water is added to spheroplasts of *E. coli* incubated in the presence of valinomycin and ubiquinone-1, menadione or fumarate, proton translocation with an $\rightarrow H^+/2e^-$ stoichiometry of close to 2.0 (Table I) is observed (*14*). The proton translocation with fumarate as acceptor is not observed in the presence of the respiratory inhibitor 2-*n*-heptyl-4-hydroxyquinoline *N*-oxide, whereas that with menadione and ubiquinone-1 is unaffected. This inhibitor probably inhibits a cytochrome *b*-dependent oxidation of the endogenous menaquinol by fumarate. A scheme for proton translocation in the H_2-menaquinone segment of the respiratory chain is shown in Fig. 1B where the translocated species is the electron and the associated proton-consuming and proton-releasing reactions occurring on opposite sides of the membrane give rise to the net proton translocation. This differs from a conformational proton pump where the translocated species is the proton. Since the $\rightarrow H^+/2e^-$ stoichiometry for the H_2-fumarate pathway is also close to 2.0, electron transfer from menaquinol to fumarate probably does not drive proton translocation.

Hydrogenase consists of only one type of subunit of molecular weight 56,000–58,000 (*9*, *10*). Thus, this enzyme is the simplest respiratory system known to effect proton translocation.

Nitrate Reductase

The nitrate reductase complex of *E. coli* is probably the best characterized bacterial respiratory system. The enzyme consists of 3 subunits, α, β, and γ, of molecular weights 142,000–155,000, 58,000–67,000, and

19,000–20,000, respectively, the reported values varying from author to author (for references, see *1*). The γ subunit is the apoprotein of the cytochrome *b*-556 (NO_3^-), the synthesis of which is co-induced with that of nitrate reductase. The γ subunit is accessible to lactoperoxidase-catalyzed radioiodination at the periplasmic aspect of the cytoplasmic membrane, while the α subunit is accessible at the cytoplasmic aspect (*23*). Subsequent studies using immunochemical methods (*24*) and covalent labeling using transglutaminase (*25*), diazotized [^{125}I]diodosulfanilic acid and diazobenzene [^{35}S]sulfonate (*26*) confirmed the location of the α subunit. Labeling with diazotized [^{125}I]diodosufanilic acid and with diazobenzene [^{35}S]sulfonate (*26*) and limited digestion with trypsin revealed that the β subunit is also accessible at the cytoplasmic aspect of the membrane (*27*). Confirmation of the transmembranous orientation of the nitrate reductase complex comes from the observation that the non-permeant diquat radical is oxidized via the nitrate reductase pathway at the periplasmic aspect of the membrane in a heme-dependent, ubiquinone-independent manner (*17*), while benzyl viologen radical (*17*) and $FMNH_2$ (*16*) are oxidized in a heme-independent manner, and nitrate is reduced at the cytoplasmic aspect (*28*, *29*).

The problem with measuring $\rightarrow H^+/2e^-$ stoichiometries in bacteria is that it is difficult to be sure which substrate is being used. To overcome this problem, we measured proton translocation in response to the addition of limiting amounts of ubiquinol-1 to spheroplasts of *E. coli* incubated either anaerobically in an excess of nitrate or aerobically. The $\rightarrow H/2e^-$ stoichiometries are given in Table I. The use of the double-quinone mutant, AN384 (*8*), showed that the proton translocation is independent of endogenous quinones. In contrast, the heme-deficient mutant A1004a (*30*) was unable to oxidize ubiquinol-1 unless the growth medium was supplemented with 5-aminolevulinic acid, which allows heme synthesis to procede. The stoichiometry with either oxygen or nitrate was close to 2.0. In contrast to mitochondria (*31*), $\rightarrow H^+/2e^-$ ratios in *E. coli* do not appear to be underestimated due to phosphate transport (*2*, *32*).

To understand the mechanism of proton translocation in the ubiquinol-1 segment of the respiratory chain it is important to know on which side of the membrane protons are consumed in the reaction

$$2H^+ + 2e^- + NO_3^- \longrightarrow H_2O + NO_2^- \quad (2)$$

To test this point we used benzyl viologen radical as electron donor; this radical was able to rapidly permeate the cytoplasmic membrane and react with nitrate reductase via the heme-independent pathway. Bulk phase pH changes lagged considerably behind the absorption changes (due to oxidation of the colored benzyl viologen radical to the colorless divalent cation) in both rate and extent unless the cytoplasmic membrane was made permeable to protons by addition of the protonophore carbonyl cyanide *m*-chlorophenylhydrazone or nigericin, which facilitates H^+/K^+ exchange (*2*). It can be concluded that the protons of reaction (2) are consumed at the cytoplasmic aspect of the cytoplasmic membrane.

Two other reductants, diquat radical and reduced *N*-methylphenazonium methosulfate, resemble ubiquinol-1 in ability to interact with the nitrate reductase pathway in a heme-dependent, ubiquinone-independent, 2-*n*-heptyl-4-hydroxyquinoline-*N*-oxide-sensitive manner (*2*, *17*). Diquat radical, an electron donor, was unable to drive proton translocation, while reduced *N*-methylphenazonium methosulfate, a hydride donor, drove proton translocation with an $\rightarrow H^+/2e^-$ stoichiometry of less than one (*2*).

Figure 2A shows a scheme to account for proton translocation in the ubiquinol-nitrate segment of the respiratory chain. This scheme is consistent with the transmembranous orientation of the complex, the internal site of proton consumption and the $\rightarrow H^+/2e^-$ stoichiometry of almost two. When ubiquinol-1 is replaced with a hydride donor the stoichiometry falls to one (Fig 2B), and with an electron donor, to zero (Fig. 2C). In each case (Fig. 2A–C) the reaction is electron translocating. The failure to detect proton translocation when diquat radical is electron donor excludes the operation of a proton pump. This contrasts with mammalian cytochrome *c* oxidase, where electron transfer alone is sufficient to drive proton translocation (*33*).

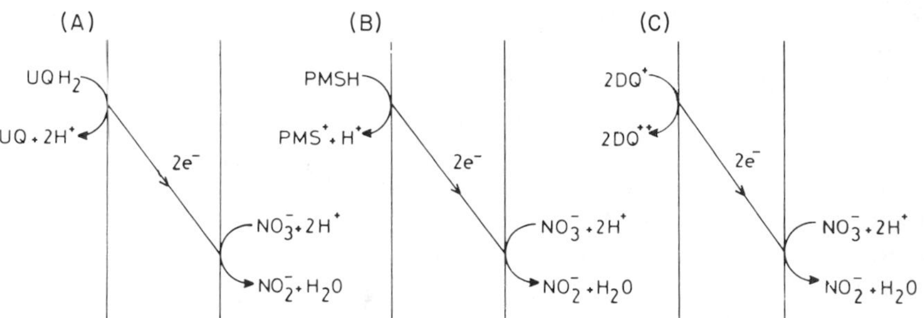

Fig. 2. Scheme for the oxidation of (A) ubiquinol, (B) reduced *N*-methylphenazonium methosulfate and (C) diquat radical via the nitrate reductase pathway of *E. coli*. See text for details. Abbreviations: UQH_2, ubiquinol; UQ, ubiquinone; PMSH, reduced *N*-methylphenazonium (methosulfate salt); PMS^+, oxidized *N*-methylphenazonium; DQ^{++} and DQ^+, respectively, oxidized and radical species of diquat.

DISCUSSION

Is Ubiquinone-1 an Accurate Model of Endogenous Ubiquinone-8?

Because of the differing side chains, it seems inevitable that ubiquinone-1 will exhibit some different biochemical properties from the endogenous ubiquinone-8. The real problem, however, is not that such differences may occur, but whether ubiquinone-1 faithfully takes the place of ubiquinone-8 in the mechanism by which substrate dehydrogenases are functionally connected to the nitrate reductase system and to cytochrome oxidase. The evidence suggests that it does. Ubiquinone-1 is able to reconstitute NADH oxidase activity and the associated quenching of atebrin fluorescence in inside–out membrane vesicles from a ubiquinone-deficient mutant (*34*). Furthermore, ubiquinone-1 is able to reconstitute formate-dependent proton translocation, with either nitrate or oxygen as acceptor (*2*), and also active transport of proline driven by lactate oxidation (*35*) in the double-quinone mutant AN 384.

Role of Ubiquinone in Proton Translocation

One of the important features of ubiquinone is that it is a hydrogen acceptor; reduction of ubiquinone to ubiquinol requires not only electrons but also protons. Thus, reduction of ubiquinone at the cytoplasmic aspect of the membrane will cause alkalinization of the cytoplasm (Fig. 1B). Oxidation of ubiquinol at the periplasmic aspect will release protons and cause acidification of the extracellular bulk phase. To behave in this manner, ubiquinone must act as a mobile hydrogen carrier in a classical redox loop as proposed by Mitchell (*36*). Figure 3 shows how ubiquinone could act in this manner in the formate–nitrate pathway to give a stoichiometry of 4.0; this fits quite well with experimental data (Table I) that gave a value as high as 3.6. Ubiquinone could act in this way to connect any dehydrogenase of appropriate redox potential to any oxidase or nitrate reductase. The overall $\rightarrow H^+/2e^-$ stoichiometry would be the sum of the stoichiometries of the dehydrogenase and oxidase or nitrate reductase. Thus with succinate as donor, the overall stoichiometry with either oxygen or nitrate is 2.0 because the succinate–ubiquinone reaction is not proton translocating (*21*).

Studies on model systems show that ubiquinone can behave as a mobile hydrogen carrier. Futami *et al.* (*37*) showed ubiquinone is able to mediate hydrogen transfer across artificial lipid vesicle membranes.

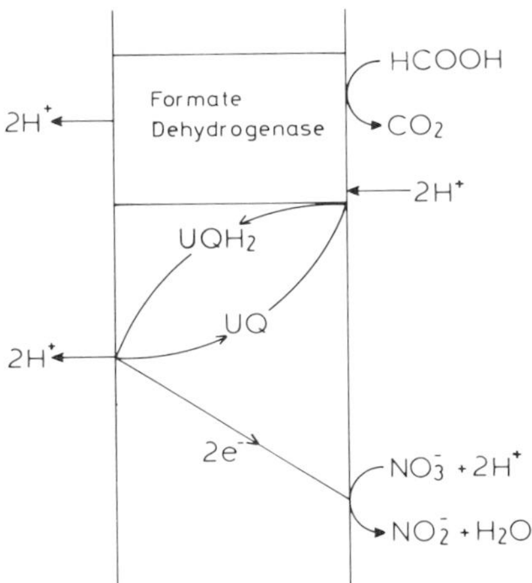

Fig. 3. Scheme for the role of ubiquinone in the proton-translocating formate to nitrate pathway in *E. coli*. The mechanism of proton translocation by formate dehydrogenase is not known. Ubiquinone (UQ) is portrayed as a mobile hydrogen carrier. Full details are given in the text.

From studies of ubiquinone in phospholipid monolayers, Quinn and Esfahani (*38*) concluded that ubiquinone would be thermodynamically capable of transbilayer movement. However, such results must always be treated with caution when extrapolating to physiological systems, since ubiquinone function is almost certainly influenced by interaction with respiratory proteins (*19*). More information is required on transmembranous mobility of ubiquinone in the *E. coli* membrane. The exact sites of ubiquinone oxidation and reduction require verification, as those shown in Figs. 1 and 2 are only implied by analogy with sites for artificial oxidoreduction donors and acceptors. We attempted to explore the topography of the quinone sites using water-soluble quinone analogs (Fig. 4) which, because of their large substituent groups, would not be expected to be able to permeate the cytoplasmic membrane. Unfortunately, neither analog was a good substrate for formate dehydrogenase, ubiquinol oxidase, or nitrate reductase. However, the potential of these compounds as specific inhibitors of oxidation and reduction of naturally occurring quinones has yet to be explored.

Downie and Cox (*39*) proposed that an $\rightarrow H^+/2e^-$ stoichiometry of 4.0 for aerobic oxidation of NADH can be explained by what they term a "Q cycle." In fact, the scheme they present is not a true Q cycle as described by Mitchell (*18*) but merely shows ubiquinone acting at two sites. The scheme of Downie and Cox (*39*) is based upon steady-state spectral measurements of the *b*-type cytochromes in membranes from

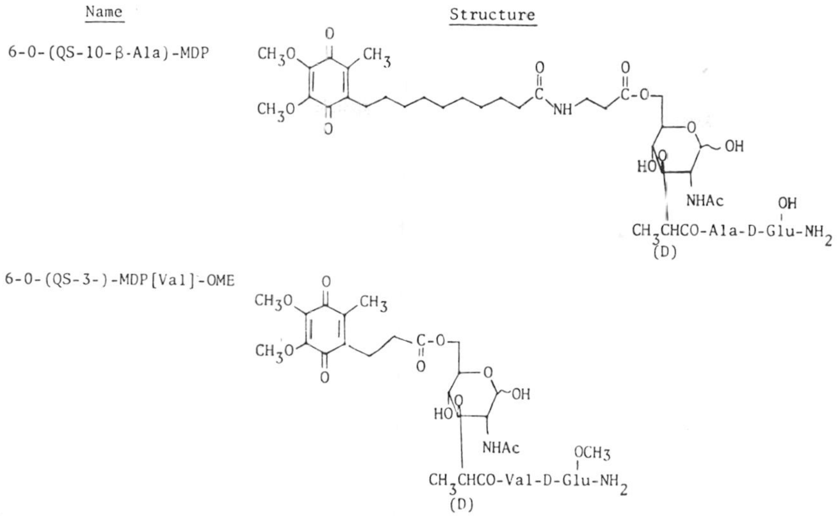

Name | Structure

6-0-(QS-10-β-Ala)-MDP

6-0-(QS-3-)-MDP[Val]-OME

Fig. 4. Water-soluble ubiquinone analogs.

ubiquinone-sufficient and ubiquinone-deficient strains. The interpretation of such data is equivocal. In their scheme there is no proton translocation in the NADH–ubiquinone segment of the respiratory chain. Our data show conclusively that proton translocation occurs in the formate-ubiquinone, hydrogen-ubiquinone, and ubiquinol-oxygen segments of the respiratory chain. Since $\rightarrow H^+/2e^-$ stoichiometries for both NADH oxidation and formate oxidation are close to 4.0 (*21*), the NADH dehydrogenase must be proton translocating. The failure of Brookman *et al.* (*40*) to detect proton translocation by NADH dehydrogenase in an ATPase-negative heme-deficient mutant may reflect a widespread effect on the respiratory chain due to the adverse conditions under which this strain must grow. Loss of proton-translocating ability of NADH dehydrogenase has been reported under conditions of sulphate limitation (*41*).

Role of Menaquinone

Menaquinone is involved in the proton-translocating reduction of fumarate. The proton-translocating step is in the reduction of menaquinone to menaquinol (Fig. 1B). The menaquinol-fumarate pathway does not appear to be proton translocating. Therefore, menaquinone does not need to move across the membrane as does ubiquinone.

Further Considerations of Proton Translocation

The evidence for chemiosmotic mechanisms of proton translocation by photosynthetic systems, both in bacterial chromatophores (*42*, *43*) and chloroplasts (*44*, *45*) is strong. There are now three well documented bacterial systems in which a transmembranous electron transfer gives rise to proton translocation: the H_2-menaquinone (*14*) and ubiquinol-nitrate (*2*) segments of the anaerobic respiratory chain of *E. coli* and the formate-fumarate pathway of *Vibrio succinogenes* (*46*, *47*). In addition electron-transfer reactions involved in the oxidation of Fe^{2+} by *Thiobacillus ferrooxidans* are reported to span the membrane (*48*) and, although not proton translocating, provide a chemiosmotic mechanism for establishing a difference in electrochemical proton actvity across the membrane (*49*).

The situation in mitochondrial systems is much more complicated. The work of Wikström and Krab (*33*) and also of Casey *et al.* (*50*) indicate that cytochrome *c* oxidase has a proton-pumping activity for which a conformational proton pump cannot be excluded. This view, however, is opposed by Mitchell and Moyle (*51*) and Lorusso *et al.* (*52*). But recent estimates of $\rightarrow H^+/2e^-$ ratios of 3–4/classical oxidative phosphorylation site (*31*) are not readily explained by present knowledge of the respiratory carriers if proton translocation is effected by a chemiosmotic mechanism.

We believe, however, that it is unjustified to extrapolate unconditionally results from bacterial systems to mitochondrial systems (*53*). Proton translocation in mitochondria probably involves vectorial electron transfer, but it is possible that additional mechanisms may exist for the actual pumping of protons across the membrane. This additional complexity may be reflected in the large number of subunits in mitochondrial as opposed to bacterial systems (for references, see *2*). For example, the NADH dehydrogenase of mitochondria contains over 20 subunits (*56*), while the equivalent enzyme in *E. coli* requires only one (*5*). Consideration should be given to the evolution of the complex mitochondrial proton-translocating systems from the simpler bacterial systems (*54*).

CONCLUDING REMARKS

The anaerobic respiratory chain of *E. coli* is an attractive system in which to study the role of ubiquinone. The evidence suggests that ubiquinone functions as a mobile hydrogen carrier and this function depends on the structural and functional orientation of the respiratory enzymes within the membrane. However, more information is needed about the transmembrane and lateral mobility of ubiquinone and the lateral mobility of the enzymes with which it reacts. We also need to know more about the iron–sulphur centers involved, as these play an important role in ubiquinone function in mitochondria (*19*). Little is known about such proteins in *E. coli* at present (*56*, *57*). More information may be forthcoming on the role of ubiquinone in aerobic respiration in *E. coli* now that work is in progress on purification of the cytochromes (*58–61*), and also on the isolation of mutants which lack specific cytochromes (*62*).

ACKNOWLEDGMENTS

The work reported from our laboratory was supported by the Science Research Council and the Medical Research Council. We thank Dr. B. Trumpower for a preprint of his manuscript. Our thanks go to Dr. I. Imada (Central Research Division, Takeda Chemical Industries Ltd, Osaka 532, Japan) for the gift of the ubiquinone analogs.

REFERENCES

1. Haddock, B. A., and Jones, C. W. (1977). *Bacteriol. Rev.* **41,** 47–99.
2. Jones, R. W., Lamont, A., and Garland, P. B. (1980). *Biochem. J.* **190,** 79–94.
3. Haddock, B. A. (1977). *Symp. Soc. Gen. Microbiol.* **27,** 95–120.
4. Old, R. W., and Primrose, S. B. (1980). "Principles of Gene Manipulation." Blackwell, Oxford.
5. Young, I. G., Jaworoski, A., and Poulis, M. I. (1978). *Gene* **4,** 25–36.
6. Cole, S. T., and Guest, J. R. (1979). *FEMS Microbiol. Lett.* **5,** 65–67.
7. Cole, S. T., and Guest, J. R. (1980). *Mole. Gen. Genet.* **178,** 409–418.
8. Wallace, B. J., and Young, I. G. (1977). *Biochim. Biophys. Acta* **461,** 84–100.
9. Adams, M. W. W., and Hall, D. O. (1979). *Biochem. J.* **183,** 11–22.
10. Graham, A., Boxer, D. H., Haddock, B. A., Mandrand-Berthelot, M.-A., and Jones, R. W. (1980). *FEBS Lett.* **113,** 167–172.
11. Enoch, H. G., and Lester, R. L. (1975). *J. Biol. Chem.* **250,** 6693–6705.
12. Clegg, R. A. (1976). *Biochem. J.* **153,** 533–541.
13. Jones, R. W. (1979). *Biochem. Soc. Trans.* **7,** 724–725.
14. Jones, R. W. (1980). *Biochem. J.* **188,** 345–350.
15. Jones, R. W. (1980). *FEMS Microbiol. Lett.* **8,** 167–171.
16. Kemp, M. B., Haddock, B. A., and Garland, P. B. (1975). *Biochem. J.* **148,** 329–333.
17. Jones, R. W., and Garland, P. B. (1977). *Biochem. J.* **164,** 199–211.
18. Mitchell, P. (1976). *J. Theor. Biol.* **62,** 327–367.
19. Trumpower, B. L. (1981). *J. Bioenerg. Biomembr.* **13,** 1–24.
20. Graham, A., and Boxer, D. H. (1981). *Biochem. J.* **195,** 627–637.
21. Garland, P. B., Downie, J. A., and Haddock, B. A. (1975). *Biochem. J.* **152,** 547–559.

22. Boonstra, J., and Konings, W. N. (1977). *Eur. J. Biochem.* **78,** 361–368.
23. Boxer, D. H., and Clegg, R. A. (1975). *FEBS Lett.* **60,** 54–57.
24. MacGregor, C. H., and Christopher, A. R. (1978). *Arch. Biochem. Biophys.* **185,** 204–213.
25. Graham, A., and Boxer, D. H. (1978). *Biochem. Soc. Trans.* **6,** 1210–1211.
26. Graham, A., and Boxer, D. H. (1980). *FEBS Lett.* **113,** 15–20.
27. Graham, A., and Boxer, D. H. (1980). *Biochem. Soc. Trans.* **8,** 331.
28. Jones, R. W., Ingledew, W. J., Graham, A., and Garland, P. B. (1978). *Biochem. Soc. Trans.* **6,** 1287–1289.
29. Kristjansson, J. K., and Hollocher, T. C. (1979). *J. Bacteriol.* **137,** 1227–1233.
30. Haddock, B. A. (1973). *Biochem. J.* **136,** 877–884.
31. Brand, M. D., Reynafarje, B., and Lehninger, A. L. (1976). *J. Biol. Chem.* **251,** 5670–5679.
32. Cox, J. C., and Haddock, B. A. (1978). *Biochem. Biophys. Res. Commun.* **82,** 46–52.
33. Wikström, M., and Krab, K. (1979). *Biochim. Biophys. Acta* **549,** 177–222.
34. Haddock, B. A., and Downie, J. A. (1974). *Biochem. J.* **142,** 703–706.
35. Stroobant, P., and Kaback, H. R. (1979). *Biochemistry* **18,** 226–231.
36. Mitchell, P. (1972). *J. Bioenerg.* **3,** 5–24.
37. Futami, A., Hurt, E., and Hauska, G. (1979). *Biochim. Biophys. Acta* **547,** 583–596.
38. Quinn, P. J., and Esfahani, M. A. (1980). *Biochem. J.* **185,** 715–722.
39. Downie, J. A., and Cox, G. B. (1978). *J. Bacteriol.* **133,** 477–484.
40. Brookman, J. J., Downie, J. A., Gibson, F., Cox, G. B., and Rosenberg, H. (1979). *J. Bacteriol.* **137,** 705–710.
41. Poole, R. K., and Haddock, B. A. (1975). *Biochem. J.* **152,** 537–546.
42. Jones, O. T. G. (1977). *Symp. Soc. Gen. Microbiol.* **27,** 151–183.
43. Crofts, A. R., and Bowyer, J. (1978). *In* "The Proton and Calcium Pumps" (G. F. Azzone, M. Avron, J. C. Metcalfe, E. Quagliariello, and N. Siliprandi, eds.), pp. 55–64. Elsevier/North-Holland, Amsterdam.
44. Trebst, A. (1974). *Annu. Rev. Plant Physiol.* **25,** 423–458.
45. Hauska, G., and Trebst, A. (1977). *Curr. Top. Bioenerg.* **6,** 151–220.
46. Kroger, A. (1978). *Biochim. Biophys. Acta* **505,** 129–145.
47. Kroger, A., Dorrer, E., and Winkler, E. (1980). *Biochim. Biophys. Acta* **589,** 118–136.
48. Ingledew, W. J., Cox, J. C., and Halling, P. J. (1977). *FEMS Microbiol. Lett.* **2,** 193–197.
49. Cox, J. C., Nicholls, D. G., and Ingledew, W. J. (1979). *Biochem. J.* **178,** 195–200.
50. Casey, R. P., Chappell, J. B., and Azzi, A. (1979). *Biochem. J.* **182,** 181–188.
51. Mitchell, P., and Moyle, J. (1979). *Biochem. Soc. Trans.* **7,** 887–894.
52. Lorusso, M., Capuano, F., Baffoli, D., Stefanelli, R., and Papa, S. (1979). *Biochem. J.* **182,** 133–147.
53. Mitchell, P. (1979). *Eur. J. Biochem.* **95,** 1–20.
54. Smith, S., and Ragan, C. I. (1980). *Biochem. J.* **185,** 315–326.
55. Raven, J. A., and Smith, F. A. (1976). *J. Theor. Biol.* **57,** 301–302.
56. Owen, P., Kaczorowski, G. J., and Kaback, H. R. (1980). *Biochemistry* **19,** 596–600.
57. Ingledew, W. J., Reid, G. A., Poole, R. K., Blum, H., and Ohnishi, T. (1980). *FEBS Lett.* **111,** 223–227.
58. Reid, G. A., and Ingledew, W. J. (1980). *FEBS Lett.* **109,** 1–4.
59. Kita, K., Yamato, I., and Anraku, Y. (1978). *J. Biol. Chem.* **253,** 8910–8915.
60. Kranz, R. G., and Gennis, R. B. (1982). *J. Bacteriol.* **150,** 36–45.
61. Miller, M. J., and Gennis, R. B. (1982). *Fed. Proc.* **42,** 894.
62. Green, G. N., Faiman, R., and Gennis, R. B. (1981). *Fed. Proc.* **40,** 1669.

The Electron-Transport Chain of Rhodopseudomonas sphaeroides

2

ANTONY R. CROFTS
STEVE W. MEINHARDT
JOHN R. BOWYER

INTRODUCTION

The cyclic photosynthetic electron-transfer chains of *Rhodopseudomonas sphaeroides* and *Rps. capsulata* involve three major proteins: the reaction center, a soluble cytochrome *c*, and a complex acting as a ubiquinol–cytochrome *c* oxidoreductase. This complex shares many of the characteristics of the mitochondrial Complex III (*1*, *2*) and additional observations have shown that this similarity is more exact than was previously realized. In this paper we present our results (*3–9*) on the redox properties of the components so far identified, and on the kinetics of electron transfer when oxidizing and reducing single equivalents are introduced to the chain by flash activation of the reaction center. The results will be discussed in terms of possible reaction pathways and involvement of distinct populations of ubiquinone.

RESULTS AND DISCUSSION

Role of the Rieske-Type Iron–Sulfur Center (RFe–S): Inhibition of Electron Transport by UHDBT[1]

We have previously reported on the inhibition by UHDBT of cytochrome c_2 rereduction following flash oxidation (*4–6*). We suggested

[1] Abbreviations: UHDBT, 5-*n*-undecyl-6-hydroxy-4,7-dioxobenzothiazole; RFe–S, Rieske type iron–sulfur center; UQ, ubiquinone-10.

Function of Quinones in Energy Conserving Systems

ISBN 0-12-701280-X

UHDBT inhibited electron transfer from a component between the antimycin-sensitive site and cytochrome c_2, and showed that the component had the characteristics of RFe–S. Also, the kinetics of electron transfer, the approximate stoichiometry, and the operating E_m value (and its dependence on pH) were evaluated, and these results are incorporated into Scheme I (next section). In collaboration with Prince and Dutton (*7*) we showed that, within the time resolution of the method (~1 sec), the EPR signal of RFe–S behaved as expected in response to flash or continuous illumination, and to addition of antimycin and UHDBT; both the spectrum and midpoint potential of RFe–S were modified by UHDBT. These results were interpreted as showing that UHDBT binds specifically at a site close to or identical with the RFe–S center (possibly a ubiquinone binding site), and prevents the center from undergoing its cycle of redox reactions. Comparable results were obtained using a succinate-cytochrome *c* reductase preparation (*10*, *11*).

Bound Cytochrome *c* (Cytochrome c_b)

We previously noted several characteristics of cytochrome *c* in chromatophores (and other membrane preparations) that suggested *c*-type cytochrome was not homogeneous, with a fraction behaving as if it were bound to the membrane (*4*, *8*, *9*). The characteristics noted were: (A) the absorbance maximum of cytochrome *c* in chromatophores (derived either by redox difference spectrophotometry or by time-resolved kinetic spectrophotometry), was displaced by 1–1.5 nm to the red of purified soluble cytochrome c_2 (λ max 550 ± 0.5 nm); (B) the kinetics of cytochrome *c* oxidation in UHDBT-inhibited chromatophores measured at 551 nm, showed two approximately equal phases of oxidation, one with $t_{1/2}$ ~5–10 μsec, the other with $t_{1/2}$ ~200 μsec; (C) intact spheroplasts, or vesicles prepared from them, showed a loss of approximately 50% of the cytochrome *c* found in cells or chromatophores; the remaining cytochrome *c* could not be washed off, but was unavailable to the reaction center for rapid photooxidation. Taken together with the previously published observation that the redox potential of cytochrome *c* in chromatophores differed from that of soluble cytochrome c_2 (*12*), our results indicated that two different populations of cytochrome *c* were normally present in cells or chromatophores. Recently, Wood (*13*, *14*), in an independent study, using more direct biochemical methods showed that two distinct cytochromes *c* are present in cells and chromatophores: soluble cytochrome

c_2 and a membrane bound cytochrome c (c_b) of approximately double molecular weight. Hauska (personal communication) found a complex resembling Complex III of mitochondria, containing the b-type cytochromes and cytochrome c_b, which could be isolated after mild detergent dissociation. In Fig. 1 we show spectra of the two cytochromes c resolved by full spectrum redox potentiometry of chromatophores (Fig. 1A) and by time-resolved kinetic spectrophotometry (Fig. 1B). The data of Fig. 1A were obtained by resolving spectra measured at different potentials during a reductive titration over the range $400 > E_h > 200$ (mV), into two components with $E_{m,7}$ values of 257 and 354 mV. The $E_{m,7}$ values were obtained by finding the best fit for two components to the absorbance change on titration at 551–542 nm (see Fig. legend for further details). Computer analysis shows: (A) the titration at 551–542 nm is better fitted by two components than by one component, (B) the component of $E_{m,7}$ ~355 mV has a different λ max (at 550.5 ± 0.5 nm) than the component titrating at $E_{m,7}$ ~260 mV (λ max at 552 ± 0.5 nm) and (C) in this preparation the components were present in approximately equal amounts. The data of Fig. 1B were obtained by measuring the absorbance change at different times after a saturating xenon flash (~20 μsec width at half-height) given to UHDBT-inhibited chromatophores. After correction for the reaction center change (*8*) the spectrum of the change measured 32 μsec after the start of the flash showed a peak at 550 ± 0.5 nm, while the change developing over the period between 32 and 532 μsec after the flash showed a peak at 552 ± 0.5 nm. From these and similar spectra we selected pairs of wavelengths at which the change due to one of the c-type cytochromes is maximal, while that of the unselected cytochrome is compensated. Figure 2 shows kinetic difference traces at such wavelength pairs (after correction for reaction-center changes) at several redox potentials from experiments similar to those of Fig. 1B except that UHDBT was absent. From these traces, it can be clearly seen that the two c-type cytochromes behave with distinctively different kinetics. The "free" cytochrome c_2 is oxidized rapidly ($t_{1/2}$ ~5–10 μsec) and, at high redox potentials or in the presence of UHDBT remains oxidized. The "bound" cytochrome c_b is oxidized much more slowly ($t_{1/2}$ ~ 150 μsec). When antimycin, but not UHDBT, is present, at E_h ~200 mV, cytochrome c_2 is oxidized, and then rereduced ($t_{1/2}$ ~200 μsec), while cytochrome c_b remains oxidized on the msec time scale. If the ambient redox potential is lowered to E_h ~ 100 mV (UQ_Z chemically reduced), preliminary experiments indicate that cytochrome c_b also becomes partly rereduced in the presence of antimycin, but more slowly than cytochrome c_2 ($t_{1/2}$ ~1 msec). The kinetic changes in the presence of UHDBT, or in the presence of antimycin at E_h ~200 mV are easily interpreted in terms of a simple

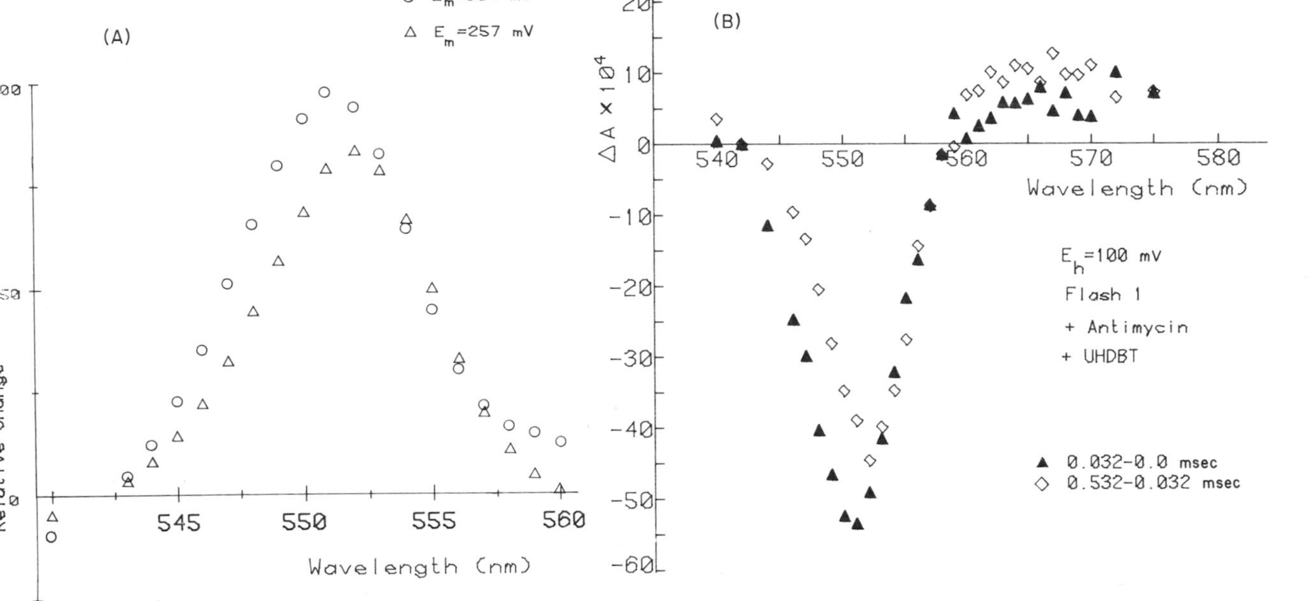

Fig. 1. Two *c*-type cytochromes of *Rps. sphaeroides*. (A) The cytochromes resolved by redox titrations. Chromatophores from *Rps. sphaeroides* Ga strain (~3 μM reaction center) were suspended in 100 m*M* KCl, 50 m*M* MOPS, pH 7.0 with 100 μM DAD, 100 μM Benzoquinone, 2 μM valinomycin, 1 μM nigericin, and ferricyanide (~100 μM). Spectra over the wavelength range 515–579 nm were measured at ~10 mV steps during reductive and oxidative titrations, with essentially similar results. The data shown are from a reductive titration over the range 380–200 mV, using dithionite as reductant. The absorbance change at 551–542 nm was analyzed for 1, 2, or 3 components. The two components fit gave $E_{m,7}$ values of 354 and 257 mV for components contributing approximately equally to the change. These values were used for further analysis to determine the contribution of each component to the change at the other points in the α-band region, then the data plotted to give the spectra shown. (B) The cytochromes separated by time-resolved spectra. Chromatophores (~0.6 μM reaction centers) were suspended in 100 m*M* KCl, 50 m*M* MOPS, pH 7.0, with 100 μM 1,2-naphthoquinone, 2 μM each of PMS, valinomycin, and nigericin, 4 μg/ml gramicidin, 10 μM antimycin, and 40 μM UHDBT, in an anaerobic redox cuvette with E_h poised at 100 ± 10 mV. Kinetic traces (average of 16, 5 msec sweep, 10 μsec filter RC, 30 sec between flashes) were accumulated at each point in the spectrum and stored for analysis (*8, 9, 33*). The spectra show the changes occurring over the times indicated.

linear kinetic scheme such as

Component	UQ_Z	⟶	RFe–S	⟶	cytochrome c_b	⟶	cytochrome c_2	⟶	P-870
Amount	0.5–4.0		0.5–0.7		0.5–0.7		0.4–0.7		1
$t_{1/2}$ for oxidation		1–2 msec		200 μsec		100 μsec		5–10 μsec	
$E_{m,7}$ (mV)	150		280		260		350		450
Inhibition by		antimycin		UHDBT		cytochrome c_2 depletion			

Although it is clear from these experiments, our previous results, and those of Wood (*14*), that cytochrome c_b and cytochrome c_2 react in series with the oxidized reaction center (P^+), and it seems clear from the effects of UHDBT that RFe–S reacts with P^+ through cytochrome c_2, we cannot be sure that cytochrome c_b is in series between RFe–S and cytochrome c_2 as shown in Scheme I (solid arrows). We will return to this point, but note the relative $E_{m,7}$ values for RFe–S (~280 mV) and cytochrome c_b (~260 mV), and the similar relation of the E_m values of the analogous components in mitochhondria (*15*) do not favor reduction of cytochrome c_b (or cytochrome c_1) by RFe–S, and parallel pathways for reduction of cytochrome c_2 by cytochrome c_b and RFe–S (dashed arrow) are as plausible as the linear scheme shown.

Electron Transfer on the Acceptor Side

Much recent work especially from Wraight and colleagues (*16–19*) has helped to clarify the reactions of the secondary quinone (UQ_{II}) involved in the two-electron gate by which electrons are transferred from the reaction center to the chain (*19*, *20*). In chromatophores, the functioning of the gate has also been observed (*3*, *21–23*) and we have recently shown a binary pattern in the extent of reduction of cytochrome *b*-50 as a function of flash number from the dark state, suggesting that electrons are available to reduce cytochrome *b*-50 only after UQ_{II} has been fully reduced (*3*). The redox dependence of the operation of the gate in chromatophores, when low concentrations of mediators are used (*3*), suggests that, at least in a fraction of centers (approximately 50%), $UQ_{II}H\cdot$ (or UQ^-_{II}) is stable in the dark at redox potentials much higher than the $E_{m,7}$ for the couple $UQ_{II}/UQ_{II}H\cdot$ found by Rutherford and Evans (*24*). This suggests a mechanism which allows $UQ_{II}H\cdot$ (or UQ^-_{II}) to exist out of equilibrium with the chain, so that experiments performed in the redox range $100 < E_h < 300$

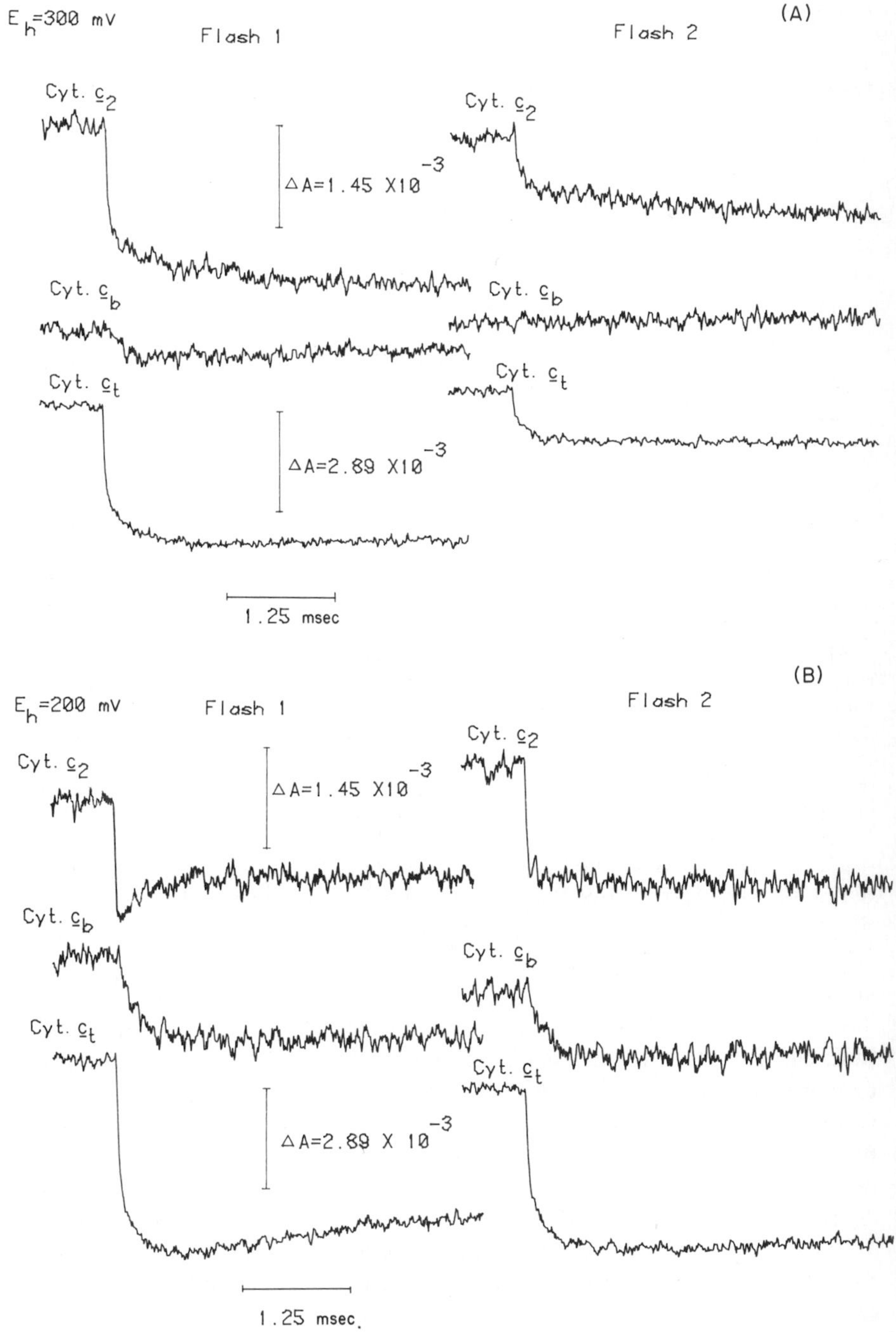

(A)

E_h=300 mV

Flash 1

Flash 2

Cyt. c_2

$\Delta A = 1.45 \times 10^{-3}$

Cyt. c_2

Cyt. c_b

Cyt. c_b

Cyt. c_t

Cyt. c_t

$\Delta A = 2.89 \times 10^{-3}$

1.25 msec

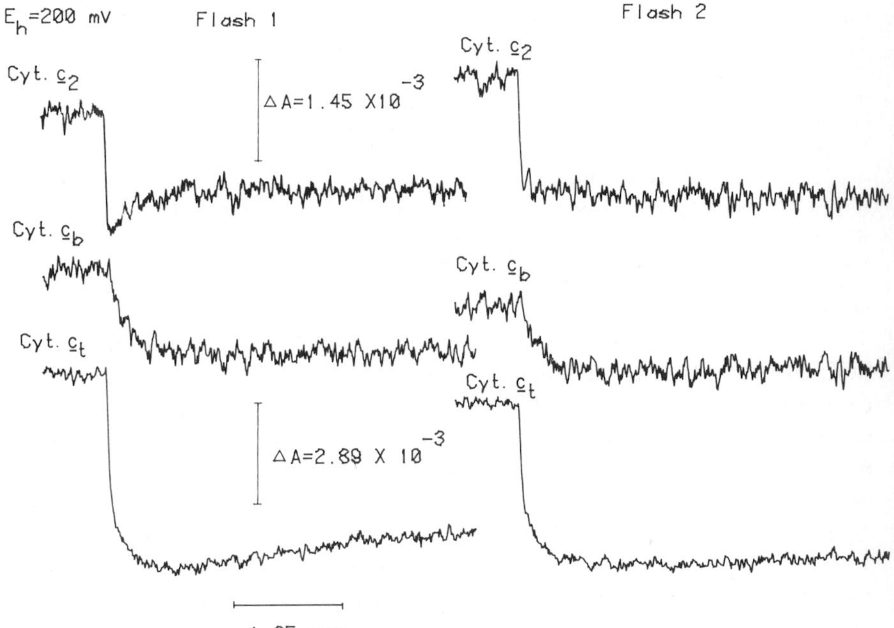

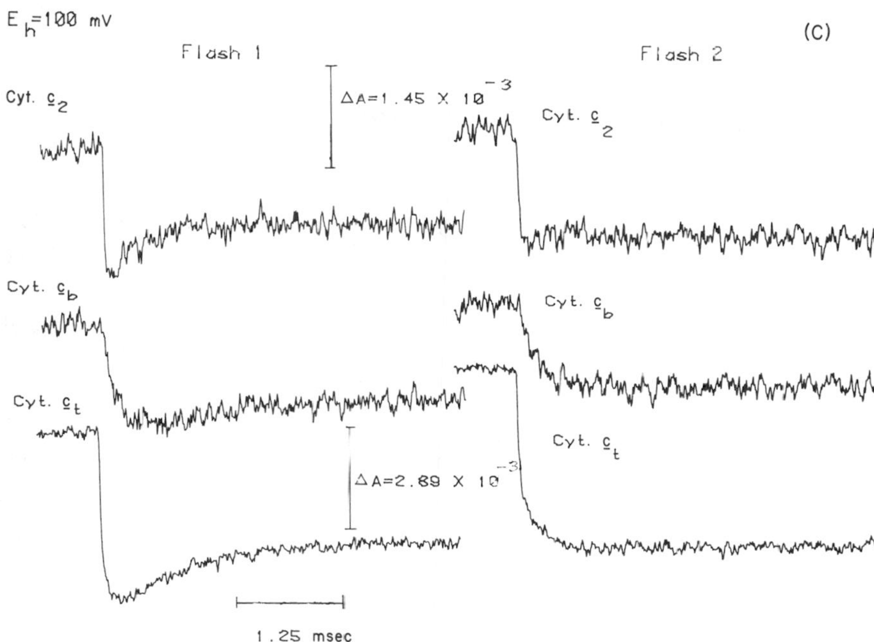

Fig. 2. Resolved kinetics of the two *c*-type cytochromes. Chromatophores (~0.6 μM reaction center) in 100 m*M* KCl, 50 m *M* MOPS, pH 7.0 were adjusted to the E_h value shown. Kinetic traces (average of 8, 5 msec sweep, 10 μsec filter RC, 75 msec between two flashes, 30 sec between groups) were accumulated at a set of wavelengths and stored for analysis. Cytochrome c_t (total cytochrome c) was measured at 551 nm; cytochrome c_b (bound cytochrome *c*) was measured at 552–548 nm; cytochrome c_2 (soluble cytochrome *c*) was measured at 550–554 nm. All changes were corrected for reaction center by subtraction of the normalized 542 nm trace (*8, 9*). Antimycin (10 μM), gramicidin (4 μg/ml), valinomycin (2 μM), and nigericin (2 μM) were present for all experiments. Mediators for Figs. 2 and 4–7 were present as follows: (A) at E_h ~300 mV, 100 μM benzoquinone, 1 μM TMPD; (B) at E_h ~200 mV, 100 μM 1,2-naphthoquinone (NQ), 1 μM TMPD; (C) at E_h ~ 100 mV, 100 μM 1,2-NQ, 1 μM TMPD, 1 μM PMS. At E_h ~ 0 mV, 100 μM 1,4-NQ, 1 μM PES, 1 μM pyocyanine, 10 μM duroquinone.

mV, using mediators at concentrations <10 μM, probably involve a heterogeneous population of chains in which a fraction of reaction centers (those with UQ_{II}^-) pass two electrons to the chain on flash excitation, and a fraction (those with UQ_{II} oxidized) pass no electrons to the chain (*3*). In confirmation of this possibility, Bowyer and Crofts (*9*) showed that in this E_h range, ametryne at 100 μM blocks electron transfer from UQ_I^- in 50% of the centers, but has no effect on cytochrome *b*-50 reduction. Stein and Wraight (personal communication) found that in reaction centers, ametryne, like *o*-phenanthroline (*18, 25*) blocks electron transfer from UQ_I^- to

UQ_{II}, but not to UQ_{II}^-. Such an effect in chromatophores would account for our observation in terms of the hypothesis of heterogeneous chains outlined above.

Role of Cytochrome *b*-(−90)

We obtained precise spectra of the three *b*-type cytochromes of *Rps. sphaeroides* in the α-band region, and this has made it possible to select appropriate wavelengths for the kinetic resolution of the redox changes of the individual cytochromes (*8*, *25a*). It was previously thought that cytochrome *b*-50 was the only *b*-type cytochrome undergoing rapid redox changes during operation of the cyclic chain (*1*, *2*). We were able to demonstrate unambiguously that, in the presence of antimycin, cytochrome *b*-(−90) is rapidly reduced following flash excitation (Figs. 3–5).

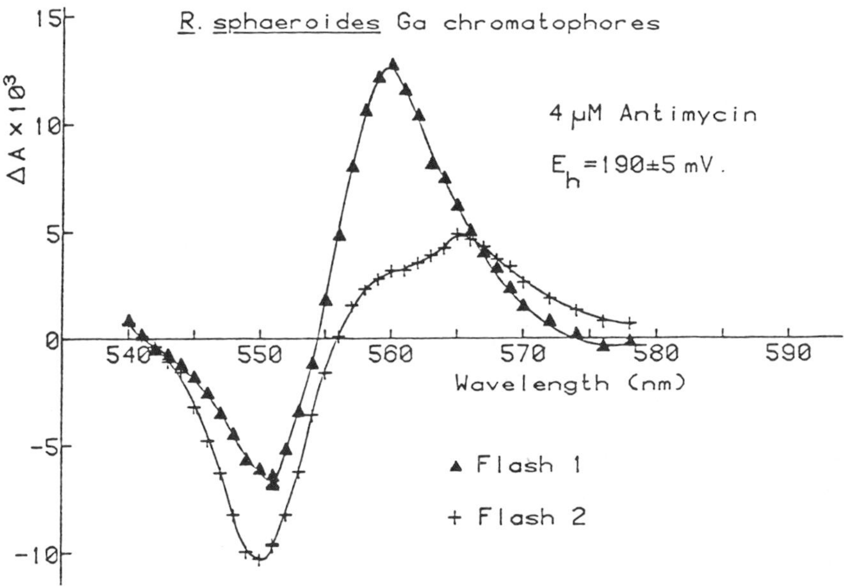

Fig. 3. Light-minus-dark difference spectra of cytochrome changes in *Rps. sphaeroides* Ga chromatophores. Chromatophores were suspended to 1.2 μM RC in pH 7.0 buffer containing 10 μM each of DAD, 1,4-NQ, 1,2-NQ, 2-OH-1,4-NQ; 1 μM valinomycin and 4 μM antimycin at E_h 190 ± 5 mV. At each wavelength, chromatophores were subjected to two flashes with 20 msec between each flash. Signals were not averaged; 33 sec elapsed between each measurement. ▲, the change 13 msec after the first flash; +, the change 5 msec after the second flash, using the point 17.5 msec after the first flash as the baseline. The contribution of the reaction center change was removed by subtraction of the appropriately normalized change recorded at 603 nm.

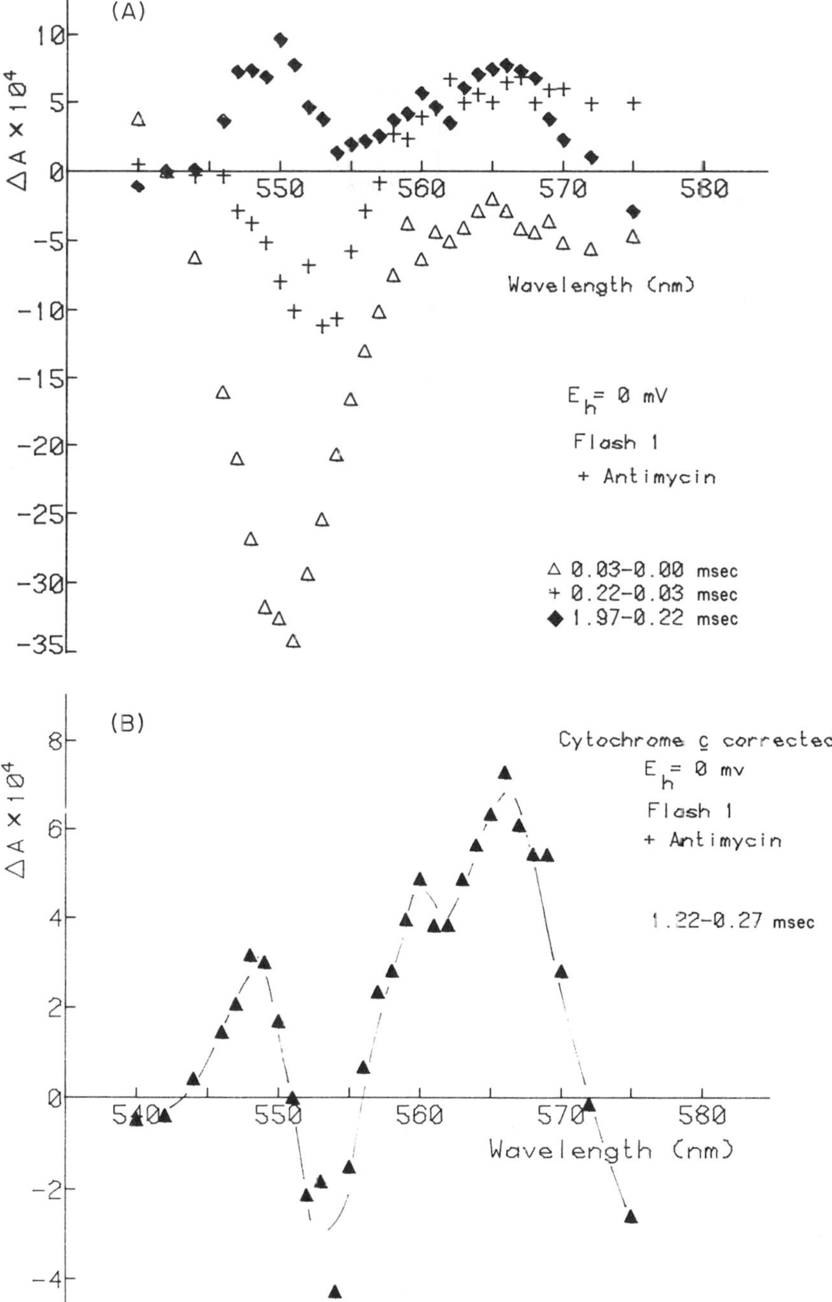

Fig. 4. Time-resolved spectra at E_h ~O mV. Chromatophores (~3 μM reaction center) in the same medium as for Fig. 2, were adjusted to E_h 0 ± 5 mV. Traces (average of 16, 5 msec sweep, 10 μsec filter RC, 30 sec between flashes) were accumulated at each wavelength and stored for analysis. (A) Spectra show the changes which occurred over the intervals indicated, and are all corrected for reaction center change. (B) Points additionally corrected for cytochrome *c* (assuming a homogeneous population).

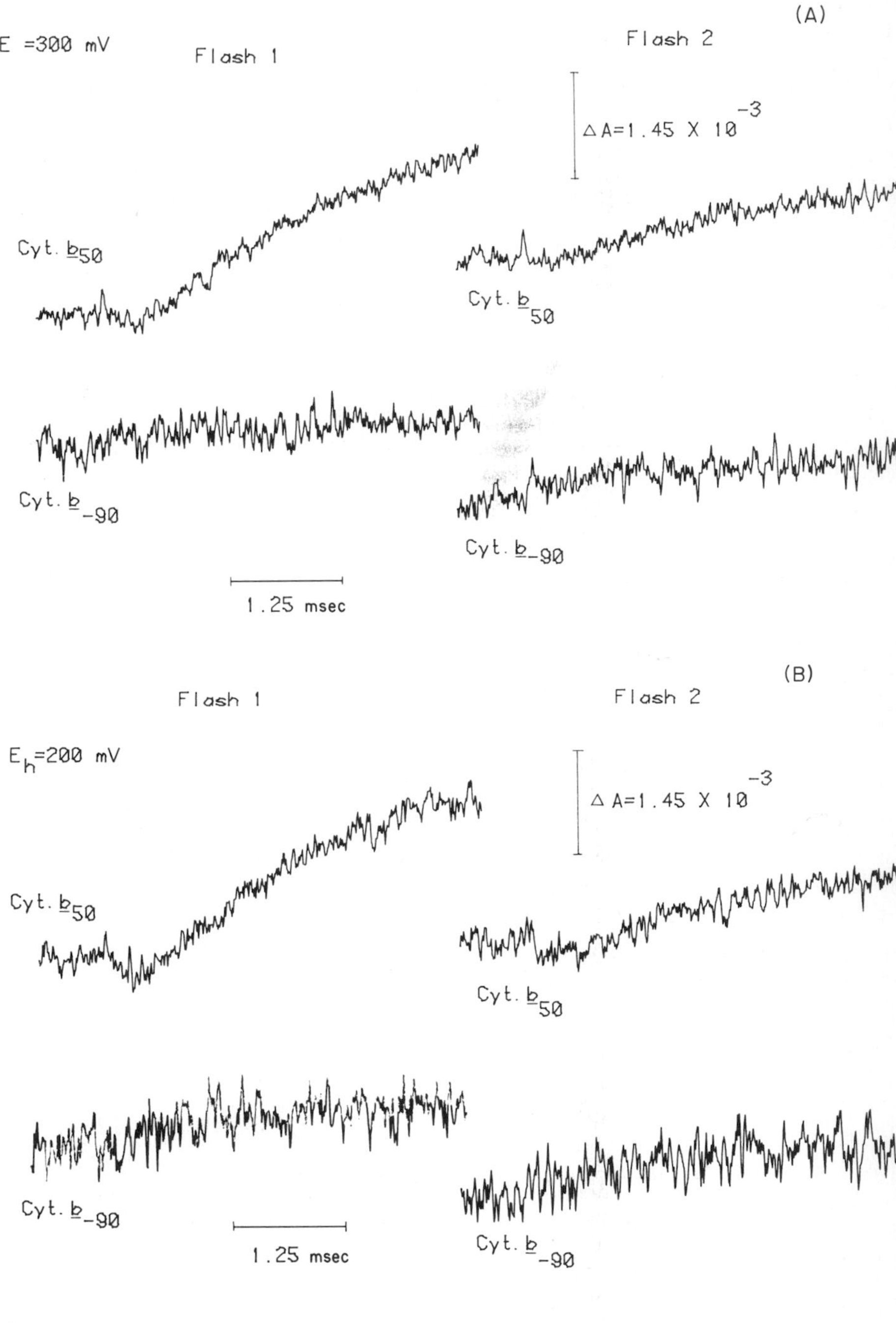
(A)
E =300 mV
Flash 1
Flash 2
ΔA=1.45 X 10^-3
Cyt. b50
Cyt. b50
Cyt. b-90
Cyt. b-90
1.25 msec
(B)
Flash 1
Flash 2
Eh=200 mV
ΔA=1.45 X 10^-3
Cyt. b50
Cyt. b50
Cyt. b-90
Cyt. b-90
1.25 msec

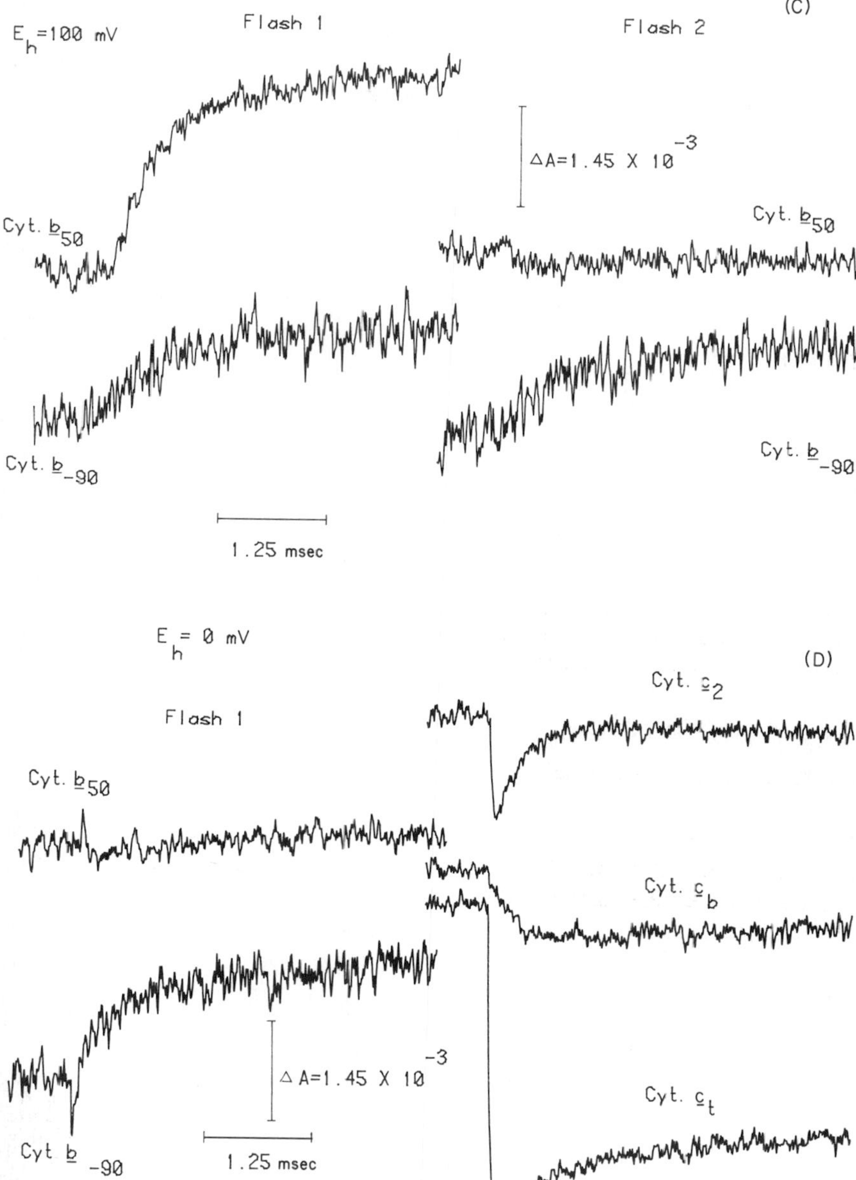

Fig. 5. Kinetics of cytochrome *b* changes in the presence of antimycin at several potentials. Conditions were as for Fig. 2. Cytochrome *b*-50 was measured at 560–570 nm; cytochrome *b*-(−90) was measured at 566–555 nm, after subtraction of changes due to cytochrome c_2 and cytochrome c_b measured as in Fig. 2, using normalization parameters from spectra like those in Fig. 1. All changes were corrected for reaction center. Cytochrome *b* changes at E_h ~300 mV (A), ~200 mV (B), and ~100 mV (C) for each of two flashes. Cytochromes *b* and *c* change at E_h ~0 mV on flash number one (D).

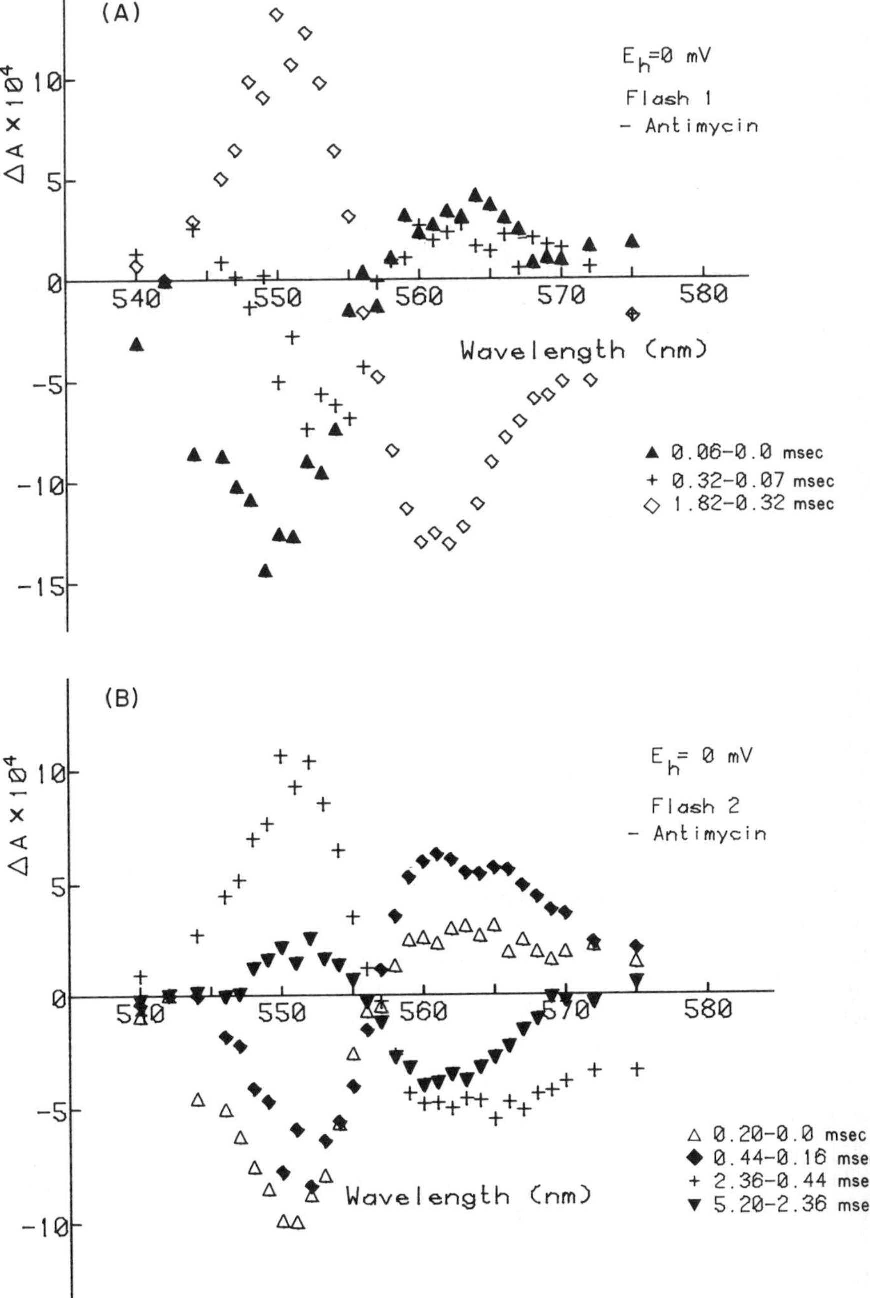
(A)
E_h=0 mV
Flash 1
- Antimycin
ΔA × 10^4
Wavelength (nm)
0.06-0.0 msec
0.32-0.07 msec
1.82-0.32 msec
(B)
E_h= 0 mV
Flash 2
- Antimycin
ΔA × 10^4
Wavelength (nm)
0.20-0.0 msec
0.44-0.16 msec
2.36-0.44 msec
5.20-2.36 msec

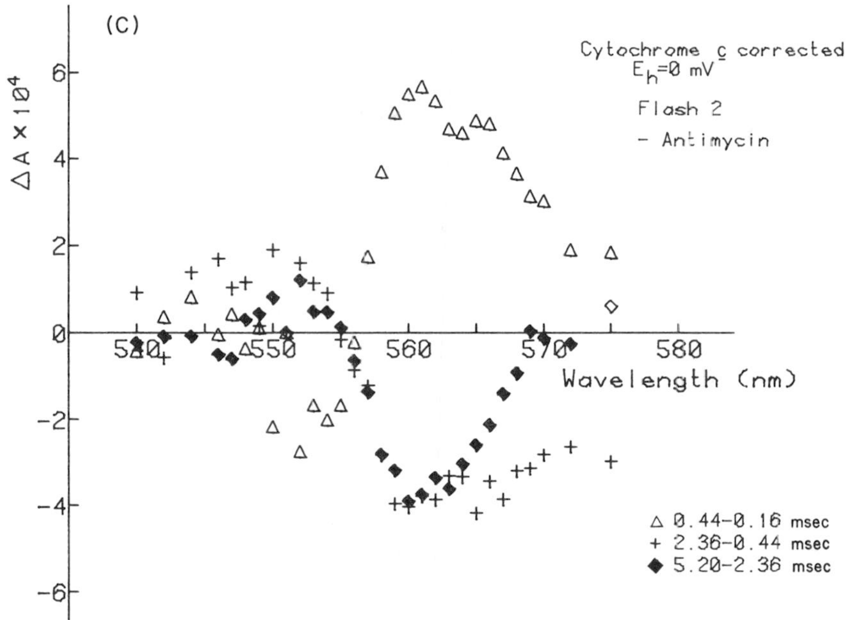

Fig. 6. Time-resolved spectra at $E_h \sim 0$ mV in the absence of antimycin. Conditions as for Fig. 2 but without antimycin. Traces (average of 16; first 200/500 points represent 2 msec with filter at 10 μsec RC, last 300/500 points represent 12 msec, with filter at 50 μsec; flashes were 8.33 msec apart, flash groups were 15 sec apart) were accumulated at the wavelengths shown and stored for analysis. The spectra show changes occurring over the times indicated, after correction for reaction center. Spectra were additionally corrected for cytochrome *c* where indicated.

Because of the large number of overlapping absorbance changes in this part of the spectrum and the broad double band of cytochrome *b*-(−90) it is difficult to resolve the spectra from those of cytochrome *b*-50, and those due to electron transfer from cytochrome c_b to cytochrome c_2. However, when cytochrome *b*-50 is reduced before excitation (either by a preilluminating flash, or by redox poising), the contribution of cytochrome *b*-(−90) can be clearly seen in time-resolved spectra (Figs. 3 and 4). By selection of a wavelength pair at which contributions due to cytochrome *b*-50 compensate, and by successive subtraction from kinetic traces of the changes due to the reaction center and the two separately resolved *c*-type cytochromes, we were able to separate the kinetic changes of cytochrome *b*-(−90) from the other known components. Although kinetic traces are rather noisy (Fig. 5), they show reduction of cytochrome *b*-(−90) on the first and second of two flashes, with a $t_{1/2} \sim 550$ μsec at $E_h \sim 100$ mV, and with no obvious delay before onset of reduction. For comparison, Fig. 5 also shows the kinetics of reduction of cytochrome *b*-50 under the same conditions (*9, 26–28*). The traces show a lag of ~200 μsec and a $t_{1/2}$ of ~500 μsec for the reduction that follows the lag. Figure 5 also shows similar kinetic traces at E_h values of zero and 200 mV. From these it can be seen that reduction of cytochrome *b*-(–90) is apparently slower and less extensive at the higher E_h, and faster ($t_{1/2}$ ~300 μsec) and more extensive at E_h ~0 mV, where cytochrome *b*-50 is largely chemically reduced before the flash.

The observation that cytochrome *b*-(−90) undergoes reduction at a rate which is rapid compared to the rate-limiting step of the cycle makes it seem probable that it is a component of the functional chain. In the absence of antimycin, kinetic traces at wavelengths appropriate for cytochrome *b*-(−90) show a transient reduction followed by oxidation, but the changes are small, and the spectra are not yet resolved sufficiently to identify cytochrome *b*-(−90) unambiguously.

A problem which immediately arises when considering this new information is the relation of cytochrome *b*-(−90) to the two-electron gate of

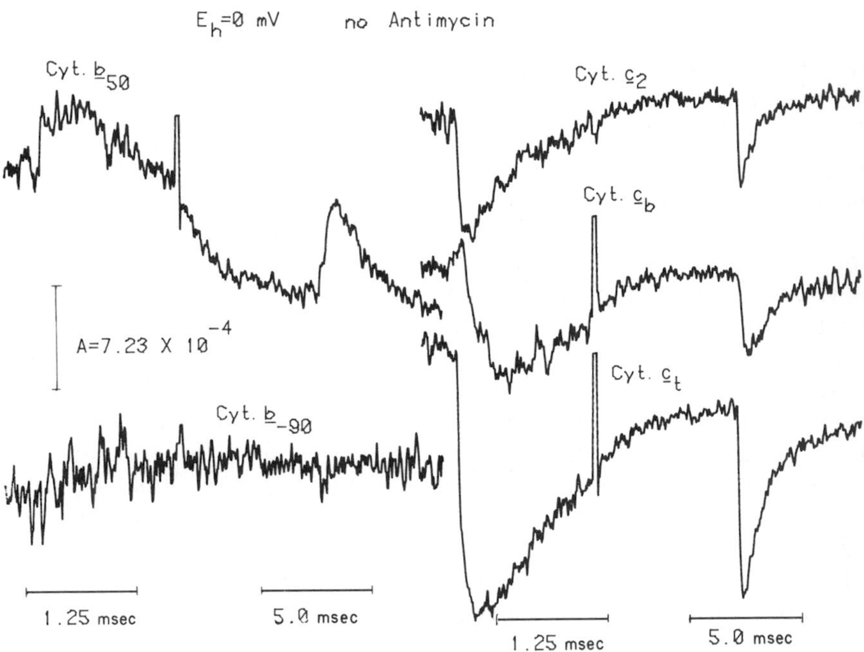

Fig. 7. Kinetics of cytochrome changes in the absence of antimycin at E_h ~0 mV. Conditions as for Fig. 6. Traces were corrected as in Figs. 2 and 5.

UQ_{II}. The $E_{m,7}$ values for the one-electron couples of UQ_{II}($E_{m,8}$ +40 mV and −40 mV for couples UQ/UQH· and UQH·/UQH_2, respectively), are too high for either of the reduced forms to act as a donor to cytochrome *b*-(−90). In collaboration with Dr. Richard Cogdell preliminary experiments using a double flash technique show a value of $E_{m,7}$ ~40 mV for the secondary acceptor (presumed to be the couple UQH·/UQH_2) that varies by −60 mV/pH unit up to a p*K* between 9 and 9.5. These values would give the couple UQH·/UQH^- a mid-potential of between −80 and −110, sufficient to reduce cytochrome *b*-(−90) to (b^-(−90H^+). However, if the low-potential couple is the reductant for cytochrome *b*-(−90), the question of the nature of the reductant for cytochrome *b*-50 still remains. In experiments in collaboration with Baccarini-Melandri, Rutherford, and Gabellini (*28a*) we found that chromatophores extracted so as to lose UQ_Z but not UQ_{II} still show a binary pattern of cytochrome *b*-50 reduction at high redox potential, suggesting that UQ_Z is not necessarily involved in the reduction.

Possible Reaction Pathways

The results presented here and in previous publications (*1–9*) show that the ubiquinone–cytochrome c_2 oxidoreductase contains the following components involved in electron transfer in the millisecond time scale: cytochrome *b*-(−90), cytochrome *b*-50, cytochrome c_b, RFe–S, and UQ_Z. This complement of components may be compared with those of the mitochondrial Complex III; cytochrome *b*-566, cytochrome b-562, cytochrome c_1, RFe–S, and UQ. This similarity suggests that the two enzymic complexes are essentially the same. The similarity extends to a comparison of redox mid-potential, and stoichiometries, in so far as these can be determined in the bacterial system, and also to the sensitivity to inhibition by antimycin and UHDBT (Table I). The similarity is further supported by the observation by Dutton *et al.* (*29*) that when mitochondrial Complex III is oxidized and reduced photochemically in a system containing the complex, cytochrome *c*, and reaction centers in solution, the flash-induced kinetics of components in the complex are similar to those observed for the analogous components in chromatophores. The similarity between the two systems has obvious evolutionary implications. More important in the present context is the probability that the similarity extends to mechanism. The kinetics of electron transport are more readily studied in the chromatophore system, and conclusions as to mechanism and reaction pathways drawn from such studies will likely be relevant to discussions of such problems in the mitochondrial system. With this in

TABLE I

Rps. sphaeroides Complex				Mitochondrial Complex III[A]			
Component	$E_{m,7}$ (mV)	λ max (nm)	Amount mole/mole RC	Component	$E_{m,7}$ (mV)	λ max (nm)	Amount (approximate mole/mole cytochrome c_1)
Cytochrome *b*-(−90)	−90	566 559	[B]	Cytochrome *b*-566	−40	566 558	1
Cytochrome-b_{50}	50	560.5	[C] 0.5–0.6	Cytochrome *b*-562	40	562	1
RFe–S	280	—	0.5–0.7	ReFe–S	280	—	1
Cytochrome c_b	260	552	0.5–0.7	Cytochrome c_1	220	551.5	1
Antimycin binding	—	—	0.5–0.7	Antimycin binding	—	—	1
Soluble cytochrome c_2	350	550	0.5–1	Soluble cytochrome *c*	270 (235)[D]	550	1

[A] Values from Dutton and Wilson (15) and literature.

[B] The extinction coefficient for cytochrome *b*-(−90) is not known. If the stoichiometry is taken to be 1 cytochrome *b*-(−90)/cytochrome c_b, then $\epsilon^{mM}_{566-555}$ is ≃ 5.7 mM^{-1} cm^{-1}.

[C] Assuming $\epsilon^{mM}_{560-570} \simeq 20$ mM^{-1} cm^{-1}.

[D] The $E_{m,7}$ value for cytochrome *c* in situ is lower than that in solution.

mind, we will discuss our results in terms of current models for electron transfer in both complexes.

Arguments for linear or Q-cycle models have been presented at length elsewhere (*1*, *2*, *9*, *29–31*). Although much of this discussion is obviously still relevant, especially for mitochondrial systems, our present results suggest that none of the simple models previously proposed accounts adequately for all our observations. The kinetic arguments we originally put forward against a simple Q cycle still stand (*27*), although details of reaction rates for new components need to be included. Nevertheless, a number of kinetic phenomena are more easily accounted for in terms of a Q cycle, than by the simple linear scheme we originally proposed (*1*, *27*). Anyone who has dabbled in potential reaction pathways will have discovered the numerous permutations of the components that can be accommodated in the basic linear and Q-cycle models.

To limit the present discussion we will make some initial limiting assumptions: (1) The mechanism of electron transfer within the complex defines a set of reaction pathways that are specific with respect to reactants, and identifiable by virtue of their kinetic parameters (rapid kinetics indicating preferred reaction paths). This assumption reflects the fact that the complexes are enzymes, and have normal biochemical characteristics. (2) The concentrations of reactants in the dark are determined by the ambient redox potential, mid-point potential, and stoichiometry of each component (i.e., the components are at equilibrium with the redox mediators). (3) The reaction pathways (as determined by rate constants), and redox characteristics of the components are not modified by either dynamic effects (conformational changes, local charge effects during turnover, etc.), or addition of inhibitors (except, of course for effects at the specific binding site).

The first of these assumptions may be expected to hold under all circumstances; the other two are probably oversimplifications, and in some circumstances are already known to be contravened.

Within the constraints set by these assumptions we can consider the experimental results. In the context of assumption (1) we may regard the following as probable: (a) the reaction pathway for donation of electrons to P-870^+ from the chromatophore complex involves soluble cytochrome c_2 as an electron shuttle between the complex and the reaction center. The characteristics of this pathway are as in Scheme I; we cannot yet distinguish between a parallel and serial electron transfer from RFe–S and cytochrome c_b to cytochrome c_2; (b) reduction of cytochrome *b*-50 depends on formation of $UQ_{II}H_2$ on flash excitation; (c) in the presence of antimycin and when UQ_Z is chemically reduced, the reduction of cytochrome *b*-50 and cytochrome *b*-(−90) occurs much more rapidly than electron transfer through the antimycin block to the high potential components. The reduction of cytochrome *b*-50 shows a pronounced lag; that of cytochrome *b*-(−90) shows no appreciable lag. In the absence of antimycin, the initial rate of reduction of cytochrome *b*-50 is the same, but the cytochrome is reoxidized. Antimycin-sensitive reoxidation occurs with kinetics similar to those of antimycin-sensitive reduction of the higher potential components of the chain, and slow phase of the carotenoid change (*9*, *27*, *32*). (d) in the absence of antimycin, and when cytochrome *b*-50 is chemically reduced before flash excitation, rereduction of the photooxidized high potential components is unmodified, and cytochrome *b*-50 is oxidized with similar kinetics (see Fig. 7). Rereduction of cytochrome *b*-50^+ occurs much more slowly. (e) from the last two points we may conclude that (1) reduction of cytochrome *b*-50 requires some other event to occur before it can proceed (as judged by the delay before reduction); (2) reduction of cytochrome *b*-(−90) precedes that of cytochrome *b*-50 with a similar half-time for the reductive process; and (3) oxidation of cytochrome *b*-50 is kinetically coupled to electron transfer through the antimycin site.

These kinetic features seem to indicate a linear pathway for electron flow, rather than a Q cycle. However, a more complicated picture emerges when we consider other phenomena, in particular the effects of UHDBT inhibition on the kinetics of electron transfer. As we previously discussed (*5*–*7*, *9*), under these conditions and over a wide range of ambient redox potentials, the system behaves as if all the reactants of the complex, apart from cytochrome c_b, were kinetically frozen. A similar pattern of behavior is seen, in the absence of UHDBT, in a mutant of *Rps. capsulata* (R 126), which lacks the UHDBT stimulation of cytochrome *c* oxidation and which might lack a functional RFe–S center; in a mutant of *Rps. capsulata* (MT 113), which lacks cytochrome *c*; and in membrane vesicles prepared from spheroplasts, which lack cytochrome c_2 (*9*, *33*). These results all suggest that the rapid reduction of cytochrome *b*-50 occurs only if some component of the complex (presumably RFe–S) is allowed to be oxidized. A possible reason for this apparent coupling of redox reactions is provided by a Q-cycle type mechanism, in which electron flow from UQH_2 to cytochrome *c* is envisaged as leading to generation of UQH·, which acts as reductant to cytochrome *b*-50. However, a careful analysis of kinetic data for the reduction of cytochrome *b*-50 and the rereduction of cytochrome *c* and P-870^+ in the presence of antimycin, shows that the electrons leaving the complex are almost entirely accounted for by the UHDBT-sensitive oxidation of RFe–S (Table II). For a Q-cycle mechanism to be credible, an additional equivalent of reductant, derived from oxidation of UQH_2, should be available at redox potentials below the $E_{m,7}$

TABLE II

Chromatophore source	E_h (mV)	Reduction cytochrome *b*-50 (mole/mole RC)	UHDBT-sensitive, antimycin-insensitive reduction (mole/mole RC)		
			cytochrome *c*	RC	Total
Rps. sphaeroides strain Ga	100	0.53	0.54	0.16	0.7
	190	0.53	0.52	0.13	0.65
Rps. capsulata strain N22	105	0.455	0.25	0.28	0.53
	180	0.215	0.22	0.2	0.42
strain Ala pho$^+$	100	0.24	0.24	0.13	0.37
	240	0.38	0.22	0.07	0.29

[A] Chromatophores prepared from the strains indicated were suspended in 100 m*M* KCl, 50 m*M* MOPS, pH 7 in an anaerobic cuvette with 4 μM antimycin, ionophores, and mediators, and the E_h adjusted to the value shown (±5 mV). For each experiment, the kinetics and extent of cytochrome c_t, cytochrome *b*-50, and P-870 (RC) were measured following one flash. UHDBT was then added, and the kinetics remeasured. A value of $\epsilon_{560-570} = 20\ mM^{-1}\ cm^{-1}$ was assumed for cytochrome *b*-50. The total reaction center concentration was measured following multiple flashes. The right-hand column shows the number of electrons leaving the complex when only RFe–S is reduced (at 240 > E_h > 180 mV) or when RFe–S and UQ_Z are both reduced (at $E_h \sim 100$ mV).

for the couple UQ_Z/UQ_ZH_2 (+150 mV), where rapid reduction of cytochrome *b*-50 is observed (*9, 26*). Although a small increase in the reduction of cytochrome *c* and P-870$^+$ is seen under these conditions, it is never enough to account for the reduction of cytochrome *b*-50. Furthermore, our recent results indicate that an approximately equal amount of cytochrome *b*-(−90) is reduced under these conditions, making the shortfall of equivalents leaving the complex even more difficult to understand in terms of a Q cycle. However, the apparent dependence of cytochrome *b*-50 reduction on redox turnover of RFe–S is not adequately accounted for by any simple linear scheme.

Arguments for Q-cycle or linear mechanisms can be tested against other phenomena with equally ambiguous results. Q-cycle type mechanisms account well for the following observations: (1) the rate of reduction of cytochrome *b*-50 accelerates tenfold over the range of E_h in which UQ_Z becomes reduced to UQ_ZH_2 ($E_{m,7} \simeq 150$ mV) (*1, 2, 9, 26, 33*). Over a similar range ($E_{m,7} \simeq 120$ mV, Ref. *24*), UQ_{II} should become reduced to $UQ_{II}H\cdot$, but our titrations are not sufficiently well characterised to allow discrimination between these two components. If reduction of UQ_Z is necessary for the effect, then a Q cycle is a plausible mechanism; if UQ_{II}, then a linear model is preferred. In any case, the effect is seen in the presence of antimycin, where the kinetic arguments against a Q cycle apply.

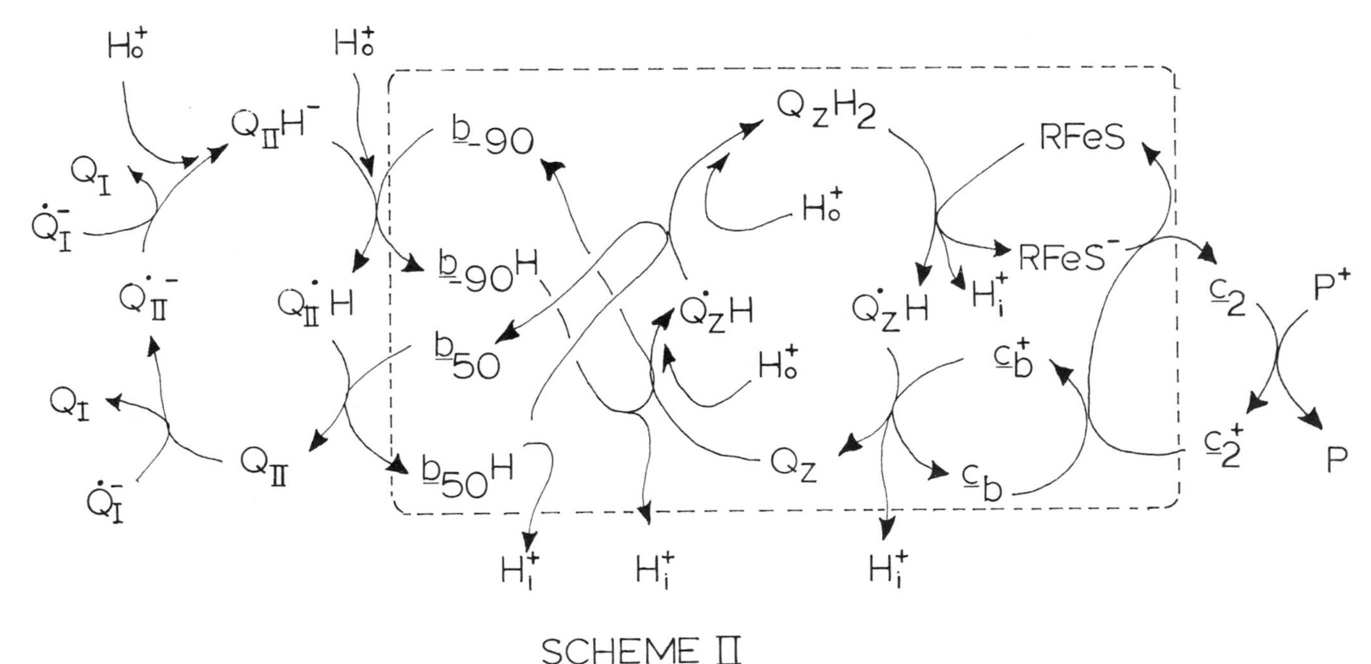
H_o^+
H_o^+
$Q_{II}H^-$
Q_I
$\dot{Q}_I^-$
b_{-90}
Q_ZH_2
RFeS
H_o^+
$b_{-90}H$
RFeS$^-$
H_i^+
$\dot{Q}_{II}^-$
$\dot{Q}_{II}H$
$\dot{Q}_ZH$
$\dot{Q}_ZH$
c_2
P^+
b_{50}
H_o^+
c_b^+
Q_I
Q_{II}
$b_{50}H$
Q_Z
c_b
c_2^+
P
$\dot{Q}_I^-$
H_i^+
H_i^+
H_i^+

SCHEME II

(2) Oxidant-induced reduction of cytochrome *b*-50 is seen much more clearly in mitochondria and isolated enzymes than in chromatophores (*10*, *11*, *34–36*). The effect has been taken as a clear indication that a Q-cycle mechanism operates, or some other similar concerted redox reaction involving oxidation of a two-electron donor by a one-electron process to leave a one-electron donor able to reduce cytochrome *b*-50 (*11*, *34–36*). Such an effect cannot be accounted for by a simple linear scheme.

It would appear that within the context of the limiting assumptions, neither a simple linear scheme nor a Q cycle can account for the observed phenomena. We would like to suggest a possible reaction scheme that resolves many of the anomalies above, but contravenes one of the limiting assumptions. We suggest that the redox potential of the cytochrome *b*-50 couple may be raised to a more positive value by interaction with the oxidized RFe–S; when RFe–S becomes oxidized, cytochrome *b*-50 becomes easier to reduce. With this assumption the mechanism of Scheme II seems quite attractive. It has the following features: (1) the complex is regarded as transferring electrons two at a time from the donor quinone pool to the cytochrome c_2 shuttle (this is achieved by a pair of low-potential cytochromes (*b*-50 and *b*-(−90) that provide electrons to the two halfreduction steps of bound UQ_Z, and a pair of high potential components (cytochrome c_b and RFe–S) that accept the electrons from the half-reduction steps.); (2) the natural redox gradients through the scheme are suitably downhill, except for the reduction of cytochrome *b*-50 (the ad hoc assumption introduced above rectifies this difficulty, and also provides an explanation for the inhibition of cytochrome *b*-50 reduction by UHDBT with the related effects discussed above, and oxidant-induced reduction of cytochrome *b*, with $UQ_{II}H\cdot$ providing the electrons); (3) the reaction scheme accounts for the known components of the complex and their stoichiometries.

ACKNOWLEDGMENTS

This work was supported by a grant from NIH-GM-2630501. We would like to thank Colin Wraight and Bill Rutherford for stimulating discussions, and Ms. Lee-Ann Oimoen for expert technical assistance.

REFERENCES

1. Crofts, A. R., and Wood, P. M. (1978). *Curr. Top. Bioenerg.* **7,** 175–244.

2. Dutton, P. L., and Prince, R. C. (1978). *In* "The Photosynthetic Bacteria" (R. K. Clayton and W. R. Sistrom, eds.), pp. 525–570. Academic Press, New York.

3. Bowyer, J. R., Tierney, G. V., and Crofts, A. R. (1979). *FEBS Lett.* **101,** 201–206.
4. Bowyer, J. R., Tierney, G. V., and Crofts, A. R. (1979). *FEBS Lett.* **101,** 207–212.
5. Bowyer, J. R., and Crofts, A. R. (1978). *In* "Frontiers of Biological Energetics" (P. L. Dutton, J. S. Leigh, and A. Scarpa, eds.), Vol. 1, pp. 326–333. Academic Press, New York.
6. Crofts, A. R. (1979). *In* "Light-Induced Charge Separation in Biology and Chemistry" (H. Gerischer and J. Katz, eds.), pp. 389–407. Dahlem Konferenzen, Berlin.
7. Bowyer, J. R. Dutton, P. L., Prince, R. C., and Crofts, A. R. (1980). *Biochim. Biophys. Acta* **592,** 445–460.
8. Bowyer, J. R., Meinhardt, S. W., Tierney, G. V., and Crofts, A. R. (1981). *Biochim. Biophys. Acta* ■■
9. Bowyer, J. R., and Crofts, A. R. (1981). *Biochim. Biophys. Acta* **636,** 218–233.
10. Trumpower, B. L., and Haggerty, J. G. (1980). *J. Bioenerg. Biomembr.* **12,** 151–163.
11. Bowyer, J. R., and Trumpower, B. L. (1980). *FEBS Lett.* **115,** 171–174.
12. Prince, R. C., and Dutton, P. L. (1977). *Biochim. Biophys. Acta* **459,** 573–577.
13. Wood, P. M. (1980). *Biochem. J.* **189,** 385–391.
14. Wood, P. M. (1980). *Biochem. J.* **192,** 761–764.
15. Dutton, P. L., and Wilson, D. F. (1974). *Biochim. Biophys. Acta* **346,** 165–212.
16. Wraight, C. A. (1979). *Biochim. Biophys. Acta* **548,** 309–327.
17. Wraight, C. A. (1979). *Photochem. Photobiol.* **30,** 767–776.
18. Wraight, C. A., and Stein, R. R. (1980). *FEBS Lett.* **113,** 273–278.
19. Wraight, C. A. (1977). *Biochim. Biophys. Acta* **459,** 525–531.
20. Vermeglio, A. (1977). *Biochim. Biophys. Acta* **459,** 516–524.
21. De Grooth, B. G., van Grondelle, R., Romijn, J. C., and Pulles, M. P. J. (1978). *Biochim. Biophys. Acta* **503,** 480–490.
22. Barouch, Y., and Clayton, R. K. (1977). *Biochim. Biophys. Acta* **462,** 785–788.
23. Fowler, C. F. (1976). *Symp. Primary Electron Transp. Energy Transduction Photosynth. Bacteria, Brussels,* Abstracts WB9.
24. Rutherford, A. W., and Evans, M. C. W. (1980). *FEBS Lett.* **110,** 257–261.
25. Vermeglio, A., Martinet, T., and Clayton, R. K. (1980). *Proc. Natl. Acad. Sci. U.S.A.* **77,** 1809–1813.
25a. Meinhardt, S. W., and Crofts, A. R. (1981)., *Biochim. Biophys. Acta,* submitted.
26. Evans, E. H., and Crofts, A. R. (1974). *Biochim. Biophys. Acta* **357,** 89–102.
27. Crofts, A. R., Crowther, D., and Tierney, G. V. (1975). *In* "Electron Transfer Chains and Oxidative Phosphorylation" (E. Quagliariello, S. Papa, E. Palmieri, E. C. Slater, and N. Siliprandi, eds.), pp. 233–241. North-Holland Publ., Amsterdam.
28. Prince, R. C., and Dutton, P. L. (1975). *Biochim. Biophys. Acta* **387,** 607–613.
28a. Baccarini-Melandri, A., Gabellini, N., Melandri, B. A., Jones, K. R., Rutherford, A. W., Crofts, A. R.,and Hurt, E. (1982). *Arch. Biochem Biophys.,* in press.
29. Dutton, P. L., Mueller, P., Packham, N. K., and Tiede, D. M. (1980). *Fed. Proc., Fed. Am. Soc. Exp. Biol.* **39,** 1802.
30. Mitchell, P. (1976). *J. Theor. Biol.* **62,** 327–367.
31. Mitchell, P. (1980). *Ann. N.Y. Acad. Sci.* **341,** 564–584.
32. Crofts, A. R., Crowther, D., Bowyer, J. R., and Tierney, G. V. (1977). *In* "Structure and Function of Energy-Transducing Membranes" (K. van Dam, and B. F. Van Gelder, eds.), pp. 139–155. Elsevier/North-Holland, Amsterdam.
33. Bowyer, J. R. (1979). Ph.D. Thesis, Univ. of Bristol.
34. Rieske, J. S. (1971). *Arch. Biochem. Biophys.* **145,** 179–193.
35. Wikstrom, M. K. F., and Berden, J. A. (1972). *Biochim. Biophys. Acta* **283,** 403–420.
36. Dutton, P. L., and Prince, R. C. (1978). *FEBS Lett.* **91,** 15–20.

Cycles and Q Cycles in Plant Photosynthesis

3

DAVID CROWTHER
GEOFFREY HIND

INTRODUCTION

The classical description of electron transfer in chloroplasts (*1*) linked photosystem II to photosystem I with a simple hydrogen-carrying plastoquinone shuttle that transferred one H^+ per electron traveling through the chain. An additional H^+/e^-, liberated inside the thylakoid with the splitting of water, provided an overall H^+/e^- stoichiometry of 2. Cyclic electron transfer around photosystem I via cytochrome *b*-563 was considered to reenter the main chain at plastoquinone, translocating one H^+ per electron completing the cycle.

Such a scheme had to be modified, however, following the discovery of a further charge translocation, observed following a single turnover flash, as a slow rise in the electric field indicating bandshift termed P518. According to the earlier scheme for electron transfer in chloroplasts, charge separation across the thylakoid membrane occurred only in the reaction centers of photosystems I and II, giving a rapid (submicrosecond) appearance of the P518 bandshift (*2*). In the last few years, however, several groups have shown a slow (millisecond) additional component of the P518 rise following single turnover flash illumination of algae (*3*) or chloroplasts (*4–6*). The maximum amplitude of this slow component is equal to that of the fast phase generated by electron transfer in photosystem I reaction centers (*7, 8*), and the appearance of the slow component is dependent on the redox poise of the electron-transfer chain (*7–10*). If this field-indicating slow rise in P518 ($P518_s$) reflects an electron crossing the thylakoid

Function of Quinones in Energy Conserving Systems

ISBN 0-12-701280-X

membrane from inside to outside in a dark step, then additional complexities must exist in the electron-transport pathway between the photosystems.

A somewhat analogous system was reported in several species of photosynthetic bacteria (*11*); in these organisms, cyclic electron flow provides ATP while NADH formation apparently occurs by energy linked reverse electron flow. They show a field-indicating bandshift analogous to P518, with one or more kinetic components reflecting electron transfer through the reaction center, and a slower rise of equal magnitude having a rise time of several milliseconds. The magnitude and rate of rise of this slow bandshift are increased by prior reduction of a special quinone molecule ("Z" or "Q_z"), (*12*, *13*) which rereduces cytochrome c_2 via the Rieske Fe–S center (*14*) and a bound cytochrome *c* (*15*). Electron flow through a cytochrome *b* of $E_{m,7} \sim +50$ mV is also greatly increased by prior reduction of Z (*12*, *13*).

The similarity between this section of the bacterial electron-transfer chain and the mitochondrial cytochrome *b*-c_1 region has been noted; it now appears that the chloroplast conforms to a similar pattern.

EXPERIMENTAL RESULTS

If electron transfer in chloroplasts is inhibited with high concentrations of DCMU[1], then cyclic electron flow around photosystem I may be induced by addition of a reductant. In aerobic samples, millimolar amounts of dithionite produce a suitable redox poise, with the effect dependent on substantial permeability barriers at the chloroplast envelope and the thylakoid membrane. When such a poising technique is used, flash-induced absorbance changes like those of Figs. 1–3 may be observed. Figure 1 shows the P518 response in untreated (left) and DCMU/dithionite treated (right) chloroplasts; the two transmembrane electrogenic steps in cyclic electron transfer around photosystem I are manifest in the DCMU/dithionite samples. Figure 2 shows separated, time-resolved spectra of absorbance changes in the cytochrome α-band region of chloroplasts similarly treated except that sufficient valinomycin was added to collapse P518 rapidly enough not to interfere with these measurements. The observed sequence of events following an actinic flash clearly was (A) cytochrome *f* oxidation, (B) cytochrome *b*-563 reduction, (C) cytochrome *f* rereduction, and (D) cytochrome *b*-563 reoxidation.

[1] Abbreviations: DCMU, 3(3,4-dichlorophenyl)-1,1-dimethylurea; DBMIB, 2,5-dibromo-3-methyl-6-isopropyl-*p*-benzoquinone

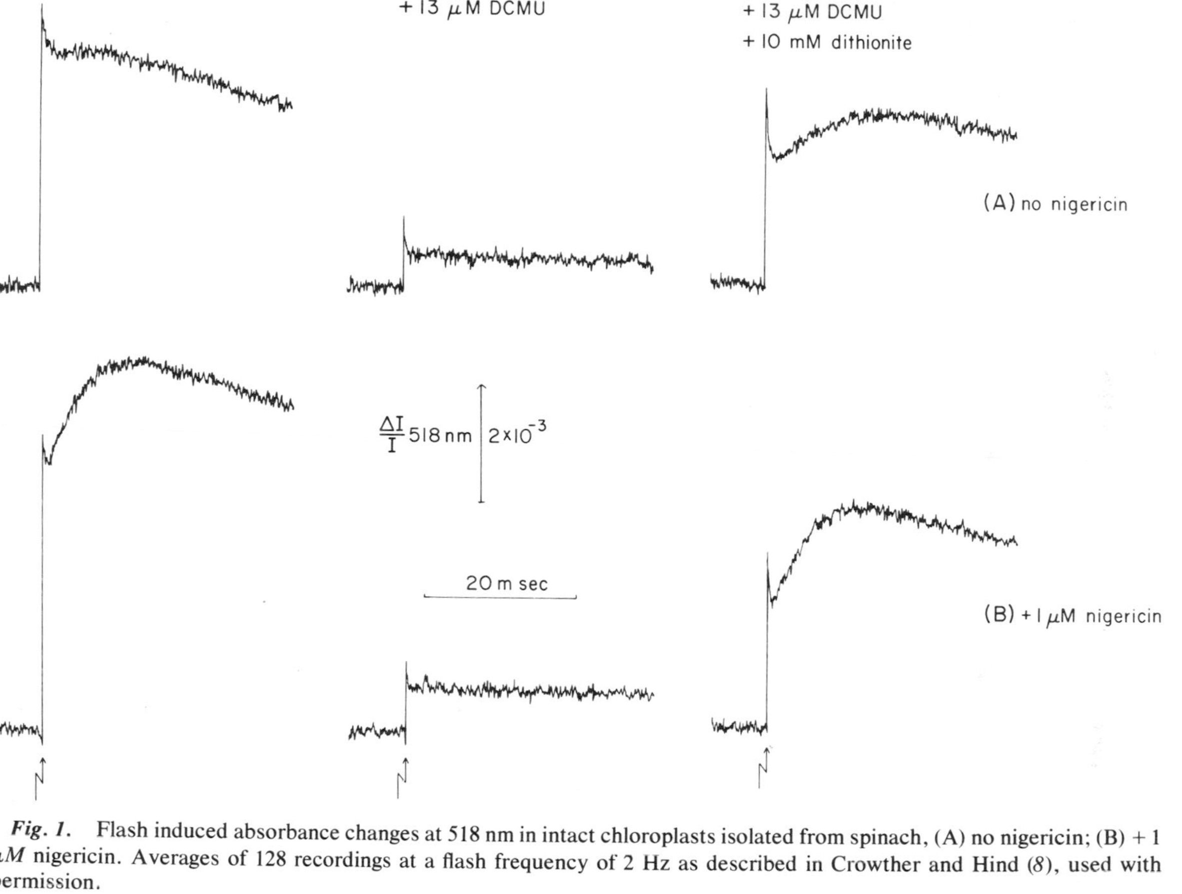

Fig. 1. Flash induced absorbance changes at 518 nm in intact chloroplasts isolated from spinach, (A) no nigericin; (B) + 1 *μM* nigericin. Averages of 128 recordings at a flash frequency of 2 Hz as described in Crowther and Hind (*8*), used with permission.

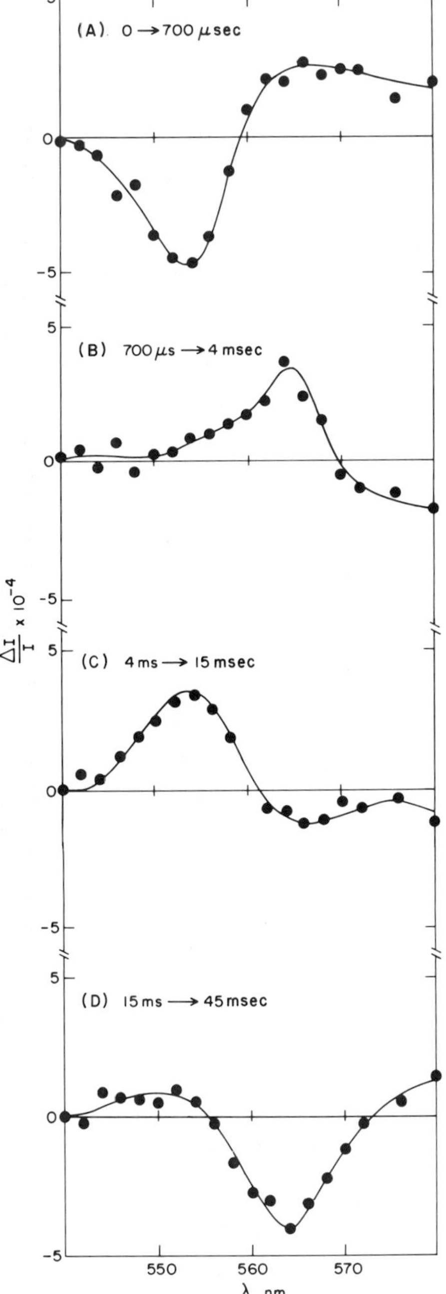

Fig. *2.* Time-resolved spectra through the cytochrome α-band region obtained during illumination of chloroplasts by repetitive (2 Hz) flashes. Time shown in (A–D) are intervals following the actinic flash. DCMU (13 μM), dithionite (10 mM), nigericin (1 μM), and valinomycin (6.7 μM) were included in the reaction medium. [For details, see Crowther and Hind (*8*), used with permission.]

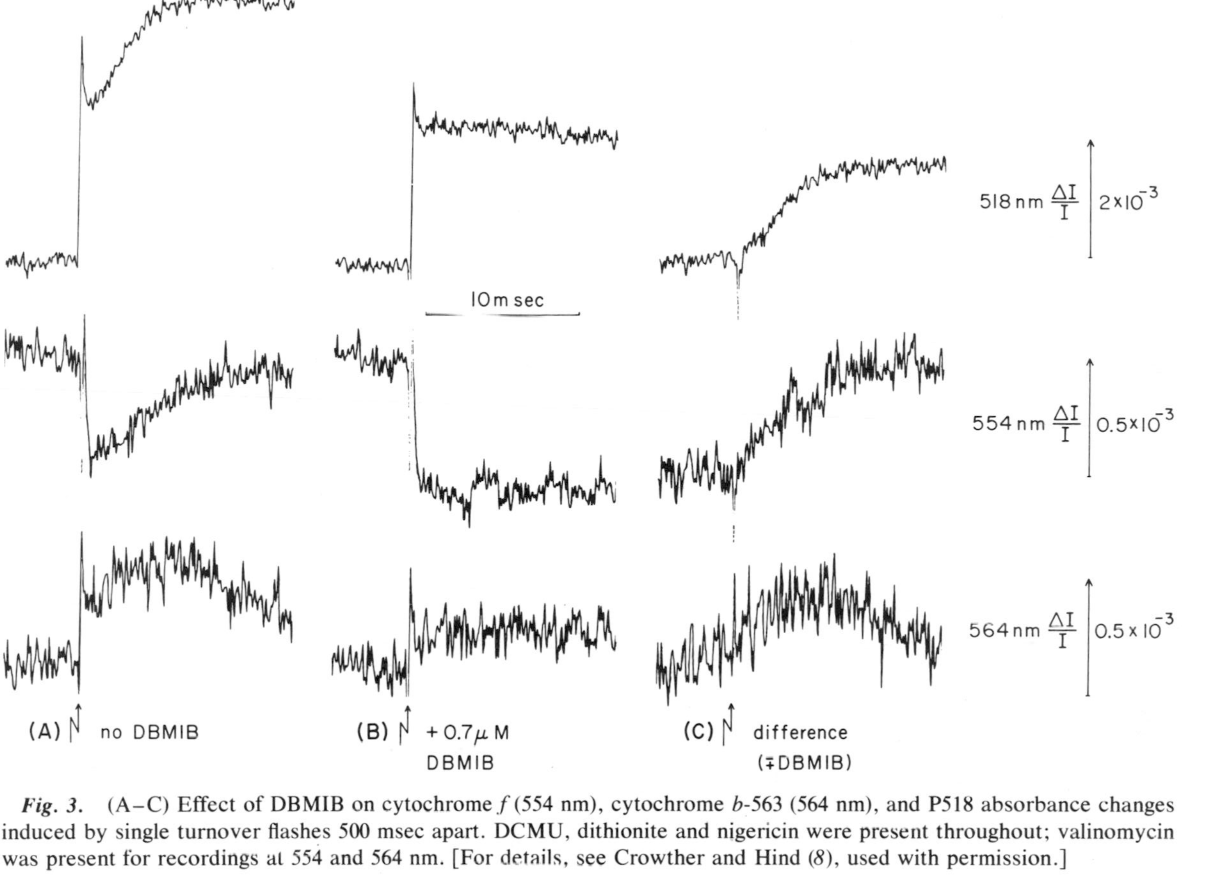

Fig. 3. (A–C) Effect of DBMIB on cytochrome *f* (554 nm), cytochrome *b*-563 (564 nm), and P518 absorbance changes induced by single turnover flashes 500 msec apart. DCMU, dithionite and nigericin were present throughout; valinomycin was present for recordings at 554 and 564 nm. [For details, see Crowther and Hind (*8*), used with permission.]

The quinone analog DBMIB is known to inhibit cytochrome *f* rereduction and generation of $P518_s$ in algae (*16*). In Fig. 3 the effects of DBMIB on cytochrome *f*, cytochrome *b*-563, and P518 kinetic changes in spinach chloroplasts are shown. Subtracted traces (with and without DBMIB) show the kinetics of $P518_s$ and cytochrome *f* rereduction are very similar, as seen for cytochrome c_2 rereduction and the slow electrochromic bandshift in *Rhodopseudomonas capsulata* (*12*). In *Chlorella* it has been claimed (*16*) that the rise of $P518_s$ is perhaps slightly slower than the rate of cytochrome *f* rereduction.

These results reinforce the view that the chloroplast electron-transfer scheme resembles that of the bacterial chromatophore in the ubiquinone/cytochrome *b*/cytochrome *c* segment. A difference is observed, though, in the kinetics and redox potential dependence of the slowest step in the cycle. In bacteria the slowest step apparently corresponds to a concerted electron movement involving cytochrome *b*-50 reoxidation, the slow electrogenic step and cytochrome c_2 rereduction, and all are activated by prior reduction of Q_z (*12*, *13*). In chloroplasts, however, the reoxidation of cytochrome *b*-563 is considerably slower than cytochrome *f* rereduction and $P518_s$ (Figs. 2 and 3), and the reoxidation is activated, along with completion of the cycle, at substantially lower potentials than are required to observe $P518_s$ and rereduction of cytochrome *f*. Evidence of this is displayed in Figs. 4 and 5 where, at controlled redox potentials in the presence of DCMU, the responses of P518 and cytochrome *b*-563 to a series of 8 flashes 250 or 500 msec apart were examined with relatively long dark intervals between flash groups. It is clear that while $P518_s$ may be seen on the first flash at potentials (around +98 mV at pH 7.6) where the plastoquinone pool ($E_{m^-7.6} \sim +44$ mV, $n = 2$ (*17*)) is mostly oxidized, it is greatly diminished on subsequent flashes. A similar conclusion was reached for *Chlorella,* where $P518_s$ appeared to titrate with and E_m about 40 mV above the E_m of the plastoquinone pool (*18*). The special quinone responsible for this effect in the chloroplast has been called U.

A similar pattern is seen when the plastoquinone pool is more than 50% reduced ($E_h \sim +36$ mV), showing that, on this time scale, the redox state of the plastoquinone pool does not substantially affect the reactions engendering the slow rise of P518. Full generation of $P518_s$ on all 8 flashes is not seen until potentials considerably below the E_m of the plastoquinone pool are reached; here the response is almost stable by about −64 mV (Fig. 4). The reoxidation of cytochrome *b*-563 shows a similar potential dependence (Fig. 5); efficient reoxidation is not seen until the ambient redox potential is well below the E_m of the plastiquinone pool.

Redox titrations of the appearance of $P518_s$ with 2 Hz repetitive flashes (Fig. 6) show that the increase in rate of the slowest step of the cycle

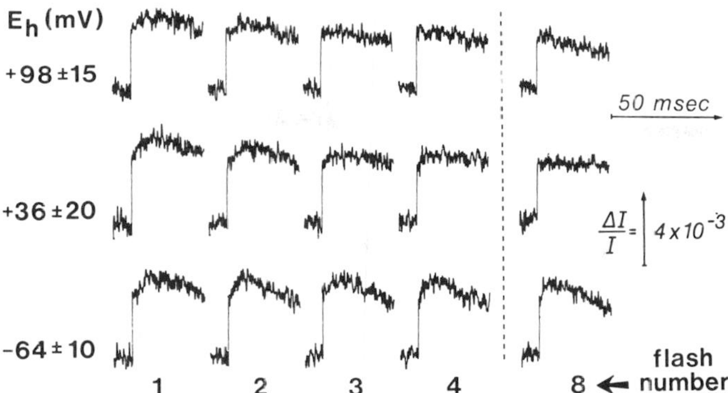

Fig. 4. Response of P518 to groups of flashes 250 msec apart following a 2 min dark period. Redox potential was poised anaerobically at the indicated values for each set. Averages of 16 groups. DCMU (15 μM) and nigericin (1 μM) were present throughout. pH of the medium was 7.6.

occurs with E_m around -55 mV ($n = 2$). This potential and n value for the component controlling the rate-limiting step may not be precise since this method effectively measured the amount of U (Q_z) rereduced 500 msec after a flash, with an equilibrium possibly disturbed from the measured potential by the action of the electron-transfer system itself. However, if

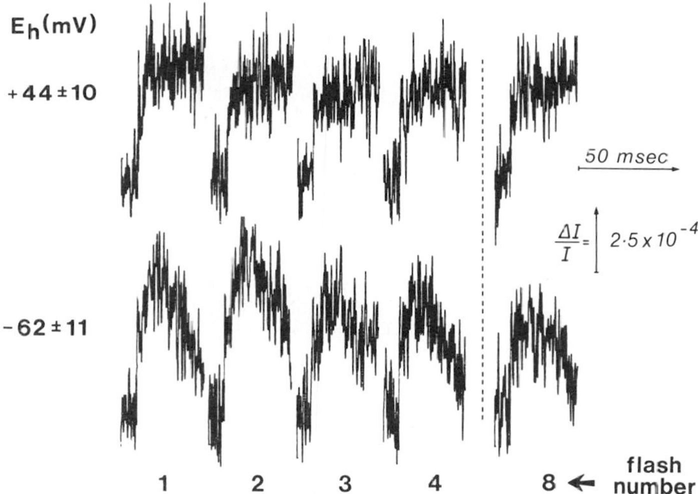

Fig. 5. Cytochrome *b*-563 response to groups of flashes 500 msec apart following 30 sec dark periods. DCMU (15 μM), nigericin (1 μM), and valinomycin (5 μM) were present throughout. Averages of 128 groups.

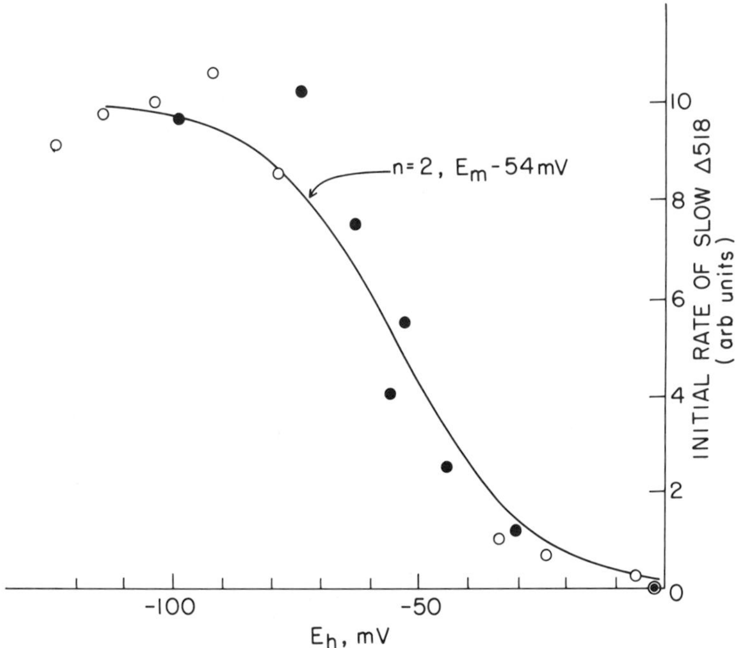

Fig. 6. Redox titration of the appearance of $P518_s$ in 2 Hz repetitive flashes. Details in Crowther and Hind (*8*), used with permission. (●), oxidative and (○), reductive titrations, respectively.

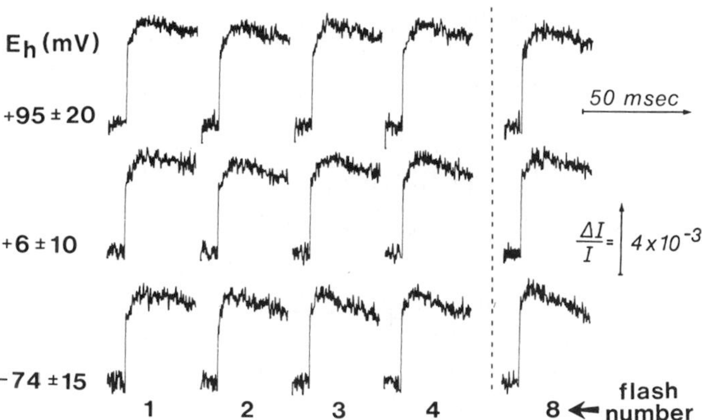

Fig. 7. Responses of P518 to flash groups with photosystem II active. Conditions as for Fig. 4 but without DCMU.

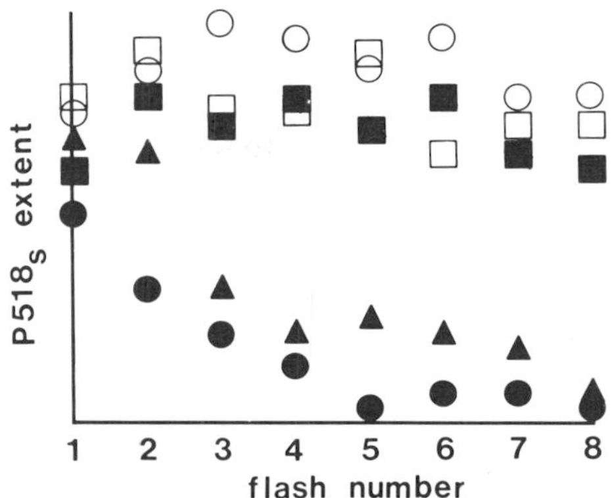

Fig. 8. Summary of P518$_S$ behavior from Figs. 4 and 7. With DCMU: (●) +98 mV, (▲) +36 mV, (■) −64 mV; without DCMU (○) +95 mV, (□) −74 mV. P518$_S$ extent was measured 12 msec after the flash.

the time required for rereduction of U (presumed to be the rate-limiting step of the cycle) were substantially greater than 500 msec at high potentials and substantially lower at the lowest potentials measured, then the E_m measured would be a good estimate of that required for efficient completion of the cycle. The results in Fig. 4 support this contention, showing the different E_h ranges required for observation of P518$_s$ on the first flash or on all 8 flashes, indicating completion of the cycle (rereduction of U).

If photosystem II is active (no DCMU present) during experiments, such as shown in Fig. 3, a slightly different pattern is seen in flash trains (Fig. 7); the loss of P518$_s$ with increasing flash number is not seen if U is reduced initially (+95 mV), showing that photosystem II may reduce U without intervention of the plastoquinone pool. This behavior is more clearly presented in Fig. 8, where the extents of P518$_s$ under the conditions of Figs. 4 and 7 are plotted against flash number.

DISCUSSION

While more work is needed to present a convincing mechanism for chloroplast electron flow, these results, and those of other workers to date (*3–10, 16, 18–20*) may most simply be described by a variant of the Q cycle proposed by Mitchell (*21*). Figure 9 shows a minimal scheme for such a mechanism. U is a special quinone (analogous to Q_z in chromatophores) which, on oxidation, can donate one electron to photosystem I

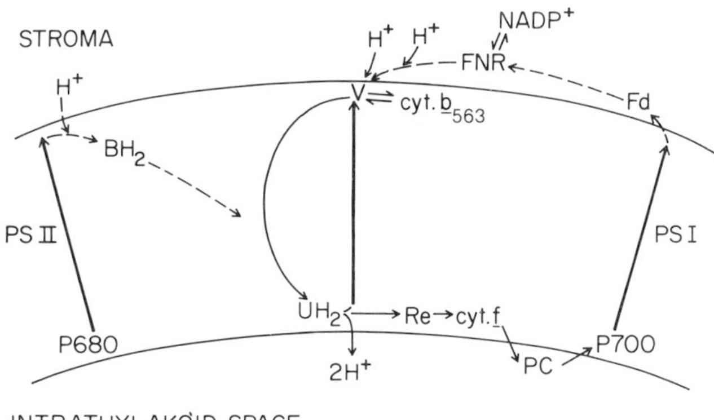

Fig. 9. Possible scheme for chloroplast electron transfer, based on Q cycle proposed by Mitchell (*21*).

via plastocyanin, cytochrome *f*, and the Rieske Fe–S center (Re), and the other via a transmembrane electrogenic reaction to a plastoquinone reductase (V) at the outer surface of the thylakoid. This plastoquinone reductase requires two electrons (and two protons) that do not arrive simultaneously; one comes from U, the other from photosystem I. Cytochrome *b*-563 may act as a "buffer" store, as its reduction and reoxidation kinetics do not match well with other observed reactions.

When two electrons are available to V, it may rereduce U via an electroneutral transmembrane hydrogen transfer on plastoquinol. Alternatively, if photosystem II is active, U may be rereduced by the bound plastoquinone B of photosystem II, from which electron transfer proceeds pairwise over a wide range of redox potentials (J. M. Bowes and A. R. Crofts, personal communication) giving rise under some conditions to weak oscillations in $P518_s$ with flash number (*18–20*). It seems unlikely that rereduction of U in the cyclic pathway around photosystem I proceeds through plastoquinone B, as photosystem I units are not stoichiometrically associated with photosystem II units throughout the thylakoid membrane (*22*); additionally, absorbance changes like those reported for photosystem I driven cyclic flow in spinach (*8*) were reported from maize bundle sheath cells, which lack photosystem II activity and grana stacks (*23*). However, plastoquinone B does seem able to rapidly reduce U without obligate use of the plastoquinone pool (Fig. 7) (*18*) showing efficient input to the cycle from photosystem II.

Convincing evidence for the operation of this (Q cycle?) extra energy conserving site in stoichiometric linear (H_2O to $NADP^+$) electron flow has not yet been presented. Our experiments (*10*) have suggested that in the presence of acceptors such as methyl viologen, $P518_s$ is not seen with repetitive flashes. That it may be seen on the first few flashes from a dark state (*6*, *18–20*) is not surprising, since under these conditions the only prerequisites for $P518_s$ observation are reduced U and flash oxidation of cytochrome *f*. However, electron input to V from photosystem I via ferredoxin: NADP reductase (*24*) seems required to cause efficient completion of the proposed Q cycle (i.e., rereduction of U and reoxidation of V) on each of a series of flashes, and thus to attain full activity of this extra coupling site. Indeed, optimal operation *in vivo* seems to occur when photosystem II is partially closed by reduction of one of its acceptors (*10*). Thus, while cyclic electron flow around photosystem I may well include a Q cycle, such a mechanism is unlikely to operate in linear electron flow.

Whereas the transthylakoid ΔpH seems to limit the overall rate of electron flow (*25*), the controlling factor in the switch between linear and cyclic electron flows in the chloroplast seems to be the reduction of a specific component in the thylakoid membrane—which could be the same as that effecting the distribution of energy between photosystems I and II (*26*, *27*). Hopefully, the pattern and mechanism of chloroplast electron flow will soon become clear.

ACKNOWLEDGMENTS

The authors would like to thank Cathy Chia and Edward Robson for technical assistance. Nigericin was kindly donated by Dr. R. L. Hamill of Eli Lilly Laboratories. This research was carried out at Brookhaven National Laboratory under the auspices of the United States Department of Energy.

REFERENCES

1. Trebst, A. (1974). *Annu. Rev. Plant Physiol.* **25,** 423–458.
2. Witt, H. T. (1979). *Biochim. Biophys. Acta* **505,** 355–427.
3. Joliot, P., and Delosme, R. (1974). *Biochim. Biophys. Acta* **357,** 267–284.
4. Slovacek, R. E., and Hind, G. (1978). *Biochem. Biophys. Res. Commun.* **84,** 901–906.
5. Horváth, G., Droppa, M., Mustardy, L. A., and Faludi-Dániel, Á. (1978). *Planta* **141,** 239–244.
6. Velthuys, B. R. (1978). *Proc. Natl. Acad. Sci. U.S.A.* **75,** 6031–6034.
7. Crowther, D., Mills, J. D., and Hind, G. (1979). *FEBS Lett.* **98,** 386–390.
8. Crowther, D., and Hind, G. (1980). *Arch. Biochem. Biophys.* **204,** 568–577.
9. Slovacek, R. E., Crowther, D., and Hind, G. (1979). *Biochim. Biophys. Acta* **547,** 138–148.

10. Slovacek, R. E., Crowther, D., and Hind, G. (1980). *Biochim. Biophys. Acta* **592,** 49515—505.

11. Wraight, C. A., Cogdell, R. J., and Chance, B. (1978). *In* "The Photosynthetic Bacteria" (R. K. Clayton and W. R. Sistrom, eds.), pp. 471–511, Plenum, New York.

12. Crofts, A. R., Crowther, D., Bowyer, J. R., and Tierney, G. V. (1977). *In* "Structure and Function of Energy Transducing Membranes (K. van Dam and B. F. van Gelder, eds.), pp. 133–155. Elsevier/North-Holland, Amsterdam.

13. van den Berg, W. H., Prince, R. C., Bashford, C. L., Takamiya, K., Bonner, W. D., and Dutton, P. L. (1979). *J. Biol. Chem.* **254,** 8594–8604.

14. Bowyer, J. R., Dutton, P. L., Prince, R. C., and Crofts, A. R. (1980). *Biochim. Biophys. Acta* **592,** 445–460.

15. Wood, P. M. (1980). *Biochem. J.* **189,** 385–391.

16. Bouges-Bocquet, B. (1977). *Biochim. Biophys. Acta* **462,** 371–379.

17. Okayama, S. (1976). *Biochim. Biophys. Acta* **440,** 331–336.

18. Bouges-Bocquet, B. (1980). *FEBS Lett.* **117,** 54–58.

19. Velthuys, B. R. (1979). *Proc. Natl. Acad. Sci. U.S.A.* **75,** 6031–6034.

20. Velthuys, B. R. (1980). *FEBS Lett.* **115,** 167–170.

21. Mitchell, P. (1976). *J. Theor. Biol.* **62,** 327–367.

22. Melis, A., and Brown, J. S. (1980). *Proc. Natl. Acad. Sci. U.S.A.* **77,** 4712–4716.

23. Leegood, R. C., Crowther, D., Walker, D. A., and Hind, G. (1981). *FEBS Lett.* **126,** 89–92.

24. Shahak, Y., Crowther, D., and Hind, G. (1981). *Biochim. Biophys. Acta* **636,** 234–243.

25. Slovacek, R. E., and Hind, G. (1981). *Biochim. Biophys. Acta Biochim. Biophys Acta* **635,** 393–404.

26. Horton, P., and Black, M. T. (1980). *FEBS Lett.* **119,** 141–144.

27. Bennett, J., Steinback, K. E., and Arntzen, C. J. (1980). *Proc. Natl. Acad. Sci. U.S.A.* **77,** 5253–5257.

Differentiation between C and M Side Reactions of Ubiquinone Using Imposed Electron–Electrochemical Potential Gradient

4

ASHER GOPHER
MENACHEM GUTMAN

INTRODUCTION

Ubiquinone is a lipid-soluble electron carrier that participates in redox reactions on the C and M sides of the mitochondrial membrane. While determination of the total redox state of the quinone is feasible (yet not by an easy, nondestructive method), the local redox state at different loci of the membrane is indeterminable. In the present study we utilized the interaction of ubiquinone with defined membrane markers, the *b*-566 and *b*-561 cytochromes, to evaluate reactions of the quinone with C and M side electron carriers. The oxidation of cytochrome *b* by ATP in antimycin-inhibited mitochondria has been recognized (*1–3*), and it has been demonstrated that the oxidized species is cytochrome *b*-561 (*2*, *3*) and that oxidation requires the presence of antimycin. While the experimental observations are clear and accurate, the mechanism leading to the oxidation is still ambiguous. According to Slater and Lee (*3*) the oxidant is an un-

Function of Quinones in Energy Conserving Systems

ISBN 0-12-701280-X

known species and two coupling sites were proposed in the dehydrogenase-quinone-cytochrome *b* junction. The interpretation by Flatmark and Pedersen (*2*) utilized the term "transducing" carrier. This nomenclature is not fully compatible with the present role given to the protonmotive force as the intermediate in coupling reactions. The Q cycle describes cytochrome *b*-561 as a carrier unique in its ability to be reduced by donors located either on the M side or on the C side of the membrane. In the absence of antimycin and in the presence of KCN and ascorbate plus TMPD (to keep *c* and *a* type cytochromes in a fully reduced state), cytochrome *b*-561 is reduced by the dehydrogenases and attains redox equilibrium with the M phase. In the presence of antimycin, the equilibrium of cytochrome *b*-561 with the M phase sided redox couple is mediated by cytochrome *b*-566, located on the C side of the membrane (*4*, *5*). Indeed, evidence for such a dual pathway for reduction of the *b*-type cytochromes was observed by Trumpower in his iron–sulfur protein depleted preparation (*6*).

As in the presence of antimycin, cytochrome *b*-561 equilibrates with the substrate (on the M side) only via cytochrome *b*-566 (located on the C side), and its redox state will be altered whenever the electrochemical potentials of the protons of the two phases are not the same. Acidification of the C phase (due to ATPase-linked proton translocation) will increase the redox potential of cytochrome *b*-566 (*7*), which will act as an oxidant with respect to cytochrome *b*-561. Similarly, according to Walz (*8*), $\Delta\Psi$ (positive on C side) will alter the electron–electrochemical potential ($\tilde{\mu}e$), causing carriers on the C side of the membrane to behave as oxidants with respect to carriers on the M side of the membrane. The effect of either mechanism will be an oxidation of cytochrome *b*-561. Consequently, interpretation of the energy dependent oxidation of cytochrome *b*-561 calls for a possible distinction between the effect of ΔpH and $\Delta\Psi$ on the redox potential of ubiquinone.

Previous studies on oxidation of cytochrome *b*-561 were executed under conditions where the redox potential of the substrate was not defined and the oxidation of cytochrome *b*-561 was not expressed in millivolts. Under such conditions the extent of oxidation is simply not related to the magnitude of the driving force. To improve our ability to interpret the results, we carried out our experiments in a titrative mode. This technique enabled us to relate the redox potential measured for cytochromes *b*-561 and *b*-566 with the redox potential of the substrate. As will be shown, such a quantitative approach led to unambiguous assignment of $\Delta\Psi$ as the driving force for oxidation of cytochrome *b*-561.

EXPERIMENTAL RESULTS

Oxidation of Cytochrome *b* by ATP in Submitochondrial Particles

Inside-out submitochondrial particles (*9*) expose the succinate binding site to the bulk of the solution. Thus, the limited permeability of fumarate across the membrane does not hamper our ability to control the redox potential by this redox couple. In the following experiments we used this sytem to investigate the mechanism of oxidation of the *b*-type cytochromes by ATP. To prevent redox changes in the *c* and *a*-type cytochromes, KCN, ascorbate, and TMPD were always present in the reaction mixture.

In accordance with Flatmark and Pedersen (*2*), we observed an enhanced reduction of the *b*-type cytochromes upon addition of ATP to ETPH pretreated with ascorbate, TMPD, KCN, and fumarate plus succinate at molar ratio of 10 (not shown). When antimycin was added prior to ATP, addition of ATP caused an oxidation of cytochrome *b* having a maxiumum at 562.5 nm (Fig. 1 and insert). This observation is more informative once the redox state of each *b*-type cytochrome is calculated according to its absorption at the two wavelengths 566 and 561 nm (*10*). The results of these calculations are presented in Fig. 2. Titration of cytochromes *b*-561 (A), or *b*-566 (B), with increasing concentrations of antimycin shifts their redox state by 40 mV, respectively, in a sigmoidal curve. The inflection point and end point of the curve are identical with those measured for antimycin inhibition of succinoxidase activity (not shown).

The effect of antimycin on the redox states of cytochromes *b*-561 and *b*-566 (in the absence of ATP) is readily explained by the known effect of antimycin on the redox mid-potential (E_m) of cytochromes *b*-561 and *b*-566 (*11*). If we calculate, for the extreme points (in the absence of, or in saturation by antimycin) the potential of cytochrome *b* ($E_h = E_m + 60 \log b_{ox}/b_{red}$), a constant value is obtained for cytochromes *b*-561 and *b*-566. This value is practically identical with the applied potential of the fumarate/succinate couple (Fig. 2).

Repeating these antimycin titrations in the presence of ATP alters the responses of the cytochromes. The effect of ATP can be studied as a function of the degree of inhibition of the electron transport system by antimycin. At low concentrations of antimycin the redox state of cytochrome *b*-561 is unaffected by ATP, but at concentrations high enough to block electron transport, cytochrome *b*-561 in the presence

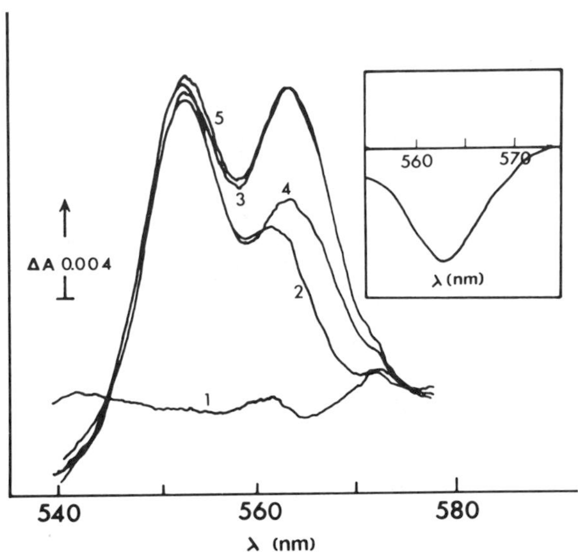

Fig. 1. The effect of ATP on the redox state of the *b* cytochromes in ETPH in the presence of antimycin. Dual wavelength spectral scanning (isosbestic reference point at 575 nm). ETPH (1.7 mg/ml) was suspended in 0.18 *M* sucrose, 50 m*M* Tris-acetate, pH 7.4, and 5 m*M* $MgSO_4$ (30 °C) and the spectra were measured after the following sequential additions: 1. oxidized baseline; 2. addition of 2 m*M* KCN, 66 μ*M* TMPD, 4 m*M* ascorbate, 10 m*M* fumarate, 1 m*M* succinate; 3. antimycin (2 nmole/mg protein); 4. ATP (1 m*M*), 5. FCCP (1 μ*M*). Insert. Difference spectra showing the effect of ATP obtained by subtracting line 4 from line 3.

of ATP, assumes a redox state some 40 mV more positive than in the absence of ATP (Fig. 2A). The response of cytochrome *b*-566 to ATP is more ambiguous (Fig. 2B). In the presence of saturating concentrations of antimycin, the redox state of cytochrome *b*-566 is that of the fumarate/succinate couple whether or not ATP is present. In the absence of antimycin, addition of ATP initiates a reverse electron flux. Thus the response of the redox state is also effected by kinetic considerations and quantitative interpretation should be avoided.

Effect of the Applied Redox Potential on the Response of the *b*-Type Cytochromes to ATP

In the presence of saturating concentrations of ATP and antimycin, the extent of oxidation of cytochrome *b*-561 is a function of the fumarate-to-succinate ratio. A Nernst plot of the redox state of cytochrome *b*-561 in the absence and in the presence of ATP is given in Fig. 3. Cytochrome *b*-561 behaves as a single electron carrier with a

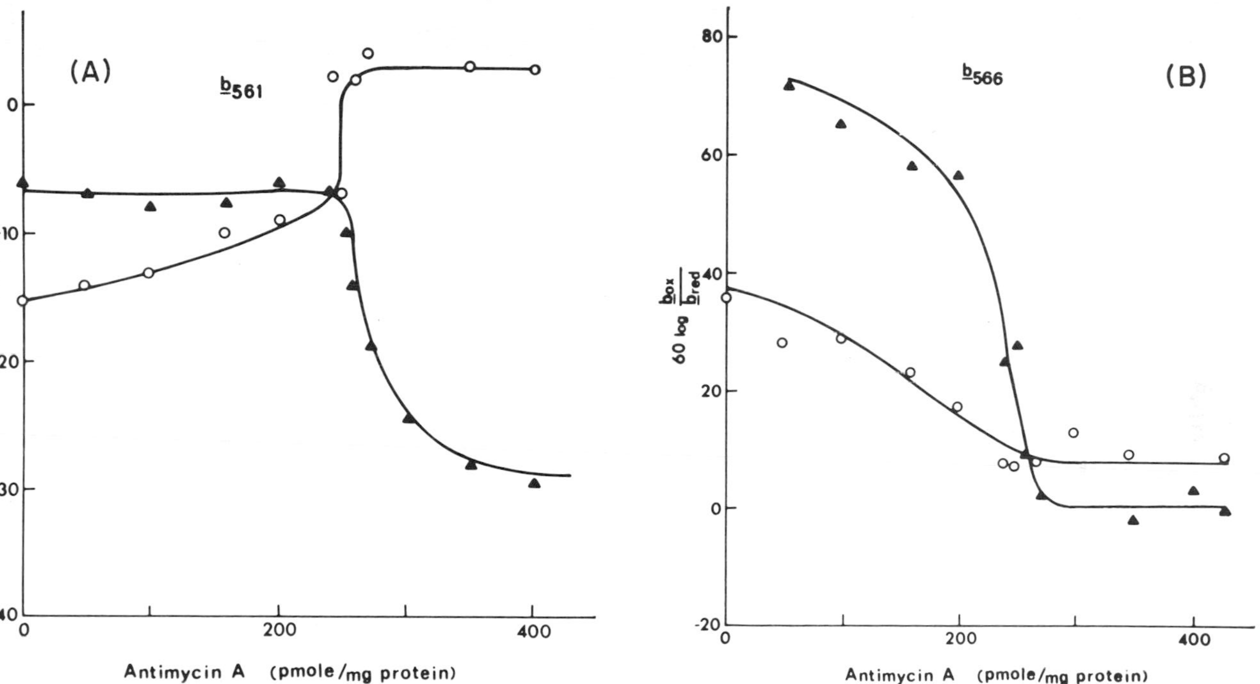

Fig. 2. The effect of antimycin concentration on the response of cytochromes *b*-561 and *b*-566 to ATP. ETPH (1.5 mg/ml) suspended in sucrose-Tris buffer (Fig. 1) and cytochrome *b* was reduced by addition of 2 m*M* KCN, 4 m*M* ascorbate, 66 μM TMPD, 10 m*M* fumarate, and 1 m*M* succinate followed by successive additions of antimycin and 1 m*M* ATP. The spectra between 540–610 nm using 575 nm as reference wavelength were recorded. The redox states of cytochromes *b*-561 (A) and *b*-566 (B) before (▲) and after (○) addition of ATP were calculated according to the following equations (*10*)

$$\Delta A_{561-575} = 0.75\ (b\text{-}^{+2}_{561}) + 0.25\ (b\text{-}^{+2}_{566}),$$

$$\Delta A_{566-575} - 0.45\ (b\text{-}^{+2}_{561}) + 0.55\ (b\text{-}^{+2}_{566}).$$

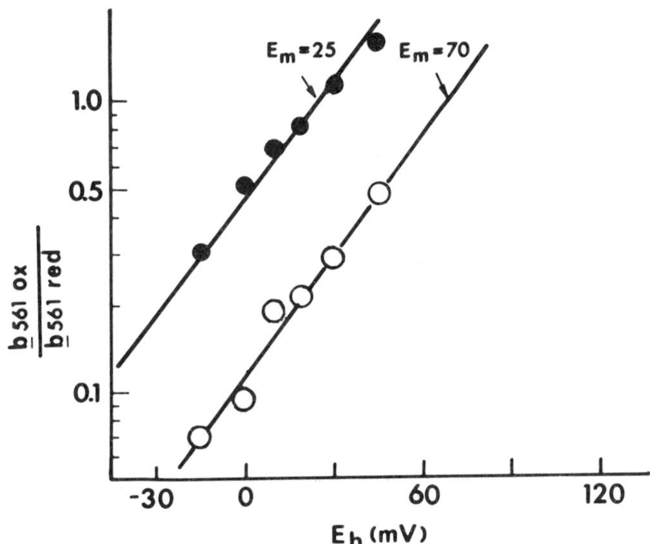

Fig. 3. The effect of ATP on the redox titration of cytochrome *b*-561 in antimycin-inhibited submitochondrial particles. Experiments as in Fig. 1 were repeated with various fumarate/succinate ratios to give the redox potentials indicated in the abscissa. The redox state of cytochrome *b*-561 was calculated before (○) and after (●) addition of ATP as described in Fig. 2.

mid-potential of +70 mV. In the presence of ATP, its mid-potential assumes a value of +25 mV. The response of cytochrome *b*-566 to the addition of ATP was marginal. In some experiments no change in the redox state of cytochrome *b*-566 was measured, while in others a small oxidation (~9%) was noted. Repeating these measurements under strict anaerobic conditions (to avoid the oxidant effect on the redox state of cytochrome *b*-566 (*12*)) or in the presence of phenazine methosulfate (to accelerate the redox relaxation (*13*)) did not change the response. The redox state of cytochrome *b*-566 was equal to that of the donor couple in the presence or in the absence of ATP.

Effect of pH on ATP-Dependent Oxidation

Experiments as described in Fig. 3, were repeated at the pH values indicated in Fig. 4. The apparent mid-potentials of cytochrome *b*-561 in the absence and the presence of ATP are a linear function of pH in the range 6.0–8.5. The slopes of the lines are practically identical and correspond to a redox reaction with stoichiometry of one proton per electron (Fig. 4).

Driving Force for Oxidation of Cytochrome *b*-561

The extent of cytochrome *b*-561 oxidation with response to ATP concentration follows a saturation curve. The ATP concentration giving 50% of maximal oxidation is 25 μM. Changing the phosphate potential of the reaction also modulates the magnitude of the oxidation. At a high phosphate potential ($\Delta G_{ATP} < -12$ kcal/mole), the oxidation reaches a maximal limiting value.

Measurable oxidation is obtained at such a low phosphate potential (−6 kcal/mole), that even commercial ADP (usually contaminated by ~0.5% ATP) causes detectable oxidation. Prior treatment of the ADP

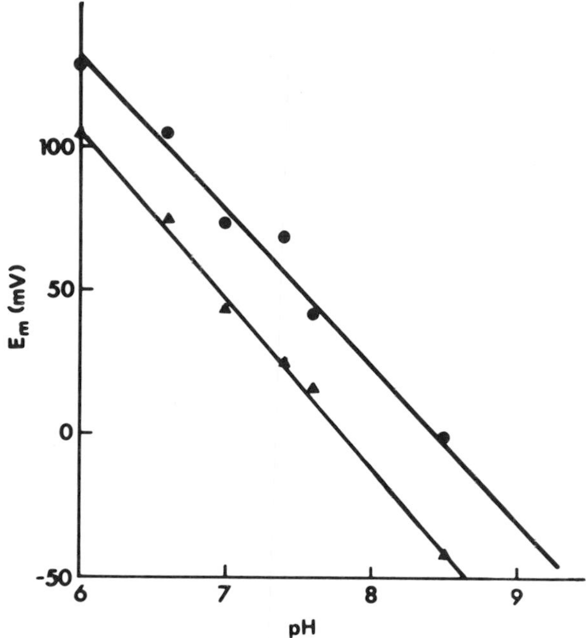

Fig. 4. The pH dependence of cytochrome *b*-561 mid-potential as measured in the presence or absence of ATP for antimycin-inhibited submitochondrial particles. The experiment described in Fig. 3 was repeated at the indicated pH values. The E_m values, measured in the absence (●) and presence (▲) of ATP, are drawn with respect to pH. The redox potential of the fumarate/succinate couple was corrected for the pH.

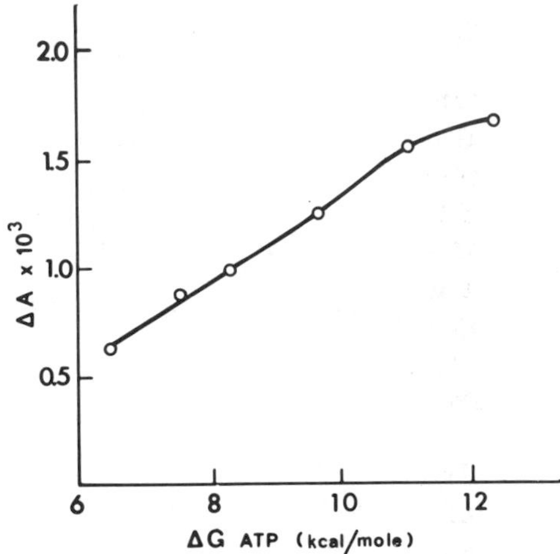

Fig. 5. The effect of phosphate potential on the oxidation of cytochrome *b*-561 in antimycin-inhibited submitochondrial particles. The experimental procedure as in Fig. 1 was repeated, only the concentration of ATP was varied and added together with ADP and phosphate to the indicated phosphate potential ($\Delta G^{o\prime}{}_{ATP} = -7.3$ kcal/mole). Cytochrome *b*-561 oxidation was measured at 562.5–575 nm.

solution with hexokinase plus glucose abolishes the oxidation measured upon addition of ADP.

To investigate the nature of the driving force for the oxidation, we tried to distinguish between $\Delta\Psi$ and ΔpH components of the protonmotive force. ΔpH was measured fluorimetrically (*14*) using 9-aminoacridine. Phosphorylating submitochondrial particles, (suspended in 30 m*M* KCl, 50 m*M* Tris-chloride and 5 m*M* $MgCl_2$, pH 7.4 KCl medium), can easily reach 2.5–3 pH units at the expense of NADH oxidation or 2.4 units during ATP hydrolysis. Addition of nigericin (1 μM) caused a complete recovery of the fluorescence. In the same buffer and in the presence of nigericin, ATP caused the normal oxidation of cytochrome *b*-561. Futhermore, addition of nigericin after addition of ATP did not reverse the oxidation of cytochrome *b*-561.

Further evidence against ΔpH as the driving force for oxidation came from studies in sucrose-Tris-Mes buffer. This system lacks a permeant anion, maintains a high $\Delta\Psi$ value, but fails to demonstrate a fluorescence quenching of 9-aminoacridine upon addition of ATP. However, normal oxidation of cytochrome *b*-561 was measured in this buffer system even in the presence of nigericin.

Identification of $\Delta\Psi$ as the Driving Force for Oxidation

To quantitate the oxidation of cytochrome *b* with respect to $\Delta\Psi$, we measured $\Delta\Psi$ by the spectral shift of safranine (*15*). In submitochondrial vesicles the magnitude of the safranine signal is a linear function of $\Delta\Psi$ in the range of 0–100 mV (positive inside, not shown). We also determined the magnitude of $\Delta\Psi$ built up during ATP hydrolysis. The same value was measured both by safranine spectral shift (100 mV) and by flow dialysis (95 ± 10 mV). Thus as long as the magnitude of the signal is kept within the linear section and adheres to the standard concentration of the reactants, protein and safranine, the magnitude of $\Delta\Psi$ can be directly measured.

The correlation between $\Delta\Psi$ and E_h is depicted in Fig. 6 where the two potentials are drawn with respect to each other. No matter which method was utilized to de-energize the membranes (FCCP, permeant

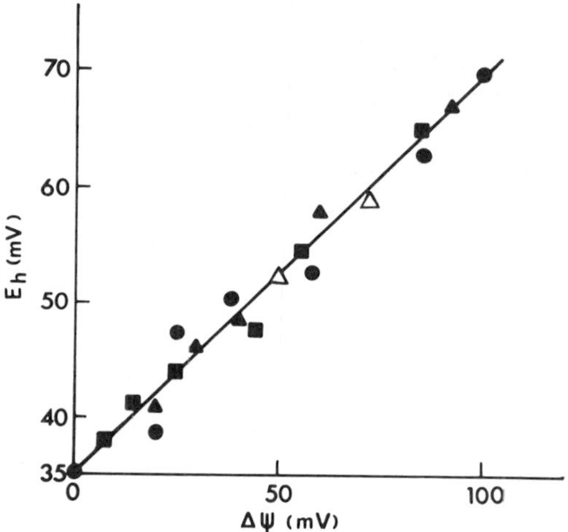

Fig. 6. Correlation between the redox potential of cytochrome *b*-561 in antimycin-inhibited submitochondrial particles and the ATPase driven $\Delta\Psi$. $\Delta\Psi$ was measured by following the ATP-induced safranine spectral shift at 511–533 nm. The highest measured value was corroborated by equilibrium dialysis. The redox potential (E_h) of cytochrome *b*561 was calculated from the redox state and the mid-potential as detailed in Fig. 2. The results represent two kinds of experiments. The two potentials when measured at steady-states in the presence of various concentrations of permeant anions (NO^-_3 (△) and SCN^- (●). The rest of the points in this figure were taken from the experiment summarized in Table I. The magnitudes of the membrane and redox potentials (at identical times after addition of 1 μM FCCP (▲) or 1μM oligomycin (■) is shown.

TABLE I

The Correlation Between the Pseudo-First-Order Rate Constants for the Change in Redox State of Cytochrome *b*-561 and the Safranine Spectral Shift during the Buildup and Collapse of $\Delta\psi$ in Antimycin-Inhibited Submitochondrial Particles.[A]

	Rate constant of response to ATP (sec^{-1})		Rate constant of response to collapse of $\Delta\psi$		
Experiment no.	Cytochrome *b*-651	Safranine spectral shift	$\Delta\psi$ Collapsing agent	Cytochrome *b*-561	Safranine spectral shift
1	0.125	0.111	FCCP	0.055	0.068
2	0.111	0.14	Oligomycin	0.083	0.125
3	0.125	0.14	Hexokinase + glucose	0.078	0.067

[A] The oxidation of cytochrome *b*-561 (measured at 561–575 nm) and the buildup of $\Delta\psi$ (measured by safranine) were measured kinetically as described in Figs. 2 and 6, respectively, showing a pseudo-first-order kinetics. The collapse of $\Delta\psi$ was achieved by FCCP (1 μM), oligomycin (1 μM) or by removal of ATP by hexokinase (100 units) and glucose (1 mM). All kinetic responses were found to follow a pseudo-first-order kinetics over 90% of the measured reaction.

anions or blocking of ATPase), there is a linear correlation between the shift in mid-potential of cytochrome *b*-561 and the measured $\Delta\Psi$.

Linear correlation of the two parameters was also observed in kinetic studies (see Table I). In these experiments, the kinetics of the formation and disappearance of oxidized cytochrome *b*-561 was measured in parallel to the kinetics of appearance and collapse of $\Delta\Psi$. $\Delta\Psi$ was induced by addition of ATP and then abolished by uncoupler, blocking of ATPase by oligomycin, or exhaustion of ATP by hexokinase plus glucose. As seen in Table I the kinetics of the decrease of E_h and $\Delta\Psi$ are characterized by the same rate constants.

DISCUSSION

The oxidation of cytochrome *b*-561 reported in this study represents the interaction of the cytochromes with a single redox couple under the influence of an externally coupled driving force, generated by ATP hydrolysis. This oxidation should not be confused with the effect of ATP on the redox state of cytochrome *b* in mitochondria, maintained in the presence of ascorbate as electron donor and fully oxidized pyridine nucleotides (*10*). While the oxidation we observed can be classified as a static head (*17*), the system described by Wilson (*10*) approximates a level flow through two coupling sites, one at the *c*-*b*-Q level and the other through NADH dehydrogenase (*18*, *19*).

Early reports on oxidation of cytochrome *b*-561 in the presence of ATP and antimycin utilized submitochondrial particles (1, 3) in the presence of mixed substrates (*1*) or NADH as a low potential donor interacting with cytochrome *c* through two coupling sites (*3*). Later studies were performed on intact mitochondria (*2*, *17*) where the presence of endogenous substrate and the enzymes of the Krebs cycle impaired the ability to maintain a known redox potential in the system. Therefore, we preferred to employ submitochrondrial particles where the oxidation of succinate occurs on the externally located M side of the membrane. Furthermore, in the absence of fumarase no low potential substrate (malate) can be formed. Thus, we could carry out our experiments under well defined redox potentials.

By poising the redox potential of the bulk over a wide range (-30 to $+30$ mV), we could monitor the effect of ATP when the initial redox state of the carriers under observation varied tenfold (see Fig. 3). This allowed quantitation of the oxidation in absolute units of the shift of redox potential expressed in millivolts.

The redox state of each *b*-type cytochrome was measured in the presence and absence of antimycin and in the presence and absence of ATP. Of these four defined states, the redox state measured in the presence of ATP and absence of antimycin is not suitable for any quantitative analysis. This state represents reverse electron transport from cytochrome c_1 (or cytochrome *c*) to the quinone level which then equilibrates with the fumarate/succinate couple. As the rate-limiting step in this sequential reaction is not defined, quantitative conclusions cannot be drawn. Alternatively, redox states measured in the presence of antimycin and ATP show no electron flux and are suitable for analysis. A state of equilibrium (or quasi-equilibrium) of the redox components is established when the electrochemical potential of the electron (*8*) is equal for all reactants. In our system there are three components and two phases: (a) the bulk phase, where the redox system is maintained constant by fumarate and succinate, (regarded as the reference), (b) the membrane phase that is permeable to electrons, either through the cytochromes or the quinone, and (c) the internal aqueous phase that can assume an electric potential of +100 mV, with respect to the bulk phase.

The correlation between the redox potential, calculated from the redox state of the cytochromes and the corresponding E_h value, of the fumarate/succinate couple is given in Fig. 7. In the absence of antimycin and ATP, cytochromes *b*-561 and *b*-566 are in redox equilibrium with the succinate/fumarate couple. In this case $\Delta\Psi = 0$. The electrical potential does not affect the electrochemical potential of the electrons, and the redox state of the carriers on both sides of the membrane are in accord with the exepected value. Once antimycin is added, the mid-potentials of cytochromes *b*-566 and *b*-561 are shifted, respectively, from −62 to +30 mV and 50 to 70 mV. This explains why the equilibrium of cytochrome *b*-566 with the donor couple becomes experimentally observable. As seen in Fig. 7 under these conditions the redox state and the corresponding E_h values measured for the *b*-type cytochromes are in accord with the potential set by the fumarate/succinate couple.

Once ATP is added the situation is changed. Cytochrome *b*-566, located on the opposite side of the membrane with respect to the reference couple (Scheme I), still assumes the redox potential of the bulk. However, cytochrome *b*-561 is closer to the bulk and reflects a major shift of the electron–electrochemical potential. The magnitude of this discrepancy is ~40 mV and is not sensitive to pH (see Fig. 4). Cytochrome *b*-561 oxidation measured under these conditions is an energy dependent reaction, and is sensitive to uncouplers (Fig. 1). The oxidation is not measured in the presence of oligomycin and the magnitude is proportional to the phosphate potential applied to the system (Fig. 5).

The driving force for cytochrome *b*-561 oxidation can be the protonmotive force or only one of its components, ΔpH or ΔΨ. Involvement of ΔpH in cytochrome *b*-566 oxidation can be eliminated for the following reasons. Cytochrome *b*-561 oxidation measured in sucrose-tris-acetate buffer, where no ΔpH was detected by 9-aminoacridine, is similar to that measured in KCl medium where ΔpH = 2.5. Addition of nigericin col-

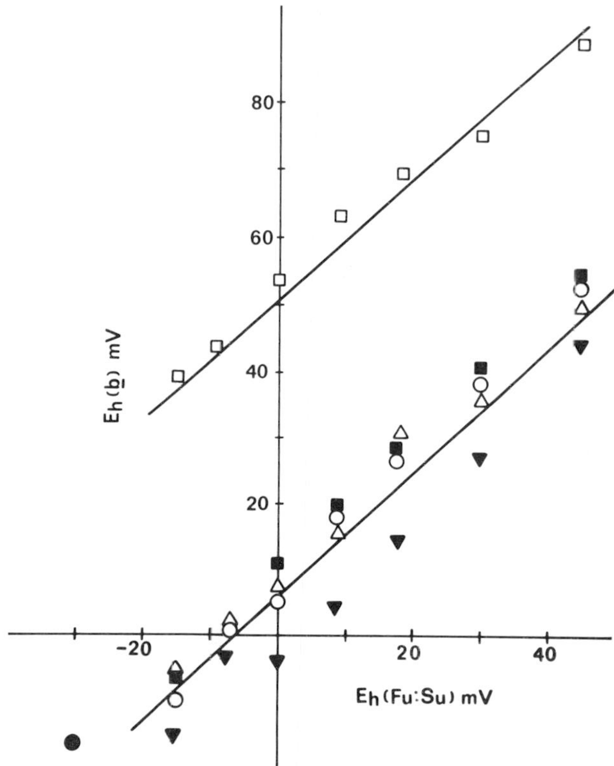

Fig. 7. The correlation between the measured potential of the *b*-type cytochromes and the redox potential of fumarate/succinate in energized and non-energized submitochondrial particles. ETPH (1.5 mg/ml) treated with 2 m*M* KCN, 4 m*M* ascorbate, and 66μ*M* TMPD. The redox states of cytochromes *b*-561 and *b*-566 were calculated as described in Fig. 2 and are drawn with respect to the redox potential of the fumarate/succinate couple. The experiments were carried out either in the absence or presence of antimycin in energized (1 m*M* ATP) or de-energized particles. ○, cytochrome *b*-561 in the absence of antimycin and ATP; △, cytochrome *b*-561 in the presence of antimycin and absence of ATP; □, cytochrome *b*-561 in the presence of ATP and antimycin; ●, cytochrome *b*-566 in the absence of antimycin and ATP; ▼, cytochrome *b*-566 in the presence of antimycin and absence of ATP; ■, cytochrome *b*-566 in the presence of ATP and antimycin.

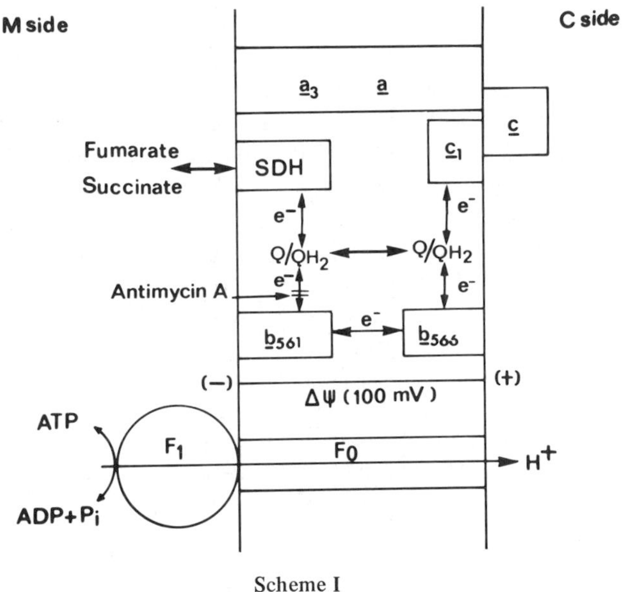

Scheme I

lapses the ΔpH built up in the KCl medium without affecting the magnitude of the oxidation when expressed either as percent of cytochrome *b*-561, or calculated in millivolts. Moreover, addition of KSCN, which replaces ΔΨ by ΔpH, completely abolishes ATP-dependent oxidation of cytochrome *b*-561.

The dependence of the oxidation on ΔΨ is demonstrated by the following facts. There is a linear correlation between ΔΨ (measured by safranine or flow dialysis) and cytochrome *b*-561 oxidation (Fig. 6). Furthermore, the kinetics of the formation and decay of ΔΨ (measured by safranine) and ΔE_h for cytochrome *b*-561 are characterized by the same time constants (Table I). Thus, identification of ΔΨ as the driving force for the oxidation is supported by both equilibrium and kinetic measurements.

Once this conclusion is accepted, explanation of the oxidation becomes a straightforward consequence of the Q cycle, or any other mechanism that stipulates that the interaction of cytochrome *b*-561 with Q is antimycin sensitive and locates cytochromes *b*-561 and *b*-566 on the M and C sides, respectively. In the absence of antimycin, no relative change in the redox state of cytochromes *b*-561 and *b*-566 is expected as both are in equilibrium with the quinone, and uncharged mobile carrier (*20*). ATP-dependent ΔΨ will change the electrochemical potential of the electron (*8*) rendering the C side cytochrome *b*-566 an oxidant with respect to cytochrome *b*-561. However, this gradient in electrochemical potential will be dissipated into electron flux by the following mechanism: Cytochrome *b*-566 is located in the membrane C side together with some of the quinone, with relative redox state unaffected by $\Delta\Psi$. Consequently, any enhanced reduction of cytochrome *b*-566 driven by $\Delta\Psi$ is paralleled by comparable reduction of some of the quinone. As ubiquinone is an uncharged mobile carrier, any gradient in its redox state (among different loci in the membrane) will dissipate by diffusion until quasi-equilibrium is established. The same considerations are applied for cytochrome *b*-561. Its redox equilibrium with M side quinone is not affected by $\Delta\Psi$ and, as the redox state of ubiquinone in different loci is equalized (or nearly so) by diffusion, cytochromes *b*-561 and *b*-566 will assume close redox potentials (see Fig. 6) reflecting the various rate constants associated with the quasi-equilibrium state.

This type of open system (equilibrated with two redox couples, fumarate/succinate and ascorbate/TMPD), functions as a coupling system converting the protonmotive force into an uphill reverse electron transport. In contrast, once antimycin is present, the oxidation of cytochrome *b*-561 by QH_2 (M side) is inhibited, and the magnitude of the oxidation in millivolts should be fully equated with the electrical gradient between cytochrome *b*-561 and its C side oxidant (most probably cytochrome *b*-566). This accounts for the oxidation of cytochrome *b*-561. Under these conditions, cytochrome *b*-566 senses an electrochemical potential of $E_{b-566} = E_{out} + 100$ mV and should react in a highly reduced state. However, because its equilibrium with Q (C side) is not affected by $\Delta\Psi$, it will equilibrate in a redox reaction with the quinone which, because of its mobile nature, maintains the redox potential of the M side. Thus we account for cytochrome *b*-561 oxidation without a corresponding shift in the redox state (or redox potential) of cytochrome *b*-566.

The effect of acidification of the inner space (C side) of the vesicles on the redox state of the *b*-type cytochromes is of interest. Generation of ΔpH will alter the mid-potential of these components facing the C side; thus at constant electrochemical potential (as poised by the succinate/fumarate couple) their redox state will vary. (Cytochrome *b*-561 faces the M side of the membrane (see Scheme I) and thus acidification of the inner space will not affect its redox state, in accord with our observations). The redox potential of cytochrome *b*-566 is pH dependent only above pH 6.9 and the E_m varies with a slope of 60 mV/pH unit (*7*). Thus, under our experimental conditions (where the initial pH = 7.4), the maximal shift in the mid-potential of cytochrome *b*-566 cannot exceed 30 mV (corresponding to 0.5 pH unit), whatever the size of $\Delta\Psi$. A 30 mV shift of mid-potential is large enough to be observed as enhanced reduction of cytochrome *b*-566. Still, whenever antimycin was present an ATP-dependent reduction of cytochrome *b*-566 was not observed. This might suggest that in submitochondrial particles cytochrome *b*-566 exchanges protons with the M side of the membrane rather than with the C side.

The last point for consideration is the ratio between $\Delta\Psi$ and the resulting oxidation by cytochrome *b*-561. As seen in Fig. 6, the redox potential applied to cytochrome *b*-561 (in the presence of antimycin) increases above the potential of the bulk (fumarate/succinate) as a linear function of $\Delta\Psi$. The slope of the linear function (0.35 ± 0.05) is constant over the entire range of measured $\Delta\Psi$ values. As $\Delta\Psi$ was measured by a double calibration system (K^+ gradient and equilibrium dialysis), the less-than-unity value of the slope indicates that only a fraction of the total electric potential gradient affects the redox equilibrium of cytochromes *b*-566 and *b*-561 (*8*). Knowledge of the electric potential profile across the membrane is presently limited. Yet, based on our measurements, it seems that cytochromes *b*-566 and *b*-561 are located in the membrane such that only a fraction of the total electric field affects their relative redox state.

REFERENCES

1. Hinkle, P. C., Butow, R. A., Racker, E., and Chance, B. (1967). *J. Biol. Chem.* **242,** 5169–5173.

2. Flatmark, T., and Pedersen, J. I. (1973). *Biochim. Biophys. Acta* **325,** 16–28.

3. Slater, E. C., and Lee, I. Y. (1973). *In* "Oxidases and Related Redox Systems" (T. E. King, R. K. Morton, and M. Morrison, eds.), Vol. 2, pp. 823–850. Univ. Park Press, Baltimore, Maryland.

4. Wikström, M. K. F. (1973). *Biochim. Biophys. Acta* **301,** 155–193.

5. Mitchell, P. (1975). *FEBS Lett.* **56,** 1–6.

6. Trumpower, B. L. (1976). *Biochem. Biophys. Res. Commun.* **70,** 73–80.

7. Wilson, D. F., Erecinska, M., Leigh, J. S., and Koppelman, M. (1972). *Arch. Biochem. Biophys.* **151,** 112–121.

8. Walz, D. (1979). *Biochim. Biophys. Acta* **505,** 279–353.

9. Hansen, M., and Smith, A. L. (1964). *Biochim. Biophys. Acta* **81,** 214–222.

10. Wilson, D. F., and Erecinska, M. (1975). *Arch. Biochem. Biophys.* **167,** 116–127.

11. Erecinska, M., and Wilson, D. F. (1976). *Arch. Biochem. Biophys.* **174,** 143–157.

12. Baum, H., and Rieske, J. S. (1966). *Biochem. Biophys. Res. Commun.* **24,** 1–9.

13. Wikström, M. K. F., and Berden, J. A. (1972). *Biochim. Biophys. Acta* **283,** 403–420.

14. Rottenberg, H., and Lee, C. P. (1975). *Biochemistry* **14,** 2675–2680.

15. Akerman, K. E. O., and Wikström, M. K. F. (1976). *FEBS Lett.* **68,** 191–197.

16. Colowick, S. P., and Womack, F. C. (1969). *J. Biol. Chem.* **244,** 774–777.

17. Kedem, O., and Caplan, S. R. (1965). *Trans. Faraday Soc.* **61,** 1897–1911.

18. Gutman, M., Beinert, M., and Singer, T. P. (1975). *In* "Electron Transfer Chains and Oxidative Phosphorylation" (E. Quagliariello, S. Papa, F. Palmieri, E. C. Slater, and N. Siliprandi, eds.), pp. 55–62. North-Holland Publ., Amsterdam.

19. Rottenberg, H., and Gutman, M. (1977). *Biochemistry* **16,** 3220–3227.

20. Hauska, G. (1977). *In* "Bioenergetics of Membranes" (L. Packer, G. C. Papgeorgiou, and A. Trebst, eds.), pp. 177–187. Elsevier/North-Holland, Amsterdam.

The Proton Translocation Function of the Ubiquinone-Cytochrome c *Oxidoreductase of Mitochondria*

5

S. PAPA
F. GUERRIERI
M. LORUSSO
G. IZZO
F. CAPUANO

INTRODUCTION

The ubiquinone-cytochrome *c* oxidoreductase of mitochondria (b-c_1 complex, or Complex III) corresponds to the second energy conserving site of the respiratory chain and is the segment of the cytochrome system in which proton pumping function is clearly established (*1–5*). Yet the mechanism by which the reducing equivalents, donated by dehydrogenases, are accepted by the redox centers of the b-c_1 complex, and transferred to cytochrome *c* (*6–8*), and transmembrane $\Delta\tilde{\mu}H^+$ is generated is still unsettled. It is currently disputed whether proton translocation results directly from vectorial ligand conduction by prosthetic groups at the catalytic centers (*9*) or involves cooperative proton-transfer reactions in apoproteins, alternatively to or in conjunction with vectorial catalysis (*10–12*).

In its original redox-loop formulation (*13*), the direct ligand conduction

Function of Quinones in Energy Conserving Systems

ISBN 0-12-701280-X

mechanism required a → $H^+/2e^-$ quotient of two and a → $q^+/2e^-$ quotient of zero for proton release from mitochondria, associated with electron flow from quinol to cytochrome *c*. However these quotients, measured independently in various laboratories, were found to be four and two, respectively (*2*–*5*). These stoichiometries, and related observations in mitochondria and other coupling membranes, were still explained in terms of direct ligand conduction by development of mechanisms like the protonmotive Q cycle (*14*, *15*) or two quinone redox loops in series (*16*).

Exhaustive scrutiny reveals that while important support is apparently given to these mechanisms by the identification of separate quinone systems (King, Chapter 1, this volume; Ohnishi *et al.*, Paper 7, Chapter IV, this volume) and quinone-binding proteins (Yu and Yu, Paper 7, Chapter V, this volume) the models, as they are presently formulated, depend on certain restrictive conditions that remain highly speculative. These models do not account for proton motive characteristics of semiquinones (Swallow, Paper 3, Chapter II, this volume), cooperative linkage existing in the *b* cytochromes and Rieske Fe–S protein between the redox state of the metal and proton motive equilibria in apoproteins. The occurrence of these linkage phenomena (which by analogy to the Bohr effects of hemoglobin (*17*) are designated redox–Bohr effects (*11*), is in fact indicated by pH dependence of the midpoint potential (*18*) of *b* cytochromes (*19*) and the Rieske Fe–S protein (*20*).

The relationship between the E_m and protolytic equilibria in the enzyme (apoprotein?) is described by Eq. 1 (*18*)

$$E_m = E_o + \frac{RT}{F} \ln \frac{[H^+] + K_{red}}{[H^+] + K_{ox}} \tag{1}$$

The equation predicts that when $[H^+] >> K_{red}$ and $K_{ox} >> [H^+]$, $E_m = E_o + 0.06(\log [H^+] - \log K_{ox})$; that is, in this pH range, the midpoint redox potential decreases 60 mV per pH unit increase.

Utilizing the p*K* values of the oxidized and reduced enzyme it is possible to calculate the ionization state of the oxidized and reduced *b* cytochromes as a function of pH, and from this the extent to which, at any given pH, the system functions as an electron or effective hydrogen carrier. The profile of the pH dependence of the extent to which *b* cytochromes and Rieske Fe–S protein function as effective hydrogen carriers is shown in Fig. 1. Quinols in organic solvents with p*K*'s close to 12 function as hydrogen carriers up to pH 11; however, semiquinones with p*K*'s around 6 (Swallow, Paper 3, Chapter 2, this volume) change from hydrogen to electron carriers as the pH rises.

We previously proposed (*3*, *10*, *11*, 21) that Bohr effects in metalloproteins of the *b*-c_1 complex can result in redox-linked vectorial H^+ translo-

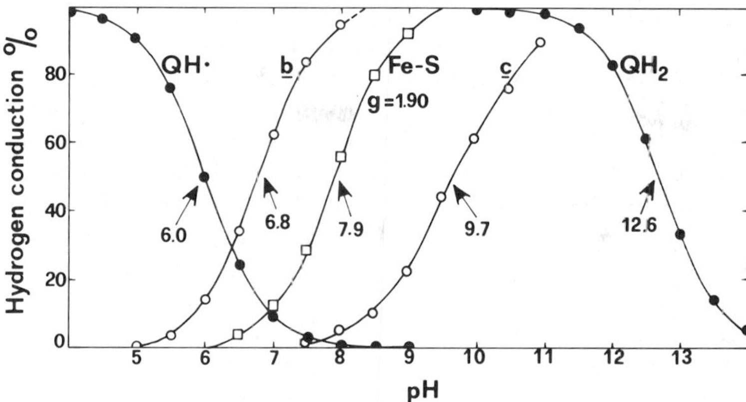

Fig. 1. Percentage of b and c cytochromes b and c, Fe–S protein (g = 1.90), ubiquinol and ubisemiquinol functioning as hydrogen carriers as a function of pH. For details, see text.

cation across the membrane, provided certain requirements are met. The first requirement is that the Bohr effects occur in the physiological pH range. This condition is met by the b cytochromes and the Rieske Fe–S protein. The second requirement is that the apoproteins be large enough to span the membrane. This seems to be true for some polypeptide subunits of Complex III (*22*). The third requirement is that the redox-linked ionized groups be arranged to constitute an H^+ channel that conducts H^+ uphill one way across the membrane.

In this paper we present direct measurements of redox–Bohr effects in the b-c_1 complex and experiments indicating direct involvement of Bohr effects and of the proton motive characteristics of semiquinones in the proton-pumping function of the b-c_1 complex.

EXPERIMENTAL RESULTS

Redox–Bohr Effects

Scalar proton-transfer reactions arising from redox–Bohr effects in electron carriers can be estimated from net pH changes associated with redox transitions (*23*) in isolated redox enzymes. Aerobic oxidation of electron carriers has to result in the consumption of an equivalent amount of H^+, and their reduction by hydrogenated reductants in the production of the same amount of H^+. Deficits in the consumption and production of H^+ are respectively a measure of H^+ release and H^+ binding to proton motive groups participating in redox–Bohr effects.

Figure 2 presents measurements of Bohr protons in isolated b-c_1 complex. The redox centers of the complex were reduced by succinate anaerobically by contaminating succinate dehydrogenase (*24*). Aerobic oxidation was then produced by purified cytochrome *c* oxidase and cytochrome *c* added in traces. Figure 2A shows that net aerobic oxidation of *c* and *b* cytochromes was accompanied, as expected, by H^+ consumption for protonation of reduced oxygen to H_2O. Net rereduction of the cytochromes by succinate, occurred after oxygen exhaustion, and

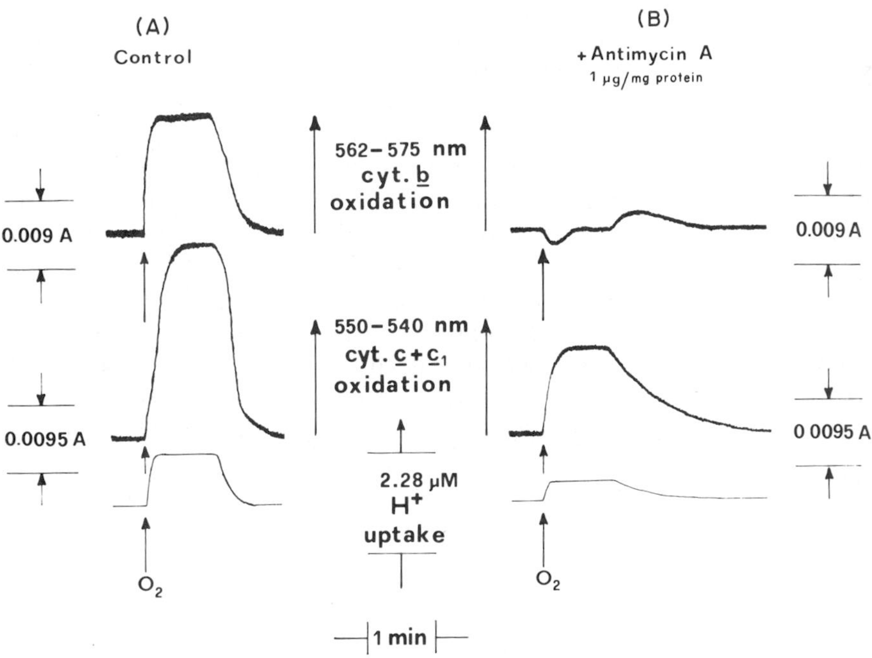

Fig. 2. Analysis of redox–Bohr effects in isolated b-c_1 complex, without (A) and with (B) Antimycin A. The b-c_1 complex (0.33 mg protein/ml) was incubated in: 200 m*M* sucrose, 30 m*M* KCl, 0.83 m*M* K-succinate, 0.5 m*M* K-malonate, 0.25 μ*M* cytochrome *c*, 0.005 mg purified cytochrome oxidase, and 4 μg/ml purified catalase, pH 7.2. Temperature 25 °C. Volume 2.29 ml. The incubation occurred under a continuous stream of Argon. After anaerobiosis, redox transitions were obtained by repetitive pulses of 1 μl of 0.1% H_2O_2. The amounts of *b* cytochromes were calculated by using $\Delta\epsilon_{mM} = 20$ at 562–576 nm. The amounts of cytochromes $c + c_1$ were calculated by using $\Delta\epsilon_{mM} = 19.1$ at 550–540 nm; the amount of cytochrome c_1 was obtained by substraction of the amount of exogenous cytochrome *c* (0.25 μ*M*). The Fe–S center g = 1.90 was taken as equivalent to the amount of cytochrome c_1.

resulted in release of the same amount of protons as was consumed on oxygenation. However, it can be seen that the amount of H^+ bound and subsequently released was less than half the sum of electron carriers oxidized. The difference gives the estimate of the protons exchanged as a result of redox–Bohr effects in the *b*-c_1 complex. Since it was shown that the midpoint redox potential of cytochrome c_1 is pH independent (*25*), the Bohr protons must be related to the redox transition of *b* cytochromes and Rieske Fe–S protein and thus exhibit an H^+/e^- coupling number of 1.15 (Table I).

In Fig. 2B a concentration of antimycin A that prevented any net redox change of *b* cytochromes was added. Under these conditions H^+ uptake was less than half the sum of electron carriers oxidized. Measurement of the Bohr protons, when related to net redox transition of the Rieske Fe–S protein, gave H^+/e^- coupling number of 1.5. The same H^+/e^- coupling number for Bohr protons in the Fe–S protein could be computed from the experiment in the absence of antimycin, when the observed oxidation of cytochrome *b* was corrected for the estimated extent of effective hydrogen conduction by this carrier as described in Fig. 1. Thus, the present data directly demonstrate the occurrence of redox–Bohr effects in the Rieske Fe–S protein. Note however that the measured H^+/e^- coupling number of 1.5 at pH 7.2 is considerably higher than expected from the reported pH dependence of the E_m of this carrier (*20*). It is possible that the extra H^+ released are derived from aerobic oxidation of protein bound QH_2 to $Q^{\overline{\cdot}}$, present in the isolated *b*-c_1 complex (*24*), according to

$$2\ QH_2 + \tfrac{1}{2}\ O_2 \rightarrow 2Q^{\overline{\cdot}} + 2\ H^+ + H_2O. \tag{2}$$

Clearly, further work is needed to characterize these important phenomena.

$\rightarrow H^+/e^-$ Stoichiometry

In an attempt to verify involvement of redox–Bohr effects of components of the *b*-c_1 complex and the protolytic reaction of semiquinones in the proton-pumping function of the complex, the effect of pH on the $\rightarrow H^+/e^-$ stoichiometry for proton release from mitochondria associated with electron flow in the cytochrome system from quinols to oxygen or cytochrome *c* was investigated.

Figure 3 illustrates the experimental procedure for measurement of the $\rightarrow H^+/e^-$ ratio for aerobic (Experiment A) and ferricyanide oxidation of duroquinol (Experiment B) (*26*). In Experiment A, duroquinol

TABLE I

Analytical Evaluation of Redox–Bohr Effects in Isolated b–c_1 complex[A]

Control				Antimycin 1 μg/mg protein			
H^+ Movements ng ions · ml^{-1}		e^- Carriers oxidation nmoles · ml^{-1}		H^+ Movements ng ions · ml^{-1}		e^- Carriers oxidation nmoles · ml^{-1}	
H^+ uptake	1.143	Cytochrome b	0.750	H^+ uptake	0.457	Cytochrome b	—
Bohr H^+	2.247	Cytochromes $c + c_1$	1.445	Bohr H^+	0.611	Cytochromes $c + c_1$	0.659
($\Sigma e^- - H^+$ uptake)		Fe–S	1.195	($\Sigma e^- - H^+$ uptake)		Fe–S	0.409
Bohr H^+/b, Fe–S = 1.15			3.390	Bohr H^+/Fe–S = 1.49			1.068

[A] For experimental conditions, see legend to Fig. 2.

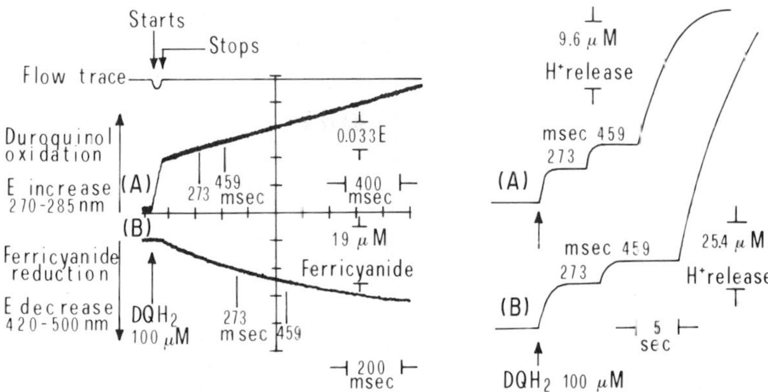

Fig. 3. Rapid kinetic analysis of proton translocation associated with aerobic (Experiment A) or ferricyanide (Experiment B) oxidation of duroquinol in beef heart mitochondria. Experimental conditions: (A), Mitochondria (2 mg of protein/ml) were preincubated for 5 min in the main syringe of the flow apparatus in an air-saturated medium containing: 150 m*M* KCl, 5 m*M* K-malonate, 0.5 μg of valinomycin/mg of protein, 0.5 μg of rotenone/mg of protein and 2 μg of oligomycin/mg of protein; pH 7.2. Mitochondria were then pulsed with 6 m*M* methanolic solution of duroquinol (final concentration in the mitochondrial suspension, 100 μ*M*). (B), mitochondria (2 mg/ml) were incubated for 10 min in a medium containing: 150 m*M* KCl, 5 m*M* K-malonate, 2 m*M* glycylglycine, 30 nmole of *N*-ethylmaleimide/mg of protein, 0.5 μg of valinomycin/mg protein and 0.5 μg of oligomycin/mg of protein. After this interval 2 m*M* KCN and 0.5 μg of rotenone/mg of protein were added and the pH adjusted to 7.2; 250 μ*M* ferricyanide was then added and after 5 min the mitochondrial suspension was pulsed with 6 m*M* methanolic solution of duroquinol (final concentration in the mitochondrial suspension, 100 μ*M*). [For other experimental details, see Papa *et al.* (*26*).]

was rapidly injected into an aerobic suspension of beef heart mitochondria, and duroquinol oxidation and proton release were monitored with a stopped-flow spectrophotometer and a continuous flow pH-meter, respectively. The rates of quinol oxidation and acidification were computed from the progress of the reaction from 273 to 459 msec. This interval was chosen because addition of the methanolic solution of duroquinol to the mitochondrial suspension caused a potentiometric artifact that added to the initial acidification caused by duroquinol oxidation, and the duroquinol solution contained traces of oxidized quinone. The $\rightarrow H^+/2e^-$ quotient so measured was, at pH 7.2, 3.6 ± 0.15 (*26*).

In Experiment B aerobic mitochondria supplemented with KCN were equilibrated with 250 μ*M* ferricyanide. Duroquinol (100 μ*M*) was then rapidly injected and the rate of ferricyanide reduction, monitored at 420–500 nm, and proton release were measured as described for Experiment A. The $\rightarrow H^+/2e^-$ ratio was 3.6 ± 0.04 (*26*). Note that in both experiments, the b-c_1 complex had already gone through a number of turnovers at the time the → H^+/e^- was measured.

Figure 4 illustrates an experiment in which electron flow in the b-c_1 complex was activated by rapidly adding ferricytochrome *c* to KCN-inhibited mitochondria supplemented with duroquinol. In this case the → H^+/e^- ratio was computed from the progress of the reaction at 45 msec after the oxidant pulse, the approximate time at which the b-c_1 complex had completed the first turnover. Ferricytochrome *c* was rapidly reduced by an antimycin-sensitive reaction associated with rapid H^+ release. The → H^+/e^- ratio computed from the antimycin-sensitive reactions was 3.6 ± 0.1 (*26*).

Figure 5 illustrates the effect of pH of the medium on the → H^+/e^- quotients, measured with the three approaches described. In the two cases of reductant pulses the → H^+/e^- ratios, measured under turning-over conditions of the b-c_1 complex, exhibited a peculiar pH dependence and were practically the same value irrespective of whether electrons were transferred to oxygen or to ferricyanide. This, incidentally, shows that cytochrome oxidase is not involved in the proton transloca-

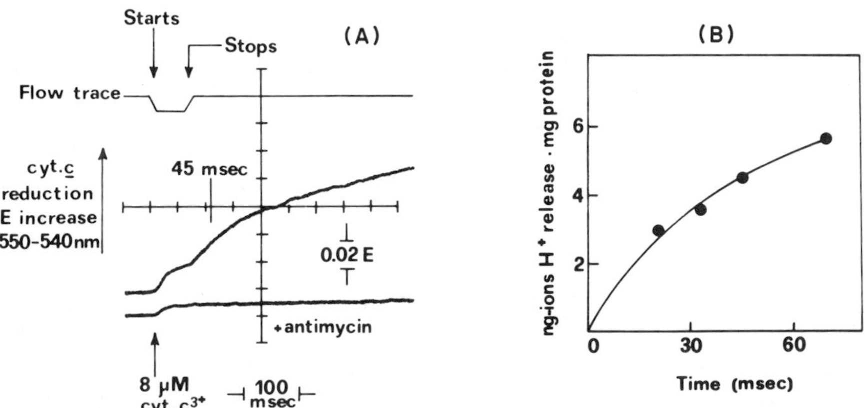

Fig. 4. (A) Reduction of exogenous ferricytochrome *c* by duroquinol, and (B) associated proton release in KCN-inhibited beef heart mitochondria. Mitochondria (1 mg of protein/ml) were incubated for 5 min in the main syringes of the flow apparatus in the medium described in Fig. 3, which contained also 1 m*M* KCN. After preincubation, a methanolic solution of duroquinol was added directly in the main syringes, to a final concentration of 100 μ*M*. After 2 min, the mitochondrial suspension was pulsed with ferricytochrome *c*, present in the side syringe as 0.48 m*M* solution in 180 m*M* KCl (the final concentration in the mitochondrial suspension was 8 μ*M*). The H^+ release shown was corrected for pH change caused by ferricytochrome *c* addition to antimycin supplemented mitochondria. [For other experimental details, see Papa *et al.* (*26*).]

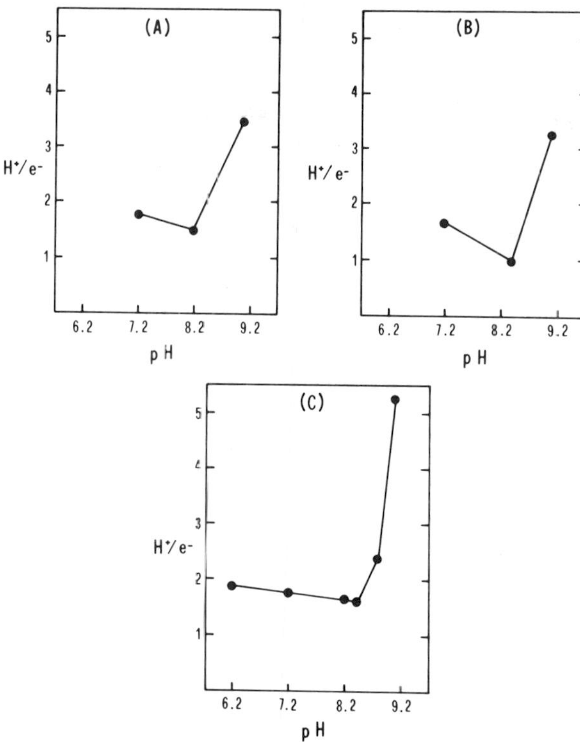

Fig. 5. pH dependence of H^+/e^- ratios for (A) aerobic oxidation of duroquinol, (B) ferricyanide oxidation of ubiquinol-1, and (C) cytochrome *c* reduction by duroquinol. Experimental conditions: (A), see Fig. 3A. (B), see Fig. 3B (the concentration of ferricyanide was 150 μM and 50 μM ubiquinol-1 was used as reductant), and (C), see Fig. 4. In Experiments A and B, the H^+/e^- ratios were corrected for nonenzymatic oxidation reaction of quinols. In Experiment C the H^+/e^- ratio was correct for the antimycin-insensitive reaction.

tion process, and is an activity of the b-c_1 complex (*26*). The $\rightarrow H^+/e^-$ ratio first decreased as the pH was raised from 7.2 to 8.2; from a value of 1.8 in both cases, to 1.4 in the case of the duroquinol $\rightarrow O_2$ reaction, and to one for the ubiquinol-1 $\rightarrow$ ferricyanide system. The more pronounced decrease observed for ubiquinol-1 is probably due to some direct reaction of ferricyanide with quinol or semiquinol. In both cases however, further increase of the pH to 9.1 raised the $\rightarrow H^+/e^-$ ratio to 3.3–3.4.

The single turnover duroquinol $\rightarrow$ ferricytochrome system produced the same pattern of pH dependence of the $\rightarrow H^+/e^-$ quotient. The values of the quotients were practically the same as those measured with the reductant pulse for pH values up to 8.1. Interestingly, the H^+/e^- quotient at pH 9 was much higher than in the reductant pulse, reaching a value around five.

DISCUSSION

The peculiar pH dependence of the → H^+/e^- stoichiometry for proton translocation associated with electron flow in the b-c_1 complex with quinols as the source of reducing equivalents, and the observation that this stoichiometry can exceed, (at alkaline pH) the value of two, cannot be explained in terms of direct ligand conduction mechanisms, such as the Q cycle (*14*, *15*) or two quinone redox-loops in series (*16*). In fact both schemes predict, in principle, a fixed → H^+/e^- quotient of two (→ $H^+/2e^-$ of 4) unless one invokes semiquinone dismutative reactions in addition to the specific one-electron transfer reaction characteristics of the two schemes. Thus, Mitchell and Moyle (Paper 7, Chapter VII, this volume) observed a → $H^+/2e^-$ stoichiometry of five in some conditions (low medium pH, or the presence of NEM), which declined to four by raising the pH. This decrease of the → H^+/e^- stoichiometry does, in fact, resemble what we observed in the pH range 7.2 to 8.2. It seems quite likely that the characteristic pH dependence of the → H^+/e^- stoichiometry described in this paper derives from the pH dependence of effective hydrogen transfer by semiquinone, *b* cytochromes and Fe–S protein of Complex III.

It is particularly significant that the → $H^+/2^-$ ratio changes from a value of two at neutral pH, to a value of three at pH 9. This could be accounted for by direct protonmotive catalysis by quinone at site 2, whereby 1 → H^+/e^- is derived directly from the scalar oxidation of quinol by electron acceptors of the b-c_1 complex, and an additional one or two H^+/e^- represent vectorial proton translocation resulting from successive transfer of reducing equivalents through a second system of protein-stabilized ubisemiquinone. At this site the → H^+/e^- stoichiometry would vary from one to two, depending on the ionization state of ubisemiquinone radical.

A simple direct ligand conduction mechanism by quinones, however, does not explain: (a) the decrease of the → H^+/e^- ratio below two at pH around eight; (b) the rise of the → H^+/e^- ratio in the pH range from eight to nine. In fact, ubisemiquinone with a pK of 6, or lower, will already be almost completely deprotonated at a pH below eight; (c) the observation that in the single turnover elicited by the cytochrome c^{3+} pulse, the → H^+/e^- ratio reaches the value of five at pH nine. All these observations

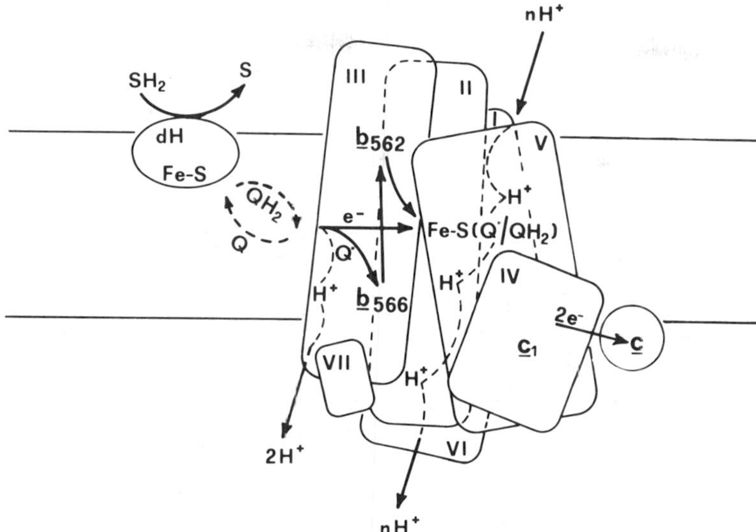

Fig. 6. Postulated mechanism of electron transfer and proton translocation in the b-c_1 segment of the respiratory chain.

taken together strongly suggest that redox–Bohr effects in the b-c_1 complex play a direct role in the proton-pumping function of the complex.

These ideas are schematically described in Fig. 6 for which we also utilized some of the available information on topology of the b-c_1 complex subunits (*22*). It is proposed that quinols are oxidized in two one-electron steps by the centers of the complex . The quinol would be oxidized by the Rieske Fe–S protein in an antimycin-insensitive reaction. The ubisemiquinone radical so produced would be oxidized by cytochrome b-566, which then transfers electrons to cytochrome b-562 followed by electron transfer to the Fe–S protein through an antimycin-sensitive reaction. Thus, the two electrons reunite at the level of the Fe–S protein, pass through ubisemiquinone associated with the Fe–S center and then pass to cytochrome c_1. The cytochrome b shunt explains the well-known phenomenon of the antimycin promoted oxidant-induced reduction of b cytochromes (*7*, *27*). Anisotropic protonation of the UQ·/UQH_2 couple from the inner M side of the membrane upon reduction, and proton liberation at the outer C side upon oxidation would result in transmembrane translocation of two to four H^+ per $2e^-$ traversing the Fe–S-UQ·-cytochrome c_1 system. In such a model, the benzenoid ring of quinone is conceived as being localized in the center of the membrane where it would oscillate laterally between the electron donor and electron acceptor. Proton access to the Fe–S-UQ· center from the matrix aqueous phase, and proton release to the outer aqueous space would occur through specific proton channels in the polypeptides (subunits V ?) of the complex; access of H^+ into the channel and release of H^+ on the opposite side would be favored by redox-linked pK shifts of ionizable groups in the channel (vectorial Bohr mechanism). The involvement of these groups would account for the high pK of the proton-pumping process, which, judging from the rise of the $\rightarrow H^+/e^-$ ratio at alkaline pH (Fig. 5), should fall between eight and nine and is thus much higher than that of the ubisemiquinone. Net oxidation of the redox centers of the Fe–S-UQ· center results in a decrease of the pK of ionizable groups, acting in series with their deprotonation as proton-transfer sites in the proton channel. This would explain the extra H^+ release/e^- observed during completion of the first turnover, in excess of the H^+ release by the turning-over enzyme, when only vectorial proton release occurs and the stoichiometry is dictated by the UQ·/UQH_2 couple.

However, the possibility that redox–Bohr effects alone are directly responsible for the proton-pumping process without intervention of the quinone system should also be considered. These schemes, and in particular the proposed involvement of specific ubisemiquinone systems and redox–Bohr effects, represent a reasonable basis for the design of further experimental work directed toward elucidation of the mechanism of proton pumps.

REFERENCES

1. Mitchell, P., and Moyle, J. (1967). *In* "Biochemistry of Mitochondria" (E. C. Slater, Z. Kaniuga, and L. Wojtczak, eds.), pp. 55–74. Academic Press, New York.
2. Leung, K. H., and Hinkle, P. C. (1975). *J. Biol. Chem.* **250,** 8467–8471.
3. Papa, S., Guerrieri, F., Lorusso, M., Izzo, G., Boffoli, D., and Capuano, F. (1977). *FEBS Symp.* No. 42, 502–519.
4. Alexandre, A., Reynafarje, B., and Lehninger, A. L. (1978). *Proc. Natl. Acad. Sci. U.S.A.* **75,** 5296–5300.
5. Pozzan, T., Miconi, V., Di Virgilio, F., and Azzone, G. F. (1979). *J. Biol. Chem.* **254,** 10200–10205.
6. Baum, H., Rieske, J. S., Silman, H. I., and Lipton, S. H. (1967). *Proc. Natl. Acad. Sci. U.S.A.* **57,** 798–805.
7. Wikstöm, M. K. F. (1973). *Biochim. Biophys. Acta* **301,** 155–193.
8. Rieske, J. A. (1976). *Biochim. Biophys. Acta* **456,** 195–274.
9. Mitchell, P. (1979). *Eur. J. Biochem.* **95,** 1–20.
10. Papa, S., Guerrieri, F., Lorusso, M., and Simone, S. (1973). *Biochimie* **55,** 703–716.

11. Papa, S. (1976). *Biochim. Biophys. Acta* **456,** 39–84.
12. von Jagow, G., and Engel, W. D. (1980). *FEBS Lett.* **111,** 1–5.
13. Mitchell, P. (1966). "Chemiosmotic Coupling in Oxidative and Photosynthetic Phosphorylation." Glynn Res., Bodmin, England.
14. Mitchell, P. (1975). *FEBS Lett.* **56,** 1–6.
15. Mitchell, P. (1976). *J. Theor. Biol.* **62,** 327–367.
16. Crofts, A. R., Crowther, D., and Tierney, G. V. (1975). *In* "Electron Transfer Chains and Oxidative Phosphorylation" (E. Quagliariello, S. Papa, F. Palmieri, E. C. Slater, and N. Siliprandi, eds.), pp. 233–241. North-Holland Publ., Amsterdam.
17. Kilmartin, J. V., and Rossi-Bernardi, L. (1973). *Physiol. Rev.* **53,** 836–889.
18. Clark, W. M. (1960). "Oxidation-Reduction Potentials of Organic Systems," Waverley Press, Baltimore, Maryland.
19. Urban, P. F., and Klingenberg, M. (1969). *Eur. J. Biochem.* **9,** 519–525.
20. Prince, R. C., and Dutton, P. L. (1976). *FEBS Lett.* **65,** 117–119.
21. Papa, S., Guerrieri, F., Lorusso, M., Izzo, G., Boffoli, D., and Stefanelli, R. (1978). *In* "Membrane Proteins" (P. Nicholls *et al.*, eds.), pp. 37–48. Pergamon, Oxford.
22. Bell, R. L., Sweetland, J., Ludwig, B., and Capaldi, R. A. (1979). *Proc. Natl. Acad. Sci. U.S.A.* **76,** 741–745.
23. Papa, S., Guerrieri, F., and Izzo, G. (1979). *FEBS Lett.* **105,** 213–216.
24. Nelson, B. D., and Gellerfors, P. (1978). *In* "Biomembranes: Part D: Biological Oxidations, Mitochondrial and Microbial Systems" (S. Fleischer and L. Packer, eds.), Methods in Enzymology, Vol. 53, pp. 80–91. Academic Press, New York.
25. von Jagow, G., Schägger, H., Engel, W. D., Hackenberg, H., and Kolb, H. G. (1978). *In* "Energy Conservation in Biological Membranes" (G. Schäfer and M. Klingenberg, eds.), Mosbach Colloquium, pp. 200–209. Springer-Verlag, Berlin and New York.
26. Papa, S., Guerrieri, F., Lorusso, M., Izzo, G., Boffoli, D., Capuano, F., Capitanio, N., and Altamura, N. (1980). *Biochem. J.* **192,** 203–218.
27. Chance, B., Wilson, D. F., Dutton, P. L., and Erecinska, M. (1970). *Proc. Natl. Acad. Sci. U.S.A.* **66,** 1175–1182.

Pathways of Respiration-Coupled H^+ Extrusion via Ubiquinone in Rat Liver Mitochondria

6

ADOLFO ALEXANDRE
ALBERT L. LEHNINGER

INTRODUCTION

It is now widely agreed that passage of a pair of electrons through site 2 of the respiratory chain of mitochondria, or the functionally equivalent portion of the electron-transport chain in photosynthetic bacteria, causes translocation of four H^+ from one side of the membrane to the other. It is also widely acknowledged that ubiquinone is a participant in the translocation of at least two and possibly all four H^+ ejected in site 2. However, several different mechanisms for the participation of ubiquinone in electron and H^+ transport have been proposed; some are considered elsewhere in this volume.

During a study of the stoichiometric relationship of H^+ ejection to electron transport during the oxidation of various substrates by rat liver mitochondria, we made some observations that raised questions about how ubiquinone functions during respiration-coupled H^+ translocation with different electron donors. In brief, we found that the oxidation of the site 2 substrates, succinate and glycerol 3-phosphate, by rotenone-treated mitochondria obtained from thyroxine-pretreated animals gave precisely identical H^+/O ratios close to 8, as well as identical $Ca^{2+}/2e^-$ uptake ratios close to 4 (*12*). It is well known that the substrate site of succinate dehydrogenase is located on the matrix or M side of the inner membrane (*7, 15, 17, 19, 21*), whereas glycerol phosphate dehydrogenase is on the C side (*9,*

Function of Quinones in Energy Conserving Systems

ISBN 0-12-701280-X

11, *23*, *25*). Therefore it was a reasonable assumption that substrate protons from succinate would be released to the M side and those from glycerol phosphate would be released to the C side. Since our stoichiometric data were identical for the two substrates, questions were raised about the fate of the substrate protons in each case. If the two H^+ derived from glycerol 3-phosphate were released to the C side, they would be scalar H^+ and should not require movement of a permeant charge-compensating cation into the mitochondria. Yet Ca^{2+} uptake data indicated that the same number of electric charges were separated during electron flow from the two substrates to oxygen.

We therefore undertook a comparative study of the stoichiometry of H^+ movements and charge separation in site 2 of the respiratory chain, with succinate and glycerol 3-phosphate as substrates and ferricyanide or ferricytochrome *c* as electron acceptor (*2*). In these experiments, rotenone was added to prevent electron flow from endogenous NAD-linked substrates and cyanide was present to prevent electron flow from cytochrome *c* to oxygen. The test systems were also supplemented with oligomycin to prevent proton movements associated with synthesis or breakdown of mitochondrial ATP, and with *N*-ethylmaleimide to block movements of H^+ on the phosphate/H^+ symporter. Proton ejection and electron flow were measured as described earlier (*3*). Electron flow was initiated with small pulses of the electron acceptor. Ca^{2+} was used in all experiments as the charge-compensating permeant cation, since Ca^{2+} is also required for maximal activation of glycerol 3-phosphate dehydrogenase. The mitochondria were obtained from rats which had been given thyroxine for a week to induce a high level of glycerol phosphate dehydrogenase activity, comparable to succinate dehydrogenase activity. Other details are in Alexandre *et al.* (*2*).

Experiments with Ferricyanide as Electron Acceptor

We readily confirmed earlier observations (*11*) that reduction of ferricyanide by glycerol phosphate in cyanide-treated rat liver mitochondria is not completely inhibited by antimycin A, indicating that ferricyanide accepts electrons to some extent from glycerol phosphate dehydrogenase or some electron carrier on the reducing side of the antimycin A block. Nevertheless, measurements of H^+ ejection coupled to the antimycin A-sensitive portion of ferricyanide reduction by glycerol 3-phosphate, found that the corrected $H^+/2e^-$ ratio for glycerol phosphate as electron donor was very close to four and equal to that obtained with succinate (*3*). However, the fairly substantial corrections required were undesirable. It was therefore necessary to seek an electron acceptor other than ferricyanide to obtain unambiguous stoichiometric data on movements of protons and electric charges during electron flow from glycerol phosphate and succinate through site 2.

Stoichiometric Measurements with Ferricytochrome *c* as Electron Acceptor

We found that ferricytochrome *c* could be used as electron acceptor with both substrates, provided that the liver mitochondria from thyroxine-treated rats were first divested of the outer membrane, to allow ferricytochrome *c* free access to the inner membrane. In control experiments, the mitoplasts (*6*) yielded $H^+/2e^-$ ejection ratios close to four with succinate as substrate, confirming earlier measurements in a number of laboratories with ferricyanide as acceptor in intact mitochondria (*3*). Moreover, under these conditions, electron flow from both succinate and glycerol 3-phosphate to ferricytochrome *c* was found to be inhibited over 97% by antimycin A. The mitoplast–ferricytochrome *c* system was therefore employed for comparative stoichiometric studies on glycerol phosphate and succinate as electron donors into site 2.

In a typical pair of experiments the $H^+/2e^-$ ejection ratio was found to be 3.70 with α-glycerol phosphate as substrate, and 3.72 with succinate as substrate. Both the reduction of ferricytochrome *c* and the ejection of H^+ were inhibited over 95% by the addition of antimycin A to both the succinate and glycerol phosphate systems. Variation of the amount of ferricytochrome *c* added in the pulses from 2 to 10 nmole per mg produced no significant change in the $H^+/2e^-$ translocation ratio with either substrate. When NEM was omitted from the test systems, the $H^+/2e^-$ ratios declined to about three with both substrates, as expected (*5*, *22*). When no substrate was added to the system the rate of reduction of ferricytochrome *c* was less than 3% of the rate in the presence of either succinate or glycerol phosphate.

In similar experiments, the uptake of the charge-compensating permeant cation Ca^{2+} was measured with a Ca^{2+} selective electrode during ferricytochrome *c* pulses in the presence of glycerol 3-phosphate or succinate as electron donors. The $Ca^{2+}/2e^-$ uptake ratios for both substrates were nearly identical at 0.96, equivalent to uptake of 1.92 positive electric charges transferred inward per $2e^-$, very close to the ratios observed in earlier measurements on the succinate-ferricyanide span (*3*). When FCCP was added to the otherwise complete system close to two H^+ appeared in the medium per $2e^-$ in the case of both substrates.

TABLE I

Observed Stoichiometric Relationships in the Span Substrate → Ferricytochrome *c*

Substrate	$H^+/2e^-$ Ejection ratio	$Ca^{2+}/2e^-$ Uptake ratio	$H^+/2e^-$ Formation ratio (FCCP)
Succinate	4	1	2
Glycerol 3-phosphate	4	1	2

To summarize the findings (Table 1), succinate and glycerol phosphate gave identical $H^+/2e^-$ ejection ratios with ferricytochrome *c* as acceptor, identical $Ca^{2+}/2e^-$ uptake ratios approaching 1.0, and identical $H^+/2e^-$ formation ratios in the presence of the protonphore FCCP. These data therefore confirm and extend our earlier observations that electron transport from either succinate or glycerol phosphate to oxygen via sites 2 and 3 give identical $H^+/2e^-$ ejection ratios close to 8.0 (*12*).

Sidedness of the Dehydrogenases

The equality of the stoichiometric ratios for site 2 for these two substrates, whose dehydrogenases were shown to be on opposite sides of the inner membrane, raises basic questions regarding the fate of both the electrons and protons removed from these substrates by their respective dehydrogenases. Since glycerol phosphate dehydrogenase, like succinate dehydrogenase, is a complex system containing flavin nucleotide and iron–sulfur centers and also passes reducing equivalents to ubiquinone and the b-c_1 complex, these two substrates in all probability feed reducing equivalents into a common pathway; therefore, both could be expected to activate respiration-coupled vectoral H^+ extrusion by common pathways.

The basic problems posed by the identical stoichiometries may now be pointed out. Since succinate dehydrogenase is located on the M side of the inner membrane, the electron pair originating from succinate must ultimately cross the membrane in order to reduce cytochrome *c*, located on the C side. However, there is no net translocation across the membrane of electrons originating from glycerol phosphate: they arise and terminate on the C side. The net movements of electrons across the membrane from the matrix to the medium in site 2 is therefore two with succinate, but zero with glycerol phosphate. Also, the sidedness of the release of the proton pairs removed from the two substrates by their dehydrogenases must be considered. The two H^+ removed from succinate by its dehydrogenase, whose active sites face the matrix, could be expected to appear in the matrix and the two H^+ removed from glycerol phosphate by its dehydrogenase, which faces the medium, could be expected to appear in the medium (Fig. 1). However, these assignments for the sidedness of substrate proton release from the two dehydrogenases must be reconciled with two sets of facts: (1) both dehydrogenases feed electrons into ubiquinone and the b-c_1 complex, and (2) both systems yield identical H^+ and Ca^{2+} stoichiometries. The models in Fig. 1 show that if substrate protons from the two dehydrogenases are released to opposite sides of the membrane, and subsequently four H^+ are translocated from the matrix to the medium while $2e^-$ from each substrate flow through site 2 of the electron-transport chain, then the stoichiometries observed with succinate would be consistent with the model. However, glycerol phosphate could be expected to cause the appearance of six H^+ in the medium (two scalar H^+ from the substrate and four vectorial H^+ from electron transport per se). Moreover, glycerol phosphate oxidation would require the uptake of two Ca^{2+} to compensate for the net loss of four H^+ from the matrix. Model 1, although consistent with the data for succinate, is clearly incorrect for oxidation of glycerol phosphate, which delivered a total of only four H^+ to the medium and caused uptake of only one Ca^{2+}.

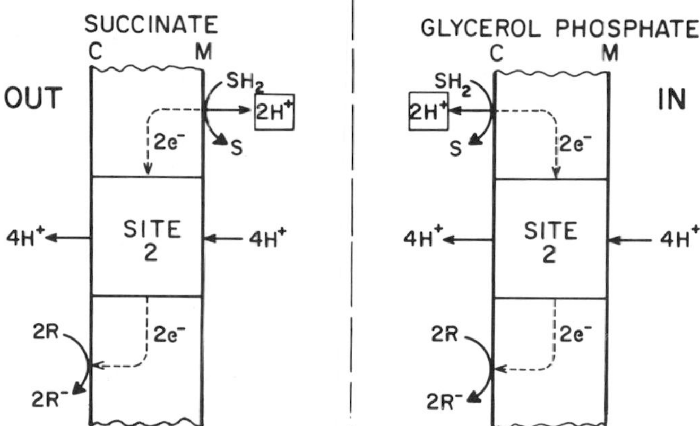

Fig. 1. Model 1 for the movement of H^+ and electron flow from succinate and glycerol phosphate to cytochrome c via energy-conserving site 2. This model assumes that the two substrate protons are discharged to the same side of the inner membrane on which the dehydrogenase active sites are located, and that subsequent flow of the corresponding electrons through site 2 causes translocation of four H^+ from the M side to the C side.

Other Models for Delivery of H^+ from Site 2

Several different models for the delivery of four H^+ into the C side coupled to electron transport from the two dehydrogenases have been considered, but only one, Model 2, can account in a simple way for the stoichiometric data (Fig. 2). It requires that succinate dehydrogenase deliver its two substrate protons into the medium rather than the matrix, that glycerol phosphate dehydrogenase also deliver its substrate protons into the medium, and that the subsequent common pathway of electron transport from both substrates to cytochrome *c* translocate vectorially only two H^+ per $2e^-$ from the matrix to the medium. In this model succinate oxidation would translocate a total of four H^+ to the outside, two H^+ arising from succinate via its dehydrogenase and two H^+ coupled to flow of $2e^-$ via the b-c_1 complex to cytochrome *c*. Glycerol phosphate dehydrogenase would deliver two H^+ directly into the medium (these would be scalar) and the subsequent electron transport via the b-c_1 complex to cytochrome *c* would translocate the other two H^+ from the matrix to the medium. Model 2 would also account for the $Ca^{2+}/2e^-$ uptake ratio of 1.0 in both cases. However, model 2 is incomplete, since it must also account for measurements indicating that reducing equivalents originating from NADH in site 1 probably also yield four H^+ on passing through site 2. The $H^+/2e^-$ ejection ratio for sites 1 + 2 together has been found to approach eight (*13*, *20*) and for sites 1 + 2 + 3 to approach 12 (*22*).

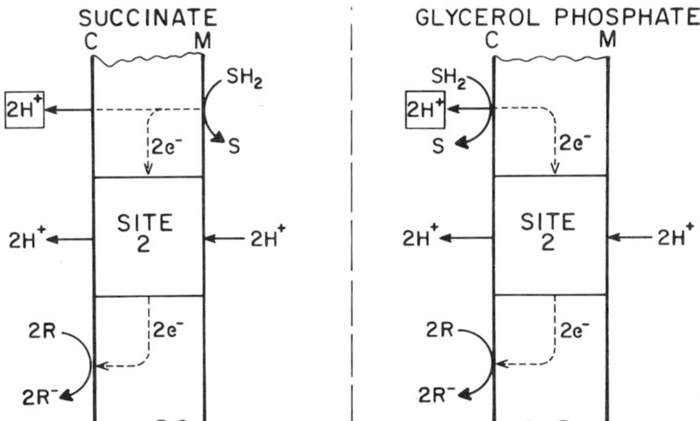

Fig. 2. Model 2 for the movement of protons and electrons from succinate and glycerol phosphate to cytochrome *c* via site 2. This model assumes that both succinate dehydrogenase and glycerol phosphate dehydrogenase deliver their substrate protons to the C side and that subsequent electron transport through site 2 translocates only two H^+ from the M side to the C side.

Experiments on the Sidedness of H^+ Extrusion by Succinate Dehydrogenase using *N,N,N,N'*-Tetramethylphenylenediamine as Electron Acceptor

Additional information on the sidedness of succinate proton delivery and the role of ubiquinone in H^+ translocation in site 2 was obtained by a different approach. The b-c_1 complex was blocked with antimycin A and electrons were conducted from succinate around the antimycin A block using *N,N,N,N'*-tetramethylphenylenediamine (TMPD) as electron carrier to ferricyanide. In this system the b-c_1 complex is bypassed, eliminating whatever vectorial proton movements are coupled to electron transport through the b-c_1 complex. In appropriate control experiments it was confirmed that TMPD can bypass the antimycin-sensitive block during electron flow from either succinate or NAD-linked substrates to oxygen. Moreover, it was determined by different methods that, under these experimental conditions, neither TMPD nor its reduction product $TMPD^+$ (Wurster's blue), acted as a transmembrane protonophore (*2*).

To determine at what point TMPD accepts electrons during oxidation of succinate, the inhibitor thenoyltrifluoroacetone (TTFA) was employed. TTFA inhibits electron flow from succinate to ubiquinone, but does not inhibit highly purified preparations of succinate dehydrogenase. ESR measurements indicate that TTFA inhibits electron flow from iron–sulfur center S-3 to ubiquinone (*1*). We found that TTFA inhibits succinate oxidation via the TMPD bypass around the antimycin A block, to the extent of 85% or more. Thus TMPD accepts electrons at some point between the site of TTFA inhibition and the site of antimycin A inhibition, as indicated in the following scheme

$$
\begin{array}{ccccccccc}
 & & & \text{TTFA} & & & & \text{Antimycin A} & \\
\text{Succinate} & \longrightarrow & [\text{Fe–S}] & \dashv\!\!\longrightarrow & \text{Q} & \longrightarrow & b & \dashv\!\!\longrightarrow & c_1 \\
 & & & & \downarrow & & & & \\
 & & & & \text{TMPD} & & & &
\end{array}
$$

While different proposals were made for the sequence of interaction of ubiquinone with members of the b-c_1 complex, under our conditions it appears most probable that TMPD accepts electrons from a reduced form of ubiquinone, as indicated in the above scheme.

Use of the TMPD bypass around the antimycin block allows determination of the side of the mitochondrial membrane from which protons from succinate are delivered by ubiquinone, whether into the M space or C space. To make such experiments unambiguous, ferricyanide was used as the terminal electron acceptor in the presence of cyanide, thus avoiding

H^+ translocation as electrons go through cytochrome oxidase. If the substrate protons of succinate are delivered directly into the C space, it would be reasonable to expect them to appear in the medium promptly and without lag, by an electroneutral process. Their appearance should not be hastened by addition of the protonophore FCCP. If the substrate protons are delivered into the M space, it could be expected that their appearance in the C space would be delayed or slowed, in which case FCCP should hasten their appearance. We therefore measured the rate of appearance of H^+ in the medium coupled to electron flow from succinate to ferricyanide via the TMPD bypass in antimycin A-treated mitochondria, as well as the effect of FCCP. In the absence of FCCP, H^+ immediately appeared in the medium without a lag period during succinate oxidation. Moreover, when FCCP was added there was absolutely no acceleration of the rate of H^+ appearance in the medium, even when the rate of electron flow was varied over a wide range by increasing the TMPD concentration. Under no conditions was a lag observed in the appearance of H^+ in the medium nor was any rate enhancement afforded by FCCP. These and other tests indicate that the H^+ are delivered into the C space during oxidation of succinate via ubiquinone (*2*).

An Explicit Model of H^+ Extrusion in Site 2

Results with the TMPD bypass are consistent with Model 2 (Fig. 2) for H^+ movements associated with succinate oxidation. They also allow postulation of a more explicit model for the role of ubiquinone in H^+ extrusion from site 2 substrates as shown in Fig. 3. This model proposes that substrate hydrogens from all incoming sources (succinate, glycerol 3-phosphate, as well as other flavin-linked site 2 substrates), are transferred within the membrane to ubiquinone, reducing it to ubiquinol. Oxidation of the ubiquinol by the b-c_1 complex would deliver the site 2 substrate H^+ to the medium; this portion of site 2 is designated site 2A. We assume that the other two H^+ that appear in the medium during passage of $2e^-$ through site 2 are translocated from the M space to the C space by electron flow through the b-c_1 complex, either by a second ubiquinone-dependent step (for review, see *8*; see also *14*, *16*) or by pumping via cytochrome *b* (*18*, *24*); this step is designated site 2B. The species of ubiquinol involved in site 2A must have appropriate sidedness, so that when it is oxidized protons appear only on the C side, whatever the sidedness of the original electron donor. Free ubiquinol is unlikely to have such sidedness, but if ubiquinol is oxidized while bound to a protein of the dehydrogenase com-

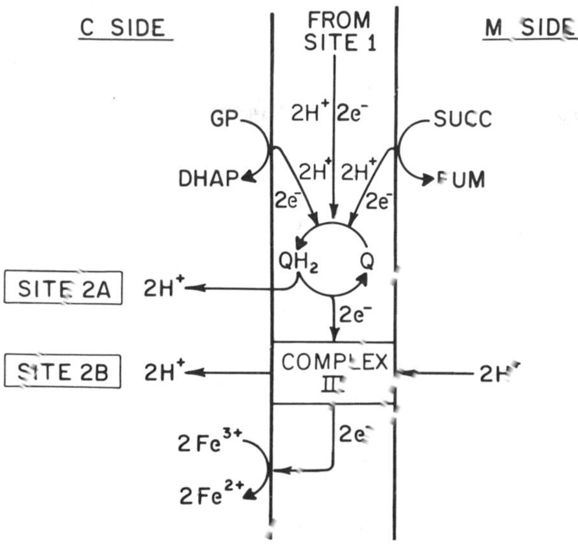

Fig. 3. An explicit model for electron flow and H^+ movements from succinate and glycerol 3-phosphate in site 2. This model, the only model that accounts for the observed stoichiometries in Table I, proposes that the two substrate protons from both succinate and glycerol phosphate are passed to ubiquinone (Q) within the membrane, in parallel with the transfer of electron pairs to Q via iron–sulfur centers associated with the dehydrogenases. The ubiquinol (QH_2) so formed is then reoxidized by the b-c_1 complex in such a way that the two quinol protons are delivered to the C side. Subsequent electron transport through the b-c_1 complex to cytochrome c causes translocation of two H^+ from the M side to the C side. The NADH dehydrogenase system of site 1 is also proposed to pass two $H^+ + 2e^-$ to ubiquinone within the membrane, as shown for succinate and glycerol phosphate. This formation differs from traditional views on the action of ubiquinone, which is usually assumed to acquire two H^+ from the M compartment on its reduction to ubiquinol.

plex or the b-c_1 complex, or to some other protein of fixed symmetry, then the appropriate sidedness of H^+ release could be assured.

This model for site 2 proton movements (*2*) must be clearly distinguished from the more traditional models for the role of ubiquninone in H^+ translocation. These traditional models propose that ubiquinone functions as a proton translocator by taking up the H^+ from the matrix space during its reduction to ubiquinol and delivers them to the medium on reoxidation to ubiquinone. The traditional Q cycle models are incompatible with our data since they cannot account for the identical H^+ and charge separation stoichiometries for both glycerol 3-phosphate and succinate oxidation. The model we proposed envisions that ubiquinone collects substrate hydrogens from within the membrane, rather than as H^+

from the matrix. As ubiquinol is reoxidized H^+ are delivered to the medium and electrons to the b-c_1 complex.

The model in Fig. 3 has other implications regarding the possible role of iron–sulfur centers in the transfer of reducing equivalents from substrates to ubiquinone within the membrane. We suggest that the pairs of substrate hydrogens are transferred within the membrane in the form of two $H^+ + 2e^-$, in such a way that two H^+ are transferred from the dehydrogenase to ubiquinone in some intramembrane domain, in parallel with the transfer of $2e^-$ from reduced flavin to ubiquinone via the iron–sulfur center(s). This scheme is not only consistent with our experimental data but also provides a feasible route for substrate hydrogen atoms to be transferred to ubiquinone without entering the matrix space.

Finally, the data and considerations developed here must also be compatible with the P/O ratios of oxidative phosphorylation with succinate and glycerol phosphate. The model in Fig. 3 is consistent with our data showing that two electric charges are separated across the membrane as reducing equivalents pass from either succinate or glycerol phosphate to cytochrome c, and implies that the P/O ratios for the oxidation of succinate and glycerol phosphate should be identical, since electron flow from cytochrome c to oxygen is in a common pathway for both substrates. The available data on the P/O ratios support this view: glycerol phosphate oxidation gives P/O values equal to those of succinate oxidation (*10*). In view of our earlier finding that electron flow from cytochrome c to oxygen separates at least five and possibly six charges (*4*), it would be expected that site 2 delivers 0.5 ATP and site 3 delivers 1.5 ATP, to make a theoretical total of two ATP per atom of oxygen reduced from either succinate or glycerol 3-phosphate. Further experiments are under way to test the pathways that are postulated in Fig. 3.

CONCLUSIONS

Most studies of H^+ extrusion coupled to mitochondrial electron transport through site 2 have employed succinate as the electron donor, which introduces reducing equivalents from the M side of the membrane, where its dehydrogenase is located. However, comparative experiments on rat liver mitochondria from thyroxinized rats with glycerol phosphate as electron donor, which furnishes reducing equivalents to ubiquinone from glycerol phosphate dehydrogenase on the C side, place important limitations on the site and mechanism of H^+ extrusion in site 2. The H^+/O ejection ratios for succinate and glycerol phosphate (sites 2 + 3) are identical at close to 8.0. Moreover, the $H^+/2e^-$ ejection ratios for antimycin-sensitive electron flow from succinate and from glycerol phosphate to ferrocytochrome *c* in mitoplasts are also identical at close to 4.0. In addition, the charge separation ratios with Ca^{2+} were also identical for the two substrates. These observations thus indicate that the site 2 H^+ extruding reactions must be organized in such a way that the *substrate* protons from both substrates appear on the C side of the membrane, even though the dehydrogenase active sites are on opposite sides. Experiments on the sidedness of H^+ release from succinate when the b-c_1 complex is inhibited by antimycin A and by-passed with TMPD show that 2 H^+ from succinate appear directly on the C side of the membrane in a TTFA-inhibited reaction. To accommodate these and other observations it is concluded that the substrate protons from both succinate and glycerol phosphate, and indeed all flavin-linked substrates, regardless of the sidedness of their dehydrogenases, are transferred to ubiquinone *within* the membrane and are then released from ubiquinol to the C side by a protolytic dehydrogenation step (site 2A). The second pair of H^+ released in site 2 are proposed to be translocated from the matrix to the medium on passage of $2e^-$ through the cytochrome b-c_1 complex (site 2B).

REFERENCES

1. Ackrell, B. A. C., Kearney, E. B., Coles, C. J., Singer, T. P., Beinert, H., Wan, Y.-P., and Kolkers, K. (1977). *Arch. Biochem. Biophys.* **182,** 107–117.
2. Alexandre, A., Galiazzo, F., and Lehninger, A. L. (1980). *J. Biol. Chem.* **255,** 10721–10730.
3. Alexandre, A., and Lehninger, A. L. (1979). *J. Biol. Chem.* **254,** 11555–11560.
4. Alexandre, A., Reynafarje, B., and Lehninger, A. L. (1978). *Proc. Natl. Acad. Sci. U.S.A.* **75,** 5296–5300.
5. Brand, M. D., Reynafarje, B., and Lehninger, A. L. (1976). *J. Biol. Chem.* **251,** 5670–5679.
6. Chan, T. L., Pedersen, P. L., and Greenawalt, J. W. (1970). *J. Cell Biol.* **45,** 291–305.
7. Chappell, J. B., and Haarhoff, K. N. (1967). *In* "Biochemistry of Mitochondria" (E. C. Slater, Z. Kaniuga, and L. Wojtczak, eds.), pp. 75–91. Academic Press, New York.
8. Crofts, A. R., and Wood, P. M. (1978). *Curr. Top. Bioenerg.* **7,** 175–244.
9. Donnellan, J. F., Barker, M. D., Wood, J., and Beechey, R. B. (1970). *Biochem. J.* **120,** 467–478.
10. Ernster, L., and Nordenbrand, K. (1967). *In* "Oxidation and Phosphorylation" (R. W. Estabrook and M. D. Pullman, eds.), Methods in Enzymology, Vol. 10, pp. 86–94. Academic Press, New York.
11. Klingenberg, M., and Buchholz, M. (1970). *Eur. J. Biochem.* **13,** 247–252.
12. Lehninger, A. L., Reynafarje, B., and Alexandre, A. (1977). *In* "Structure and Function of Energy-Transducing Membranes" (K. van Dam and B. V. van Gelder, eds.), pp. 95–106. Elsevier/North-Holland, Amsterdam.

13. Lehninger, A. L., Reynafarje, B., Alexandre, A., and Villalobo, A. (1979). *In* "Membrane Bioenergetics" (C. P. Lee, G. Schatz, and L. Ernster, eds.), pp. 393–404. Addison-Wesley, Reading, Massachusetts.

14. Malviya, A. N., Nicholls, P., and Elliott, W. B. (1980). *Biochim. Biophys. Acta* **589,** 137–149.

15. Merli, A., Capaldi, R. A., Ackrell, B. A. C., and Kearney, E. B. (1979). *Biochemistry* **18,** 1393–1400.

16. Mitchell, P. (1976). *J. Theor. Biol.* **62,** 327–367.

17. Palmieri, F., Prezioso, G., Quagliariello, E., and Klingenberg, M. (1971). *Eur. J. Biochem.* **22,** 66–74.

18. Papa, S. (1976). *Biochim. Biophys. Acta* **456,** 39–84.

19. Papa, S., Lofrumento, N. E., Loglisci, M., and Quagliariello, E. (1969). *Biochim. Biophys. Acta* **189,** 311–314.

20. Pozzan, T., Miconi, V., Di Virgilio, F., and Azzone, G. F. (1979). *J. Biol. Chem.* **254,** 10200–10205.

21. Quagliariello, E., and Palmieri, F. (1968). *Eur. J. Biochem.* **4,** 20–27.

22. Reynafarje, B., Brand, M. D., and Lehninger, A. L. (1976). *J. Biol. Chem.* **251,** 7442–7451.

23. Scott, D. M., Storey, B. T., and Lee, C.-P. (1978). *Biochem. Biophys. Res. Commun.* **83,** 641–648.

24. von Jagow, G., and Engel, W. D. (1980). *FEBS Lett.* **111,** 1–5.

25. Wohlrab, H. (1977). *Biochim. Biophys. Acta* **462,** 102–112.

Protonmotive Mechanisms of Quinone Function

7

PETER MITCHELL
JENNIFER MOYLE

INTRODUCTION

It is now generally agreed that conservation of energy by redox chain and photoredox chain systems in mitochondria, bacteria, and chloroplasts depends on their protonmotive function. Therefore, our considerations of the function of quinones in electron-transfer and energy-conserving systems center on the role they play in coupling electron transfer to proton translocation (*1*, *2*).

In this paper, our main object is to obtain some new insights concerning the molecular mechanisms of the protonmotive ubiquinol–cytochrome *c* oxidoreductase of mitochondria (*3–5*), the ubiquinol–cytochrome c_2 oxidoreductase of photosynthetic bacteria (*5–8*), and the plastoquinol–plastocyanin oxidoreductase of chloroplasts (*8–10*); all appear remarkably similar in structure and function and are conveniently described collectively as QH_2 dehydrogenases.[1] For brevity, we may also refer to these

[1] Abbreviations: Q, quinone; QH, semiquinone; Q·, semiquinone or semiquinone anion; UQ, ubiquinone; PQ, plastoquinone; Fe–S, iron–sulfur center; PC, plastocyanin; d, dehydrogenase or electron donor; deH, dehydrogenase; succ, succinate; Ch., choline; Gly-Gly, glycylglycine; EGTA, ethyleneglycol-*bis*aminoethyltetraacetic acid; val, valinomycin; FCCP, carbonylcyanide *p*-trifluoromethoxyphenylhydrazone; rot, rotenone; ant, antimycin; NEM, *N*-ethylmaleimide; HQNO, 2-heptyl-4-hydroxyquinoline *N*-oxide; UHDBT, 5-*n*-undecyl-6-hydroxy-4,7-dioxobenzothiazole; DBMIB, 2,5-dibromo-6-isopropyl-3-methylbenzoquinone; BAL, 2,3-dimercaptopropanol; pH_o, outer pH; pH_i, inner pH; $\leftarrow H/e^-$ and $\rightarrow e^-/e^-$, numbers of hydrogen atoms and electrons, respectively, translocated per electron

Function of Quinones in Energy Conserving Systems

ISBN 0-12-701280-X

complex protonmotive osmoenzymes as *c* reductase, c_2 reductase, and PC reductase, respectively. They have been difficult to recognise as physically discrete protein molecule species, because their insolubility in water has presented special technical problems in separation and purification. However, it now appears legitimate to regard *c* reductase as a well-defined protein molecule species that is plugged through the fluid–lipid bilayer membrane in such a way that its central hydrophobic and lipophilic region acts as a specific osmotic barrier domain that is continuous with the hydrocarbon osmotic barrier domain B of the primary membrane (*3–5, 11–14*), as illustrated crudely in Fig. 1. The monomeric unit of *c* reductase contains four redox-functional centers (*3–5*): the Rieske iron–sulfur center (Fe–S), and cytochromes c_1, *b*-566, and *b*-562. It also contains UQ. However, this may not be a permanent prosthetic group, but may represent UQ that is bound reversibly as substrate in catalytic Q-reactive centers (*15*) that stabilize transition states (*1, 16*), such as the quinol anion QH^- (*17*), the semiquinone QH, or the semiquinone anion Q^- (*18–22*), perhaps in the iron–sulfur protein and in the cytochrome *b* subunits (*4, 21, 22*). The c_2 reductase and the PC reductase resemble the *c* reductase in containing the Rieske iron–sulfur center and cytochrome *b* (*7, 11, 23*). In *Rhodopseudomonas sphaeroides,* c_2 reductase has been shown by Wood (*24*) to contain a high molecular weight cytochrome *c*, resembling cytochrome c_1, whereas the PC reductase contains cytochrome *f* instead of cytochrome c_1 (*9–11*). It is clear that the c_2 and PC reductases are strikingly similar to *c* reductase in composition and structure, but they are not yet sufficiently well characterized to permit quantitative comparisons.

Both the fully reduced and the fully oxidized forms of the UQH_2/UQ and PQH_2/PQ couples are so hydrophobic and lipophilic that they are largely confined to the hydrocarbon B domain of the membrane, where their average concentration is some 20 times that of the cytochromes c_1 or *f* of the QH_2 dehydrogenases. Recent work by Hauska and colleagues (*25–27*) and by Quinn and Esfahani (*28*) has shown that, despite the large bulk of the polyisoprenoid chains, UQH_2/UQ and PQH_2/PQ couples are remarkably mobile in the B domain of the membrane. Thus, they may not only function as the laterally mobile pool of hydrogen in the form of the reducing substrate for the QH_2 dehydrogenases in the membrane, as originally conceived by Kröger and Klingenberg (*29*), but they may also act as good conductors of hydrogen from one side of the B domain of the mem-

transferred: *a*, *b*, *c*, and *f*, cytochromes identified also by suffixes; P, N, and B, protonically positive, protonically negative and osmotic barrier domains, respectively; o and i, Q-reactive centers on the protonic output or P, and protonic input or N sides, respectively, of QH_2 dehydrogenases.

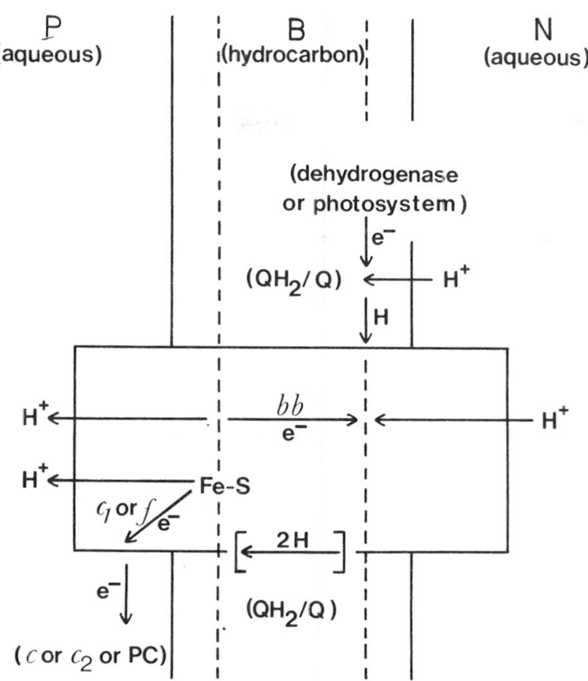

Fig. 1. Transmembrane structural and functional topology of QH_2 dehydrogenases. The dashed lines represent the surfaces of the nonaqueous osmotic barrier domain B, which is continuous between the hydrocarbon domain of the primary membrane and the QH_2 dehydrogenase represented by the rectangular box plugged through it. The protonically positive and negative aqueous domains are denoted by P and N, respectively. Italic letters represent cytochromes, and Fe–S and PC represent the Rieske iron–sulfur center and plastocyanin, respectively.

brane to the other. Particularly important consequences may stem from this special localization and specific hydrogen-conducting function of the electron-donor substrates for the QH_2 dehydrogenases. Likewise, it appears to be particularly significant that, as indicated in Fig. 1, electron acceptors for the QH_2 dehydrogenases, cytochromes c, c_2, or PC, are located at the P surface of the membrane, where they may be able to undergo rapid translational and/or rotational diffusion at the membrane surface between their electron donor and their electron acceptor, cytochromes c_1 or f, located at the P pole of the QH_2 dehydrogenase molecule (*30–33*). To facilitate comparisons between mitochondrial, chromatophore, and chloroplast QH_2 dehydrogenases, we use the symbols P and N to refer to the aqueous proton-conducting domains that are normally at relatively positive and negative protonic potential energies, respectively, on either side of the hydrocarbon B domain of the membrane.

All the research groups studying the chemiosmotic stoichiometry of the QH_2 dehydrogenases of mitochondria (*14*) and chromatophores (*6–8*) agree that, under the usual experimental conditions, for each hydrogen atom (or $e^- + H^+$) donated to the reductase, and for each electron accepted by cytochromes *c* or c_2, two protons appear on the P side, one proton disappears on the N side, and one electron is conducted from the P to the N side of the B domain (Fig. 1). In other words, the $\leftarrow H/e^-$ and $\rightarrow e^-/e^-$ ratios are two and one, respectively. There is increasing experimental support for the view that the $\leftarrow H/e^-$ and $\rightarrow e^-/e^-$ ratios may likewise be two and one, respectively, for the QH_2 dehydrogenase of chloroplasts (*1*, *34–38*).

Figure 1 is designed to show the net transmembrane translocation processes per electron transferred, and the general topological arrangement of the QH_2 dehydrogenases without specifying particular ligand-conduction connections through the B domain surfaces denoted by the dashed lines. The diagram emphasizes that the essential process of conduction of hydrogen from N to P (denoted by the arrow in the square brackets), could occur either via specifically bound Q that is conformationally mobile in the polypeptide system of the B domain of the QH_2 dehydrogenase, or via the Q pool, (denoted by QH_2/Q), in the hydrocarbon B domain of the primary membrane, or both (*1*, *14*, *15*). There is much less uncertainty about the electron-translocating function of the QH_2 dehydrogenases, which may reasonably be attributed to the *b* cytochromes, denoted by *bb*, that are known to be located in the hydrophobic B domain of the molecule, where their redox state appears to be influenced by the electric potential difference between the P and N domains (*39*).

Our representation of Fe–S as the electron donor for cytochromes c_1 or *f* in Fig. 1 is based on work by Trumpower and colleagues on the extraction and reconstitution of the Rieske iron–sulfur protein in cytochrome *c* reductase preparations (*40*), on recent work by Bowyer and colleagues (*41*) in which the UQ analog 5-*n*-undecyl-6-hydroxy-4,7-dioxobenzothiazole (UHDBT) inhibited electron transfer in bacterial chromatophores between Fe–S and cytochrome c_2, and on Malkin and Posner's observation (*42*) that a mutant of *Lemna perpusilla* with a block of electron transfer between PQ and cytochrome *f* lacked the $g = 1.89$ paramagnetic resonance signal of the Rieske iron–sulfur protein.

The fundamental question of chemiosmotic molecular mechanism in the QH_2 dehydrogenase is: how does the topology and chemistry of the ligand-conducting components couple the scalar redox throughput of each reducing equivalent to the vectorial conduction of 2 H from N to P and of

1 e^- from P to N? Our attempts to answer this question were based on two closely related ligand-conduction concepts: the linear redox loop and the Q cycle (*1*). Since the only known hydrogen conductor in the QH_2 reductases appeared to be UQ or PQ, various linear Q-loop and Q-cycle schemes were suggested as being the most promising for describing the molecular mechanism (*1*, *16*).

The ligand-conduction diagrams in Fig. 2 illustrate that the essential difference between linear and cyclic types of scheme for QH_2 dehydrogenases is that the Q-loop type of scheme (A) is conceived as translocating the two hydrogen atoms through two consecutive hydrogen conductors that act as electron-proton symporters, connected in series through an electron conductor; whereas the Q-cycle type of scheme (B) is conceived as translocating the two hydrogen atoms in parallel through a single 2H conductor that is connected to a closed electron-conducting loop through which one of the electrons recirculates, so that the system catalyzes the symport of two protons with one electron. Diagrams (C) and (D) correspond to (A) and (B), respectively, and include possible proton-translocating redox species of Q. In these diagrams, and elsewhere in this paper, we represent the semiquinone radical as QH and the anion as Q^- (without a dot because it is generally known that these molecular species contain an unpaired electron). In the Q-cycle diagram of Fig. 2D, Q-reactive redox centers are denoted by o and i. The o and i originally stood for outer and inner centers in the *c* reductase of mitochondria, but they may also be used to represent Q-reactive protonic output and input centers, respectively, in all QH_2 dehydrogenases.

The tightness of coupling of electron transfer to proton translocation in the Q-loop and Q-cycle schemes described by Fig. 2 depends on the topo-

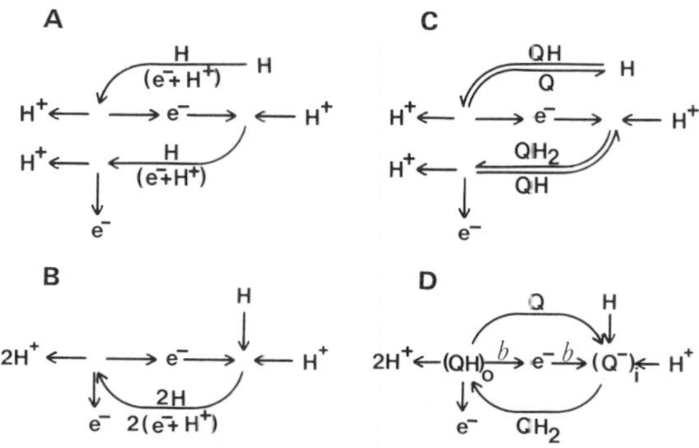

Fig. 2. Ligand-conduction diagrams of linear Q loop (A and C) and Q cycle (B and D) mechanisms in QH_2 dehydrogenases.

logical connectedness of four one-electron redox reactions corresponding to oxidation and reduction of the couples Q^{2-}/Q^- and Q^-/Q, that interact via specific ligand-conducting pathways. To facilitate comparison, the Q-loop and Q-cycle schemes are shown in Fig. 3 as alternative ligand-conducting connections between the same set of four redox components. The d represents a dehydrogenase or other electron donor system. The specific binding of cytochromes *b*-562 or b_i, and inhibition of cytochrome *b* oxidation by antimycin (*3–8, 18–20, 34, 35, 43, 44*), has been, and continues to be, extremely useful for distinguishing between linear-loop and Q-cycle schemes, because it greatly facilitiates observations on the so-called oxidant-induced reduction of cytochrome *b* (*45, 46*) that is diagnostic of the assumed interdependence of the reduction of oxidized Fe–S and cytochrome b_o at center o of the Q cycle (*1, 4, 46*). Recently, Bowyer and Trumpower (*47*) directly demonstrated the rapid reduction of cytochrome c_1 during cytochrome *b* reduction in mitochondrial cytochrome *c* reductase in the presence of antimycin, and showed that this reaction, which is diagnostic of center o of the Q cycle, is specifically inhibited by the Q analog UHDBT (*48*). Correspondingly, Surkov and Konstantinov (*49*) observed that the degree of oxidation of cytochrome *b* by the autoxidizable Q analog 2,5-dibromo-6-isopropyl-3-methylbenzoquinone (DBMIB) is enhanced by antimycin, which presumably inhibits cytochrome *b* reduction at center i under these conditions. Collectively, these results would be very difficult to explain in terms of a linear Q-loop scheme.

The ligand-conduction diagrams of Fig. 3 show that although the overall chemiosmotic stoichiometry of the Q-loop and Q-cycle systems is the same, and both ligand-conduction systems could operate alternatively in a

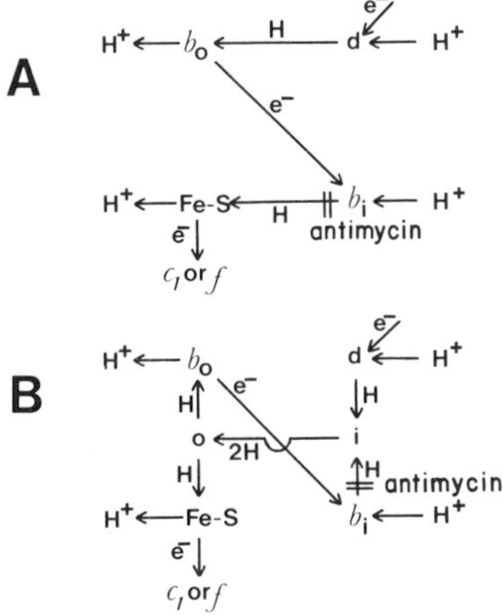

Fig. 3. Comparative ligand-conducting connections in linear Q loop (A) and Q cycle (B) schemes. Italic letters represent cytochromes, and Fe–S and d represent the Rieske center and a dehydrogenase or other electron (hydrogen) donor, respectively. The site of inhibition by antimycin is marked antimycin. The symbols o and i represent Q-reactive centers on the protonic output and input sides of the QH_2 dehydrogenase (corresponding to P and N), respectively. Cytochromes *b* have suffixes o and i to denote their location.

given QH_2 dehydrogenase molecule, they could not operate simultaneously, or alternately in rapid succession, without uncoupling or change of stoichiometry (*16*). For example if, in the Q-cycle scheme of Fig. 2D or Fig. 3B, half the QH produced by oxidation of QH_2 at center o escaped from the P to the N side of the B domain (in accordance with the Q-loop scheme) and was reduced to QH_2 at center i, the resulting dismutation of QH across the B domain would give rise to $\leftarrow$ H/e^- and $\rightarrow$ e^-/e^- ratios of 1.5 and 0.5 instead of two and one, respectively.

The understanding of the somewhat novel problems of electrostatic effects in chemiosmotic redox systems (*50*), such as those posed by the ligand-conduction diagrams of Fig. 3, was greatly facilitated by consideration of the total potential energies of the electrons, protons and hydrogen atoms ($\bar{\mu}e^-$, $\bar{\mu}H^+$, and $\bar{\mu}H$) that tend to flow toward equilibrium through the specific ligand-conduction pathways (*51–53*). We dwell on it somewhat here, because, as emphasized recently by Walz (*54*), the remarkable simplicity and power of this method still seems to be insufficiently appreciated.

At uniform temperature and pressure, and near chemiosmotic equilibrium, it is convenient to define the local total potentials, and the total protonic potential difference $\Delta\bar{\mu}H^+$ between the P and N domains, as follows

$$\bar{\mu}e^-(\alpha) = -E_h(\alpha) - \psi(\alpha) \tag{1}$$

$$\bar{\mu}H^+(\alpha) = \psi(\alpha) - ZpH(\alpha) \tag{2}$$

$$\bar{\mu}H(\alpha) = \bar{\mu}e^-(\alpha) + \bar{\mu}H^+(\alpha) = -E_h(\alpha) - ZpH(\alpha) \tag{3}$$

$$\Delta\bar{\mu}H^+ = \bar{\mu}H^+(P) - \bar{\mu}H^+(N) \tag{4}$$

where E_h means the local redox potential in mV on the standard hydrogen scale, ψ means the local electrostatic potential in mV, pH means $-\log_{10}$ (local H^+ ion activity) defined statistically (*55*), the bracketed symbols denote the domain of a given potential, and Z is the conventional factor 2.3 RT/F, which is about 60 at 25°.

It can be readily shown by use of equations (1–4) that, near chemiosmotic equilibrium, for both the linear Q-loop and the Q-cycle scheme of Fig. 3, the total electronic potential difference between the electron donor d and the electron acceptor c_1 or f would be equal to twice the total protonic potential difference across the B domain, or

$$\bar{\mu}e^-(d) - \bar{\mu}e^-(c_1 \text{ or } f) = 2\Delta\bar{\mu}H^+. \tag{5}$$

Also, the total electronic potentials of the b cytochromes would be the same, so that

$$E_h(b_i) - E_h(b_o) = \psi(b_o) - \psi(b_i). \tag{6}$$

However, for the linear Q-loop system the total electronic potentials and the corresponding E_h values of the b cytochromes would be fixed relative to the donor and acceptor potentials according to the relationships

$$\bar{\mu}e^-(b_o) - \bar{\mu}e^-(c_1 \text{ or } f) = \Delta\bar{\mu}H^+ \tag{7}$$

or

$$E_h(c_1 \text{ or } f) - E_h(b_o) = \psi(b_o) - \psi(c_1 \text{ or } f) + \Delta\bar{\mu}H^+, \tag{8}$$

and

$$\bar{\mu}e^-(d) - \bar{\mu}e^-(b_i) = \Delta\bar{\mu}H^+ \tag{9}$$

or

$$E_h(c_1 \text{ or } f) - E_h(b_0) = \psi(b_0) - \psi(c_1 \text{ or } f) + \Delta\mu H^+, \tag{10}$$

But, for the Q-cycle system described in Fig. 3B, the total electronic potentials and corresponding E_h values of the b cytochromes would not be thermodynamically fixed relative to the donor and acceptor potentials. However, as discussed previously (*1*), the rapid net oxidation of the Q pool by QH_2 dehydrogenases implies that electrons donated to Q at center i are virtually in equilibrium with the QH_2/Q couple at center i, or on the N side of the membrane, through mechanisms such as those illustrated in Fig. 4 (from *1*). In other words, when we include this condition in the Q-cycle scheme of Fig. 3B, it follows that

$$\bar{\mu}e^-(b_i) = \bar{\mu}e^-(d), \tag{11}$$

and therefore;

$$\bar{\mu}e^-(b_o) - \bar{\mu}e^-(c_1 \text{ or } f) = 2\Delta\bar{\mu}H^+ \tag{12}$$

or

$$E_h(c_1 \text{ or } f) - E_h(b_o) = \psi(b_o) - \psi(c_1 \text{ or } f) + 2\Delta\bar{\mu}H^+. \tag{13}$$

Thus, comparing equations (7) or (8) with (12) or (13), it is evident that the linear Q-loop and Q-cycle schemes predict very different poises of the redox components of the QH_2 dehydrogenases. The Q-cycle scheme accounts reasonably well for the observed poises of the redox components in the ubiquinol–cytochrome *c* oxidoreductase system of mitochondria. But the linear Q-loop scheme was not able to do so without new ad hoc assumptions (*1*).

It is especially noteworthy that the acid–base and redox chemistry of ubiquinone and plastoquinone enables these lipophilic quinones to play a key part in the process of specific electron–proton symport or hydrogen conduction through the B domain of Q-loop or Q-cycle systems. They exhibit what may aptly be called a direct or substrate-level membrane Bohr effect that is much more biochemically simple and explicit than the linked-function Bohr effect invoked by proponents of conformational or other indirect chemiosmotic reaction mechanisms (*5*, *56*). The quinols QH_2 are relatively weak acids that are almost completely protonated at pH 7, but their oxidation products QH_2^+ and QH_2^{2+} are relatively strong acids that deprotonate to give the semiquinone QH, the semiquinone anion Q^- and the quinone Q. The semiquinone has a pK_a near six (for relevant data, see *1*, *2*, *4*, *17*). We suggest that the expected surface-active property and high surface conductance of the semiquinone anions UQ^- and PQ^- may be functionally important for rapid lateral electron transfer over the relatively alkaline N surface of the membrane. By contrast, the hydrophobic and lipophilic properties of the components of the UQH_2/UQ and PQH_2/PQ couples enable them to catalyze tightly coupled $e^- + H^+$ conduction through the hydrocarbon B domain of the primary membrane; and evidence that semiquinones UQH and PQH are also hydrophobic may possibly allow the QH_2/QH and QH/Q couples, as well as the QH_2/Q couple, to catalyze tightly coupled $e^- + H^+$ conduction through the hydrocarbon *B* domain of the primary membrane, or perhaps through specific Q-binding amphipathic or hydrophobic proteins.

The proposed parallel translocation of two H^+ per e^- in the Q-cycle scheme is peculiarly compatible with the redox chemistry of ubiquinone and plastoquinone. Referring to Fig. 2D, the relatively high energy (and low stability) of the semiquinone QH and its anion Q^- favors both the input of the second hydrogen atom to give QH_2 from Q^- at center i, and the output of the second hydrogen atom to give $Q + 2H^+$ at center o. As discussed previously (*1*, *16*), it is energetically feasible for the acceptance of the first electron from QH_2 at center o by cytochromes c_1 or *f*, at an E_h around +300 mV, to drive the second electron from QH through center o into the P side of cytochrome *b* at a sufficiently negative E_h (meaning a sufficiently high electronic energy) to reduce the heme nucleus of cytochrome *b*; the reduced cytochrome *b* can subsequently reoxidize as the electron returns, again via center i, to its lower energetic state in QH_2. These considerations are, of course, related to the important phenomenon of oxidant-induced reduction of cytochrome *b* in QH_2 dehydrogenases.

Figure 4 (from *1*) represents attempts to explain why cytochrome *c* reductase can catalyze the rapid antimycin-sensitive acceptance of *x* electrons from the Q pool independently of the rapid antimycin-insensitive donation of *y* electrons to the Q pool by a dehydrogenase; and we suggest that these considerations may be generally applicable to QH_2 dehydrogenases. Tight stoichiometric coupling of proton translocation to electron transfer requires that, at center o, the two electrons should be transferred

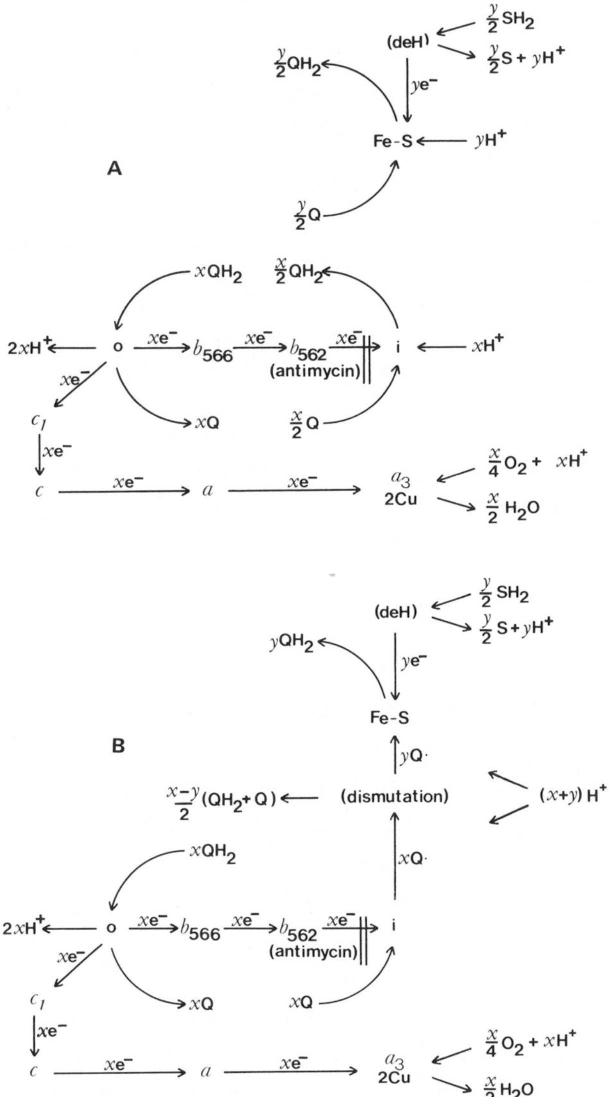

Fig. 4. Suggested Q-cycle mechanisms for independent oxidation and reduction of the Q pool in the mitochondrial cytochrome system [from Mitchell (*1*)]. (A) Chemically separate centers for reduction of Q to QH_2 in the dehydrogenase (deH) and in the cytochrome *c* reductase containing Q-reactive centers o and i. (B) A cooperative duplex center for reduction of Q to QH_2, in which successive one-electron transfers are catalyzed by center i of the cytochrome *c* reductase and by the iron–sulfur center (Fe–S) of the dehydrogenase, the two one-electron transfers being connected by dismutation of the semiquinone or semiquinone anion represented by Q·. Electron translocation by cytochrome oxidase is shown at the bottom of both diagrams. The numbers of electrons accepted by O_2 and donated by the dehydrogenase are represented by *x* and *y*, respectively. Other conventions as in Figs. 1 and 3.

separately, and at different energy levels, from QH_2 to cytochrome c_1 (via Fe–S) and from QH to cytochrome *b*-566, and that QH dismutation should be suppressed. This presumably would require very specific kinetic and/or spatial channeling of the electron transfer from QH to *b*-566, but only a moderate binding and stabilization energy of QH at center o. Conversely, as indicated in Fig. 4, net reduction of the Q pool would require the equilibration of the energies of the two electrons transferred from cytochrome *b*-562 at center i, either (A) by the transfer of the electrons alternately to Q and QH, thus producing QH_2 at center i, or (B) by the transfer of the electrons only to Q at center i, followed by dismutation of the Q^- or QH so produced. To bring the two electron transfers to about the same energy level (or E_h) at center i, the effective stability constant of QH would have to be raised from its normal value of around 10^{-10} in nonpolar media to near unity; this would require a very high binding and stabilization energy of Q^- or QH at center i. The relative binding affinities of QH_2, QH^- and QH at center o and of Q, Q^-, and QH at center i may be especially relevant to the experimental observations described in this paper.

EXPERIMENTAL RESULTS

We have attempted to elucidate the chemiosmotic molecular mechanism of cytochrome *c* reductase by measuring the chemiosmotic stoichiometry of the reaction in suspensions of rat liver mitochondria under various experimental conditions, using oxidant pulse methods and analogous rate methods developed in our laboratory (*14*, *57–59*). In these methods, a suspension of mitochondria was equilibrated anaerobically in a closed glass vessel equipped with both H^+ and K^+ sensitive electrode systems that produce continuous records of the effective H^+ and K^+ content of the suspension medium H_o^+ and K_o^+. Valinomycin was added to make the mitochondrial cristae membrane specifically permeable to K^+ in media such that the K^+ was virtually the only permeant ion species present. Thus, inward electron translocation caused a corresponding electrophoretic import of K^+ into the mitochondria, and could be measured by the fall in K_o^+, while outward hydrogen translocation from internal mitochondrial reductants could be measured by the rise in H_o^+. Under these conditions, when a pulse of respiration was induced by injecting a known quantity of standard O_2 or anaerobic ferricyanide solution into the anaerobic mitochondrial suspension, the $\leftarrow H/2e^-$ and $\rightarrow e^-/2e^-$ ratios could be estimated from the recorded excursions of H_o^+ and K_o^+.

TABLE I

Chemiosmotic Stoichiometry of Cytochrome *c* Reductase of Rat Liver Mitochondria Under Various Conditions[A]

Medium	Additions	Oxidant	$\leftarrow H/2e^-$	$\rightarrow e^-/2e^-$
(a) 150 m*M* KCl, 1 m*M* EDTA, 3.3 m*M* Gly-Gly	—	O_2	4.06 ± 0.11 (*11*)	
	—	$Fe(CN)_6^{3-}$	3.80 ± 0.03 (*4*)	
	Succ	O_2	4.10 ± 0.15 (*9*)	
	NEM	O_2	4.98 ± 0.04 (*5*)	
	NEM	$Fe(CN)_6^{3-}$	3.88 ± 0.11 (*4*)	
	Succ + NEM	O_2	4.98 ± 0.09 (*14*)	
	Succ + NEM	$Fe(CN)_6^{3-}$	4.04 ± 0.15 (*4*)	
(b) 240 m*M* Sucrose, 10 m*M* $MgSO_4$, 1 m*M* EGTA, 3.3 m*M* Gly-Gly	Succ	O_2	4.10 ± 0.11 (*7*)	1.95 ± 0.10 (*7*)
	Succ + NEM	O_2	4.89 ± 0.05 (*6*)	2.90 ± 0.10 (*6*)
(c) 240 m*M* Sucrose, 10 m*M* Ch-Cl, 1 m*M* EDTA, 3.3 m*M* Gly-Gly	—	O_2	5.00 ± 0.05 (*4*)	2.92 ± 0.08 (*4*)
	NEM	O_2	4.95 ± 0.05 (*4*)	2.90 ± 0.10 (*4*)
(d) 150 m*M* Ch-Cl, 1 m*M* EDTA, 3.3 m*M* Gly-Gly	—	O_2	5.10 ± 0.10 (*6*)	3.05 ± 0.10 (*6*)
	—	$Fe(CN)_6^{3-}$	3.85 ± 0.06 (*12*)	1.82 ± 0.07 (*6*)
	Succ	O_2	5.06 ± 0.06 (*6*)	3.02 ± 0.09 (*6*)
	NEM	O_2	5.10 ± 0.08 (*2*)	3.10 ± 0.10 (*2*)

[A] Rat liver mitochondria were suspended (about 6 mg of protein/ml) in anaerobic media in a closed, stirred, reaction vessel of 3.3 ml capacity at 25° and at pH_0 7.0–7.1. All media, in addition to the components listed contained carbonic anhydrase (20 μg/ml). Rotenone (0.4 μ*M*) was introduced after addition of the mitochondria. Where indicated, 0.5 m*M* succ and 0.2 m*M* NEM were present. Valinomycin (10 or 100 μg/g of mitochondrial protein in media containing 150 or about 1 m*M* K^+, respectively) or 1 μ*M* FCCP was also present. After 20 min anaerobic preincubation, pulses of O_2 as air-saturated saline, or pulses of anaerobic potassium ferricyanide solution, were injected into the reaction vessel. For general experimental methods, see Mitchell *et al.* (*57, 58*). In calculating the $\leftarrow H/2e^-$ and $\rightarrow e^-/2e^-$ values for cytochrome *c* reductase when oxygen was used as oxidant, the values previously found experimentally for cytochrome oxidase (*14, 56*) were subtracted from the overall values observed. (For abbreviations, see footnote 1.)

The quantity of oxidant injected was generally less than the quantity of ubiquinol in the mitochondrial Q pool. The mitochondrial cytochrome *c* was present on the external surface of the cristae membrane and reacted rapidly with ferricyanide, which did not permeate the cristae membrane. The protonmotive NADH dehydrogenase was very specifically and completely inhibited by rotenone. Therefore, we used ferricyanide pulses in suspensions of rotenone-treated mitochondria to measure the $\leftarrow H/2e^-$ and $\rightarrow e^-/2e^-$ ratios of cytochrome *c* reductase, either in the presence or absence of a Q-reducing substrate such as succinate or choline. We also used oxygen pulses under similar conditions, but in that case a value of 2.0 was subtracted from the $\rightarrow e^-/2e^-$ values to correct for the observed purely electron translocating function of cytochrome oxidase. In Table I we assembled the results of a large number of experiments done on rotenone-treated rat liver mitochondria at pH_0 7.0–7.1 and 25° in four different media and with various additions. The results show that with O_2 as oxidant, under certain conditions, the $\leftarrow H/2e^-$ ratio and, where measurable, the $\rightarrow e^-/2e^-$ ratio for the *c* reductase were near four and two, respectively. But with O_2 as oxidant, in the presence of *N*-ethylmaleinide (NEM), or in media (c) or (d), the ratios were near five and three, respectively, revealing the translocation of an odd proton. However, when the $\leftarrow H/2e^-$ and $\rightarrow e^-/2e^-$ ratios were near five and three, respectively, with O_2 as oxidant, they were near four and two, respectively, with ferricyanide.

In these oxidant pulse experiments, the initial concentration of ferricyanide was only about 20 μM, and consequently the rate of electron transfer with ferricyanide was only about 25% of that with O_2. Therefore, we tested the effect of slowing electron transfer with O_2 as oxidant by titrating with 2-heptyl-4-hydroxyquinoline *N*-oxide (HQNO). Figure 5 shows that with O_2 as oxidant, the translocation of the fifth proton was

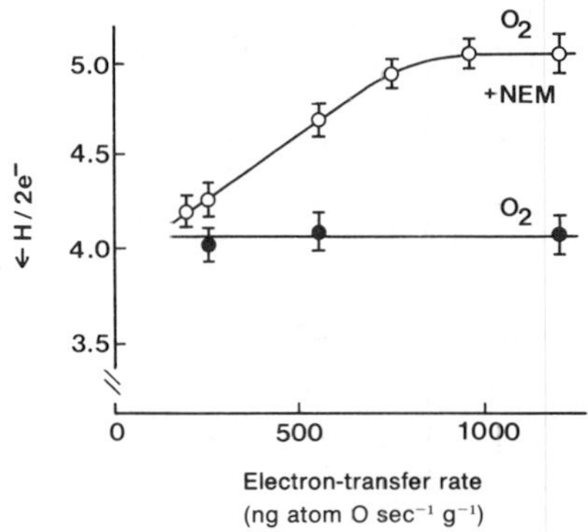

Fig. 5. Suppression of the odd proton translocation by slowing the electron-transfer rate in O_2 pulse experiments with rotenone-treated rat liver mitochondria in medium (a) of Table I at pH_0 7.0–7.1 and 25°. The electron-transfer rate was varied by the presence of different concentrations of HQNO up to a maximum of 1.5 μmole/g of mitochondrial protein. (○), experiments done in the presence of 0.2 mM NEM. For other details see Table I and text.

suppressed when the electron-transfer rate was decreased to about 25% of the normal rate; but under conditions such that the ← $H/2e^-$ ratio was normally four, as in medium (a), this ratio was independent of the electron-transfer rate. Thus, the inability of ferricyanide to give the odd proton translocation could be accounted for simply by the relatively low electron-transfer rate with ferricyanide.

The choline chloride media (c) and (d) of Table I are known to depress pH_I, the internal mitochondrial pH, by as much as 0.8 pH unit. Therefore we observed the effect of varying pH_I by measuring the ← $H/2e^-$ ratio in rotenone-treated mitochondria with succinate as substrate and O_2 as oxidant at various pH_O values, as illustrated in Figs. 6B–C. The odd proton translocation was inhibited as pH_O was increased from six toward nine both in the absence and presence of NEM. In Fig. 6B, a KCl medium was used, whereas in Fig. 6C, a choline–chloride medium (with depressed pH_I) was used. The steepest slope of the curves occurred at a lower pH_O in Fig. 6B than in Fig. 6C confirming our interpretation that the translocation of the odd proton is directly influenced by pH_I (*58*).

Figure 6A shows the ← $H/2e^-$ ratio in experiments done without rotenone in medium (b), using endogenous NADH as substrate, with O_2 as oxidant, either in the absence or presence of NEM. It is especially noteworthy that these and other supporting experiments show that the odd proton translocation is not seen when the Q pool and/or the cytochrome *c* reductase are reduced by NADH dehydrogenase.

Summarizing, it seems that the translocation of the odd proton in the

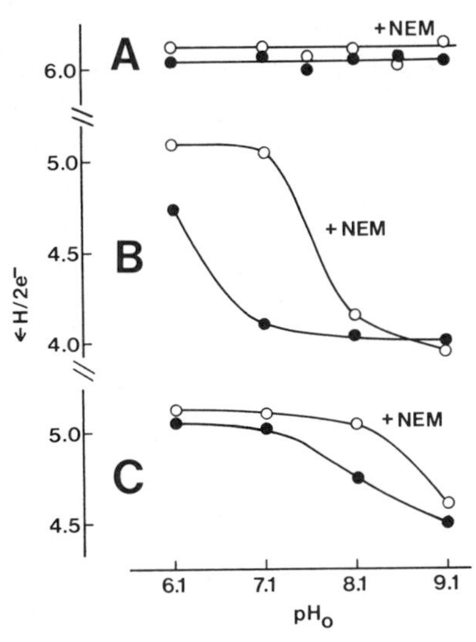

Fig. 6. Suppression of odd proton translocation by high pH_O and by NADH oxidation in O_2 pulse experiments. In (A) the medium was (b) of Table I, and no rotenone was present. In (B) the medium was 140 m*M* KCl, 10 m*M* $MgSO_4$, 1 m*M* EGTA and 3.3 m*M* glyclyglycine; and in (C) the medium was (c) of Table I. In (B) and (C) the substrate was 0.5 m*M* succinate, and 0.4 μ*M* rotenone was present. The pH_O was adjusted with HCl and KOH in (B), and with HCl and choline hydroxide in (A) and (C). For other details see Table I and text.

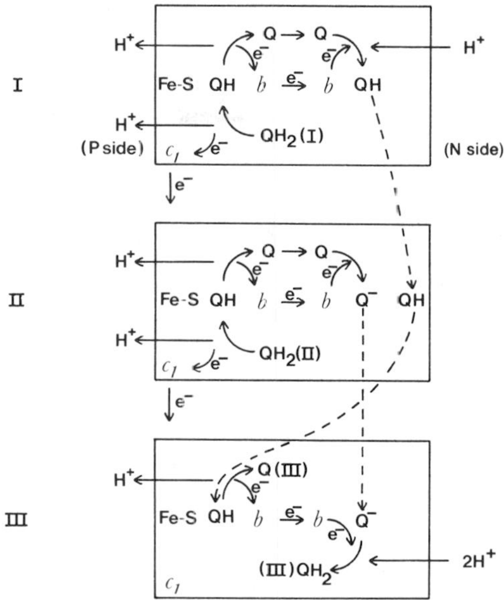

Fig. 7. Suggested molecular mechanism for the odd proton translocation in cytochrome *c* reductase. For explanation see text.

← $H/2e^-$ ratio of five in rotenone-treated mitochondria depends on a high rate of electron transport through cytochrome *c* reductase relative to the rate of some unidentified reaction inhibited by NEM, and on conditions that depress pH_I. Therefore, the odd proton translocation might be explained on the basis of the Q-cycle scheme shown in Fig. 4 (*16*). The mechanism proposed is represented in three stages in Fig. 7. In stage I, a QH_2 [represented as QH_2(I)] is oxidized to QH, with the ejection of $2H^+$ on the P side of the cytochrome *c* reductase. In stage II, a second QH_2 [represented as QH_2(II)] is oxidised to Q^-, with the ejection of a further $2H^-$ on the P side. In the third stage, there is a trans-barrier migration of QH, and the odd H^+ is ejected on the P side as a result of the trans-barrier dismutation between this QH and the Q^- produced on the N side in stage II. In the three stages, two ubiquinols, QH_2(I) and QH_2(II), would be taken up and one ubiquinol, QH_2(III), and one ubiquinone, Q(III), would be produced. Thus, one QH_2 would be oxidized to Q by the transfer of $2e^-$ to cytochrome *c* in the overall process. All three stages might conceivably be catalyzed in a monomeric *c* reductase. Alternatively, stages I and II might occur in neighboring cytochrome *c* reductase monomers, and stage III might then follow, after trans-barrier migration of QH from the N side of one monomer to the P side of the other.

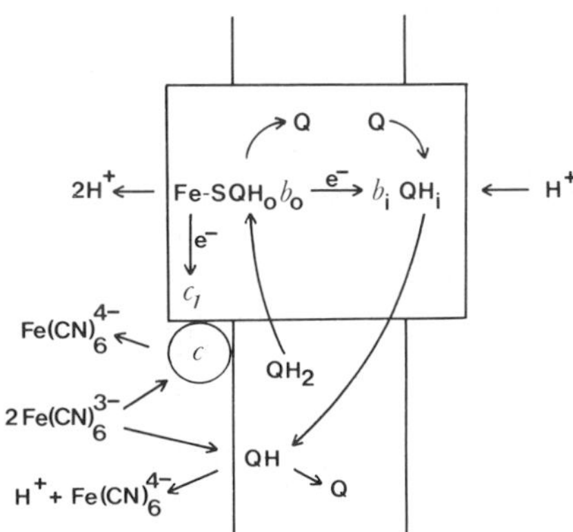

Fig. 8. Suggested molecular mechanism for ← $H/2e^-$ ratio of three given by cytochrome *c* reductase with ferricyanide as oxidant. Conventions as in Figs. 1 and 3.

If it is true that the odd proton translocation arises from outward diffusion of QH from center i, followed by trans-barrier QH dismutation, it might be possible to interrupt the process by using a high concentration of an impermeant external oxidant, like ferricyanide, that could directly oxidize QH at the external surface of the B domain. Figure 8 illustrates that such an interruption of the trans-barrier dismutation of QH would be expected to lower the ← $H/2e^-$ ratio of the cytochrome *c* reductase from about five to about three. Therefore, using a rate method for measuring the ← $H/2e^-$ ratio in the suspensions of rotenone-treated mitochondria in medium (d) of Table I at pH_O 7.0–7.1 and 25°, we observed the effect of raising the ferricyanide concentration from the 20 μM used in the oxidant pulse experiments towards 1 m*M*. As the ferricyanide concentration was increased, the rate of electron transfer was increased. As shown in Fig. 9,

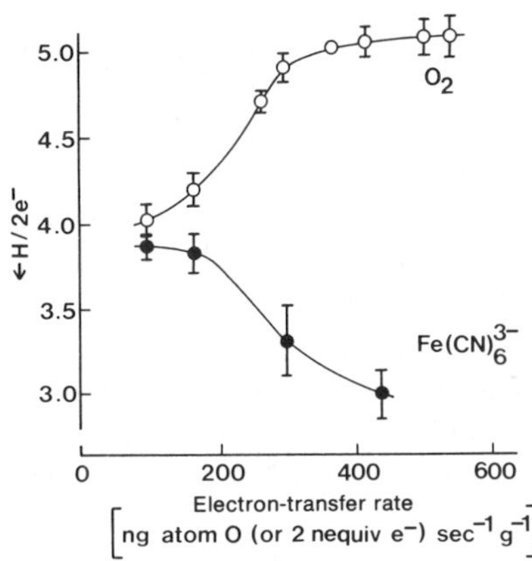

Fig. 9. Divergent dependence of ← $H/2e^-$ ratio on electron-transfer rate in O_2 pulse and ferricyanide pulse experiments, using rotenone-treated rat liver mitochondria at pH_O 7.0–7.1 and 25° in medium (d) of Table I. In the O_2 pulse experiments (○), the electron-transfer rate was varied with HQNO, as in Fig. 5. In the ferricyanide pulse experiments (●), the electron-transfer rate was varied by using different quantities of ferricyanide in the pulse, up to a maximum giving an initial concentration of 1m*M*. For other details see Table I and text.

to distinguish between the direct effect of the change of ferricyanide concentration and the indirect effect of the change of electron-transfer rate, we plotted the ← $H/2e^-$ values given by various concentrations of ferricyanide against the observed rate of electron transfer, and compared them with the ← $H/2e^-$ values given in the same medium by O_2 pulse experiments in which the electron-transfer rate was varied by titrating with HQNO. As the electron-transfer rate increased, the ← $H/2e^-$ ratio went from four to five in control experiments with O_2 as oxidant and from four to three in experiments with ferricyanide as oxidant. However, when similar experiments were done in medium (a) that gave a ← $H/2e^-$ ratio near four, independent of the electron-transfer rate with O_2 as oxidant (Fig. 5), the ← $H/2e^-$ ratios were also near four with ferricyanide as oxidant, and independent of both the electron-transfer rate and the ferricyanide concentration. The results of these, and other related experiments beyond the scope of the present paper, are consistent with the mechanism suggested in Fig. 8, and support the view that the odd proton translocation arises from the outward diffusion of QH from center i.

DISCUSSION

The mechanism proposed to account for the odd proton translocation by cytochrome *c* reductase (*16*) is obviously somewhat speculative, but has the considerable merit of being implicit in the Q-cycle schemes reproduced in Fig. 4 (from *1*). The phenomenon of the odd proton translocation may therefore provide evidence about the normal Q-cycle mechanism.

The appearance of the odd proton would presumably be promoted by a low pH_I, as observed, because the ubisemiquinone head-group would be much less likely to pass into and through the B domain in the hydrophilic anionic form Q^- than in the protonated form QH. The notion (*1*, *16*) that ubisemiquinone may be stabilized as the anion Q^- at center i (at or near the antimycin-binding site) of cytochrome *c* reductase is consistent with observations of Yu and colleagues (*19*) on the antimycin-sensitive ubisemiquinone EPR signal in purified cytochrome *c* reductase preparations reduced by catalytic amounts of succinate dehydrogenase in the presence of succinate. The signal indicated that, under favorable conditions, more than 30% of the ubiquinone in the system was present as the radical. But accumulation of the Q radical(s) occurred only when the pH was at or above 7.4 (*19*). Somewhat analogous observations on succinate–cytochrome *c* reductase by Ohnishi and Trumpower (*20*) likewise support the view that center i may bind Q^- rather than QH.

Referring to Fig. 7, it was suggested (*16*) that, under normal conditions with the N side of cytochrome *c* reductase at high pH in stage I, QH_2(I) might be oxidized to give unprotonated Q^- at center i (side *N*). Hence this Q^- could be reduced through to QH_2 in stage II, while QH_2(II) was being oxidized to Q, in accordance with Fig. 4A. Alternatively, 2Q might be produced in stages I + II, so that local dismutation to QH_2 and Q^- could occur at the N side, in accordance with Fig. 4B. The effect of NEM in promoting the appearance of the odd proton might therefore be explained by destabilization of Q^- and enhancement of Q^- protonation at center i, or by inhibition of QH dismutation on the N side, or both. The promotion of the odd proton translocation by relatively high rates of electron transfer, and therefore of Q^- and QH production at center i, might likewise be explained by the greater accumulation of QH on the N side, and by the resulting increased rate of diffusion of QH to the P surface of the B domain.

It is particularly noteworthy that the odd proton translocation can occur when cytochrome *c* reductase is reduced by the Q pool and/or by succinate dehydrogenase and/or by choline dehydrogenase. But it cannot be seen when the Q pool and/or cytochrome *c* reductase is reduced by NADH dehydrogenase. This may imply that NADH dehydrogenase prevents accumulation of Q^- or QH originating from center i of the cytochrome *c* reductase, perhaps by reducing it directly to QH_2 in accordance with Fig. 4B.

There is still much uncertainty about the precise physical and chemical interactions between the Q-reducing iron–sulfur flavoprotein dehydrogenases and the QH_2-oxidizing cytochrome *c* reductase in mitochondria (*14–16*). It is important to discover, for example, whether the Q-binding proteins of the iron–sulfur dehydrogenases, such as those of succinate dehydrogenase (*15*, *60*), and/or the terminal Fe–S centers of these dehydrogenases, participate in electron and hydrogen transfer between the Q pool and cytochrome *c* reductase, and whether hydrogen atoms or electrons are transferred to cytochrome *c* reductase (*14–16*). Perhaps advances in the knowledge of the primary and secondary electron-conducting and hydrogen-conducting quinones of photoredox chains (*61*) may help to answer such questions of electron and hydrogen transfer from the donor systems to the Q pool and to the QH_2 dehydrogenases in mitochondria as well as in photosynthetic bacteria and chloroplasts. In this context, it is noteworthy, for example that the observations of Binder and Selman (*62*) on artificial quinol-mediated cyclic photoredox activity in chloroplasts indicated that the quinol anion QH^- may donate electrons directly to the plastoquinone pool, whereas the protonated quinol QH_2 must donate electrons via the bound plastoquinone B.

As the experimental knowledge about the mitochondrial cytochrome *c* reductase has become more detailed and sophisticated, it has obviously become increasingly difficult to reconcile it with a linear Q-loop type of scheme. On the other hand, the predictions of the Q-cycle type of scheme have stood up to experimental scrutiny remarkably well so far, at least for mitochondrial cytochrome *c* reductase. But the position has been more controversial in the case of the cytochrome c_2 reductase of photosynthetic bacteria. Crofts and colleagues (*8*, *63–65*) generally favored a linear–loop scheme, while Dutton and colleagues (*6*, *66–69*) generally preferred a Q-cycle scheme. Both groups invoked multiple sites of action of antimycin (*8*, *63*, *64*, *66*) pending subsequent research (*43*, *65*, *70*). However, work by Dutton and colleagues revealed oxidant-induced reduction of cytochrome *b*-50 in antimycin-treated chromatophores (*67*). Further studies (*43*) indicated linked antimycin-insensitive reduction of cytochromes c_2 and *b*-50 by protein-complexed QH_2 (described as ZH_2 in *63*), and the binding of antimycin to a single site per cytochrome c_2 reductase where it caused a cytochrome *b*-50 blue shift and blocked oxidation of cytochrome *b* by QH at the N side (center i), but not the reduction of cytochrome *b* by QH at the P side (center o) of the cytochrome c_2 reductase. These observations clearly favored a Q-cycle type of scheme. In contrast, Crofts and colleagues observed that the half-time of cytochrome *b* reduction, following a saturating flash in the absence of antimycin, was significantly shorter than that of cytochrome c_2 reduction (*63*, *64*), which appeared incompatible with synchronous reduction of cytochromes *b* and c_2 at center o. A similar phenomenon was observed by Hind and colleagues in chloroplasts, where reduction of cytochrome *b*-563 occurred faster than that of cytochrome *f* after a saturating flash in the absence of antimycin (*35*). However, the work of Trumpower (*4*, *22*) suggested that the delay of cytochrome c_2 or *f* reduction may be explained by the redox participation of the Rieske iron–sulfur center between ubiquinol or plastoquinol and cytochromes c_2 or *f*, if QH migrating from side N can rapidly reduce cytochrome *b* at center o (*16*). Some confirmation of this view is provided by the recent observations of Bowyer et al. (*41*) that, in chromatophore cytochrome c_2 reductase, UHDBT but not antimycin inhibited cytochromes c_2 and *b* reduction, and antimycin but not UHDBT inhibited cytochrome *b* oxidation. They did not consider it to be clear how the observed interaction of UHDBT with the iron–sulfur center could explain the other effects of this inhibitor, such as prevention of cytochrome *b* reduction and elimination of the relatively slow antimycin-sensitive uptake of the second proton from side N after a flash. They suggested that these effects may reflect additional sites of action of the inhibitor. It seems to us, however, that most of their experimental findings may be rather easily explained on the basis of a Q-cycle mechanism if we accept the thesis that antimycin is a specific inhibitor of events at center i (*1*), whereas UHDBT is a specific inhibitor of events at center o (*22*); and if we also consider that the redox function of UHDBT may possibly cause complications in the pattern of redox flows and poises like those described by Surkov and Konstantinov (*49*) for DBMIB. In our view, the effects of antimycin and DBMIB on the PC reductase of chloroplasts (*41*) may likewise be explained on the basis that antimycin and DBMIB are specific inhibitors of events at centers i and o, respectively, in a Q-cycle mechanism.

The experimental work of Malviya *et al.* (*71*) on the redox reactions of the *b* cytochromes in cytochrome *c*-deficient mitochondria and submitochondrial particles seems to us to be consistent with the normal Q-cycle scheme, in which cytochromes *b*-566 and *b*-562 are electronically connected in series between centers o and i, although they prefer a different interpretation. However, their work confirms that antimycin inhibits oxidation of cytochrome *b*-562 (at center i), and also suggests that 2,3-dimercaptopropanol (BAL), like UHDBT, is an inhibitor of events at center o, presumably by inactivating or dislocating the Rieske protein. It is a welcome development that the "Slater factor" (*72*) may have been identified at last. As discussed previously (*16*), the observation of de Vries and colleagues (*73*) that the EPR signal of the Rieske center is split may possibly be explained by the presence of a Q-binding site (center o) at or near the Fe–S center. Further studies of the EPR spectra of the protonmotive QH_2 dehydrogenases are likely to be valuable for extending our knowledge about their molecular mechanisms.

To conclude our commentary on possible protonmotive QH_2 dehydrogenase mechanisms, we consider the significance of some newly discovered facts about the chemistry of quinol oxidation from the work of Rich and Bendall (*17*), together with recent knowledge of the location and electron-transfer function of the Rieske iron–sulfur center in cytochrome *c* reductase from the work of Trumpower and colleagues (*4*, *22*, *40*). We suggest that in protonmotive QH_2 dehydrogenases generally, the function of the Rieske center may be to facilitate deprotonation and oxidation of QH_2 at center o by relatively tight complexation of the reactive anionic species QH^- when the Rieske center is in the oxidized state, and by relatively loose complexation and donation to ferricytochrome b_o of the QH or Q^- radical when the Rieske center is reduced by accepting the electron from the tightly complexed QH^-. After donation of the QH or Q^- radical to ferricytochrome b_o, the reduced Rieske center would donate an electron to ferricytochrome c_1 or *f* and return to the oxidized state. This function of the Rieske center might require controlled conformational articulations as well as specific spatial interrelationships between the Rieske iron–sulfur protein and cytochromes c_1 or *f*, and b_o in the dehydrogenases. The specific transfer of one electron from QH^- to the Rieske center, and the other electron from QH or Q^- to cytochrome *b*, as required at center o of the Q cycle, would presumably depend on discriminatingly high binding constants and rate constants, not only for reaction of QH^- with the oxidized Rieske center, but also for reaction of QH and/or Q^- with oxidized cytochrome b_0. For that reason, these proposals for the normal operation of the Q-cycle mechanism may also account for participation of a linear Q-loop type of ligand conduction under certain conditions. For example, as discussed previously (*16*), when cytochrome *b* was completely reduced in anaerobic mitochondrial suspensions, at the commencement of respiration by a pulse of O_2, the QH released from the Rieske center would presumably escape across the B domain to center i until cytochrome b_0 became sufficiently oxidized to bind and oxidize the QH at center o. Trumpower (*22*) has suggested that a reversible structural dislocation of the Rieske protein occurs when cytochrome *c* reductase is highly reduced (*22*) and this might also favor transient Q-loop behavior in cytochrome *c* reductase. It is noteworthy that this type of Q-loop activity gives a $\leftarrow H/2e^-$ ratio of 2.0, compared with 4.0 for the Q cycle; and it is conceivable that steady operation of this ''low gear'' activity could be varied to give $\leftarrow H/2e^-$ ratios between four and two, according to physiological circumstances. We do not believe that good experimental evidence is yet available for variable chemiosmotic stoichiometry in cytochrome *c* reductase or cytochrome c_2 reductase. But, as hinted by Rich and Bendall (*17*), it is tempting to suggest that some such mechanism might account for the apparent variability of protonmotive gearing in chloroplasts.

ACKNOWLEDGMENTS

We thank Robert Harper and Sue Rush for expert technical assistance and help in preparing the manuscript, and we are indebted to Derek Bendall, John Bowyer, and Bernie Trumpower for sending us advance information about their work, and for useful discussions on the subject of this paper. We thank Glynn Research Ltd. for general financial support.

REFERENCES

1. Mitchell, P. (1976). *J. Theor. Biol.* **62,** 327–367.
2. Crane, F. L. (1977). *Annu. Rev. Biochem.* **46,** 439–469.
3. Rieske, J. S. (1976). *Biochim. Biophys. Acta* **456,** 195–247.

4. Trumpower, B. L., and Katki, A. G. (1979). *In* "Membrane Proteins in Energy Transduction" (R. A. Capaldi, ed.), pp. 89–200. Dekker, New York.
5. von Jagow, G., and Sebald, W. (1980). *Annu. Rev. Biochem.* **49,** 281–314.
6. Dutton, P. L., and Prince, R. C. (1978). *In* "The Bacteria" (I. C. Gunsalus, ed.), Vol. 6, pp. 523–584. Academic Press, New York.
7. Crofts, A. R., Meinhardt, S. W., and Bowyer, J. R. (1982). Chapter 34, this volume.
8. Crofts, A. R., and Wood, P. M. (1978). *Curr. Top. Bioenerg.* **7,** 175–244.
9. Nelson, N., and Neumann, J. (1972). *J. Biol. Chem.* **247,** 1817–1824.
10. Rich, P. R., and Bendall, D. S. (1980). *Biochim. Biophys. Acta* **591,** 153–161.
11. Rich, P. R., Heathcote, P., Evans, M. C. W., and Bendall, D. S. (1980). *FEBS Lett.* **116,** 51–56.
12. Weiss, H., Wingfield, P., and Leonard, K. (1979). *In* "Membrane Bioenergetics" (C. P. Lee, G. Schatz, and L. Ernster, eds.), pp. 119–132. Addison-Wesley, Reading, Massachusetts.
13. Engel, W. D., Schägger, H., and von Jagow, G. (1980). *Biochim. Biophys. Acta* **592,** 211–222.
14. Mitchell, P. (1980). *Ann. N.Y. Acad. Sci.* **341,** 564–584.
15. King, T. E. (1978). *FEBS Symp.* No. 45, 17–31.
16. Mitchell, P. (1982). *In* "Oxidases and Related Redox Systems (T. E. King, *et al.*, eds.), Pergamon, Oxford, in press.
17. Rich, P. R., and Bendall, D. S. (1980). *Biochim. Biophys. Acta* **592,** 506–518.
18. Konstantinov, A. A., and Ruuge, E. K. (1977). *FEBS Lett.* **81,** 137–141.
19. Yu, C. A., Nagaoka, S., Yu, L. and King, T. E. (1980). *Arch. Biochem. Biophys.* **204,** 59–70.
20. Ohnishi, T., and Trumpower, B. L. (1980). *J. Biol. Chem.* **255,** 3278–3284.
21. Yu, C. A., and Yu, L. (1980). *Biochem. Biophys. Res. Commun.* **96,** 286–292.
22. Trumpower, B. L. (1981). *J. Bioenerg. Biomembr.* **13,** 1–24.
23. Malkin, R., and Bearden, A. J. (1978). *Biochim. Biophys. Acta* **505,** 147–181.
24. Wood, P. M. (1980). *Biochem. J.* **189,** 385–391.
25. Futami, A., Hurt, E., and Hauska, G. (1979). *Biochim. Biophys. Acta* **547,** 583–596.
26. Futami, A., and Hauska, G. (1979). *Biochim. Biophys. Acta* **547,** 597–608.
27. Hauska, G., and Hurt, E. (1982). Paper No. 1, Chapter III, this volume.
28. Quinn, P. J., and Esfahani, A. (1980). *Biochem. J.* **185,** 715–722.
29. Kröger, A., and Klingenberg, M. (1973). *Eur. J. Biochem.* **34,** 358–368.
30. Roberts, H., and Hess, B. (1977). *Biochim. Biophys. Acta* **462,** 215–234.
31. Swanson, M., Speck, S. H., Koppenol, W. H., and Margoliash, E. (1982). *In* "Interactions between Iron and Proteins in Oxygen and Electron Transport" (Q. H. Gibson and C. Ho, eds.), Vol. 2 pp. 51–56. Elsevier, North-Holland.
32. Overfield, R. E., and Wraight, C. A. (1980). *Biochemistry* **19,** 3328–3334.
33. Plesničar, M., and Bendall, D. S. (1973). *Eur. J. Biochem.* **34,** 483–488.
34. Mills, J. D., Slovacek, R. E., and Hind, G. (1978). *Biochim. Biophys. Acta* **504,** 298–309.
35. Slovacek, R. E. Crowther, D., and Hind, G. (1979). *Biochim. Biophys. Acta* **547,** 138–148.
36. Fairneau, J., Garab, G., Horváth, G., and Faludi-Dániel, A. (1980). *FEBS Lett.* **118,** 119–122.
37. Velthuys, B. R. (1980). *FEBS Lett.* **115,** 167–170.
38. Velthuys, B. R. (1982). Paper No. 1, Chapter VI, this volume.
39. Mitchell, P. (1972). *FEBS Symp.* No. 28, 353–370.
40. Trumpower, B. L., Edwards, C. A., and Ohnishi, T. (1980). *J. Biol. Chem.* **255,** 7487–7493.

41. Bowyer, J. R., Dutton, P. L., Prince, R. C., and Crofts, A. R. (1980). *Biochim. Biophys. Acta* **592,** 455–460.
42. Malkin, R., and Posner, H. B. (1978). *Biochim. Biophys. Acta* **501,** 552–554.
43. van den Bergh, W. H., Prince, R. C., Bashford, C. L., Takamiya, K., Bonner, W. D., and Dutton, P. L. (1979). *J. Biol. Chem.* **254,** 8594–8604.
44. Roberts, H., Smith, S. C., Marzuki, S., and Linnane, A. W. (1980). *Arch. Biochem. Biophys.* **200,** 387–395.
45. Rieske, J. S. (1971). *Arch. Biochem. Biophys.* **145,** 179–193.
46. Wikström, M. K. F., and Berden, J. A. (1972). *Biochim. Biophys. Acta* **283,** 403–420.
47. Bowyer, J. R., and Trumpower, B. L. (1981). *J. Biol. Chem.* **256,** 2245–2251.
48. Bowyer, J. R., and Trumpower, B. L. (1980). *FEBS Lett.* **115,** 171–174.
49. Surkov, S. A., and Konstantinov, A. (1980). *FEBS Lett.* **109,** 283–288.
50. Mitchell, P. (1969). *Theor. Exp. Biophys.* **2,** 159–215.
51. Mitchell, P. (1963). *Biochem. Soc. Symp.* No. 22, 142–168.
52. Mitchell, P. (1976). *Symp. Soc. Gen. Microbiol.* **27,** 383–423.
53. Mitchell, P. (1979). *In* "Membrane Bioenergetics" (C. P. Lee, G. Schatz, and L. Ernster, eds.), pp. 361–372. Addison-Wesley, Reading, Massachusetts.
54. Walz, D. (1979). *Biochim. Biophys. Acta* **505,** 279–353.
55. Mitchell, P. (1967). *Nature (London)* **214,** 400.
56. von Jagow, G., and Engel, W. D. (1980). *FEBS Lett.* **111,** 1–5.
57. Mitchell, P., Moyle, J., and Mitchell, R. (1979). *In* "Biomembranes, Bioenergetics, Oxidative Phosphorylation" (S. Fleischer and L. Packer, eds.), Methods in Enzymology, Vol. 55, pp. 627–640. Academic Press, New York.
58. Mitchell, P., and Moyle, J. (1979). *Biochem. Soc. Trans.* **7,** 887–894.
59. Mitchell, P. (1982). *In* "Oxidases and Related Redox Systems" (T. E. King, et al., eds.), Vol. 3, Pergamon, Oxford, in press.
60. Yu, C. A., and Yu, L. (1980). *Biochemistry* **19,** 3579–3585.
61. Wraight, C. (1982). Paper 2, Chapter IV, this volume.
62. Binder, R. G., and Selman, B. R. (1980). *Biochim. Biophys. Acta* **592,** 314–322.
63. Crofts, A. R., Crowther, D., and Tierney, G. V. (1975). *In* "Electron Transfer Chains and Oxidative Phosphorylation" (E. Quagliariello, S. Papa, F. Palmieri, E. C. Slater, and N. Siliprandi, eds.), pp. 233–241. North-Holland Publ., Amsterdam.
64. Crofts, A. R. Crowther, D., Celis, H., DeCelis, S. A., and Tierney, G. V. (1977). *Biochem. Soc. Trans.* **5,** 491–495.
65. Bowyer, J. R., and Crofts, A. R. (1978). *In* "Frontiers of Biological Energetics" (P. L. Dutton, J. S. Leigh, and A. Scarpa, eds.), Vol. 1, pp. 326–333. Academic Press, New York.
66. Petty, K. M., Jackson, J. B., and Dutton, P. L. (1977). *FEBS Lett.* **84,** 299–303.
67. Dutton, P. L., and Prince, R. C. (1978). *FEBS Lett.* **91,** 15–20.
68. Prince, R. C., van den Bergh, W. H., Takamiya, K., Bahsford, L., and Dutton, P. L. (1978). *In* "Frontiers of Biological Energetics" (P. L. Dutton, J. S. Leigh, and A. Scarpa, eds.), Vol. 1., pp. 201–209. Academic Press, New York.
69. Petty, K., Jackson, J. B., and Dutton, P. L. (1979). *Biochim. Biophys. Acta* **546,** 17–42.
70. Bowyer, J. R., and Crofts, A. R. (1980). *Biochim. Biophys. Acta* **591,** 298–311.
71. Malviya, A. N., Nicholls, P., and Elliott, W. B. (1980). *Biochim. Biophys. Acta* **589,** 137–149.
72. Slater, E. C. (1949). *Biochem. J.* **45,** 14–30.
73. de Vries, S., Albracht, S. P. J., and Leeuwerik, F. J. (1979). *Biochim. Biophys. Acta* **546,** 316–333.

INDEX

D

E

G

H

I

L

M

N

O

P

Q

R

S

T

U

V